ADDITIVE MANUFACTURING

Innovations, Advances, and Applications

Captions for Figures on Cover

(a) Photograph of the developed micro-stereolithography system. (From Chapter 4, Figure 4.4A, Authors: *Jae-Won Choi, Yanfeng Lu, and Ryan B. Wicker*)

(b) Microstructures of powder bed sample after etching.
  - (A) Powder with different sizes (optical micrograph)
  - (B) Single particle (optical micrograph)
  - (C) Single particle (scanning electron micrograph)
  (From Chapter 7, Figure 7.9A–C, Authors: *Xibing Going, James Lydon, Kenneth Cooper, and Kevin Chou*)

(c) Key low thermal budget annealing technologies for advanced material and device development. (From Chapter 5, Figure 5.6, Authors: *Pooran C. Joshi, Teja Kuruganti, and Chad E. Duty*)

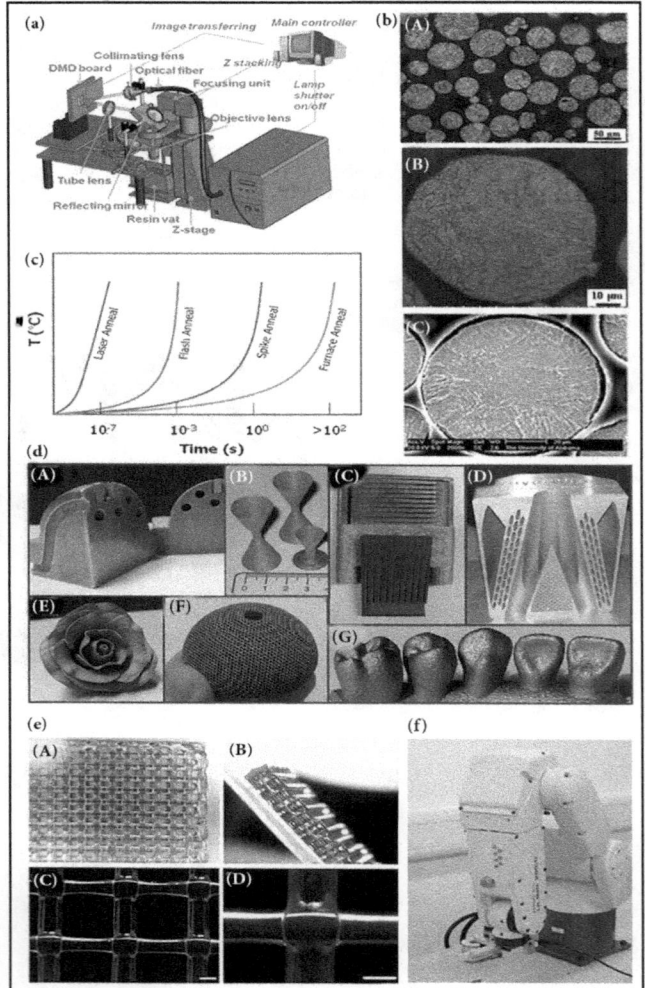

(d) Some complex SLM parts.
  - (A) Stainless steel mold with conformal cooling to enhance the productivity in injection molding
  - (B) Ti6Al4V thin wall structures
  - (C) 316L stainless steel heating plate for aerospace industry
  - (D) Advanced nozzle with internal cooling system from AlSi10Mg alloy
  - (E) Stainless steel artistic flower
  - (F) Ti6Al4V biomedical acetabular cup with advanced cellular porosity for improved biocompatibility
  - (G) CoCr dental parts
  (From Chapter 3, Figure 3.2 A–G, Authors: *Jean-Pierre Kruth, Sasan Dadbakhsh, Bey Vrancken, Karolien Kempen, Jef Vleugels, and Jan Van Humbeeck*)

(e) (A) Macroscopic images of a four-layered hydrogel scaffold: Top view
  - (B) Macroscopic images of a four-layered hydrogel scaffold: Side view
  - (C) Optical micrographs of scaffold microarchitecture (needle Ø 250 µm, fiber spacing 1 mm); scale bar: 250 µm
  - (D) Higher magnification of the intersection between fibers; scale bar: 250 µm
  (From Chapter 16, Figure 16.2, Authors: *Sara Maria Giannitelli, Pamela Mozetic, Marcella Trombetta, and Alberto Rainer*)

(f) Photograph of experimental apparatus. Soldering iron and wire feeder attached to 6 axis robotic arm and used to deposit lead-free solder tracks from a substrate mounted in the sample holder (From Chapter 2, part of Figure 2.2, Author: *Abinand Rangesh*)

ADDITIVE MANUFACTURING
Innovations, Advances, and Applications

Edited by

T.S. Srivatsan • T.S. Sudarshan

CRC Press
Taylor & Francis Group
Boca Raton London New York

CRC Press is an imprint of the
Taylor & Francis Group, an **informa** business

CRC Press
Taylor & Francis Group
6000 Broken Sound Parkway NW, Suite 300
Boca Raton, FL 33487-2742

First issued in paperback 2020

© 2016 by Taylor & Francis Group, LLC
CRC Press is an imprint of Taylor & Francis Group, an Informa business

No claim to original U.S. Government works

ISBN-13: 978-1-4987-1477-8 (hbk)
ISBN-13: 978-0-367-73778-8 (pbk)

Library of Congress Cataloging-in-Publication Data

Additive manufacturing : innovations, advances, and applications / editors, T.S. Srivatsan and T.S. Sudarshan.
    pages cm
Includes bibliographical references and index.
ISBN 978-1-4987-1477-8 (alk. paper)
1. Three-dimensional printing. I. Srivatsan, T. S. II. Sudarshan, T. S., 1955-

TS171.95.A44 2015
621.9'88--dc23   2015016369

Visit the Taylor & Francis Web site at
http://www.taylorandfrancis.com

and the CRC Press Web site at
http://www.crcpress.com

Contents

Preface

The innovation of creating a three-dimensional object layer by layer using computer-aided design (CAD) was originally termed *rapid prototyping*, a valuable technique that was developed in the early 1980s for the purpose of manufacturing. In its early stages, rapid prototyping was typically used to create models and prototype parts and offered quick realization of what engineers had envisioned. Rapid prototyping was one of the preliminary processes that eventually culminated in additive manufacturing (AM), which allows the production of actual printed parts, in addition to models. The most notable advances the process offers are the development and production of products with a noticeable reduction in both time and cost, facilitated by increased human interaction and optimization of the product development cycle, thus making it possible to create almost any shape that would otherwise be difficult to machine using conventional techniques. With the emergence of additive manufacturing, scientists, engineers, and even students can rapidly build and analyze models for the purpose of theoretical comprehension and related studies. In the medical profession, doctors have been able to build models of various parts of the body to analyze injuries or disease and to plan appropriate medical procedures. Additive manufacturing has also made it possible for market researchers to gather the opinions of potential buyers of newly developed products and for artists to explore their creativity.

The gradual growth and eventual transition of rapid prototyping to three-dimensional (3D) printing have allowed the process to gain ground as a valuable process for making prototype parts among manufacturers of many types. In recent months, several companies have implemented the actual use of 3D printing to make prototype parts. Major manufacturers such as Boeing, Airbus Industries, General Electric, and even Siemens are using AM to develop a number of high-value production parts. Siemens has been using additive manufacturing to produce over 100 different types of spare parts and other gas turbine components, resulting in a reduction in repair time of as much as 90% in some cases. Siemens predicts that 3D printing will revolutionize the availability and supply of spare parts. Normally, spare parts have been mass produced, stored, and then shipped as needed; however, additive manufacturing has allowed Siemens to print them as required. Likewise, the aerospace industry is using additive manufacturing to produce lighter parts with reduced material waste as compared to traditional subtractive machining. Most recently, a United Kingdom–based automotive and aerospace parts maker, in collaboration with Airbus, developed a titanium bracket that was 3D printed in 40 minutes vs. 4 hours by conventional machining, cutting material usage by well over 50%. The ability to utilize additive manufacturing near the point of use allows on-demand manufacturing and drastically reduces both inventory and wasted time. This has made possible the rapid growth of additive manufacturing since its initiation in 1988. Over the next two decades, the annual growth rate of worldwide revenues of all additive manufacturing products and services was 25.4%, but from 2010 to 2013 the growth rate was 27.4%.

This book consists of 17 chapters. To meet the needs of different readers, each chapter provides a clear, compelling, and complete discussion of the subject matter. The first chapter introduces the reader to the techniques that are viable for metallic materials while highlighting the advantages of each technique with specific reference to technological applications. In the next two chapters, the contributing authors present and discuss additive manufacturing of metals using the techniques of free space deposition (Chapter 2) and selective laser melting (Chapter 3). In Chapters 4 and 5, the contributing authors provide an overview of specific technologies related to additive manufacturing.

The next three chapters discuss various aspects pertinent to powder-based additive manufacturing: the application of radiometry in laser powder deposition-based additive manufacturing (Chapter 6), the use of powders in electron beam melting (Chapter 7), and advanced concepts aimed at studying and understanding the simulation of powder-based additive manufacturing processes (Chapter 8). Chapter 9 then presents the influence of process parameters on microstructure and the properties of laser-deposited materials, and Chapter 10 addresses the integration of gas-permeable structures. The use of additive manufacturing for components made from ceramic materials is discussed in Chapter 11. With specific reference to polymeric materials, Chapter 12 provides a comprehensive overview of key aspects related to reactive inkjet printing of nylon materials for the purpose of additive manufacturing. The next several chapters on biomedical applications of additive manufacturing were written by renowned experts in their fields. Chapter 13 provides a comparison between additive manufacturing materials and human tissues, Chapters 14 and 15 address the use of additive manufacturing for medical devices, and the use of additive manufacturing for drug and cell delivery is presented in Chapter 16. The relevance of additive manufacturing to rare earth magnets is the focus of Chapter 17. In each chapter, the contributing authors have made an attempt to present applications of the particular additive manufacturing technologies being discussed, the future prospects and far-reaching applications of those technologies, and developments to be made in areas that have been considered to be either impossible or uneconomical in the past.

Overall, this text on additive manufacturing provides a solid background for understanding the immediate past, the ongoing present, and emerging trends, with an emphasis on innovations and advances in its use for a wide spectrum of manufacturing applications, including the human healthcare system. This text can very well serve as a single reference book or even as textbook for

1. Seniors in undergraduate programs in the fields of materials science and engineering, manufacturing engineering, and biomedical engineering
2. Beginning graduate students
3. Researchers in both research and industrial laboratories who are studying various aspects related to materials, products, and additive manufacturing
4. Engineers seeking technologically novel and economically viable innovations for a spectrum of both performance-critical and non-performance-critical applications

We anticipate that this bound volume will be of much interest to scientists, engineers, technologists, and entrepreneurs.

MATLAB® is a registered trademark of The MathWorks, Inc. For product information, please contact:

The MathWorks, Inc.
3 Apple Hill Drive
Natick, MA 01760-2098 USA
Tel: 508 647 7000
Fax: 508-647-7001
E-mail: info@mathworks.com
Web: www.mathworks.com

Acknowledgments

The editors gratefully acknowledge the understanding and valued support they received from the authors of the various chapters contained in this text. Efforts made by the contributing authors to present and discuss the different topics greatly enhance the scientific and technological content and are very much appreciated. The useful comments and suggestions made by the referees on each chapter further helped to elevate the technical content and merit of the final version of each chapter.

Our publisher, CRC Press, has been very supportive and patient throughout the entire process, beginning with the conception of this intellectual project. We extend an abundance of thanks, valued appreciation, and gratitude to the editorial staff at CRC Press. Specifically, we must mention Allison Shatkin, senior acquisitions editor for materials science and chemical engineering, and Amber Donley, project coordinator, editorial project development, for their sustained interest, involvement, attention, and energetic assistance stemming from understanding coupled with an overall willingness to help both the editors and the contributing authors. They ensured timely execution of the numerous intricacies related to smooth completion of this volume, from the moment of its approval and up until compilation and publication. At moments of need, the editors greatly appreciated Amber Donley's support while she remained courteous, professional, and enthusiastically helpful.

Special thanks inlaid with an abundance of appreciation are also extended to Dr. K. Manigandan, research scholar (research associate) at The University of Akron, Ohio, for his almost ceaseless, relentless, and tireless efforts to ensure proper formatting and layout of the chapters. Of course, most importantly and worthy of recording, is that the timely compilation and publication of this bound volume would not have been possible without the understanding, cooperation, assistance, and patience of the authors and the positive contributions of the peer reviewers.

Editors

T.S. Srivatsan, PhD, professor of materials science and Engineering in the Department of Mechanical Engineering at The University of Akron, earned his master's of science degree in aerospace engineering in 1981 and his doctoral degree in mechanical engineering in 1984 from the Georgia Institute of Technology. Dr. Srivatsan joined the Department of Mechanical Engineering faculty at The University of Akron in 1987. Since then, he has instructed undergraduate and graduate courses in the areas of advanced materials and manufacturing processes, mechanical behavior of materials, fatigue of engineering materials and structures, fracture mechanics, materials science and engineering, mechanical measurements, design of mechanical systems, and mechanical engineering laboratory. His research areas currently include the fatigue and fracture behavior of advanced materials, including monolithic, intermetallic, and nano-materials and metal–matrix composites; processing techniques for advanced materials and nanostructure materials; relationships between processing and mechanical behavior; electron microscopy; failure analysis; and mechanical design. Dr. Srivatsan has authored or edited 57 books in such areas as cross-pollinating mechanical design; processing and fabrication of advanced materials; deformation, fatigue, and fracture of ordered intermetallic materials; machining of composites; failure analysis; and technology of rapid solidification processing of materials. He serves as co-editor of the *International Journal on Materials and Manufacturing Processes* and is on the editorial advisory board of several journals within the domain of materials science and engineering. He has delivered over 200 technical presentations at national and international meetings and symposia, technical/professional societies, and research and educational institutions. He has authored or co-authored over 700 archival publications, including articles in international journals, chapters in books, proceedings of national and international conferences, reviews of books, and technical reports. In recognition of his efforts, contributions, and impact on furthering science, technology, and education, Dr. Srivatsan has been elected a fellow of the American Society for Materials, International; a fellow of the American Society of Mechanical Engineers; and a fellow of the American Association for the Advancement of Science. He has also been recognized as outstanding young alumnus of Georgia Institute of Technology and outstanding research faculty of the College of Engineering at The University of Akron, in addition to receiving the Dean Louis Hill Award for exceptional dedication and service. He has also consulted with the U.S. Air Force and U.S. Navy, national research laboratories, and industries related to aerospace, automotive, power generation, leisure-related products, and applied medical sciences.

T.S. Sudarshan, PhD, earned his bachelor of technology degree in metallurgy from the Indian Institute of Technology, Madras, and later completed his master's of science and doctoral degrees in materials engineering science from Virginia Polytechnic Institute and State University. Dr Sudarshan is currently the president and CEO of Materials Modification, Inc., which is at the forefront of research, development, and commercialization of advanced materials utilizing novel processing techniques. He has demonstrated technological leadership for well over three decades and has worked extensively throughout his career in the areas of nanotechnology and surface engineering, for which he is well known throughout the world. Through his leadership, over $60 million in funding has been raised for the primary purpose of very high-risk, high-payoff advanced technology-related programs in several non-traditional areas. Dr. Sudarshan has published well over 170 papers in archival journals and has edited 29 books on surface modification technologies and advanced materials. He is currently the editor of two international journals and holds numerous patents. In recognition of his efforts, contributions, and impact on furthering science, technology, and its far-reaching applications, he has been elected as a fellow of the American Society for Materials, International; a fellow of the International Federation for Heat Treatment and Surface Engineering; and a fellow of the Institute of Materials, Minerals and Mining. He was conferred with the distinguished alumni award of the Indian Institute of Technology on Institute Day in 2014.

Contributors

Richard Bibb
Loughborough Design School
Loughborough University
Loughborough, United Kingdom

Jae-Won Choi
Department of Mechanical Engineering
The University of Akron
Akron, Ohio

Kevin Chou
Mechanical Engineering Department
The University of Alabama
Tuscaloosa, Alabama

Kenneth Cooper
Additive Manufacturing Laboratory
Marshall Space Flight Center
Huntsville, Alabama

Sasan Dadbakhsh
Department of Mechanical Engineering
University of Leuven (KU Leuven)
Leuven, Belgium

Chad E. Duty
Oak Ridge National Laboratory
Oak Ridge, Tennessee

Mohsen Eshraghi
Department of Mechanical Engineering
California State University, Los Angeles
Los Angeles, California

Saeed Fathi
Additive Manufacturing Research Group
Wolfson School of Mechanical and
    Manufacturing Engineering
Loughborough University
Loughborough, United Kingdom

Sergio D. Felicelli
Department of Mechanical Engineering
The University of Akron
Akron, Ohio

Sara Maria Giannitelli
Department of Engineering, Tissue
    Engineering Unit
Università Campus Bio-Medico di Roma
Rome, Italy

Xibing Gong
Mechanical Engineering Department
The University of Alabama
Tuscaloosa, Alabama

Joshua J. Hammell
Department of Mechanical Engineering
South Dakota School of Mines and Technology
Rapid City, South Dakota

Pooran C. Joshi
Oak Ridge National Laboratory
Oak Ridge, Tennessee

Karolien Kempen
Department of Mechanical Engineering
University of Leuven (KU Leuven)
Leuven, Belgium

and

Department of Mechanical Engineering
    Technology
Katholieke Universiteit Leuven
Geel, Belgium

Christoph Klahn
Product Development Group Zurich pdlz
ETH Zurich
Zurich, Switzerland

Vemuru V. Krishnamurthy
School of Physics, Astronomy, and
    Computational Sciences
George Mason University
Fairfax, Virginia

Jean-Pierre Kruth
Department of Mechanical Engineering
University of Leuven (KU Leuven)
Leuven, Belgium

Teja Kuruganti
Oak Ridge National Laboratory
Oak Ridge, Tennessee

Michael A. Langerman
Department of Mechanical Engineering
South Dakota School of Mines and Technology
Rapid City, South Dakota

Yanfeng Lu
Department of Mechanical Engineering
The University of Akron
Akron, Ohio

James Lydon
Additive Manufacturing Laboratory
Marshall Space Flight Center
Huntsville, Alabama

K. Manigandan
Department of Mechanical Engineering
The University of Akron
Akron, Ohio

James D. McGuffin-Cawley
Department of Materials Science and
    Engineering
Case Western Reserve University
Cleveland, Ohio

Mirko Meboldt
Product Development Group Zurich pd|z
ETH Zurich
Zurich, Switzerland

Pamela Mozetic
Department of Engineering, Tissue
    Engineering Unit
Università Campus Bio-Medico di Roma
Rome, Italy

Deepankar Pal
Department of Industrial Engineering
University of Louisville
Louisville, Kentucky

Jayanthi Parthasarathy
MedCAD
Dallas, Texas

Alberto Rainer
Department of Engineering, Tissue
    Engineering Unit
Università Campus Bio-Medico di Roma
Rome, Italy

Abinand Rangesh
Lumi Ventures, LLC
Brookline, Massachusetts

T.S. Srivatsan
Department of Mechanical Engineering
The University of Akron
Akron, Ohio

Brent Stucker
Department of Industrial Engineering
University of Louisville
Louisville, Kentucky

T.S. Sudarshan
Materials Modification, Inc.
Fairfax, Virginia

Chong Teng
Department of Industrial Engineering
University of Louisville
Louisville, Kentucky

Darren Thompson
Regional Medical Physics Department
Newcastle upon Tyne Hospitals NHS Trust
Newcastle upon Tyne, United Kingdom

James L. Tomich
Department of Materials Engineering and
    Science
South Dakota School of Mines and Technology
Rapid City, South Dakota

Marcella Trombetta
Department of Engineering, Tissue
    Engineering Unit
Università Campus Bio-Medico di Roma
Rome, Italy

Brett L. Trotter
Department of Mechanical Engineering
South Dakota School of Mines and Technology
Rapid City, South Dakota

Jan Van Humbeeck
Department of Materials Engineering
University of Leuven (KU Leuven)
Leuven, Belgium

Jef Vleugels
Department of Materials Engineering
University of Leuven (KU Leuven)
Leuven, Belgium

Bey Vrancken
Department of Materials Engineering
University of Leuven (KU Leuven)
Leuven, Belgium

Ryan B. Wicker
W.M. Keck Center for 3D Innovation
The University of Texas at El Paso
El Paso, Texas

John Winder
Health and Rehabilitation Sciences Research
    Institute
University of Ulster
Northern Ireland, United Kingdom

1 Additive Manufacturing of Materials
Viable Techniques, Metals, Advances, Advantages, and Applications

T.S. Srivatsan, K. Manigandan, and T.S. Sudarshan

CONTENTS

ABSTRACT

Additive manufacturing (AM) has acquired the status of a mainstream manufacturing process. The technique facilitates the building of parts through the addition of materials one layer at a time using a computerized three-dimensional (3D) solid model. Additive manufacturing does not require cutting tools, coolants, or other auxiliary resources used in conventional manufacturing. The technique allows for optimization of design coupled with the ability to produce customized parts on demand. The numerous advantages over conventional manufacturing have captured the imagination of the public, resulting in additive manufacturing being referred to as the "third industrial revolution." The governing fundamentals and working principles of additive manufacturing offer a spectrum of advantages, including near-net shape capabilities, superior design, geometric flexibility, innovations in fabrication using multiple materials, reduced tooling and fixturing, shorter cycle times for both design and manufacturing, and overall savings in both energy and costs while concurrently enabling production on a global scale. This chapter provides an overview of the technologically viable techniques for metallic materials and explores the key ingredients essential for both applying and advancing this technology so as to make it feasible at all levels. The most commonly used engineered metallic materials and their short-term mechanical properties are examined, and the emerging and far-reaching applications of this revolutionary technique to education, health, and well-being are briefly summarized.

1.1 INTRODUCTION

Manufacturing is essential for both the creation of wealth and improvements in the quality of life. The intricacies of manufacturing are applied to (1) system design and organization, (2) technological logistics, and (3) operational planning and control. Manufacturing technology can be classified as either conventional or non-conventional processes. The conventional processes have been in existence since prior to 1950, and the non-conventional processes are those that have been both adopted and implemented since the 1950s.

Additive manufacturing (AM) can be described as the process of joining or adding materials with the primary objective of making objects from three-dimensional (3D) model data using the layer-by-layer principle.[1] The layer manufacturing (LM) technologies are known among both engineering and scientific communities as rapid manufacturing[2] or rapid prototyping (RP).[3] Unlike conventional manufacturing techniques, such as machining and stamping, which tend to fabricate products by removing material from a larger stock or sheet metal, the technique of additive manufacturing creates final shapes by the addition of materials. It makes effective and efficient use of available raw materials and produces minimal waste while offering satisfactory accuracy in the geometry of the finished parts.[1–3] Additive manufacturing allows a design in the form of a computerized three-dimensional model to be easily transformed to a finished end product without the use or aid of any additional fixtures and cutting tools. This opens up the possibility of producing parts that have a complex geometry and would be quite difficult to obtain using conventional material removal processes. The benefits of additive manufacturing can lead to novel innovations in design, a necessary process for the manufacturing and assembly of any product.

Additive manufacturing has also made possible the development of environmentally friendly products. Unlike traditional manufacturing processes, which tend to impose various constraints on the design of a product, the flexibility intrinsic to additive manufacturing allows manufacturers to optimize production schedules, thus eliminating potential waste.[4] The ability of additive manufacturing to construct parts and/or products having a complex geometry implies that the previously separated parts can be easily consolidated into a single object. The "topologically" optimized designs that additive manufacturing is capable of realizing can increase the functionality of a product, thereby reducing both the amount of energy that is consumed and the fuel or natural resources required for its operation.[5]

Rapid strides were made in the development, progressive growth, and eventual commercialization of additive manufacturing technology in the early 1980s, and additive manufacturing has experienced exponential growth in several sectors of the industry related to commercial products, aerospace technology, and even health and medicine.[6] A considerable amount of progress has been made, resulting in the creation of an expectation that additive manufacturing will revolutionize manufacturing while providing benefits to society at large. A few of these benefits as conceived and put forth in the 1980s include the following:

1. Reduced usage of raw materials and consumption of energy, key contributing factors to environmental sustainability
2. On-demand manufacturing, which presents a novel opportunity to reconfigure the manufacturing supply chain so as to offer less expensive products to end users or customers and in the process utilize minimal resources
3. Customized healthcare products to meet the specific needs of individual customers, which in turn could be expected to have a significant influence on improving the overall well-being of the population

Over the last several decades, periodic reviews summarizing the progress and outcomes of additive manufacturing have been published.[7-40] A sizeable number of these papers have focused on the roles that various processing technologies can play in the engineering of a product designed for a specific purpose. The objective of this chapter is to present information pertinent to the additive manufacturing techniques being used for metallic materials, with specific reference to innovations in the use of viable techniques, advances being made with specific reference to outcomes, and potential far-reaching applications. This chapter is organized as follows:

Section 1.2 provides a cursory overview of the background of additive manufacturing pertaining to the emergence, key characteristics, and noteworthy advantages and disadvantages of the most viable and technologically relevant techniques.

Section 1.3 presents the most viable and applicable techniques available for engineering the manufacturing of products made from metallic materials.

Section 1.4 briefly summarizes use of the cold spray technique for the purpose of additive manufacturing.

Section 1.5 presents an overview of the key ingredients required for ensuring success in the application of additive manufacturing.

Section 1.6 offers a glimpse into the most significant advantages of putting to effective use additive manufacturing technology.

Section 1.7 provides a compelling overview of the technological advances that have resulted from the emergence and use of additive manufacturing.

Section 1.8 briefly surveys the published literature with specific reference to the mechanical properties of user-friendly additive manufactured materials.

Section 1.9 provides a useful record of the potential far-reaching applications that can result from the use of additive manufacturing.

Section 1.10 presents a short overview of the emergence and use of additive manufacturing in space.

Section 1.11 addresses the impact of additive manufacturing on education.

Section 1.12 takes an informative look at the impact of additive manufacturing on the overall well-being and health of the populations it serves.

Section 1.13 provides a short overview of the standards relevant to additive manufacturing.

Section 1.14 briefly summarizes the possible future of additive manufacturing.

1.2 TYPES OF ADDITIVE MANUFACTURING TECHNIQUES

The technology of additive manufacturing in essence consists of three basic steps:

1. A computerized three-dimensional (3D) solid model is developed and converted into a standard AM file format, such as the traditional Standard Tessellation Language (STL)[41] or a more recent AM format.[42]
2. This file is then sent to an additive manufacturing machine where it is manipulated to change both the position and orientation of the part or to simply scale the part.
3. The part is then built layer by layer on an additive manufacturing machine.

In 1988, a new approach to additive manufacturing based on application of the layer-by-layer principle was demonstrated in the form of stereolithography (SLA). Initially, the main drawbacks of the technique of stereolithography were poor dimensional accuracy and the fact that its application was essentially restricted to "touch" and "feel" models. With incremental yet noticeable advances in technology and resultant improvements in accuracy, AM became an accepted method for prototyping. The term *rapid prototyping* appeared and soon after became universally accepted. Over the years, other forms of the layer-manufacturing processes were developed, and a few studies have been conducted to compare the various processes involved in these new manufacturing technologies.[43,44] In the mid-1990s, continuing research efforts resulted in rapid prototyping processes becoming indispensable components of engineering the rapid development of a product.

Various terms have been suggested for these types of technology, including (1) *material ingress manufacturing* or *additive manufacturing*, (2) *freeform fabrication*, (3) *layer manufacturing*, or (4) *3D printing*. The term *rapid prototyping*, however, became the one most widely accepted and used.[45] Rapid prototyping was defined to be direct fabrication of a physical part from a 3D model with the aid of additive manufacturing, much in the same way as an office printer places two-dimensional (2D) digital files onto pieces of paper. According to the ASTM Committee F12,[46]

1. Additive manufacturing is defined as a process of joining materials to make objects from 3D model data, usually layer upon layer as opposed to traditional processes based on subtractive manufacturing technologies.
2. 3D printing is the fabrication of objects through the deposition of a material using a print head, nozzle, or other printing technologies. 3D printing is a process of using successive layers of printed material to form solid 3D objects of virtually any shape from a digital model.

In 2009, ASTM Committee F42 was formed to develop standards for additive manufacturing. An important contribution of Committee F42 has been the development of a terminology standard that defines the different processes required to build 3D parts from computer-aided design (CAD) files. More recently, due to the further development and concurrent emergence of AM technologies, ASTM has specified other commonly used synonyms for additive manufacturing, including (1) *additive fabrication*, (2) *additive processes*, (3) *additive techniques*, (4) *additive layer manufacturing*, (5) *layer manufacturing*, and (6) *freeform fabrication*. An overall overview of the working principle of additive manufacturing is shown in Figure 1.1. This manufacturing method has attracted the attention of U.S. industry and research universities as a direct result of the creation of America Makes, formerly known as the National Additive Manufacturing Innovation Institute (NAMII). America Makes is a government-led consortium that seeks to address and attend to issues specific to additive

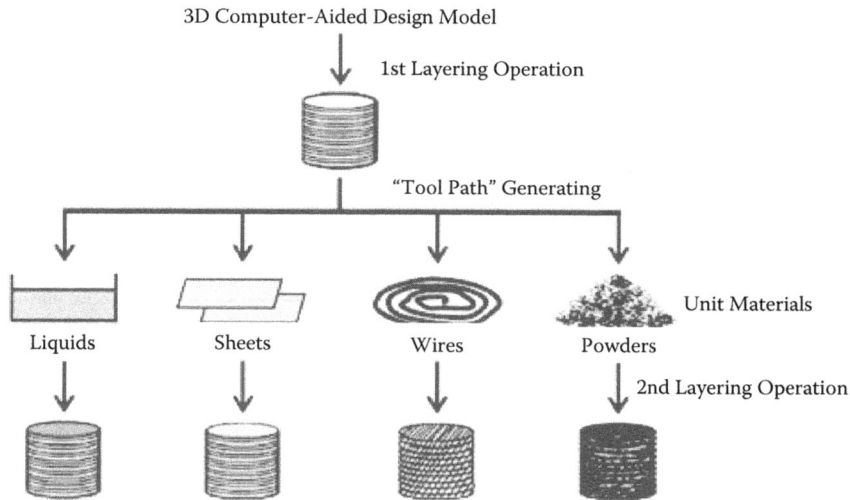

3D Computer-Aided Design Model

1st Layering Operation

"Tool Path" Generating

Liquids   Sheets   Wires   Powders   Unit Materials

2nd Layering Operation

FIGURE 1.1   A representation of the working profiles of additive manufacturing: the layer-by-layer fabrication of a product using various unit materials.[75]

manufacturing. Both the National Aeronautics and Space Administration (NASA) and the Air Force Research Laboratory (AFRL) support America Makes, and the program can play an important role in coordinating actions while concurrently avoiding the duplication of efforts. Similar developments are also taking place within the European Union (EU). The EU agenda on additive manufacturing began in the late 2010s with the adoption of an additive manufacturing platform.

1.2.1   RAPID PROTOTYPING

Rapid prototyping (RP), one of the earlier additive manufacturing processes, allows for the creation of printed parts, not just models. A few of the noticeable advances this process offered to product development and production include the following: (1) reductions in both time and costs, (2) enhanced human interaction, (3) the possibility of creating almost any shape that would otherwise be very difficult to machine, and (4) a shortened product development cycle.[47,48] It soon came to be referred to as direct fabrication of a part using a 3D model by way of additive manufacturing. As there are no tool setting, workpiece fixture, and tool wear compensation requirements, the process was categorized as being the most flexible and direct form of digital manufacturing. In industry circles, the process began to be referred to as *additive digital manufacturing* (ADM) rather than rapid prototyping. In the mid-1990s, the rapid prototyping processes became firmly established as indispensable components for the rapid development of a product. With the emergence and use of rapid prototyping, both scientists and students have been able to build and analyze models for the purpose of theoretical analysis, comprehension, and study. Medical doctors can build models of a diseased or injured human body for the purpose of analysis and determining the various test procedures to be used. With the aid of rapid prototyping, market researchers can receive quick feedback on what the ultimate end users think of a new product.

The key steps involved in product development using rapid prototyping are shown in Figure 1.2. It is apparent in this figure that creating models more quickly can save a lot of time, and there are advantages to be gained by being able to test a number of models to eliminate flaws in their design and make any changes required for improved performance. The technology of rapid prototyping has been used not just for the creation of models but also for the creation of finished products constructed of plastic materials.[48] Subsequently, this technology has come to be termed 3D printing (3DP), although in reality it originated as rapid prototyping.[49,50]

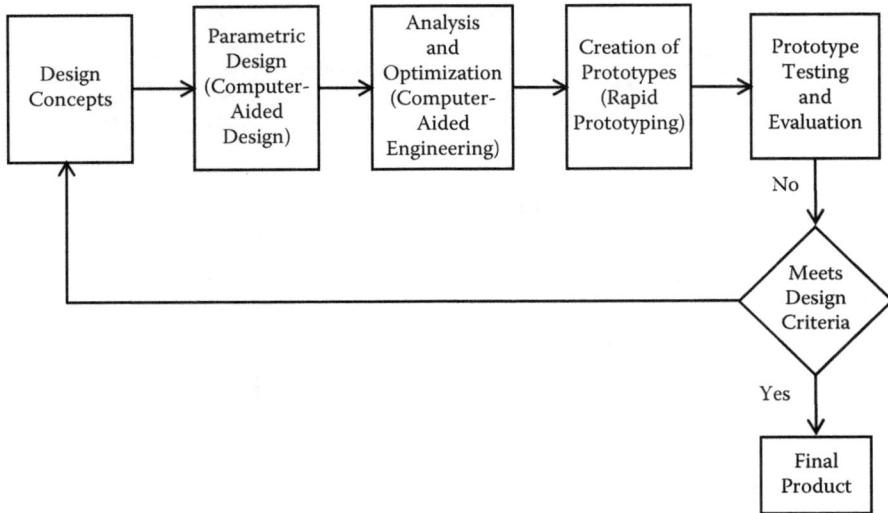

FIGURE 1.2 Pictorial depiction of the product development cycle.[19]

The advent of rapid prototyping and the concurrent emergence and use of the technologies of computer-aided design (CAD), computer numerical control (CNC), and computer-aided manufacturing (CAM) resulted in the evolution of rapid manufacturing (RM). The CAD, CNC, and CAM technologies, when combined together, made possible the printing of three-dimensional objects.[48,50,51] Rapid prototyping was found not to be the best solution in all cases, especially when dimensions of the part are larger than the available additive manufacturing printers. This raised the need for the use of CNC machining processes. The materials often chosen and used for rapid prototyping are still very limited. It became clear following years of trials that it is possible to print parts made from both metals and ceramics but not from all of the available and commonly used manufacturing materials.[51]

Different additive manufacturing processes build and eventually consolidate the layers in different ways. A few of the processes depend on thermal energy from laser or electron beams that is directed via optics to melt or sinter (form a coherent mass without initiating melting) metal powder or plastic powder together. Figure 1.3 provides a pictorial overview of the various additive manufacturing processes. In this figure, the processes are classified as liquid-based, solid-based, or powder-based.[51] The processes discussed in this chapter are considered to be most relevant and promising with regard to the overall future of this rapidly evolving technology for a broad spectrum of materials. Additive processes, which generate parts in a layered manner, first appeared in the 1980s with stereolithography (SLA). Since then, several new ideas have been put forth, many patents have been approved, new processes have been invented, and a few have eventually been commercialized. An overview of the different processes is given in Table 1.1. The processes considered and briefly discussed in this section are

1. Stereolithography (SLA)
2. Fused deposition modeling (FDM)
3. Laminated object manufacturing (LOM)
4. 3D printing (3DP)
5. Selective laser sintering (SLS)
6. Laser Engineered Net Shaping (LENS™)
7. Electron beam melting (EBM)

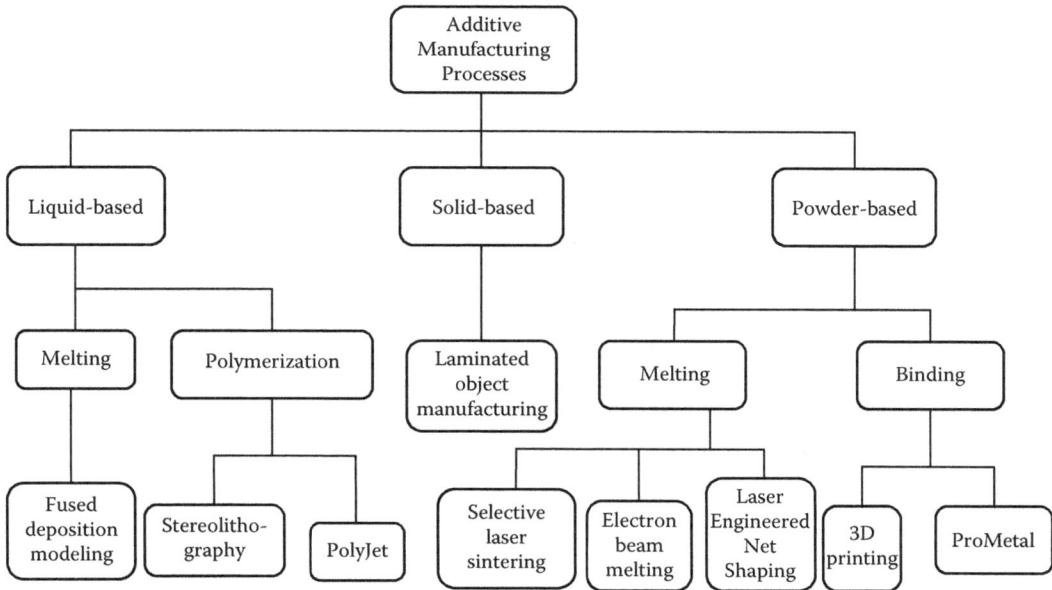

FIGURE 1.3 Pictorial overview of various additive manufacturing processes.[51]

The liquid-based and powder-based processes were found to be more promising than the solid-based processes, such as laminated object manufacturing (LOM). In 2004, the techniques of electron beam melting (EBM), Laser Engineered Net Shaping (LENS), ProMetal, and PolyJet became non-existent.

1.2.2 STEREOLITHOGRAPHY

Stereolithography, initially developed by 3D Systems, Inc. (Rock Hill, SC), was the first and most widely applied rapid prototyping process. The patent for SLA (U.S. Patent 4575330) was awarded in 1986. In essence, it is a liquid-based process that consists of the curing or solidification of a photosensitive polymer when an ultraviolet laser makes contact with the resin. The process begins with construction of a model using CAD software. The model is then translated to a STL file in which the pieces are cut into "slices," with each slice containing the information required for each layer. The thickness of each layer as well as its resolution depend on the equipment used. A platform is created

TABLE 1.1

Layer Manufacturing Technologies, Acronyms, and Years of Development[63]

Name	Acronym	Years of Development
Stereolithography	SLA	1986–1988
Solid ground curing	SGC	1986–1988[a]
Laminated object manufacturing	LOM	1985–1991
Fused deposition modeling	FDM	1988–1991
Selective laser sintering	SLS	1987–1992
3D printing	3DP	1985–1997

[a] Discontinued in 1999.

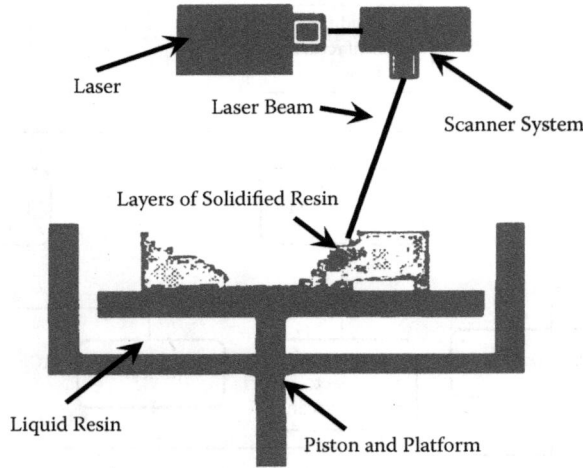

FIGURE 1.4 The technique of stereolithography and the key components involved.[19]

to hold the piece and to support any overhanging structures. An ultraviolet laser applied to the resin initiates solidification at specific locations of each layer. When the layer is finished the platform is lowered. When the process is finally done any excess resin is drained and subsequently reused.[48,50,51] Sustained progress through the years resulted in the development of an advancement of this process that offered much higher resolution and was known as *microstereolithography*. In this process, a layer thickness of less than 10 microns is easily achieved.[52]

The technique of microstereolithography helps provide three-dimensional microstructures. Promising applications of this novel technology are ultra-dense routing for three-dimensional structural electronic substrates having width, thickness, and space dimensions extending down to the level of microns. The technique has made possible many new applications in the domains spanning consumer, medical, and defense electronics. A few examples include the following:

1. Camouflaged sensors that can be hidden in plain sight for homeland security purposes
2. Dynamically adaptive prosthetics in which comfort and fit can be monitored and adjusted
3. Electronics embedded in structural components of a vehicle or building
4. Implantable electronics consisting of biocompatible materials
5. Microsystem packaging that requires high spatial resolution to create intricate cavities
6. Physically small, yet electronically large, 3D antennas that offer improved performance
7. Wearable electronics that are made to fit a specific industry

The key components or parts of a stereolithography machine are shown in Figure 1.4.

The basic principle governing the process of stereolithography is photopolymerization. This is essentially a process where a liquid monomer or a polymer is converted into a solid by the application of ultraviolet light. The ultraviolet light acts as a catalyst for the reactions; the process is also known as ultraviolet curing. It has also been found to be applicable for powders of a ceramic suspended in a liquid.[53]

Errors can be made during the process of stereolithography. A significant error that often occurs during processing is over-curing, which tends to occur in overhanging parts because of the lack of fusion with a bottom layer. An error in the scanned line shape can be introduced during the scanning process. Because the resin used is a high-viscosity liquid, the thickness of the layer is often the only variable and errors can occur during position control. Other errors can occur when the part or component must have a specific surface finish that under normal circumstances would be provided manually.[54] When sustained research and development efforts eventually led to the possibility of using different materials to build a piece, the process became known as *multiple material stereolithography*. In order to print with different materials, all of the resin has to be drained and new material provided when the process reaches the layer where the change must take place. This is essential even if the first material is to be used again as it is only possible to print consecutive layers.

1.2.3 FUSED DEPOSITION MODELING

The patent for fused deposition modeling (FDM) (U.S. Patent 5121329) was awarded in 1992. In this technique, liquid thermoplastic material is easily extruded from a movable FDM head and then deposited in ultra-thin layers onto a substrate. Basically, the material is heated to one degree above its melting point so that it solidifies immediately following extrusion and also easily cold welds to the previous layers. The materials that have been used for this additive manufacturing technique include polycarbonate (PC), acrylonitrile butadiene styrene (ABS), polyphenyl sulfone (PPSF), PC–ABS blends, medical-grade polycarbonate, wax, metals, and even ceramics.[3] In recent years, machines having two nozzles have been developed; one nozzle is for the part material and the other for the support material, which is often less expensive and can easily break away from the parent without impairing its surface finish.[55] The noticeable disadvantages of this technique include the following:[56]

1. A seam line between the layers
2. The need for supports
3. Longer build-up time, sometimes requiring days to build complex parts
4. Delamination arising as a direct consequence of fluctuation in temperature
5. Low resolution of the Z-axis compared to other additive manufacturing techniques

The observable advantages of this process include the following:[48,50]

1. No chemical post-processing required
2. Less expensive equipment due to no resins having to be cured
3. Use of materials resulting in an overall cost-effective process

A schematic of the basic fused deposition modeling process is shown in Figure 1.5.

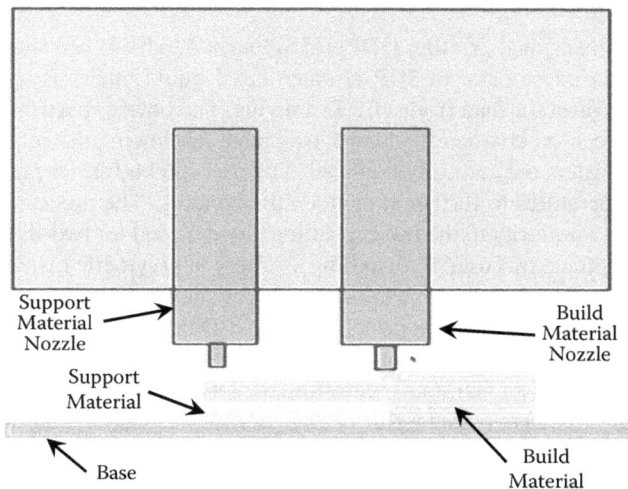

FIGURE 1.5 The key components of fused deposition modeling.[19]

1.2.4 LAMINATED OBJECT MANUFACTURING

The patent for laminated object manufacturing (documented in U.S. Patent 4752352) was awarded in 1988.[56] The simplest LOM technique uses adhesive-coated sheet materials. The adhesive, which is either precoated onto the materials or deposited on the surface immediately prior to bonding, allows the sheets to be attached to each other. The 3D parts are then manufactured by sequential lamination and the cutting of two-dimensional (2D) cross-sections using a laser beam. The velocity and focus of the laser beam are adjusted so the cutting depth corresponds to the thickness of the layer. By careful selection of both the velocity and focus, damage to the underlying layers is either minimized or avoided.

A variety of materials, including paper, metals, plastics, fabrics, synthetic materials, and even composites, have been tried. Most importantly, this technique is relatively inexpensive, and toxic fumes are not generated. It offers the advantage of being automated, thus requiring minimal operator assistance. The primary limitation of this technique can be associated with accuracy problems along the Z-axis resulting in instability in dimensions. This technique can also cause internal cavities, thus influencing the quality of the end product. To totally reduce waste, post-production is necessary, and in a few cases secondary processes are necessary to generate parts that are functional.[58] The noticeable advantages of this technique include the following:

1. Low cost
2. No need for post-processing or support structures
3. No deformation or phase change during the process
4. The possibility of building large parts

Four observable disadvantages of this technique are

1. Fabrication material is subtracted, thus causing wastage.
2. The surface has low definition.
3. The material is directionally dependent with regard to both machinability and mechanical properties.
4. Complex internal cavities are difficult to build.

1.2.5 THREE-DIMENSIONAL PRINTING

The patent for three-dimensional printing (3DP) (U.S. Patent 5204055) was awarded in 1993.[59] This process is an MIT-licensed process. In 3DP, a water-based liquid binder is applied in a jet onto a starch-based powder to print the data from a CAD drawing. The powder particles lie on a powder bed and become glued to each other when the binder is applied. Following the sequential application of layers, the unbound powders are carefully removed. The part can be further processed by subjecting it to firing at high temperatures to further strengthen the bonding. The process is referred to as 3DP primarily because of its similarity to the inkjet printing process used for two-dimensional printing on paper. Initially, the technique was used for handling a variety of polymeric materials.[48,57] This method can also be applied for the production of metal, ceramic, and metal–ceramic composite parts. Various materials can be dispensed using different print heads, thus 3D printing can be used to exercise control over the local material composition. In fact, proper placement of the droplets can produce surfaces that have a controlled texture and can facilitate controlling the internal microstructure of the printed part.

This technique can be used to create a part of any geometry from the material chosen while concurrently facilitating controlling both the microstructure and resultant properties of the part being produced. The layering process is repeated until the part is completely printed. Subsequent to heat treatment the unbound powder is removed, resulting in the fabricated part. The sequence of operations is neatly summarized in Figure 1.6. Through sustained research and development efforts, 3DP

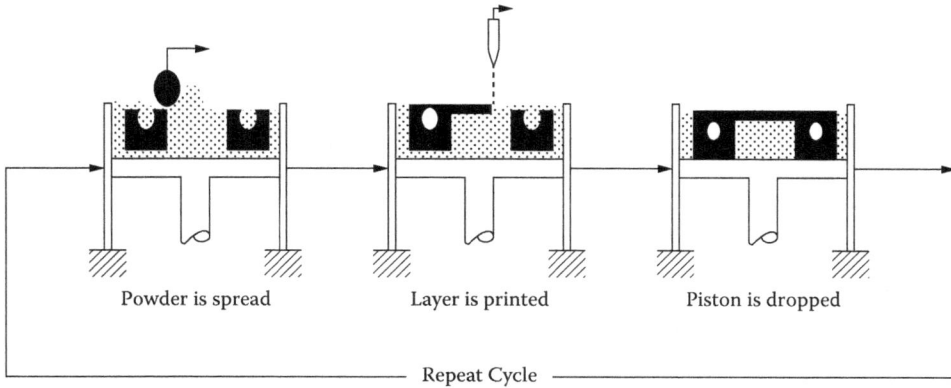

Powder is spread   Layer is printed   Piston is dropped

———————————— Repeat Cycle ————————————

FIGURE 1.6 The sequence of operations that take place in 3D printing.[58]

has been used to create parts having any geometry, including internal cavities, as long as there is a hole for the loose powder particles to escape. The support provided by the powder bed implies that parts having overhangs, undercuts, and internal cavities can be created with ease. Overall, the additive approach provides for great flexibility in addressing complex geometries. The flexible nature of 3DP is summarized in Figure 1.7. In essence, the technique of 3DP offers the advantage of speedy fabrication and reduced material cost.[60] Currently, it can easily be considered to be one of the fastest growing of all of the commercially available additive manufacturing processes.

Desktop 3D printers commercially available for consumer use allow objects such as toys and even household decorative items to be printed from the comfort of home. In foundries, this technology has seen two distinct applications. The first is for making molds. In a typical foundry, both metal and wax patterns are used to create molds at various stages. Depending on the life of the part, this historically time-consuming step was usually found to be justified; however, to print very short runs of parts, such tooling can be skipped by directly printing sand molds and cores. Very short-run, or one-off, parts can be produced with relative ease on a 3D printer if the material requirements can be met.

Three-dimensional printing is also being increasingly used in medical devices and aircraft manufacturing and is fast becoming a standard in some manufacturing processes. The development, growth, and emergence of 3D printers gradually led to their use in (1) life sciences, (2) medical devices, and (3) aerospace, with inroads being made into its use for hobbies and even personal products.[61,62] The rapid development of consumer 3D printers began with the emergence of printer kits, which eventually evolved into neatly wrapped, useable-out-of-the-box printers.[62]

Combined with 3D printing, additive manufacturing has the potential to allow the manufacture of individual, customized products anywhere in the world. Rather than manufacturing goods in one location and shipping them all over the world, 3D printing makes it possible to send design blueprints instantaneously so the goods can be produced at the actual location where they are needed.

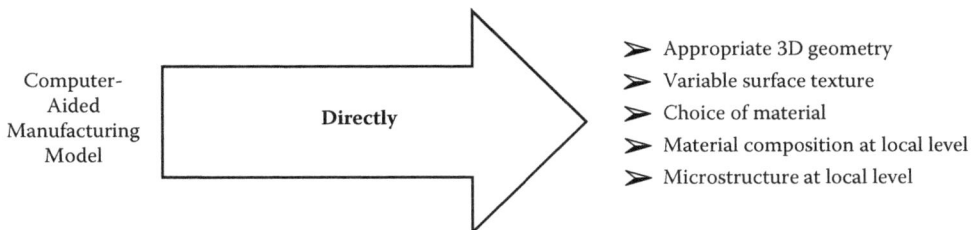

Computer-Aided Manufacturing Model

Directly

➤ Appropriate 3D geometry
➤ Variable surface texture
➤ Choice of material
➤ Material composition at local level
➤ Microstructure at local level

FIGURE 1.7 The key components of 3D printing.[58]

Traditional manufacturing has almost always required a significant investment in both factories and machinery, but 3D printing can vastly reduce such overhead while also reducing transportation costs and the costs associated with logistics. 3D printing allows just-in-time logistics to be replaced by just-in-time manufacturing, giving businesses the opportunity to be more competitive and agile.

Over time, 3D printing has grown significantly more sophisticated. The newer generation of 3D printers can create increasingly complex objects, even ones that have different component parts. Advances in the processing of metals, such as metal sintering, have made it possible for 3D printers to no longer be restricted to generic plastic materials. 3D printing also offers the potential to manufacture products with multiple working components, thus creating a product that is operative immediately without further assembly. These and numerous other capabilities will grow with time, significantly enhancing the power of 3D printing.

In the high-performance aerospace industry, functional parts have been 3D printed and tested for use on Tornado fighter jets used by the Royal Air Force (RAF).[60] In North America, General Electric Aviation is producing a key fuel injector nozzle using 3D printing. In fact, the aerospace industry is putting 3D printing to effective use to make lighter parts with reduced material waste compared to traditional subtractive machining. Although several companies are using 3D printing to make prototype parts, major manufacturers such as Airbus, General Electric, and Siemens are currently using the technology to develop high-quality production parts. Siemens has used the technology to make spare parts and important gas turbine components, resulting in a reduction in repair times by as much as 90%. The ability to manufacture parts when they are needed minimizes costs arising from storage, as well as the lead time required to replace a critical part when it fails. In 2013, General Electric Aviation acquired Morris Technologies and its direct metal laser sintering (DMLS) technology. The company is using the additive manufacturing technology for the manufacture of fuel nozzles. Airbus has put additive manufacturing to effective use, along with Rolls-Royce and Pratt & Whitney. Aerojet Rocketdyne is also working on the use of 3D printing for components of a rocket engine.

In the consumer-oriented automotive sector, 3D printing has been used for a few components, such as mirror mounts, to reduce costs. 3D printing does have some limitations with regard to its use in this sector, such as a rough surface finish, limitations in size, and generally overall high costs to produce the end part or product. A commercially available 3D printing machine is shown in Figure 1.8.

FIGURE 1.8 Commercially marketed 3D printing machine.

TABLE 1.2
Summary of Worldwide Demand for 3D Printers

Region	Demand ($ million)				Annual Growth (%)	
	2007	2012	2017	2022	2007–2012	2012–2017
North America	361	900	2285	5120	20.0	20.5
United States	335	825	2085	4610	19.8	20.4
Canada	26	75	200	510	23.6	21.7
Western Europe	194	495	1225	2750	20.6	19.9
Asia Pacific	183	445	1170	2930	19.4	21.3
China	56	155	500	1450	22.6	26.4
Japan	81	165	330	660	15.3	14.9
Other Asia-Pacific	46	125	340	820	22.1	22.2
Central and South America	7	25	75	22	29.0	24.6
Eastern Europe	13	35	103	290	21.9	24.1
Africa/Middle East	17	50	142	390	24.1	23.2
Total	775	1950	5000	11,700	20.3	20.7

Source: Leigh, S.J. et al., *PLoS ONE*, 7(11), e49365, 2012.

Based on the growth in the use of 3D printers and 3D printing technology, world demand for 3D printers, in conjunction with all of the related or relevant materials and software, is projected to rise by about 20% every year, reaching $5 billion by the year 2017 (see Table 1.2). Professional use, such as prototyping, will continue to account for the majority of the demand for 3D printers, but one can definitely expect to see growth in consumer-related applications. Currently, 3D printers are being preferentially used for the manufacture of direct production parts and finished goods for a spectrum of applications. A few other applications of 3D printing are presented later in Section 1.9.

The drive to develop multi-material printing, full-color printing, and fabrication-grade printing is continuing to grow at a steady pace. In recent years, some of the most noticeable improvements made to 3D printers have occurred in the domain of materials. Stratasys has become involved in the use of both nylons and polycarbonates, an important development as these materials are commonly chosen for use in various manufacturing processes. In fact, the overall range of materials being printed in 3D has experienced progressive growth or expansion. Current 3D systems can print more than 100 different materials, including food ingredients, waxes, ceramics, plastics, and of course metallic materials. The ever-increasing possibility of adapting more materials for 3D printing is generating considerable excitement in the technical world. Improved industrial 3D printers are needed that are adequately streamlined so as to fit within standard office spaces, thus providing a range of affordable print solutions for users within, for example, the professional, medical and dental, and jewelry sectors. Major growth in future 3D printing technology is expected within the manufacturing sector. Although analysts have predicted that real growth will come from the personal printing arena, sustained growth in manufacturing will occur as companies look at this technology to customize or semi-customize their products commensurate with technological advances.

At its current rate of growth and implementation, 3D printing has the potential to revolutionize defense and even foreign policy, by not only making possible incredible new designs but also turning the defense industry and the entire global economy on its head. The current billion-dollar defense industry can be safely considered to be at the "bleeding" edge of this innovation. The U.S. military is already investing heavily in efforts to print uniforms, synthetic skin to treat battle wounds, and even food. By effectively undermining sanctions 3D printing can gradually change foreign policy. Although the United States has sanctioned everything from spare parts for a fighter jet to oil equipment, the emergence and prudent use of 3D printing can turn sanctions, which have been a crucial part of foreign policy for decades, into an antiquated notion.

Numerous patents on the original technology are expiring soon, which will lead to increased competition to enhance quality while lowering costs. All of the developments and concomitant applications have inspired scientists at the Massachusetts Institute of Technology to develop "4D" printing by creating materials that gradually change when they come into contact with elements, such as water.

1.2.6 SELECTIVE LASER SINTERING

The patent for selective laser sintering (SLS) (U.S. Patent 4863538) was awarded in 1989.[64,65] This additive manufacturing process uses a high-power laser to fuse small particles of the build material (metal, ceramic, polymer, glass, or any material that can be easily pulverized). The fabrication powder bed is heated to just below the melting point of the material with the primary objective of minimizing thermal distortion and to concurrently facilitate fusion to the previous layer. Each layer is then drawn on the powder bed using a laser to sinter the material. The sintered material forms the part while the unsintered powder remains in place to support the structure; it can be cleaned away and recycled once the part of interest has been built. This process of selective laser sintering offers the freedom to build complex parts that are often more durable and provide better functionality compared to other existing and preferentially used additive manufacturing processes. Further, in this process no post-curing is required, and the build time is noticeably fast. A great variety of materials can be used. The materials normally chosen include (1) plastics, (2) metals, (3) combination of metals, (4) combination of metals and polymers, and (5) combination of metals and ceramics.[52,65,66] Other materials that have been used are composite materials and reinforced polymers.[67] For the specific case of metals, it is necessary to have a binder. The binder is often made of a polymer, which is removed during heating or mixing of metals that have very different melting points.[65–67] This particular technique has been used to build parts of alumina having high strength, with polyvinyl alcohol being used as the organic binder.[48] Overall, this additive manufacturing technique can be categorized as complicated, as several build variables must be carefully controlled. The observable disadvantages specific to this process include the following:

1. Accuracy is strictly limited by the size of the particles of the chosen material.
2. Oxidation must be avoided by executing the process in an inert atmosphere or vacuum.
3. The process must occur at a constant temperature near the melting point of the chosen material.

Over time, this process has come to be referred to as direct metal laser sintering (DMLS).

1.2.7 LASER ENGINEERED NET SHAPING

Laser Engineered Net Shaping (LENS) was initially developed at the Sandia National Laboratories (SNL) in collaboration with Pratt & Whitney and was subsequently licensed to Optomec, Inc., in 1997.[68] A patent for this technique (U.S. Patent 6046426) was awarded in 2000. In Laser Engineered Net Shaping, a part is fabricated by focusing a high-powered laser beam onto a substrate with the primary objective of creating a molten pool into which metal powder particles are injected to gradually build each layer. The substrate is moved gently below the laser beam to deposit a thin cross-section and thereby create the desired part geometry. Consecutive layers are then sequentially deposited to build a 3D part. By careful control of the (1) fabrication parameters, (2) desired geometric properties including both accuracy and surface finish, and (3) material properties, such as strength and ductility, a complex part can be easily built.[69] The LENS technique has been used to

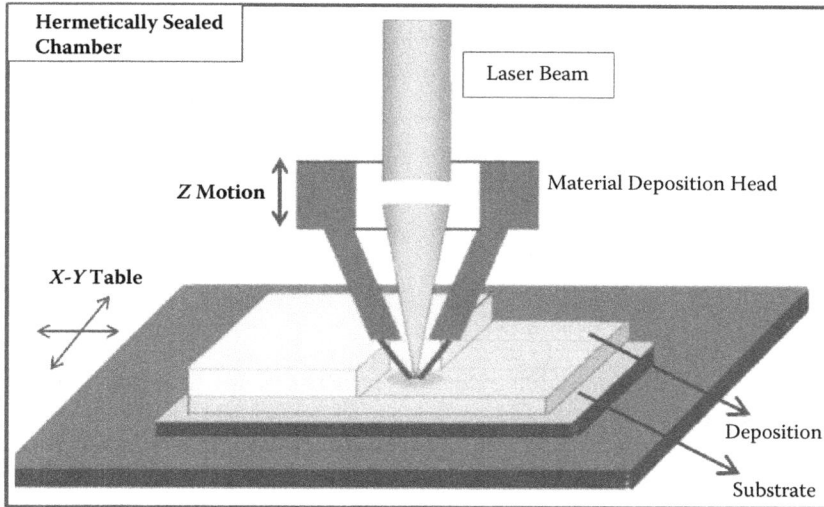

FIGURE 1.9 Key LENS components and their roles in the process.[75]

repair existing parts and structures, but it is primarily used to fabricate new ones. This technique does require a post-production process in that the finished part must be cut from the substrate. The final part has a rough surface finish, which requires subsequent machining or polishing. A noticeable problem with this process is the presence of residual stresses caused by uneven heating and cooling processes. This can result in some amount of distortion and is most certainly not desirable. The residual stresses can be noticeably significant in high-precision processes such as the repair of turbine blades.[70–72] A schematic of this process is shown in Figure 1.9.

1.2.8 ELECTRON BEAM MELTING

The process of electron beam melting (EBM) is quite similar to selective laser sintering. This relatively new and innovative additive manufacturing process has grown rapidly as a result of its technological significance and resultant far-reaching applications. In this process, an electron laser beam is used to melt the powder. The laser beam is powered by a high voltage of the order of 30 to 60 kV. To minimize or eliminate environmental interactions, such as oxidation, this process is conducted in a high-vacuum chamber when building metal parts. This process has been used extensively to build metal parts, and it has the ability to process a wide variety of pre-alloyed metal powders. A potential future application of this technique is manufacturing in outer space,[73,74] which is particularly achievable as the technique is often done within a high-vacuum chamber.

1.3 ADDITIVE MANUFACTURING TECHNIQUES USED FOR METALLIC MATERIALS

Additive manufacturing began with organic materials but gradually gained applicability for the manufacture of metallic materials when the technique was able to meet the following requirements:

1. Desired level of performance
2. Efficiency necessary for the manufacturing process
3. Endurance or sustainability
4. Energy savings
5. Cost savings

TABLE 1.3

Selected Metals and Alloys Used Commercially in Additive Manufacturing

Titanium	Aluminum	Tool Steels	Super Alloys	Stainless Steel	Refractory
Ti-6Al-4V	Al-Si-Mg	H13	IN625	316 and 316L	MoRe
ELI titanium	6061	Cermets	IN718	420	Ta-W
CP titanium	—	—	Stellite	347	Co-Cr
Gamma-titanium aluminide	—	—	—	PH 17-4	Alumina

Source: Frazier, W.E., *J. Mater. Eng. Perform.*, 23, 1917–1928, 2014.

A few of the commercially viable additive manufacturing techniques that have been developed and eventually consolidated as a direct consequence of the numerous benefits they offer were discussed in the preceding section. Some of these techniques have been developed or adapted specifically for metallic materials as extensions of the techniques used for polymeric materials. The additive manufacturing system for metallic materials can be categorized or classified in terms of

1. Material feed stock
2. Source of energy used
3. Build volume

The most common metallic materials used in additive manufacturing processes are summarized in Table 1.3, and Table 1.4 provides an overview of existing manufacturers and their specific equipment. The manufacturing systems can be divided into three broad categories:

TABLE 1.4

Representative Additive Manufacturing Equipment, Processes, and Sources of Energy

System	Equipment	Process	Energy Source
Powder bed	ARCAM	Electron beam melting	Electron beam
	EOS	Direct metal laser sintering	Yb-fiber laser
	Concept laser cusing	Selective laser melting	Fiber laser
	MTT	Selective laser melting	Fiber laser
	Phoenix System Group	Selective laser melting	Fiber laser
	Renishaw	Selective laser melting	Laser
	Realizer	Selective laser melting	Laser
	Matsuura	Selective laser melting	Fiber laser
Powder feed	Optomec (LENS)	Laser engineered net shaping	Fiber laser
	POM DMD	Direct metal deposition	Disk laser
	Accufusion laser consolidation	Laser cutting	Nd:YAG laser
	Irepa laser	Laser deposition	Laser cladding
	Trumpf	Laser deposition	—
	Huffman	Laser deposition	CO_2 cladding
Wire feed	Sciaky	Electron beam deposition	Welding
	MER plasma transferred arc	Plasma-transferred arc selected free-form fabrication	Plasma transferred arc using two 350A DC power supplies

Source: Frazier, W.E., *J. Mater. Eng. Perform.*, 23, 1917–1928, 2014.

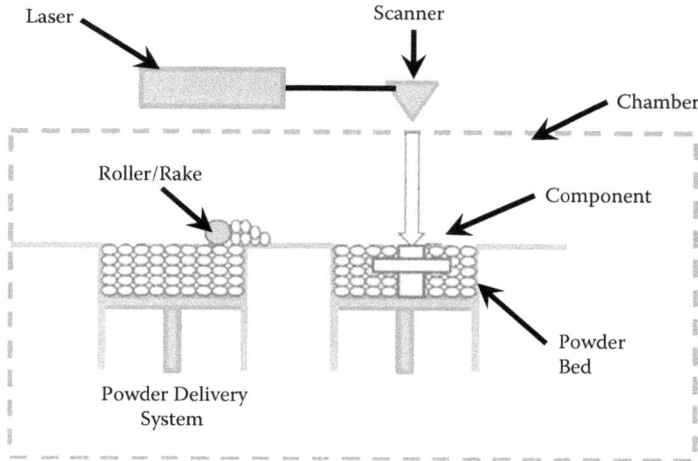

FIGURE 1.10 The key components and their role in the powder bed system used in additive manufacturing of metallic materials.[20]

1. Powder bed fusion system
2. Powder feed system
3. Wire feed system

Overall, the source of energy (arc, electron beam, or laser beam) for each feed system is the key to governing operation of the system.

1.3.1 POWDER BED FUSION

The powder bed fusion technology was initiated from selective laser sintering (SLS) and has gradually evolved into various techniques that have similar working principles but use different mechanisms to bind the powders and the layers. Methods based on a combination of laser beams and powder beds include the original selective laser sintering and the subsequent and preferentially used direct metal laser sintering (DMLS). Replacing or substituting the laser beams with electron beams resulted in the technique of electron beam melting (EBM). A synergism of an inkjet head and a powder bed system resulted in the emergence of 3D printing. A schematic of the generic powder bed system used in additive manufacturing is shown in Figure 1.10.

The key ingredients of this process consist of a (1) laser scanning system, (2) powder delivery system, (3) roller, and (4) fabricated piston. Prior to start or initiation of fabrication, the powder delivery piston moves up and the fabrication piston moves down by one layer thickness. The powder is spread and subsequently lightly compressed by a roller over the surface of the fabrication piston. A laser beam is then driven over the bed so as to selectively melt the powder under the guidance of a scanner system. Upon completion of a layer the fabrication piston moves down another layer thickness and a new layer of powder is spread over it. The process is repeated until the entire part or a solid three-dimensional component is constructed. Upon completion, the fabrication piston moves up and elevates the final object or part. The excess powder, if any, is brushed away and can be reused following proper treatment.[75,76] A distinct advantage of the powder bed fusion system is that support structures are not required. Quite often supports are added to provide thermal pathways to facilitate rapid dissipation of the heat and to concurrently exercise better control over the geometry of the part being produced. An overview of the direct metal laser sintering technique is provided in Figure 1.11.

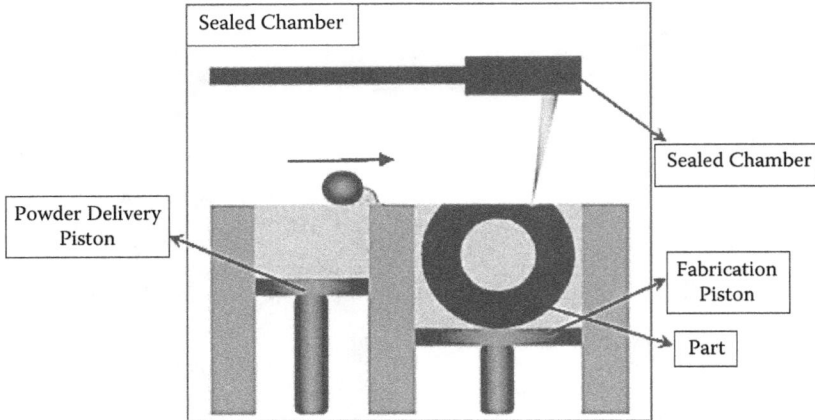

FIGURE 1.11 Visualization of the various parts involved in direct metal laser sintering (DMLS).[75]

1.3.2 POWDER FEED SYSTEM

The build volume of this system is generally large. The powder feed system generally lends itself to building a larger scale volume compared to powder bed units. In this system, the metal powder particles are conveyed through a nozzle onto the build surface. A laser is then used to melt a mono-layer or more of the powder into the shape desired. The process is repeated to create a solid three-dimensional component. There are two variations of this system:

1. The work piece remains stationary while the deposition head moves.
2. The deposition head remains stationary while the work piece is moved.

The two most noticeable advantages of using this system are the following:

1. Large build volume
2. The ability to refurbish both worn and damaged components

An illustration of the powder feed system is provided in Figure 1.12. The use of lasers allows this technique to process a wide range of metallic alloys, including titanium, nickel-based superalloys, stainless steels, and tool steels. These materials are commonly available in the powder form required by the process.[76,77] Sustained research and development efforts have culminated in two major well-developed techniques:

1. Laser Engineered Net Shaping (LENS), which was initially developed at Sandia National Laboratories (Albuquerque, NM) in the 1990s
2. Direct metal deposition (DMD), which was developed by the POM Group, Inc. (Auburn Hills, MI)

The key difference between the two techniques lies in the details specific to machine control and implementation.[78] Direct metal deposition allows for processing in an open atmosphere with local shielding provided to the molten metal. The metal powders used in this technique are both delivered and distributed using an inert gas carrier to shield the pool of molten metal from oxidation while allowing layer-to-layer adhesion to ensure a better overall wetting of the surface.[79]

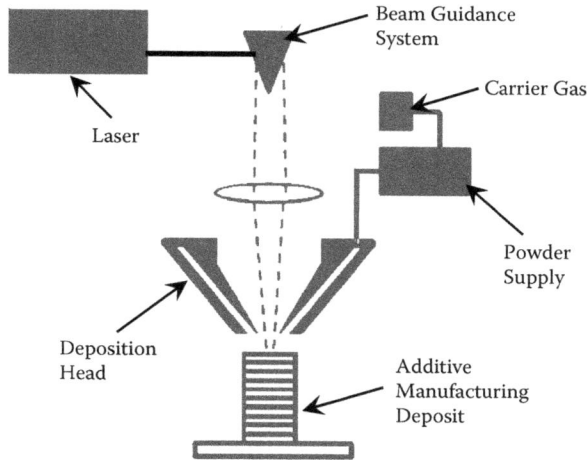

FIGURE 1.12 Key components of the powder feed system used for additive manufacturing of metallic materials.[20]

1.3.3 Directed Energy Deposition: Wire-Based Method

Directed energy deposition (DED) is another class of additive manufacturing technique. This technique makes use of a solid wire feed stock instead of metal powders. A schematic of the wire feed system is shown in Figure 1.13. A key technique belonging to this class that has gained in both significance and approval for use is electron beam freeform fabrication (EBF3). The technique of electron beam freeform fabrication was developed in 1999 by Lockheed Martin (Bethesda, MD) and disclosed to the public in 2002.[81] The technology was subsequently studied and improved upon at the NASA Langley Research Center (Hampton, VA) for the production of unitized structures using such aerospace materials as[81]

1. Alloys of aluminum
2. Alloys of titanium
3. Nickel-based superalloys
4. Titanium–aluminides
5. High-strength steel
6. Metal–matrix composites

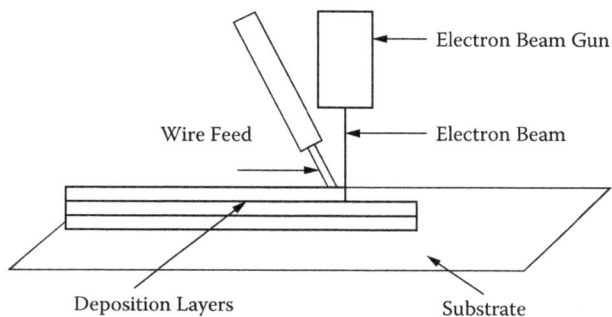

FIGURE 1.13 Simple line diagram illustrates the principle behind the wire feed system used for additive manufacturing of metallic materials.[20]

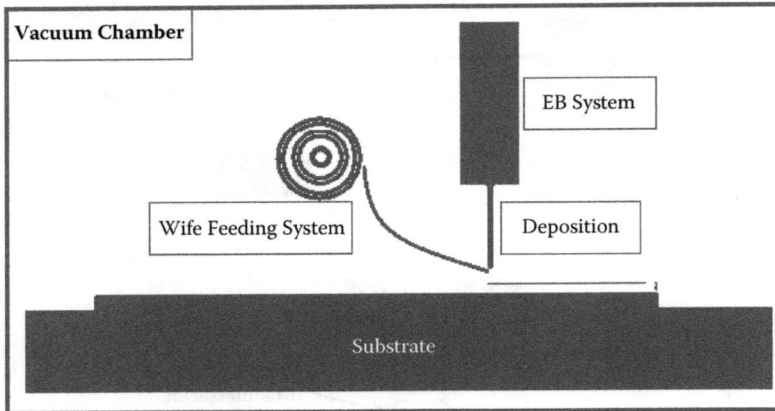

FIGURE 1.14   Key components and their roles in the electron beam freeform fabrication technique.[75]

The technique of EBF3 is similar to laser engineered net shaping except that electron beams are used as the heat source. This technique is often performed in a vacuum environment (10^{-4} torr or lower) and incorporates a metal wire feed system to deliver feedstock to the molten pool. The electron beam can be controlled and deflected very precisely and can synchronize well with highly reflective materials. Sustained research and development efforts have fine-tuned the process to give efficiencies as high as 100% in the consumption of wire feed stock and at least 95% in the use of power.[81] A schematic of the electron beam freeform fabrication technique is shown in Figure 1.14.

The National Aeronautics and Space Administration (NASA) has two types of EBF3 systems:

1. *Ground-based*—The ground-based system has a dual-wire feed system that can be loaded with either fine or coarse wire to achieve various feature definitions. Alternatively, two different alloys can be used to produce compositional gradients or components made from multiple materials.
2. *Portable*—The portable system has a single wire feeder and can be used for finer metal wire; it has high precision in positioning compared to the ground-based system, making it ideal for the fabrication of smaller parts having intricate detail.

NASA also has two portable EBF3 machines. The first machine has been flown on a microgravity research plane, and the second machine is being used for a spectrum of activities related to developing potentially viable in-space manufacturing applications.

1.4   USE OF COLD SPRAY FOR ADDITIVE MANUFACTURING

The technique of cold spray is beneficial for use in applications that put to effective use heat-sensitive substrate materials or those having difficult-to-reach spray areas. One such example is spraying inside a small cylinder, heat-sensitive tubes, or boxes with the primary intent of providing good to improved corrosion resistance. Cold spray produces deposits that are not only oxide free but also fully dense and offer acceptable mechanical properties. Basically, the process requires heating a pressurized carrier gas, typically nitrogen or air, that is passed through a convergent–divergent nozzle. The divergent section of the nozzle creates a supersonic gas jet as the carrier gas gradually expands toward the exit of the nozzle. The spray material, usually in powder form, is injected into the gas jet either upstream or downstream of the nozzle.

Based on the temperature of the process, each material requires a specific minimum particle velocity in order to form a well-bonded and dense deposit. Forming a dense deposit depends on the material's ability to plastically deform upon impact. The less ductile the material used in the spray, the greater the particle velocity required to facilitate bonding. The cold spray powder mixture must contain at least one material that can easily deform upon impact with the surface of the substrate.

As the intricacies specific to a manufacturing process gradually advance, the technique of cold spray is becoming more attractive as an enabling technology for both 3D printing and other additive manufacturing. Whereas traditional manufacturing has almost always relied on substructure manufacturing techniques that require systematic removal of material from bulk shapes by either cutting or drilling to arrive at the final shape desired, additive manufacturing builds shapes by precisely adding and immediately consolidating layers following a 3D digital model. The geometric quality of a complex 3D shape is dictated by the overall resolution of the spot size. The smaller the spot size the better, which is why laser beams on powder beds are the preferred method for producing intricate geometries comprised of special metal alloys. The smallest available spot size for cold spray deposition is 4 mm, which is sufficient for 3D restoration of metallic components in the world of remanufacturing or rapid prototyping; however, a much smaller and finer cold spray footprint is often required to produce the desired finished shape or component. With rapid strides being made in the domain of additive manufacturing, the technique of cold spray is becoming both more reliable and practical for the 3D printing of engineering components even at low temperatures.[83,84]

1.5 KEY INGREDIENTS FOR APPLYING AND ADVANCING THE TECHNOLOGY OF ADDITIVE MANUFACTURING

Significant progress has been made in the further development of techniques intrinsic to additive manufacturing, but some of the challenges that remain to be addressed are listed below:[85,86]

1. Limited amount of materials available for use in additive manufacturing processes
2. Relatively poor accuracy of the part or component caused by the "stair-stepping" effect
3. Insufficient repeatability and consistency in the end part that is produced
4. A lack of both qualification and certification methodologies for additive manufacturing processes

Products having complex shapes can now be fabricated without tools, dies, or molds in an efficient and inexpensive manner and will meet all of the necessary functional requirements. By shortening the fabrication time, production costs are appreciably reduced. Research is currently in progress with the primary objective of expediting the transformation of 3D printing from rapid prototyping to additive manufacturing using advanced materials that have the following qualities to offer:

1. Material flexibility
2. Intrinsic ability to generate fine features (less than 100 μm)
3. High throughput

The key elements that exert an influence on additive manufacturing technology are shown in Figure 1.15. These elements are discussed in the ensuing subsections with particular emphasis on existing gaps and current needs.

1.5.1 DESIGN METHODS AND STANDARDS

The unique capabilities of technologies related to additive manufacturing include the intrinsic ability to (1) fabricate complex shapes, (2) tailor materials and properties, and (3) provide functional complexities. These attributes give designers the freedom to explore both novel and innovative

```
                              ┌──────────────────────────┐
                         ┌────│          Design          │
                         │    │   method and standards   │
                         │    └──────────────────────────┘
                         │    ┌──────────────────────────┐
┌───────────────────┐    ├────│   Modeling, monitoring,  │
│     System         │    │    │    control, and process  │
│   integration      │────┤    └──────────────────────────┘
│   and cyber        │    │    ┌──────────────────────────┐
│  implementation    │    ├────│  Materials development   │
└───────────────────┘    │    │     and evaluation       │
                         │    └──────────────────────────┘
                         │    ┌──────────────────────────┐
                         └────│   Characterization and   │
                              │      certification       │
                              └──────────────────────────┘
```

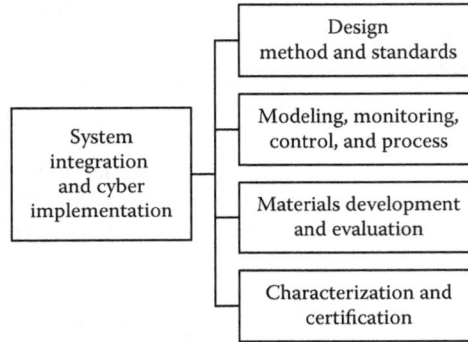

FIGURE 1.15   Key elements essential to ensuring overall effective functioning of additive manufacturing of metallic materials.[85]

applications for the technology, but there is a need for new design tools that can better represent both the functions and material interactions of additive manufacturing. Below are a few of the design tools that must be developed:

- Tools that can further aid designers in exploring designs made possible by additive manufacturing, particularly with regard to representations of shapes, properties, processes, and other related variables
- Methods for simultaneous product and process design coupled with multifunctional design
- Methods to assess life-cycle costs and the impact of additive manufacturing on both the components and products produced

The emphasis on design also requires that CAD systems be modified so as to overcome the limitations imposed by parametric boundary representations and solid modeling in representing complex geometries and multiple materials. One other important factor to be considered is that additive manufacturing calls for simulation capabilities for primitive shapes, materials, material compositions, and even functionally graded materials, to name a few. Additive manufacturing also requires multiple-scale modeling and inverse design capabilities to assist in navigating through complex process–structure–property relationships and improved finite element analysis software to put to effective and efficient use all of the available capabilities.

1.5.2   Modeling, Monitoring, Control, and Process Innovation

In light of recent technological developments and advances, the modeling, sensing, and control of additive manufacturing techniques or processes can easily be considered to rank among the highest priorities for realizing the potential far-reaching applications of this technology. The modeling of additive manufacturing processes presents significant challenges; for example, to understand the transport phenomenon in additive manufacturing processes it is essential to model the temperature, stress, and composition history. Further, it is difficult to predict the microstructures and resultant fatigue properties arising as a direct consequence of the additive manufacturing process used. This difficulty can be ascribed to the extreme heat and cooling rates, which create fundamentally new regimes of material transformation. For the case of polymeric materials, melting and recrystallization have not as yet been adequately understood to allow the robust development of mathematical models. To do so, the physics of polymeric materials be better understood with the primary objective of achieving better effective modeling. The development and availability of supercomputing have had a great impact on modeling efforts and related studies. The complex process models must often be reduced to lower-order models for use with real time parameters and resultant control of the additive manufacturing process. As of now, there exists a noticeable gap between high-fidelity modeling research and real-time efforts aimed at online process control.

Two of the noticeable challenges that must be considered with regard to additive manufacturing sensor processes include a lack of access to the build chamber and the need for intense computing power. The sensing of additive manufacturing processes requires the following:

1. *In situ* measurement of temperature, cooling rate, and residual stresses
2. Calibration of all optical sensors used for the purpose of high-accuracy measurements
3. In-process monitoring of not only the geometric dimensions but also the surface quality of the finished layers

High-speed infrared thermography has primarily been used to obtain imaging data for the prediction of microstructures. This is done by determining the grain size that results from characteristics of the melt pool. For the purpose of process control, to put to effective use such information the images from the additive manufacturing processes have to be processed at a speed of at least 30 kHz. If this is to be achieved, then there exists the challenge of using the information gained for the following purposes:

1. Online process control for both material composition and phase transformation
2. Repair of existing defects, such as the presence of pinholes or porosity, microscopic cracks, and segregation

Careful integration of the control algorithms with the existing equipment being used for additive manufacturing through the control unit of the machine creates a significant barrier to cost-effective implementation of real-time process control of additive manufacturing. Also, addressing the need for improved throughput and the need for multi-material additive manufacturing fabrication capabilities requires the following:

1. Development of multiple-nozzle array print heads and machines capable of integrating multiple additive manufacturing processes
2. Subtractive and finishing processes intrinsic to the technique of conventional manufacturing

A recent and noticeable additive manufacturing process innovation is in the domain of 3D bioprinting. The challenges related to the printing of 3D tissue scaffolds include the following:

1. Biophysical requirements specific to the scaffold's structural integrity, strength stability, and degradation, as well as cell-specific pore shape, size, porosity, and interarchitecture
2. Biological requirements pertinent to both cell loading and spatial distribution, as well as the attachment of cells, their growth, and the eventual formation of new tissues
3. Mass transport considerations with specific reference to topology and interconnectivity
4. Anatomical requirements with specific regard to anatomical compatibility and geometric fitting
5. Manufacturability requirements with reference to the overall ability of the process (e.g., availability of biomaterials coupled with the feasibility of printing) and effects of the process (e.g., warping, distortion, and overall structural integrity)

Ongoing studies have shown that printing *in vitro* biological constructs imposes a requirement for the following:

1. Development of a new generation of biomaterials in the form of bioink for (a) dispensing with cells (a cell delivery medium), (b) growing with cells (support as an extracellular matrix), and (c) functioning with cells (in the role of biomolecules)
2. Developments in engineering (vs. developments in biology) to fill the gap created by a lack of knowledge in biology
3. Commercialization of bioprinting tools so as to be able to make 3D heterogeneous structures in a viable, reliable, and reproducible manner
4. Four-dimensional bioprinting models (obtained by embedding time into the 3D bioprinting models), including stem cells having a controlled release of the biochemical molecules for exercising control of (a) complex tissues, (b) organs, (c) cellular machines, and (d) human-on-a-chip devices

1.5.3 MATERIALS DEVELOPMENT AND EVALUATION

Intense research and development in the specific domain of materials are necessary to

1. Broaden the selection of suitable or appropriate materials.
2. Prepare a database on the mechanical properties of the parts fabricated by additive manufacturing.
3. Determine any and all of the existing interactions occurring between materials and process parameters.

In the fields of metallurgical and materials engineering, it can take well over a decade to develop a new alloy and determine its various mechanical properties, such as mechanical strength, fatigue strength, and analysis. This time frame also applies to engineering new materials for the purpose of additive manufacturing—materials that have a variety of features to offer, ranging from high mechanical strength to high or improved corrosion resistance. Even with the currently available and preferred materials, noticeable advances are still needed in the preparation of raw material feedstock in the form of powder. Overall efforts in the field of materials development should include the development of both metallic and non-metallic inks that have the desired rheological properties so as to enable fairly good resolution on the order of submicrometers. For paste-extrusion-based additive manufacturing processes, the development of ceramic slurries is essential. Similarly, the development of a new generation of biomaterials to serve as cell delivery media or biomolecules is needed in the field of bioprinting.

1.5.4 CHARACTERIZATION AND CERTIFICATION

Actual production practices are much more rigorous than those traditionally used for prototyping purposes; therefore, certification is highly critical in a production-based environment, including certification of (1) equipment, (2) materials chosen and used, (3) personnel, (4) quality control, and (5) logistics. For a manufacturing process to be readily and easily adopted in industry circles, both repeatability and consistency of the manufactured parts are not only important but also essential. With gradual advances in additive manufacturing technology, these attributes have become both necessary and required over the entire build volume and between builds for each machine. Currently, the inability of additive manufacturing to guarantee material properties for a given process is inhibiting its adoption and restricting its use in industry. This is primarily because many of the companies involved lack confidence that the resulting manufactured parts will provide the mechanical properties and dimensional accuracy required to meet the needs of a specific application. The primary reason for this problem is that the prevailing additive manufacturing systems are based on a rapid prototyping machine architecture, which is quite different from the actual requirements for fabricated parts having functional use.

A lack of well-established standards has contributed to the following:

1. The material data reported by the various companies are often not comparable.
2. Various users of the additive manufacturing technology utilize different process parameters when operating their equipment consistent with their own preferences and needs.
3. There appears to exist minimal repeatability of results between suppliers and the service bureaus.
4. Only a few specifications are available that end users can reference.

1.5.5 INTEGRATION OF VARIABLES AND THEIR IMPLEMENTATION

Successful implementation of additive manufacturing technology requires prudent integration of all of the available interdisciplinary knowledge. A good example is bioprinting, for which understanding the intrinsic interactions between materials and processes is essential to bringing about effective cooperation among the engineers and biologists. Such cooperation aids in improving our knowledge of the interactions between cells and the environment in a predominantly structural environment. In addition to noticeable advances in the modeling of biological structures in three dimensions, the fourth dimension of time can be incorporated into models to predict cellular behavior. New 3D bioprinting equipment and tools will be necessary for the reliable and reproducible creation of heterogeneous biological structures.[57,59,60]

The utilization of online resources has become essential for both small- and medium-sized companies. Ensuring successful completion of the many manufacturing tasks requires the support of both suppliers and business partners. Further, there is a need to connect the various manufacturers so they are in a position to share their resources and put them to effective use. The emergence of "cloud manufacturing" has made it possible to share a pool of manufacturing resources, such as the machines commonly used in additive manufacturing; however, readily available cyber–physical systems that can be easily used for "cloud-based" additive manufacturing are in limited supply.[58,59]

1.6 ADVANTAGES AND DISADVANTAGES OF ADDITIVE MANUFACTURING

When compared with conventional manufacturing processes, few of the additive manufacturing processes have seen commercial realization stemming from the ongoing research and development, but additive manufacturing has the following advantages to offer:

- *Efficiency in material use*—Unlike conventional manufacturing based on subtractive principles wherein a large amount of material must be removed, additive manufacturing puts the raw materials to efficient and effective use by building parts layer by layer. The materials that are left over are often reused with minimum processing.
- *Efficiency in the use of resources*—Conventional manufacturing processes often require auxiliary resources such as jigs, fixtures, cutting tools, and coolants, in addition to the primary machine tool; however, additive manufacturing does not require such additional resources. As a result, a variety of parts can be made by small manufacturers that are located close to their customers. This scenario presents an opportunity for improved supply-chain dynamics.
- *Part flexibility*—Because additive manufacturing has no tooling constraints, parts having complex features can often be manufactured in a single piece. This has made it possible to not sacrifice the functionality of a part to ensure ease and repeatability in manufacturing. Also, it is currently possible to build a single part with varying mechanical properties, such as a part being flexible at one end and stiffer at the other end. This has opened up new opportunities for novel design innovations.

- *Flexibility in production*—Additive manufacturing machines do not require costly setups, so they can be economical for the production of small batches. The overall quality of the part is more dependent on the process used rather than skill-set of the operator, which allows production to be synchronized with customer demand. Also, problems associated with line balancing and production bottlenecks are eliminated because complex parts can be easily produced as a single piece.

Additive manufacturing, however, cannot fully compete with conventional manufacturing, especially in the domain of mass production, primarily because of the following:[59]

- *Limitations in size*—Additive manufacturing processes often use a liquid polymer or a powder that is inlaid with resin or plaster to build object layers. The use of such materials for additive manufacturing means that large objects cannot be produced, primarily due to a lack of material strength. Further, large objects become impractical when considering the extended amount of time required to complete the build process.
- *Presence of imperfections*—The parts produced using additive manufacturing processes often have a rough and ribbed surface finish. This appearance is primarily due to the plastic beads or large powder particles being stacked on top of each other. This unfinished look requires further surface preparation by either machining or polishing.
- *Cost*—The initial cost of additive manufacturing equipment can be high. Entry-level 3D printers average approximately $5000 and can go as high as $50,000 for higher end models. This does not include the cost of accessories, resins, and other materials required for operation.

In recent years, researchers at universities, government research laboratories, and a variety of industries have been working on improving additive manufacturing processes to overcome these shortcomings; however, it is unlikely that additive manufacturing will make traditional manufacturing processes obsolete. Further, it is reasonable to expect that several of the additive manufacturing techniques, if not all, will play increasingly important roles in manufacturing as complementary technologies that will allow for the production of replacement parts for equipment currently in use for extended periods of time.

1.7 TECHNOLOGICAL ADVANCES RESULTING FROM ADDITIVE MANUFACTURING

From the standpoint of technological innovations and advancements, additive manufacturing provides useful control over the composition, shape, and functionality of the fabricated end products while facilitating a high degree of personalization. Referred to as the "third industrial revolution,"[86] additive manufacturing has the potential to realize cost-effective mass customization and production of complex products that could not be easily manufactured using any one of the conventionally available techniques. With the aid of additive manufacturing, products can be easily manufactured

1. In a broad range of sizes (from nanometer to micrometer scale to tens of meters)
2. From a variety of materials (metals, polymers, ceramics, composites, and even biological materials)
3. With numerous functionalities (such as load-carrying brackets, energy conversion structures, and tissue-growing scaffolds)

The science and technological capabilities of additive manufacturing have made possible its use in a wide spectrum of applications, including the following:

- Cellular machines
- Custom medications
- Flexible electronics
- High-strength, lightweight aerospace structures having material gradients
- High-power, high-energy-density microbatteries
- Human organs
- Multifunctional houses
- Products having embedded multi-material sensors and actuators
- Turbine blades having internal cavities

These applications represent only a few of the opportunities offered by the technological capability and power of additive manufacturing to print complex shapes that have a controlled composition and functionality.[85,86]

Additive manufacturing has the potential to produce complex parts using functionally graded materials (FGMs). The ability to deliver different materials to building areas is necessary to build components using FGMs. This is an advantages of additive manufacturing that cannot be easily realized using any of the conventional manufacturing methods.[87] Furthermore, additive manufacturing offers the intrinsic flexibility to control the composition of a part or component to provide multiple functionality. Two noticeable examples are[88]

1. Grading tungsten carbide for the primary purpose of enhancing its erosion resistance
2. Grading cobalt for the purpose of enhancing its ductility

Additive manufacturing processes can control and optimize the properties of the part being built. A practical example of this is a pulley that contains more carbide near the hub and rim to make it harder and more wear resistant. It contains less carbide in other areas for enhanced compliance.[87] One other notable example is the nose of a metallic missile cone made using ultra-high-temperature ceramic graded to a refractory metal from the outside to inside surface of the cone so it can sustain extreme external temperatures and still be attached easily to the metallic missile cone.[88,89]

The rapidly evolving capabilities of additive manufacturing include several functions related to printing by careful programming of the behavior of the active materials. This allows the manufactured part (1) to both sense and react and (2) to compute and behave as expected. Active systems can be manufactured that are comprised of both active and passive substructures, an ability that has applications within the biomedical field. Additive manufacturing has been put to both effective and efficient use to fabricate tissue scaffolds that are bioabsorbable, biocompatible, and biodegradable and for *in vitro* biological constructs containing living cells and biological compounds.

Many of the advancements in both consumer and engineering manufacturing have been related to the following:

- Technological intersections of low-cost computer-based computational capability
- The ease of using CAD software
- Low-cost, high-precision platforms and controllers
- Availability of a variety of technologically viable additive manufacturing techniques

Much of the technological progress that is being made, in both methods and materials, is a direct consequence of ongoing research in industry, government laboratories, and academia and is now being expanded to develop a consumer base. Low-cost additive manufacturing technology is becoming more feasible and will be able to satisfy the needs of a diverse end-user community. Additive manufacturing is leading to new developments in the fields of materials science, materials engineering, and manufacturing technology, as well as in the design and development processes specific to the needs of end users.[88,89]

3D Printing and Additive
Manufacturing State of the Industry,
Annual Worldwide Progress Report

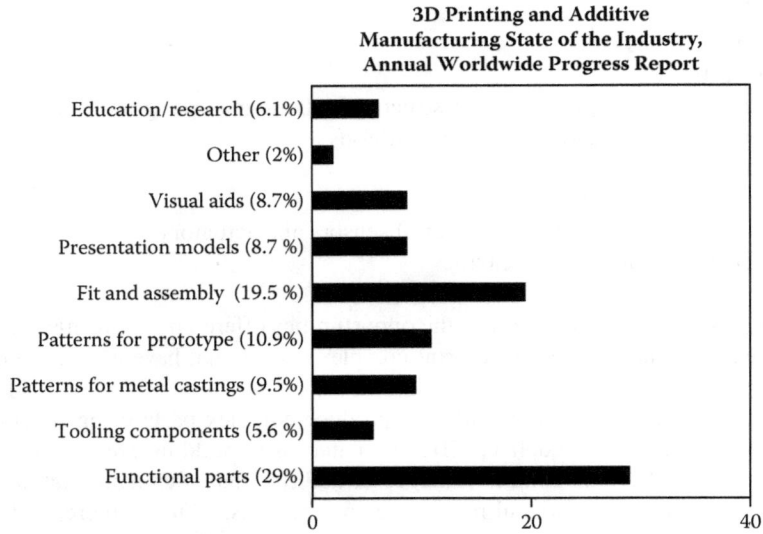

FIGURE 1.16   Bar graph showing the uses of the additive manufacturing systems in 2013. (From Wohlers, T.T., *3D Printing and Additive Manufacturing State of the Industry, Annual Worldwide Progress Report*, Wohlers Associates, Inc., Fort Collins, CO, 2014.)

Despite the many advantages additive manufacturing has to offer, it will not replace conventional manufacturing methods within the near future, but it is bringing about revolutionary changes to the manufacturing industry through its integration with conventional manufacturing technologies. A good example is the integration of laser metal deposition with computer numerical control (CNC) machining to form a hybrid process.[89] For the high-volume production of parts, additive manufacturing offers the promise and capability of providing technological support, such as the fabrication and repair of dies and molds used in various manufacturing processes (e.g., injection molding).

Currently, at least 63 companies worldwide are offering more than 66,000 professional-grade additive manufacturing systems for eight industrial sectors.[90] The largest sector is the production of consumer products and electronics at 21.8%, followed by the fabrication of parts for motor vehicles at 18.6%, medical and dental uses at 16.4%, industrial and business machines at 13.54%, and aerospace at 10.2%. An overview of the dominant products resulting from use of this technology is provided in Figure 1.16. Additive manufacturing products have transitioned from industrial prototypes to actual parts used in engineering applications, leading to the development of new and broader engineering standards in the additive manufacturing industry. The sale of additive manufacturing machines for metal manufacturing in 2013 increased by 76% over the previous year, and overall the market for 3D printing products and services grew to more than $3 billion in 2013, representing a growth of 35% over 2012.

Inspired and driven by the promise of inexpensive, highly customizable manufacturing, growth in 3D printer manufacturing has surged over the last several years. This growth has been made possible by rapid technological developments, falling costs, and new and improved applications for 3D printing technology. The 3D printing industry grew at a very rapid pace, reaching about $2 billion worldwide in 2012 and $3 billion in 2013.[91] In the United States alone, the 3D printer market in the year 2014 was estimated to be close to $4 billion with a compound annual growth rate (CAGR) of 22.8% over the period from 2009 to 2014.[91] The primary buyers of 3D printing machines come from the medical/dental and aerospace industries, and they are being used for both prototype and manufacturing purposes. The aerospace companies are using 3D printers for testing and even certification as they gear up for manufacturing on a noticeably large scale.[91,92]

With regard to its overall impact on the economy and continued sustainability, additive manufacturing offers several advantages over conventional manufacturing techniques, including (1) fabrication of structures not possible using the conventional techniques, (2) just-in-time manufacturing, (3) reduced material waste and energy consumption, and (4) shortened time to market. Most importantly, the intrinsic capability of adding materials layer by layer to create 3D objects reduces material wastage. The traditional manufacturing of aerospace components made from an alloy of titanium that are machined down to size from a large block generates up to 90% waste material, which cannot be put to use or reused immediately. The various additive manufacturing processes can reduce such waste generation, thereby reducing the energy required for the production of titanium materials and parts made from titanium alloys.[93] Complex shapes can be easily produced without tools, dies, or molds, which shortens the fabrication time and cuts production costs. A new era of digital additive manufacturing (DAM) will allow manufacturers to adapt new product designs without the limitations imposed by investing in new physical tools associated with conventional manufacturing processes.

1.8 MECHANICAL PROPERTIES OF ADDITIVE MANUFACTURED MATERIALS

There is a paucity of information in the published literature covering aspects specific to the mechanical properties, performance, and overall behavior of materials manufactured using one of the additive manufacturing techniques. Far and above the lowered cost of production, reduced energy consumed, and advantages arising from sustainability, knowledge of the properties and performance (i.e., mechanical behavior) of additive manufactured end products is essential to ensuring their sustained presence, selection, and use for a spectrum of critical applications spanning the domains of aerospace, ground transportation, and far-reaching applications in the medical field. In 2010, the Edison Welding Institute (EWI) organized the first Additive Manufacturing Consortium (AMC) in the United States.[93] This event was well attended by representatives from industry, government agencies, nonprofit research organizations, and universities whose primary purpose for attending was to record, incorporate, and implement ideas that would further nurture, foster, and promote innovation. The key aspects identified as being necessary to engineer improvements are listed below:[93]

1. Availability of material property databases
2. Process-model compensation to account for distortion and accuracy
3. Process sensing, control, and nondestructive evaluation
4. Clear and affordable paths for certification and quantification
5. Bigger, faster, and more capable equipment

Financial investments to support such endeavors include the recent initiation of the National Additive Manufacturing Innovation Institute (NAMII), funded jointly by the U.S. Department of Commerce, Department of Defense, Department of Energy, National Science Foundation, and National Aeronautics and Space Administration, along with a few other partners from industry, nonprofit organizations, and academia.

Because the mechanical properties and resultant performance of a product are greatly influenced by the internal characteristics of the chosen material, an understanding of the microstructural evolution of the various additive manufacturing processes is essential. A few studies have been conducted on selected additive manufactured metallic materials.[95–100] In these studies, the microstructure of the additive manufactured part was found to be noticeably different when compared to the conventionally manufactured materials. These differences included the presence of distinct layer patterns, heat-affected zones (HAZs), directional grains, and other fine or intrinsic microstructural features within each grain.

TABLE 1.5

Static Mechanical Properties of Processed Titanium Alloy Ti-6Al-4V[101]

Property	Typical Wrought	Orientation			
		Hot Isostatic Pressed (HIP) + Solution Heat Treated (SHT)		Hot Isostatic Pressed (HIP)	
		X–Y	Z	X–Y	Z
Yield strength (MPa)	828	887	946	848	841
Ultimate tensile strength (MPa)	897	997	1010	946	946
Elongation (%)	15.0	11.4	13.9	13.2	13.9

1.8.1 ALLOYS OF TITANIUM

For the titanium alloy (Ti6Al4V) fabricated using Laser Engineered Net Shaping (LENS), the grains parallel to the direction of deposition were found to be columnar in nature as a consequence of heat extraction from the substrate. At the interfaces between the LENS deposit and the substrate, macroscopic heat-affected zones were easily observed. Depending on the specific AM process used, the microscopic heat-affected zones had observable coarse characteristics between the layers. This finding was attributed to the reheating of previous layers as a consequence of subsequent and additional deposition.[95] It is the presence of fine microscopic heat-affected zones that give the additive manufactured microstructure a layered appearance. The presence of directional grains in conjunction with both the macroscopic heat-affected zones and microscopic heat-affected zones did result in an overall non-uniform microstructure of the additive manufactured materials. This was responsible for differences in properties along the different orientations.[95,97–100] In one independent study, it was observed that the yield strength and ultimate tensile strength of the additive manufactured processed titanium alloys exceeded the typical values for a wrought Ti6Al4V alloy. Further, the ductility of the additive manufactured alloy was found to be slightly lower than the wrought counterpart, and anisotropy in yield strength, ultimate tensile strength, and ductility was associated with the build direction. The static properties of additive manufactured Ti6Al4V was found to be comparable with the wrought product.[101,102] Some of the values reported for both the wrought and hot-isostatic processed (HIP) titanium alloys are summarized in Table 1.5.

In a recent study, it was found that the properties and microstructure of the Ti6Al4V alloy that was produced using electron beam powder bed (EBPB) and laser beam powder bed (LBPB) revealed appreciable differences.[101,102] The morphology and severity of porosity induced in the final product by the two techniques differed significantly. The researchers observed that the LBPB-processed material exhibited irregularly shaped pores, while porosity in the EBPB alloy was essentially spherical. The presence of surface defects was found to be detrimental to the high-cycle fatigue resistance of the alloy.[102] In terms of fatigue crack growth performance, the presence of porosity was found not to be the dominant influencing factor but the microstructure of the alloy was. The use of hot isostatic pressing (HIPing) to induce closure of the porosity was observed not to have a significant influence on fatigue crack growth resistance.[102]

1.8.2 NICKEL-BASED SUPERALLOYS

In the as-fabricated condition, the microstructure of nickel-based alloys was found to be columnar with grains up to 20 μm in width. Subsequent to hot isostatic pressing, the columnar grains tended to recrystallize, while the metastable γ' precipitates (Ni_3Nb) gradually dissolved. Overall, depending on the additive manufacturing technique used, unusual microstructures and

TABLE 1.6

Static Mechanical Properties of Additive Manufactured Nickel-Based Superalloy IN625[103,104]

Process	Yield Strength (MPa)	Tensile Strength (MPa)	Elongation (%)
Wrought (annealed)	450	890	44
As-fabricated (EBM)	410	750	44
Electron beam melt + hot isostatic pressed	330	770	69
Wrought (cold worked)	1100	—	18

microstructural architectures were generated, which opened the door to microstructural design. The static mechanical properties of an additive manufactured nickel-based superalloy IN625 are summarized in Table 1.6.[103] In the as-fabricated condition, the ductility of IN625 was equivalent to that of the wrought and annealed counterparts, and the yield strength was only marginally lower. Upon hot isostatic pressing, the yield strength was observed to decrease by a good 26%, with a concurrent improvement in ductility of well over 57%. The mechanical properties of the shaped metal deposition (SMD) IN718 alloy were found to be superior to those of the as-cast counterpart, but were noticeably inferior to those that were engineered using the additive manufacturing technique. The tensile strength and elongation for the entire additive manufactured processed materials were found to exceed that of the as-cast counterpart.[104] The mechanical properties of this alloy are summarized in Table 1.7.

Compared to the parts produced by conventional manufacturing processes, additive manufactured components demonstrate promising properties for most metallic materials;[93,105–108] however, the presence of observable amounts of shrinkage, porosity, and residual stresses and their mutually interactive influences can affect both the static and dynamic properties, depending on the additive manufacturing techniques used and the parameters governing the processing.[105,106] This finding has led to an examination of post-AM processed materials. Both hot isostatic pressing and heat treatments are considered to be effective methods for relieving residual stress, eliminating porosity, and recovering ductility. As a direct consequence of the specific processing condition used and the resultant characteristic microstructure, conventional post-processing treatments may not always result in the expected properties and behaviors for the materials prepared using any one of the additive manufacturing techniques.[108,109] Overall, achieving the desired properties and performance of additive manufactured materials and the resultant components requires the following factors:

1. Judicious selection of the appropriate additive manufacturing process
2. Tailoring the fabrication condition, post-processing conditions, and key control parameters

The surface finish given to an additive manufactured part is influenced by the type of equipment used, the direction of build, and the process parameters. Martukanitz and Simpson[109] reported that both the build data and feature definitions can be linked to the overall quality of the surface. In general, as the build rate increases, the feature quality or resolution decreases; thus, for fatigue-critical

TABLE 1.7

Mechanical Properties of Additive Manufactured Nickel-Based Superalloy IN718[104]

Process	Yield Strength (MPa)	Tensile Strength (MPa)	Elongation (%)
Shaped metal deposition	473	828	28
As-cast	488	786	11
Laser	552	904	16
Electron beam	580	910	22

parts that are fabricated using high-deposition-rate additive manufacturing processes, the need for post-process surface finishing becomes necessary. When properly processed, the static mechanical properties of additively manufactured metallic materials are comparable to the conventionally fabricated metallic components or counterparts. The relatively high cooling rates achieved during additive manufacturing tend to reduce partitioning while concurrently favoring reduced grain size. Most additive manufacturing fabricated materials tend to exhibit both microstructure and mechanical property anisotropy along the Z direction, or through the thickness. This direction was found to be the weakest. The dynamic properties of additive manufactured materials are noticeably influenced by the presence of macroscopic, fine microscopic (microporosity), and surface finish defects. However, when properly processed, hot isostatic pressed, and even finish machined, the additive manufactured alloys can exhibit better fatigue properties as compared to their wrought counterparts.[20,109]

1.9   POTENTIAL FAR-REACHING APPLICATIONS OF ADDITIVE MANUFACTURING

Additive manufacturing processes have been successfully used within a wide spectrum of industries spanning aerospace, automotive, biomedical, energy conversion, consumer products, and sporting goods. The additive manufacturing field can best be represented by the tree model shown in Figure 1.17, the development of which led to a Roadmap for Additive Manufacturing (RAM) workshop being held in 2009.[110] The base of the tree consists of the various feasible additive manufacturing processes. The trunk is representative of the research and development efforts that often result from use of these processes. The branches clearly represent the direct outcomes and benefits culminating from these efforts. With time, in synergism with a rapid pace of research and development, both new and emerging applications and their benefits are expected to gradually grow and attract greater

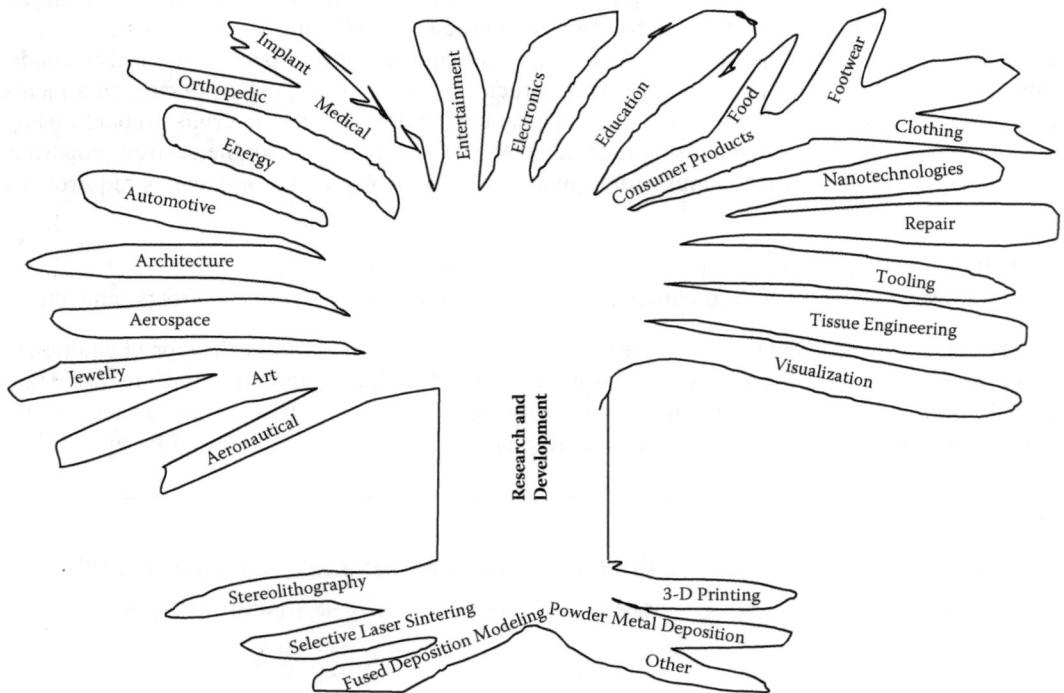

FIGURE 1.17   A simple overview of the field of additive manufacturing and the many domains for research.[110]

attention. In recent years, the use of additive manufacturing for a variety of purposes has expanded as more advanced additive manufacturing techniques are being developed and getting approved. An early application of additive manufacturing was the production of tools that had specific channels to facilitate plastic injection molding. Today, additive manufacturing has been put to effective use to make a variety of products, including (1) medical implants, (2) orthopedic and dental parts, (3) hearing aids, (4) forming tools, (5) aerospace components, (6) military and automotive components, (7) electronics, (8) art, (9) jewelry, (10) commercial lighting, (11) videogame avatars, and (12) engineered foods. Much of the current and ongoing research that is actively being pursued is focused on biomedical applications, particularly the generation of living tissues.[111]

Several of the aerospace components are noted for their complex geometries and are often made from advanced materials such as alloys of titanium, nickel-based superalloys, specialty steels, ultra-high-temperature ceramics, and even metal–matrix composites. Several of these materials are often difficult to work with and it can be time consuming to manufacture an intricate part using conventional manufacturing techniques. Further, production runs for the aerospace industry are usually small and limited to a maximum of a few thousand parts. This makes the technology of additive manufacturing both suitable and appropriate. Following tests on additive manufactured parts, BAE Systems approved a replacement part that was made entirely by additive manufacturing. The part was a plastic window breather pipe for the 146 regional jets of BAE. Around the same time, Optomec, Inc. (Albuquerque, NM) used the LENS process to fabricate complex metal components for use in satellites, helicopters, and jet engines.[112] Arup, Inc. (London), another rapidly growing small business venture, developed a 3D printing technique for creating structural steel elements that can be used in construction. Although laser sintering has traditionally been used, researchers at Arup believe that the prudent use of 3D printing to create steel elements will eventually reduce the usage of energy and the overall costs and waste so often involved with construction

In the commercial sector, research performed at General Electric Aviation's Additive Manufacturing Technology Center based in Cincinnati, OH, led to novel breakthroughs in design that encouraged additional investments in additive ma nufacturing technologies. This successful research into building fuel nozzles using additive manufacturing techniques led to the construction of a high-volume additive manufacturing factory in Auburn, AL.

Most recently, a joint research effort by researchers at NASA's Jet Propulsion Laboratory, the California Institute of Technology, and Pennsylvania State University resulted in the development of a valuable method to build gradient metal objects that will allow for the design and production of parts in which the metal within a single part changes based on specific requirements.[113]

Within the realm of space exploration, a custom 3D printer that enables the fabrication of all necessary tools and components required for a space mission will be sent to the International Space Station. The printer was developed by Made-in-Space (Mountain View, CA). The new printer successfully completed all flight certification and acceptance testing in 2014, and engineers at NASA are hoping to demonstrate that a 3D printer can function as expected in space to print components of similar quality to those fabricated on Earth.[114,115]

Optomec, Inc., successfully leveraged the advantages of LENS 3D printing and the expertise of the world's leading aerospace companies and industry organizations to advance a reliable and cost-effective approach to replace conventional repair processes, such as manual welding. Optomec's LENS process can add metal onto any existing substrate of almost any 3D shape, making it well suited for repair operations.[112] The European Space Agency, working with a cross-section of industrial partners, demonstrated in 2013 that 3D printing using lunar material was feasible, at least in principle. The shielding against radiation provided by a 3D printed block of simulated lunar regolith was evaluated and provided important input for future designs. The agency is also planning to investigate another lunar 3D printing method involving the harnessing of concentrated sunlight to melt the regolith rather than use a binding liquid. Each small step put forward in research makes the possibility of future lunar colonization a little more imaginable.[111,115]

Within the automotive industry, additive manufacturing has been explored as a viable tool in the design and development of automotive components primarily because it can shorten the development cycle while reducing not only manufacturing costs but also product costs. Additive manufacturing processes have also been used to make small quantities of structural and functional components, such as engine exhausts, braking systems, drive shafts, and gear box components for luxury and low-volume vehicles. Unlike widely used passenger cars, vehicles used in motor sports require lightweight alloys (such as alloys of titanium and aluminum) that often have complex structures and low production volumes. Additive manufacturing has been successfully applied to manufacture components that are functional for use in racing vehicles, and the technology has been put to effective use to replace the time-consuming and often expensive sand-casting and die-casting processes that have been traditionally used to make large metal components.[111,114]

Local Motors (Phoenix, AZ) has 3D printed a fully working car called the Strati. The car was printed at the International Manufacturing Technology Show held in 2014. The entire printing process took 44 hours. The Strati was the outcome of a 3D printed car design challenge that attracted more than 200 entrants from more than 30 countries. The company claims that it is the first time the main portion of a car has been printed in one piece using direct digital manufacturing. In addition to the 44 hours to print it, the car required 1 day to mill and 2 days to assemble, resulting in a 5-day build event.[116] Local Motors is a growing company in the 3D printed car arena.

German electric vehicle manufacturer Street Scooter recently completed the prototype of its C16, the exterior components of which were created using the Stratasys Object1000 3D production system. The 3D printed parts of the car include (1) its front and back panels, (2) door panels, (3) bumper systems, (4) side skirts, (5) wheel arches and lamp masks, and (6) some smaller interior items. Although many parts of the production version of the C16 will be built using more conventional methods, the 3D printing approach allowed for the prototype to be constructed inexpensively and within a time frame of 12 months.[117] The resultant 3D printed car was able to perform in strenuous testing environments at the same level as a vehicle that was made using traditionally manufactured parts.

In the biomedical field, several recent applications have enabled the use of additive manufacturing for the fabrication of (1) custom-shaped orthopedic prostheses and implants, (2) medical devices, (3) biological chips, (4) skull and jaw implants, (5) custom-molded mouthpieces for individuals suffering from sleep apnea, (6) tissue scaffolds, (7) living constructs, (8) drug-screening models, (9) surgical planning and training apparatus, and (10) 3D printed spine cages. The potential of 3D printing within the domain of spinal fusion surgery lies in its ability to be tailored to a particular patient's anatomy.[111] Recently, an orthopedic implant manufacturer (Medicrea; Neyron, France) was able to use custom software and imaging techniques to produce a spine cage made out of polymeric material and customized to perfectly fit to the patient's vertebral plates.[118] The process is patent pending, but Medicrea is hopeful that a breakthrough will pave the way for further development of implantable devices that either replace or reinforce damaged parts of the spine. The ambitious vision to devise a developmental biology-enabled, scaffoldless technique to fabricate living tissues and organs by printing living cells was realized in 2013, the 15th year of printing cells. A typical cell printing process consists of the following three distinct steps:

1. *Preprocessing*—Creating tissue-specific or organ-specific CAD models for each patient using CT scan data
2. *Processing*—Using any of the viable additive manufacturing processes to deposit living cells onto 3D biological constructs
3. *Postprocessing*—Incubating printed tissues or organs to encourage both tissue fusion and maturation

In recent years, additive manufacturing has been successfully applied in the prosthetics industry to design and fabricate lightweight and low-cost robotic parts, such as hands and wrists, thus making it possible to consolidate complex assemblies into a single unit that is functional.[111] In 2013,

Brightwake Limited (Nottinghamshire, U.K.) developed a blood recycling machine called Hemosep using the Stratasys Dimension 1200es 3D printer. The Hemosep can recover the blood that is lost following open-heart or major trauma surgery so it can be transfused back into the patient. This process, referred to as autotransfusion, reduces the volume of donor blood that is required and the problems associated with transfusion reactions.[119] The prototype device has some Stratasys 3D printed parts, including the main filtration and cooling systems. Clinical trials of well over 100 open-heart surgeries in Turkey confirmed the ability of Hemosep to significantly reduce the need for blood transfusions.[119] Medicine is certainly a field where 3D printing has made a splash.

The 3D printing techniques have allowed companies to move from an idea to a product concept design to a functional prototype much more quickly and more accurately than previous traditional prototyping methods allowed.[119] Stratasys has introduced a new 3D printer for dental applications. The printer will become the standard for prototyping and developing dental products and devices. The same company is now working with suppliers to develop integral scanners that will allow for the production of crowns, bridges, and even veneers. The Stratasys 3D printing technology promises to reduce patient time in the dental chair while making the experience of getting dental surgery more pleasant and more cost effective.

In 2014, a group of researchers at Harvard University (Boston, MA) solved one of the most difficult problems with growing artificial human organs. The team used a 3D printer to make human tissue that included rudimentary blood cells. The success of their preliminary project inspired the researchers to undertake an ambitious project to make fully functional kidneys. The researchers made significant progress by fabricating the rudimentary versions of structures in kidneys called nephrons. These artificial nephrons will help drug companies to quickly screen potential medications and will help scientists to understand kidneys at a more detailed level.[120]

At around the same time, 3D-printed syringe pumps were being produced at Michigan Technological University (Houghton) and made available to the public at a cost as low as $50 apiece. These pumps can be used in laboratories to administer small amounts of liquid for the purpose of drug delivery or chemistry-related research. Scientists can customize the design of a pump to suit their study needs just by making minor modifications in the control software.[121]

When a medication is consumed, it eventually enters the bloodstream and ends up being concentrated in the liver, whose primary function is to cleanse the blood. This means that if a drug is going to have an adverse effect on any part of the body, chances are it will be the liver. Therefore, should a pharmaceutical company want to test the safety of its products, it would be convenient to have some miniature human livers on which to experiment. The biotech firm Organovo (San Diego, CA) has begun selling such a device.[122] Known as Vive3D, the three-dimensional liver model measures just a few millimeters across and is created using a 3D bioprinter. The device incorporates two print heads; one deposits a support matrix while the other precisely places human liver cells in the matrix. The resulting models are composed of living human liver tissue and incorporate hepatocytes, stellate cells, and endothelial cells, just like an actual, full-sized liver. It is also possible to produce liver proteins such as albumin, fibrinogen, and transferrin and to synthesize cholesterol. Additionally, the cells are arranged in a 3D orientation relative to one another, as would occur naturally. By contrast, the liver cell cultures currently used to test pharmaceuticals are two dimensional and may not always function in the same manner as the actual organ.[122]

No industry is as poised to benefit from this burgeoning technology as the field of medicine. Replacing cancerous vertebrae, delivering cancer-fighting drugs, and even assisting in spinal fusion surgery are just a few of the recent advances being made. A recent groundbreaking application involved a cancer patient whose entire upper jaw was replaced with the help of 3D printing.[123] OSTEO 3D (Bangalore, India) used a CT scan to create a 3D reconstruction of the patient's face. A replica of the patient's mouth, complete with lower jaw, upper jaw, and teeth, was printed. Using the 3D printed replica as a template, a wax model was produced and adjusted for proper fit. The jaw assembly was subsequently hardened and fitted with teeth.[123] Harvard Apparatus Regenerative Technology (HART) is a biotechnology company that develops regenerative organs for the purpose of transplantation, with an initial focus on the trachea. In collaboration with academia, HART originally developed equipment that could produce hollow organs such as the trachea and gradually expanded into the regeneration of all types of hollow organs using bioreactive technology.[124]

In the industry sector, medium and large quantities of polymer-based components are being manufactured using injection molding. It is quite difficult for the techniques of additive manufacturing to compete with injection molding in producing these components; however, the additive manufacturing processes can be used to manufacture injection molds (i.e., rapid tooling) to reduce the time and costs involved in the development of new tools. Similarly, metallic parts can be easily cast from molds or dies made using additive manufacturing. An example of a company that has embraced the use of additive manufacturing in the United States is Direct Manufacturing, Inc. (Austin, TX). The company has added more machines to its shop floor as many of its customers have gradually moved toward the use of additive manufacturing and away from traditional castings, forgings, assembly of multicomponents, and subtractive manufacturing or machining. Although some machining may still be required for complex parts, the technique of additive manufacturing is considerably faster and noticeably more economical when compared to the other manufacturing methods currently being used. LENS-based 3D metal printing has been used to develop a reliable and cost-effective method to replace conventional repair processes such as welding which required the following:

- Precise definition of powder feedstock characteristics
- Improvements in process monitoring and control
- Recommendations for repair of the part or component

The potential benefits resulting from the use of any one of the emerging additive manufacturing techniques to repair high-value metal components include lower costs, higher quality, longer life, and quicker return to service.[125]

The Stratasys Objet500 Connex 3D printer was used in connection with an 85-ton injection molding press to create a 3D injection molding machine. This machine allowed for 3D printing of the injection molds to manufacture small quantities of final prototypes using the same materials, components, and even capabilities as traditional 3D molding.[126] Molds can be printed in a couple of hours with 3D printing, and tens and even hundreds of prototypes can be produced within a few days. The two biggest constraints to extensive use of the process are (1) print speeds and (2) the type of materials used. The machine must be able to print at speeds 50 to 100 times their current rates to allow for the printing of actual medical devices. Importantly, such printing could eliminate most logistics, switching, labor, and even assembly costs.[118]

Another field that can reap the benefits of implementing additive manufacturing is high-performance ceramics. The widespread or extensive use of parts made from ceramics is often limited by high costs and the tedious and laborious procedures involved in the fabrication of both prototypes and test parts. The primary factor within this process chain is fabrication of the mold. Some of the technologies of additive manufacturing offer numerous advantages that will have a positive impact on the fabrication of parts made from high-performance ceramics. With particular reference to ceramic injection molding (CIM), after giving due consideration to the observed lead times and costly tooling involved, additive manufacturing can serve as a capable complement to this means of production. The opportunity to produce both accurate and cost-efficient prototypes is of some benefit to ceramic injection molding. Once the given design has been chosen and the prototypes have been evaluated by the end user, the respective part can be fabricated and the means of production changed to ceramic injection molding so as to ensure a high throughput. With the use of additive manufacturing technology, each part costs approximately the same in production; however, in ceramic injection molding, the cost depends significantly on the size of the lot. This relationship is shown in Figure 1.18. The fabrication of ceramic parts by ceramic injection molding requires a mold, which is a significant expense, particularly if that expense is spread out over the production of only a few components.

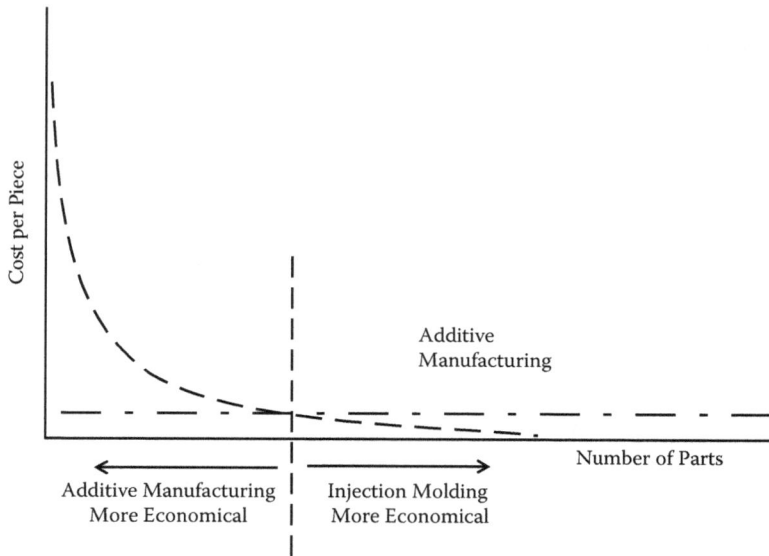

FIGURE 1.18 Comparison of the overall cost effectiveness of additive manufacturing to ceramic injection molding.[111,118]

Thus, from an economical perspective, ceramic injection molding is not favorable for the production of a few parts or even individual parts. The only prerequisite for additive manufacturing is a CAD drawing, which can be applied to any number of parts. When there exists a need to produce a small number of parts, ceramic injection molding is not the optimum choice.[111,127]

Three-dimensional printers offered by Aurora Laboratories (Palm Beach, FL) can print multiple metals at the same time, including 316 and 420 stainless steel, Inconel 625 and Inconel 718 alloys, HASTELLOY© C, brass, bronze, and mild steel, along with ceramics and plastics. These machines are a result of innovative breakthroughs in technology and design. Aurora Laboratories claims that one day a 10,000-pound thrust rocket motor could be built. The control system for the printer feeds power to a motor at a controlled speed, and the powder is propelled onto a high-energy beam. It is then melted and fused with the substrate (i.e., base material). The build area often requires an inert gas. Depending on the materials chosen for the part being built, either nitrogen or argon can be used.[125]

The military has put the many advantages offered by additive manufacturing to good use for a multitude of purposes, including turbine engine blades, heat exchangers, repairs to blades and dies, cooling ducts, and wing brackets for commercial airplanes. However, additive manufacturing must undergo the tedious, expensive, and often time-consuming process of being certified for use in the aerospace, military, and medical fields, which have emerged as markets with demanding structural applications and high-dollar-value components.[111,118]

Within the domain of consumer goods such as jewelry, precious metal designs raise unique challenges when compared to the more common alloys used in additive manufacturing due to the highly polished surfaces often desired by customers. Additive manufacturing, however, allows for the creation of designs that have never before been achievable, providing both artists and designers insight into new levels of geometrical detail. The art and industry sectors have been able to produce components that are no longer available by entering the appropriate data into an additive manufacturing machine and building the part new.

A 3D scanner and printer have been used to produce busts of individuals. In 2014, the Smithsonian Institution scanned President Barack Obama and generated a 3D bust of him that was subsequently added to the National Portrait Gallery.[128] Bakeries can print figures of the bride and groom as wedding cake toppers.[128] The arrival of 3D scanners has contributed to noticeable advances being made in the domain of 3D printing. 3D scanners share several traits with cameras. Like a camera, the 3D scanner has a cone-like field of view and can collect information about surfaces that are not completely obscured. In essence, a 3D scanner can collect distance information about surfaces within its field of view. The resulting "picture" produced by a 3D scanner is a composite of the distances measured to each point on the surface being scanned. Advances in 3D scanning continue to follow advances in camera development, including the stability and quality of the optical component. 3D scanners can provide complete measurement information on objects in the fastest way possible, which can save considerable time when modeling existing objects using CAD software. Noticeable advances in 3D scanning have resulted in time and cost reductions in the process of designing, enhancing, and testing products.[129]

The Italian firm WASP is the latest to explore the potential of additive manufacturing in the area of blending creative architecture with new, sustainable technologies. Their super-size 3D printer is capable of building low-cost housing out of mud. The company is looking to bring these benefits to a bigger stage by providing a means to quickly create shelter in developing regions where traditional forms of construction might not be possible. The idea behind the company's approach is that housing can be built on location using the materials found on site at zero cost.[130]

In the food sector, additive manufacturing is also gaining a foothold. Recently, students at the Massachusetts Institute of Technology were able to link a 3D printer to an ice-cream maker to provide on-demand soft-serve ice cream. When the device is marketed commercially to ice-cream parlors, customers will be able to order custom ice-cream treats.

The U.S. Postal Service could also benefit from 3D printing with regard to the delivery of lightweight packages. As more consumers purchase 3D printed objects, the USPS can expect to see a substantial increase in its business. Also, onsite 3D printers could be used to produce replacement parts for postal service equipment.[131]

A technique using ultraviolet light and a photosensitive polymer resin, referred to as *microstereolithography*, has allowed researchers to create ultralight but very strong materials on a microscopic scale. The light burns a 3D image into the liquid resin, layer by layer; the resin subsequently hardens when struck by ultraviolet light.[132] The excess resin is then removed, leaving behind the object. Teams of scientists, chemists, and even electrical engineers are using this approach to construct microlattices, geometric building blocks similar to cells in the body. When bonded together, the resultant material is among the lightest weight solids on Earth but is strong enough to withstand thousands of times its own weight. This material has the potential for use in transportation, aerospace, automotive, and other high-performance applications. An offshoot of this technology employs a suspension of ceramic, polymer, and metal nanoparticles that are charged so they assemble and stick to a pattern on the surface material whenever light comes in contact with the surface. This process, referred to as *electrophoretic deposition*, can create super-lightweight armor that would allow troops to move more easily on the battlefield while currently being less expensive and 20% more effective than high-strength steel in protecting the body.

1.10 EMERGENCE AND USE OF ADDITIVE MANUFACTURING IN SPACE

Both NASA and the U.S. Air Force have explored opportunities to use additive manufacturing in space because it offers the following two capabilities:

1. Reducing the volume of launch vehicles when compared to equivalent spacecraft
2. Tailoring launch vehicles that deliver materials to orbit

Both of these factors can improve the economics related to launching, and additive manufacturing can also do the following:[115]

1. Contribute to the design and manufacture of new materials and novel parts that can function well in zero gravity but not necessarily in a terrestrial environment.
2. Transform operations and logistics planning due to the ability to launch a broad category of materials that can be used for the *in situ* manufacture of a range of parts offering a wide variety of functionality.
3. Contribute to the development of space hardware and robotic systems that will allow small spacecraft to be fully manufactured in space to suit specific needs.

To a large extent, the overall pace of implementation of the different additive manufacturing technologies will depend on the following:

1. Development of new engineering and testing protocols
2. Evaluation and approval of the protocols by professional organizations
3. Emergence of new engineering and management opportunities in both the aerospace industry and government

Any new designs arising from the unique materials, structures, and manufacturing processes of additive manufacturing will have to prove their durability and safety for the applications for which they are targeted. It is clear that the inclusion of additive manufacturing in all aspects of space operations could span several decades, during which the various AM techniques for hardware production can be studied, tested, and evaluated in myriad ways.[115] Today, few of the additive manufacturing techniques are mature enough to be applied to the production of aircraft and aerospace components that can be manufactured on the ground. Thus, for now, the applications of additive manufacturing in space are limited and primarily experimental.[115]

1.10.1 POTENTIAL USER REQUIREMENTS AND ANTICIPATED TECHNOLOGIES

The National Aeronautics and Space Administration is interested in utilizing additive manufacturing to develop new aeronautical and related technologies and to discover more cost-effective ways to conduct their missions. The U.S. Air Force has applied additive manufacturing to operating and sustaining a fleet of 55 spacecraft in five separate locations for the following purposes:[111,118]

1. Protected communication
2. Wideband communication
3. Missile warning
4. Navigation
5. Space situation awareness
6. Weather-related information

Because the Air Force is actively developing new systems to maintain its space superiority, the cost and speed of innovation are critical to maintaining a competitive advantage over potential adversaries, and additive manufacturing can certainly be of some assistance in this regard. With reference to cost, aerospace companies are currently involved in pursuing projects that are aimed at developing a better understanding of the value of additive manufacturing as a viable alternative to lower the cost of tooling and as a production tool for the manufacture of key aircraft components.

1.10.2 GROUND-BASED ADDITIVE MANUFACTURING FOR SPACE

In 2011, Lockheed Martin used additive manufactured brackets for microwave communication posts for NASA's Juno spacecraft, which is currently on a mission to planet Jupiter. Other companies are exploring the use of additive manufacturing for spacecraft products or development projects in an effort to reduce not only the time required but also the costs involved. Aerojet Rocketdyne (Sacramento, CA) used additive manufacturing to build and test a rocket engine injector. The company is now offering four different space-qualified thruster systems that are produced by additive manufacturing.[113,115] In 2012, Aurora Flight Sciences (Manassas, VA) built and flew a thermoplastic drone system constructed using fused deposition modeling on a commercial 3D printer. This aircraft was built for the purpose of evaluating its aerodynamic design before going into volume production. The approach also permitted the inclusion of innovative structural arrangements that were not possible with traditional methods of construction.

At the Jet Propulsion Laboratory (Pasadena, CA), efforts are underway to use additive manufacturing to create metal objects having custom compositional gradients and specific properties at the local level. This research has made possible previously unattainable material combinations and compositions.[113]

Much research is also in progress to develop more sophisticated hybrid additive manufacturing systems that combine additive manufacturing machines with direct-write machines and other manufacturing technologies to embed electronic components and circuitry in three dimensions during fabrication. Sustained research and development efforts at the ground level in the additive manufacturing industry have focused on building machines capable of adding additive manufacturing materials to pre-constructed metallic and other substrates. Doing so would provide considerable savings in the time required for the construction of specific items when a casting or other manufactured item is used as the foundation for a complex structure.[132]

Conceived and developed by students at California Polytechnic State University—San Luis Obispo and Stanford University, the first CubeSat was launched in 2003 aboard a Russian rocket. Today, more than 100 of these miniaturized satellites have been placed in orbit for a variety of technical applications. The basic unit measures 10 cm by 10 cm by 10 cm.[133]

1.11   ADDITIVE MANUFACTURING AND ITS IMPACT ON EDUCATION

Educating the general public about additive manufacturing will eventually empower them to build upon their dreams. Further, disseminating additive manufacturing technology to potential end users will tend to democratize manufacturing while providing novel incentives for innovation. Formal education in the specific domain of additive manufacturing has already been integrated into the academic curricula at different levels of education at universities. Material related to rapid prototyping has for quite some time been part of the manufacturing engineering courses at a few engineering colleges, and rapid prototyping is addressed in the classical manufacturing textbook *Manufacturing Processes for Engineering Materials*,[114] which provides valuable information on the predecessor to additive manufacturing. The number of additive manufacturing-specific courses being offered at both the undergraduate and graduate level is growing. Two noticeable examples are

1. The "Solid Freeform Fabrication" course at the University of Texas in Austin
2. The "Rapid Prototyping in Engineering" course at the Georgia Institute of Technology in Atlanta

Community colleges have served as excellent gateways for exposing students to the many marvels of additive manufacturing. Courses offered in community colleges cover the most recent trends and developments specific to this area. As a result of the Advanced Technological Education (ATE) initiative aimed at two-year colleges, the National Science Foundation has been promoting and supporting academic curricula through its Technician Education in Additive Manufacturing (TEAM) program. ATE centers in Washington and California are focusing on the development of competencies and curricula to serve as models for future expansion of the program. Only a few educational institutions have provided access to the books, instructional guides, and other educational materials necessary for additive manufacturing courses and laboratory-related activities.[85,116]

Courses in additive manufacturing have primarily been at the level of higher education in both universities and community colleges; however, a recent secondary-level education initiative is the Defense Advanced Research Projects Agency (DARPA) program on Manufacturing Experimentation and Outreach (MENTOR). This program provides 3D printers to high schools and is part of DARPA's Adaptive Vehicle Make (AVM) portfolio of programs addressing revolutionary approaches to the design, verification, and manufacturing of complex defense systems and vehicles. The primary objective of the MENTOR program is to expose emerging generations of designers and innovators to the principles of foundry-style digital manufacturing through design challenges and competitions. The program has successfully passed its initial phase and is anticipated to be rolled out on a larger scale over the next several years.[118]

Various online resources offer additive manufacturing education and training that can reach a broad audience. MakerSpace is an online community that combines manufacturing equipment, community, and education for the purpose of enabling members to design, prototype, and create manufactured parts.[111] MakerSpace represents the democratization of design, engineering, and fabrication, as well as education. Users have already begun to produce projects having a significant impact at the national level. A goal of MakerSpace is to find low-cost options to branch out into such places as community centers, middle schools, and high schools. It is expected that such an effort could lead to an integration of design and collaboration tools that would lead to economical options for physical workspaces.

A number of additive manufacturing course textbooks and reference books have been published by researchers working in the field. Two notable ones are *Additive Manufacturing Technologies: Rapid Prototyping to Direct Digital Manufacturing*[111] and *Understanding Additive Manufacturing: Rapid Prototyping, Rapid Tooling, Rapid Manufacturing*.[134] These books attempt to

- Provide a comprehensive overview of additive manufacturing technologies along with descriptions of the support technologies, such as software systems and post-processing approaches.
- Discuss a wide variety of both new and emerging applications, such as microscale additive manufacturing, medical applications, direct write electronics, and direct digital manufacturing of end-use components.
- Introduce and provide a systematic solution for the selection and design of additive manufacturing processes.

1.12 IMPACT ON WELL-BEING AND HEALTH

Easy access to vaccines, the increased availability of medicines, and advances in both minor and major surgical procedures and therapeutic techniques have improved the overall quality of life all over the world since the end of World War II. A reduction in the mortality rate and an increase in life expectancy have occurred not only in developed countries but also in developing countries. The global population has steadily increased and the aging population continues to grow. In the year 2006 about 500 million people worldwide were 65 years of age or older. This number is projected to grow to 1 billion by 2030.[135–137] Caring for an aging population has put a noticeable strain on the budgets of governments all over the world. An Economic Cooperation and Development study found that the share of total healthcare-related expenditures attributable to individuals over 65 years of age ranged from 32 to 42%, but they represented only 12 to 18% of the population.[137] Under these circumstances, developing and delivering high-quality and economically efficient health care with the primary intent of improving the health and overall well-being of the entire population are key challenges facing society today.[137,138] The current approach to delivering high-quality, efficient health care is personalized care tailored to the particular characteristics and needs of the patient including long-term care for the elderly[139] and using the patient's biological data to determine the best course of treatment or therapy. Additive manufacturing is suited to producing customized products that can meet individual needs, thus playing a significant role in personalized health care. Importantly, it has been used to produce customized surgical implants and assistive devices for the improved heath and overall well-being of the prevailing population.[137–139]

Additive manufacturing can also be used to make surgical implants of either a solid or resorbable material. There is a growing need for surgical implants. For example, in 2001, over 20,000 individuals in the United States underwent chin augmentation surgery, a 70% increase over the previous year.[122] Similarly, lip augmentation surgeries increased by more 49% and cheek implants increased 47%.[140] Computed tomography can be used to obtain patient data to devise custom surgical implants. A solid model of the required implant is developed using reverse engineering. The model is then used to build a customized implant for the patient using the appropriate materials. This approach was found to produce accurate implants that functioned well while providing aesthetic appeal, and it can shorten both the design cycle and delivery lead time required for custom-fit surgical implants.[140] Custom implants produced using additive manufacturing also include skull implants, knee joints, elbows, and hip joints.[136,137] The most common applications can be found in the domain of dentistry. A variety of commercial dentistry products based on additive manufacturing are now available. Hearing aids are another area where additive manufacturing has made a significant contribution.

Additive manufacturing has also been used to make custom-fit safety equipment out of lightweight materials. Safety equipment, ranging from helmets to protective clothing, is an important aspect of occupational safety and health. These devices are essential to protect athletes, construction workers, firefighters, and policemen from potential harm. Safety equipment produced using additive manufacturing has the potential to provide excellent protection without sacrificing personal comfort so users can achieve a high level of performance while being assured of maximum protection.[140] One such project initiated and promoted at a university in the United Kingdom was to tailor protective specific sports garments for individual athletes by taking into account variations in the size and shape of the different individuals. The garments were often produced in one piece that fit the body perfectly without the need for seams and joints.[140]

1.13 ADDITIVE MANUFACTURING STANDARDS

The global entity that is responsible for the development and publication of international, consensus technical standards is the American Society for Testing and Materials (ASTM). Through the years more than 12,000 ASTM standards have been developed and put to active use primarily for the purpose of enhanced product safety, quality, market access, trade, and the confidence of end users or consumers. In addition to ASTM, the International Organization for Standardization (ISO) has also been involved in developing standards for technical products. The standards specific to ASTM are being jointly developed by ASTM Committee F42[141] and ISO Technical Committee 261. The development of standards must address (1) terminology, (2) processes and materials, (3) test methods, and (4) design and data formats. This has resulted in the emergence of six different areas related to raw materials, processes and equipment, and finished products. To effectively deal with the complexity intrinsic to additive manufacturing, the concerns listed below are of the highest priority and are being investigated by ASTM and ISO with the primary objective of developing relevant standards:

1. Methods for qualification and certification
2. Guidelines for design
3. Test methods for the characterization of raw materials
4. Test methods to determine the mechanical properties of finished additive manufacturing parts and components
5. Guidelines for recycling materials
6. Standard protocols for round-robin testing
7. Standard test artifacts
8. Requirements for purchased additive manufacturing parts

Until these standards have been developed and accepted by both ASTM and ISO and by manufacturing organizations, the extensive use of additive manufacturing in domains spanning engineering and product development is not likely to occur. As they exist today, the current standards do not contribute to making additive manufacturing a widely and fully accepted manufacturing process.

1.14 CONCLUDING REMARKS: THE FUTURE OF ADDITIVE MANUFACTURING

The significant advantages to be gained by fast and precise manufacturing combined with positive environmental impacts have led to additive manufacturing being referred to as the "third industrial revolution." To meet and exceed the standards set by traditional manufacturing techniques, especially for critical structural applications, continued improvements and quantification studies are essential and necessary. A comprehensive investigation aimed at systematically evaluating the static, dynamic, and high-temperature properties of additive manufactured materials must be performed with the primary objectives of

- Establishing fundamental knowledge
- Identifying process–microstructure–property relationships

Such studies would allow manufacturers not only to optimize additive manufacturing materials and techniques but also to develop effective methods for inspecting their products. The studies conducted to date have resulted in the use of additive manufacturing for the manufacture of critical components, such as turbine blades, medical devices, and even complex structural parts. Enterprising fabricators are producing 3D printed functional clocks, guns, robots, and even parts for a 3D printer. A most important and developing trend for additive manufacturing is its use for personal consumer purposes.

REFERENCES

1. ASTM. (2010). *Standard Terminology for Additive Manufacturing Technologies*, F2792-12a. ASTM International, West Conshohocken, PA.
2. Levy, G.N., Schindel, R., and Kruth, J.P. (2003). Rapid manufacturing and rapid tooling with layer manufacturing (LM) technologies: state of the art and future perspectives. *CIRP Ann. Manuf. Technol.*, 52: 589–609.
3. Kruth, J.P., Leu, M.C., and Nakagawa, T. (1998). Progress in additive manufacturing and rapid prototyping. *CIRP Ann. Manuf. Technol.*, 47: 525–540.
4. Gero, J.S. (1995). Recent advances in computational models of creative design. In: *Design Research in the Netherlands* (Oxman, R., Bax, T., and Achten, H., Eds.), pp. 56–57. ETH, Eindhoven.
5. Chu, C., Graf, G., and Rosen, D.W. (2008). Design for additive manufacturing of cellular structures. *Comput. Aided Des. Appl.*, 5: 686–696.
6. Kruth, J.P. (1991). Material ingress manufacturing by rapid prototyping techniques. *CIRP Ann. Manuf. Technol.*, 40: 603–614.
7. Flowers, J. and Moniz, M. (2002). Rapid prototyping in technology education. *Technol. Transfer*, 62(3): 7.
8. Wohlers, T. (2012). Additive manufacturing advances. *Manuf. Eng.*, 148(4): 55–56.
9. Bourell, D.L., Beaman, Jr., J.J., Leu, M.C., and Rosen, D.W. (2009). A Brief History of Additive Manufacturing and the 2009 Roadmap for Additive Manufacturing: Looking Back and Looking Ahead, paper presented at RapidTech 2009, Erfurt, Germany, May 26–27.
10. Lou, A. and Grosvenor, C. (2012). *Selective Laser Sintering: Birth of Industry*. University of Texas, Austin.
11. Jacobs, P.F. (1992). *Rapid Prototyping & Manufacturing: Fundamentals of StereoLithography*. Society of Manufacturing Engineers, Dearborn, MI.
12. Pham, D.T. and Dimov, S.S. (2001). *Rapid Manufacturing: The Technologies and Applications of Rapid Prototyping and Rapid Tooling*. Springer Verlag, New York.
13. Hauser, C., Sutcliffe, C.S., Egan M., and Fox, P. (2005). Spiral Growth Manufacturing (SGM): A Continuous Additive Manufacturing Technology for Processing Metal Powder by Selective Laser Melting, paper presented at 16th Solid Freeform Fabrication Symposium, Austin, TX, August 13.

14. Herderick, E. (2011). Additive Manufacturing of Metals: A Review, paper presented at Materials Science and Technology (MS&T) 2011, Columbus, OH, October 16–20.

15. Scott, J., Gupta, N., Weber, C., Newsome, S., Wohlers, T., and Caffrey, T. (2012). *Additive Manufacturing Status and Opportunities*. IDA, Science and Technology Policy Institute, Washington DC.

16. Vilaro, T., Colin, C., and Bartout, J.D. (2011). As-fabricated and heat-treated microstructures of the Ti-6Al-4V alloy processed by selective laser melting. *Metall. Mater. Trans. A*, 42A: 3190–3199.

17. Wang, F., Williams, S.W., Colegrove, W.P., and Antonysamy, A.A. (2013). Microstructure and mechanical properties of wire and arc additive manufactured Ti-6Al-4V. *Metall. Mater. Trans. A*, 44A: 968–977.

18. Hon, K.B.B., Han, C., and Edwardson, S.P. (2006). Investigation of new scanning pattern for stereo-lithography. *Annals CIRP*, 55(1): 217–220.

19. Wong, K.V. and Hernandez, A. (2012). A review of additive manufacturing. *ISRN Mech. Eng.*, 2012: 1–10.

20. Frazier, W.E. (2014). Metal additive manufacturing: a review. *J. Mater. Eng. Perform.*, 23: 1917–1928.

21. Milewski, J.O., Dickerson, P.G., Nemec, R.B., Lewis, G.K., and Fonseca, J.C. (1999). Application of a manufacturing model for optimization of additive processing of Inconel Alloy 690. *J. Mater. Process. Technol.*, 91: 28–28.

22. Akula, S. and Karunakaran, K.P. (2006). Hybrid adaptive layer manufacturing: an intelligent asset of direct metal rapid tooling process. *Robot. Comput. Integr. Manuf.*, 22: 113–123.

23. Santua, E.C., Shiomi, M., Osakada, K., and Laoui, T. (2006). Rapid manufacturing of metal components by laser forming. *Int. J. Machine Tools Manuf.*, 46: 1459–1468.

24. Roberts, I.A., Eng, C.J., Esterlein, R., Stanford, M., and Mynorts, D.J. (2009). A three-dimensional finite element analysis of the temperature field during laser melting of metal powders in additive layer manufacturing, *Int. J. Machine Tools Manuf.*, 49: 916–923.

25. Baufeld, B., van der Biest, O., and Gault, R. (2010). Additive manufacturing of Ti-6Al-4V components by shaped metal deposition: microstructure and mechanical properties. *Mater. Des.*, 31: S106–S111.

26. Levy, G.N. (2012). The role and future of the laser technology in additive manufacturing environment. *Phys. Proc.*, 5: 65–80.

27. Sriraman, M.R., Babu, S.S., and Short, M. (2010). Bonding characteristics during very high power ultrasonic additive manufacturing of copper. *Scripta Mater.*, 62: 560–563.

28. Ramirez, D.A., Murr, L.E., Martinez, E., Hernandez, D.H., Martinez, J.L., Machado, B.I., Medina, F., Frigola, P., and Wicker, R.B. (2011). Novel precipitate microstructural architecture developed in the fabrication of solid copper components by additive manufacturing using electron beam melting, *Acta Mater.*, 59: 4088–4099.

29. Bibb, R., Thompson, D., and Winder, J. (2011). Computed tomography characterization of additive manufacturing materials. *Med. Eng. Phys.*, 33: 590–596.

30. Baufeld, B., Brandl, E., and van der Biest, O. (2011). Wire-based additive layer manufacturing: comparison of microstructure and mechanical properties of Ti-6Al-4V fabricated by laser beam deposition. *J. Mater. Process. Technol.*, 211: 1146–1158.

31. Zaeh, M.F. and Ott, M. (2011). Investigations on heat regulation of additive manufacturing processes for metal structures. *CIRP Ann. Manuf. Technol.*, 60: 259–262.

32. Martina, F., Mehnen, J., Williams, S.W., Colegrove, P., and Wang, T. (2012). Investigation of the benefits of plasma deposition for additive layer manufacture of Ti-6Al-4V. *J. Mater. Process. Technol.*, 212: 1377–1386.

33. Brandl, E., Heckenberger, U., Holzinger, V., and Buchbinder, D. (2012). Additive manufactured AlSi10Mg samples using selective laser melting: microstructure, high cycle fatigue and fracture behavior. *Mater. Des.*, 34: 159–169.

34. Friel, R.J. and Harrys, R.A. (2013). Ultrasonic additive manufacturing: a hybrid production process for novel functional products. *Proc. CIRP*, 6: 35–40.

35. Suter, M., Weingartner, E., and Wegener, K. (2012). MHD print-head for additive manufacturing of metals. *Proc. CIRP*, 2: 102–106.

36. Boisselier, D. and Sankare, S. (2012). Influence of powder characteristics in laser direct metal deposition of SS316L for metallic parts manufacturing, *Phys. Proc.*, 39: 455–463.

37. Vayre, B., Vignat, F., and Villeneuve, F. (2012). Designing for additive manufacturing, *Proc. CIRP*, 3: 632–637.

38. Foster, D.R., Daoino, M.J., and Babu, S.S. (2012). Elastic constants of ultrasonic additive manufactured Al 3003-H18. *Ultrasonics*, 53(1): 211–218.

39. Niendorf, T. and Brenne, F. (2013). Steel showing twinning induced plasticity processed by selective laser melting: an additive manufactured high performance material. *Mater. Charact.*, 85: 57–63.

40. Cooper, D.E., Blundell., N., Maggs, S., and Gibbons, G.J. (2013). Additive layer manufacturing of Inconel 625 metal matrix composites reinforcement material evaluation. *J. Mater. Process. Technol.*, 213: 2191–2200.
41. Kumar, V. and Dutta, D. (1997). An assessment of data formats for layered manufacturing. *Adv. Eng. Software*, 28: 151–164.
42. ASTM. (2011). *Standard Specification for Additive Manufacturing File Format*, F2915-11. ASTM International, West Conshohocken, PA.
43. Juster, N.P. and Childs, T.H.C. (1994). A comparison of rapid prototyping processes. In: *Proceedings of the Third European Conference on Rapid Prototyping and Manufacturing*, University of Nottingham, July 6–7, pp. 35–52.
44. Chua, C.K., Choi, S.M., and Wong, T.S. (1998). A study of state-of-the art rapid prototyping technologies. *Int. J. Adv. Manuf. Technol.*, 14(2): 146–152.
45. Jacobs, P.F. (1992). *Rapid Prototyping & Manufacturing: Fundamentals of StereoLithography*. Society of Manufacturing Engineering, Dearborn, MI.
46. ASTM. (2014). *Standard Specification for Additive Manufacturing Titanium-6 Aluminum-4 Vanadium with Powder Bed Fusion*, F2924. ASTM International, West Conshohocken, PA.
47. Ashley, S. (1991). Rapid prototyping systems. *Mech. Eng.*, 113(4): 34.
48. Cooper, K. (2001). *Rapid Prototyping Technology*. Marcel Dekker, New York.
49. Kochan, A. (1997). Features rapid growth for rapid prototyping. *Assembly Automation*, 17(3): 215–217.
50. Noorani, R. (2006). *Rapid Prototyping: Principles and Applications*. John Wiley & Sons, New York.
51. Kruth, P.P. (1991). Material incress manufacturing by rapid prototyping techniques. *CIRP Ann. Manuf. Technol.*, 40(2): 603–614.
52. Halloran, J.W., Tomeckova, V., and Gentry, S. (2011). Photo-polymerization of powder suspensions for shaping ceramic materials. *J. Eur. Ceramic Soc.*, 31(14): 2613–2619.
53. Pham, D.T. and Ji, C. (2000). Design for stereolithography. *Proc. Inst. Mech. Eng.*, 214(5): 635–640.
54. Kim, H., Jae Won, C., and Wicker, R. (2010). Scheduling and process planning for multiple material stereolithography. *Rapid Prototyping J.*, 16(4): 232–240.
55. Pham, D.T. and Gault, R.S. (1998). A comparison of rapid prototyping technologies. *Int. J. Machine Tool Manuf.*, 38: 1257–1287, 1998.
56. Feygin, M. and Hsieh, B. (1991). Laminated Object Manufacturing (LOM): A Simpler Process, paper presented at Second Freeform Fabrication Symposium, Austin, TX, August 6–8.
57. Skelton, J. (2008). Fused deposition modeling. *3D Printers and 3D-Printing Technologies Almanac*, February 8, http://3d-print.blogspot.com/2008.
58. Kamrani, A.K. and Nasr, E.S. (2010). *Engineering Design and Rapid Prototyping*. Springer, New York.
59. Sachs, E., Cima, M., and Cornie, J. (1990). Three-dimensional printing: rapid tooling and prototypes directly from a CAD model. *CIRL Ann. Manuf. Technol.*, 39: 201–204.
60. Miller, R. (2014). Additive manufacturing (3D printing): past, present and future. *Indust. Heating*, May: 39–41.
61. Rooks, A. (2014). Is it a.m. for AM? *Cutting Tool Eng.*, February: 8.
62. Hock, L. (2014). 3-D printing: a new manufacturing staple. *R&D Mag.*, April.
63. Beaman, J.J., Barlow, J.W., Bourell, D.L., Crawford, R.H., Marcus H.L., and McAlea, K.P. (1996). *Solid Freeform Fabrication: A New Direction in Manufacturing*. Springer, New York.
64. Deckard, C. and Beaman, J.J. (1988). Process and control: issues in selective laser sintering. *ASME PED*, 33: 191–197.
65. Hwa Hsing, T., Ming Lu, C., and Hsiao Chuan, Y. (2011). Slurry-based selective laser sintering of polymer-coated ceramic powders to fabricate high strength alumina parts. *J. Eur. Ceramic Soc.*, 31(8): 1383–1388.
66. Salmoria, G.V., Paggi, R.A., Lago, A., and Beal, V.E. (2011). Microstructure and mechanical characterization of PA12/MWCNTs nanocomposite manufactured by selective laser sintering. *Polym. Testing*, 30(6): 611–615.
67. Skavko, D. and Matic, K. (2010). Selective laser sintering of composite materials technologies. *Ann. DAAAM Proc.*, 1527.
68. Mrudge, R.P. and Wald, N.R. (2007). Laser engineered net shaping advances in additive manufacturing and repair. *Weld. J.*, 86: 44–48.
69. Griffith, M.L., Schlierger, M.E., and Harwell, L.D. (1999). Understanding thermal behavior in the LENS process. *Mater. Des.*, 20: 107–113.
70. Xiong, Y. (2009). Investigation of the Laser Engineered Net Shaping Process for Nanostructured Cermets, doctoral dissertation, University of California, Davis.

71. Balkla, V.K., Bose S., and Bandyopadhyay, A. (2008). Processing of bulk alumina ceramics using laser engineered net shaping. *Int. J. Appl. Ceramic Technol.*, 5(3): 234–242.

72. Liao, Y.S., Li, H.C., and Chiu, Y.Y. (2006). Study of laminated object manufacturing with separtely applied heating and pressing. *Int. J. Adv. Manuf. Technol.*, 27(7–8): 703–708.

73. Murr, L.E., Gaytan, S., and Ramirez, D. (2012). Metal fabrication by additive manufacturing using laser and electron beam melting technologies. *J. Mater. Sci. Technol.*, 28(1): 1–14.

74. Semetay, C. (2007). Laser Engineered Net Shaping: Modeling Using Welding Simulation Concepts, doctoral dissertation, Lehigh University, Bethlehem, PA.

75. Zhai, Y. and Lados, D.A. (2012). *WPI Bi-annual Progress Report*. Integrative Materials Design Center (iMdc), Worcester, MA.

76. Castle Island. (2013). *Worldwide Guide to Rapid Prototyping: Laser Sintering*, http://www.additive3d.com/sls.htm.

77. Optomec. (2014). *LENS Technology*, www.optomec.com/?s=lens+technology.

78. Castle Island. (2013). *Direct Additive Fabrication of Metal Parts and Injection Molds*, http://www.additive3d.com/tl_221a.htm.

79. Castle Island. (2013). *Laser Powder Forming*, http://www.additive3d.com/lens.htm.

80. Brice, C.A. and Henn, D.A. (2002). Rapid Prototyping and Freeform Fabrication Via Electron Beam Welding Deposition, paper presented at International Institute of Welding Conference, Copenhagen, Denmark, June 26–28.

81. Taminger, K.M. and Hafley, R.A. (2006). Electron Beam Freeform Fabrication for Cost Effective Near-Net Shape Manufacturing, paper presented at NATO AVT-139 Specialists Meeting on Cost Effective Manufacture Via Net Shape Processing, Amsterdam, May 15–19.

82. Gibson, I., Rosen, D.W., and Stucker, B. (2010). *Additive Manufacturing Technologies: Rapid Prototyping to Direct Digital Manufacturing*. Springer, New York.

83. Villafuerte, J. (2014). Considering cold spray for additive manufacturing. *Adv. Mater. Process.*, May: 50–52.

84. Xue, L. and Islam, M.U. (2000). Free form laser consolidation for producing metallurgically sound and functional components. *J. Laser Appl.*, 12(4): 160–165.

85. Scott, J., Gupta, N., Weber, C., Newsome, S., Wohlers, T., and Caffrey, T. (2012). *Additive Manufacturing Status and Opportunities*. Science and Technology Policy Institute, Washington, DC.

86. Anon. (2012). The third industrial revolution. *Economist*, April 21, http://www.economist.com/node/21553017.

87. Jackson, T.R., Liu, H., Partrikalakis, N.M., Sachs, E.M., and Cima, M.J. (1999). Modeling and designing functionally graded material components for fabrication with local composition control. *Mater. Des.*, 20(2–3): 63–75.

88. Guo, N. and Leu, M.C. (2013). Additive manufacturing technology: applications and research needs. *Front. Mech. Eng.*, 8(3): 215–243.

89. Liu, F., Slattery, K., Kinsella, M., Newkirk, J., Chou, H.N., and Landers, R. (2007). Applications of a hybrid manufacturing process for fabrication of metallic structures. *Rapid Prototyping J.*, 13(4): 236–244.

90. Wohlers, T.T. (2014). *3D Printing and Additive Manufacturing State of the Industry, Annual Worldwide Progress Report*. Wohlers Associates, Inc., Fort Collins, CO.

91. Wohlers Assoc. (2014). Metal Additive Manufacturing Grows by Nearly 76% [press release], May 21, Wohlers Associates, Inc., Fort Collins, CO (http://wohlersassociates.com/press64.html).

92. NIST. (2013). *Measurement Science Roadmap for Metal-Based Additive Manufacturing*. National Institute for Standards and Technology, Gaithersburg, MD.

93. Additive Manufacturing Consortium, http://ewi.org/additive-manufacturing-consortium/.

94. Zhai, Y. and Lados, D.A. (2012). *Integrative Materials Design Center Biannual Progress Report*. Worcester Polytechnic Institute, Worcester, MA.

95. Murr, L.E., Gaytan, S.M., Ramirez, D.A., Martinez, E., Hernandez, J. et al. (2012). Metal fabrication by additive manufacturing using laser and electron beam melting technologies. *J. Mater. Sci. Technol.*, 28(1): 1–14.

96. Amano, R.S. and Rohatgi, P.K. (2011). Laser engineered net shaping process for SAE 4140 low alloy steel. *Mater. Sci. Eng.*, 528(22–23): 6680–6693.

97. Wu, X., Liang, J., Mei, J., Mitchell, C., Goodwin, P.S., and Voice, W. (2004). Microstructures of laser-deposited Ti-6Al-4V. *Mater. Des.*, 25(2): 137–144.

98. Thijs, L., Verhaeghe, F., Craeghs, T., van Humbeeck, J., and Kruth, J.-P. (2010). A study of the micro-structural evolution during selective laser melting of Ti-6Al-4V. *Acta Mater.*, 58(9): 3303–3312.

99. Kempen, K., Yasa, E., Thijs, L., Kruth, J.-P., and van Humbeeck, J. (2011). Microstructure and mechanical properties of selective laser melted 18Ni-300 steel. *Phys. Proc.*, 12(1): 255–263.

100. Rangers, S. (2012). Electron Beam Melting Versus Direct Metal Laser Sintering, paper presented at Midwest SAMPE's Direct Part Manufacturing Workshop, Wright State University, Fairborn, OH, November 13–14.

101. Ramosoeu, M.K.E., Booysen, G., Ngoda T.N., and Chikwanda, H.K. (2011). Mechanical Properties of Direct Laser Sintered Ti-6Al-4V, paper presented at Materials Science and Technology (MS&T) 2011, Columbus, OH, October 16–20.

102. Greitemeir, D., Schmidtke, K., Holzinger, V., and Donne, C.D. (2013). Additive Laser Manufacturing of Ti-6Al-4V and Scalmalloy© Fatigue and Fracture, paper presented at 27th Symposium of the International Committee on Aeronautical Fatigue, Jerusalem, Israel, June 5–7.

103. Murr, L.E., Martinez, E., Gaytan, S.M., Ramirez, D.A., Machado, B.I. et al. (2011). Microstructure and mechanical properties of a nickel base superalloy fabricated by electron beam melting. *Metall. Trans. A*, 42: 3491–3508.

104. Baufeld, B. (2012). Mechanical properties of Inconel 718 parts manufactured by shaped metal deposition. *J. Mater. Eng. Perform.*, 21(7): 1416–1421.

105. Ganesh, P., Kaul, R., Paul, C.P., Tiwari, P., Rai, S.K., Prasad, R.C., and Kukreja, L.M. (2010). Fatigue and fracture toughness characteristics of laser rapid manufactured Inconel 625 structures. *Mater. Sci. Eng.*, 527(29–30): 7490–7497.

106. Chlebus, E., Kuznicka, B., Kurzynowski, T., and Dyba, B. (2011). Microstructure and mechanical behavior of Ti-6Al-7Nb alloy produced by selective laser melting. *Mater. Charact.*, 62(5): 488–495.

107. Blackwell, P.L. and Wisbey, A.J. (2005). Laser-aided manufacturing technologies: their application to the near-net shape forming of a high-strength titanium alloy. *Mater. Proc. Technol.*, 170: 268.

108. Vrancken, B., Thijs, L., and Kruth, J.P. (2012). Heat treatment of Ti6Al4V produced by selective laser melting: microstructure and mechanical properties. *J. Alloys Compd.*, 541: 177.

109. Martukanitz, R., Simpson, T., and The Center for Innovative Materials Processing through Direct Digital Deposition. (2013). *Brief at the Technology Showcase.* ARL Information Center, Penn State, State College, PA.

110. Bourell, D.L., Leu, M.C., and Rosen, D. (2009). *Roadmap for Additive Manufacturing: Identifying the Future of Freeform Processing.* University of Texas, Austin.

111. Gibson, I., Rosen, D.W., and Stucker, B. (2010). *Additive Manufacturing Technologies: Rapid Prototyping to Direct Digital Manufacturing.* Springer, New York.

112. Optomec, http://www.optomec.com/.

113. Hofmann, D.C., Borgonia, J.P.C., Dillon, R.P., Suh, E.J., Mulder, J.L., and Gardner, P.B. (2013). *Applications for Gradient Metal Alloys Fabricated Using Additive Manufacturing*, NASA Technical Brief. Jet Propulsion Laboratory, National Aeronautics and Space Administration, Pasadena, CA.

114. Kalpakjian, S. and Schmid, S.R. (2008). *Manufacturing Processes for Engineering Materials*, 5th ed. Prentice Hall, Englewood Cliffs, NJ.

115. National Research Council. (2011). *3D Printing in Space.* National Academies Press, Washington DC.

116. Coxworth, B. (2014). Report on prototype electric car made with a 3D printer. *Stratasys*, November 19.

117. Roberts, S., www.youtube.com/watch?v=daiowikH7Z1.

118. Scott, J., Gupta, N., Weber, C., Newsome, S., Wohlers, T., and Caffrey, T. (2012). *Additive Manufacturing: Status and Opportunities.* Science and Technology Policy Institute, Washington DC.

119. Hock, L. (2014). 3D printing for blood recycling, medical developments. *Prod. Des. Dev.*, October 7, http://www.pddnet.com/articles/2014/10/3d-printing-blood-recycling-medical-developments.

120. Bullis, K. (2014). EmTech: 3-D printing complex kidney components. *MIT Technol. Rev.*, September 24, http://www.technologyreview.com/news/531106/emtech-3-d-printing-complex-kidney-components/.

121. Lavars, N. (2014). 3D-printed syringe pumps could cut the cost of scientific research. *Gizmag*, September 17, http://www.gizmag.com/3d-printed-syringe-pump-scientific-research/33863/.

122. Coxworth, B. (2014). Organovo now selling tiny 3D printed human livers. *Gizmag*, November 19, http://www.gizmag.com/organovo-exvive3d-liver-models/34843/.

123. Lavars, N. (2014). 3D printing helps build upper jaw prosthetic of cancer patient. *Gizmag*, October 17, http://www.gizmag.com/3d-printing-upper-jaw-prosthetic-cancer/34303/.

124. Laird, J. (2014). "Spinning" a solution for tracheal surgery. *Med. Des.*, October 20, http://medicaldesign.com/prototyping/spinning-solution-tracheal-surgery.

125. Aurora Laboratories. (2014). Metal 3D printer—yours for under US$5,000. *Mater. Today*, October 9, http://www.materialstoday.com/additive-manufacturing/news/metal-3d-printer-yours-for-under-us5000/.

126   Mraz, S. (2014). Developing medical devices with 3D IM. *Machine Des.*, October 9, http://machinedesign.com/3d-printing/developing-medical-devices-3d-im.

127.   Anon. (2014). Market trends: additive manufacturing on the rise. *Ceram. Indust. Mag.*, May 1, http://www.ceramicindustry.com/articles/93901-market-trends-additive-manufacturing-on-the-rise.

122.   McCormick, S. (2012). Chin Surgery Skyrockets among Women and Men in All Age Groups [press release]. American Society of Plastic Surgeons, Arlington Heights, IL (http://www.eurekalert.org/pub_releases/2012-04).

128.   Svensson, P. (2014). The new family portrait? 3D-printed statue selfies. *Washington Times*, October 9, http://www.washingtontimes.com/news/2014/oct/9/the-new-family-portrait-3d-printed-statue-selfies/.

129.   Hock, L. (2014). Scanning products into 3-D: 3-D scanning technology reduces costs and speeds up product development with new technology advancements. *R&D Mag.*, August 6, http://www.rdmag.com/articles/2014/08/scanning-products-3-d.

130.   Lavars, N. (2014). Wasp's 3D printers produce low-cost houses made from mud. *Gizmag*, October 20, http://www.gizmag.com/wasp-3d-printers-house-mud/34340/.

131.   Newman, J. (2014). Rapid ready tech: could 3D printing save the US Postal Service? *Desktop Eng.*, July 14, http://www.rapidreadytech.com/2014/07/could-3d-printing-save-the-us-postal-service/.

132.   Leigh, S.J., Bradley, R.J., Purssell, C.P., Billson, D.R., and Hutchins, D.A. (2012). A simple, low cost conductive composite material for 3D printing of electronic sensors. *PLoS ONE*, 7(11): e49365.

133   Davis, S. (2011). *Construction of a CubeSat Using Additive Manufacturing*, SAE Technical Paper 2011-01-2568. SAE International, Warrensdale, PA.

134.   Gebhardt, A. (2012). *Understanding Additive Manufacturing: Rapid Prototyping, Rapid Tooling, Rapid Manufacturing*. Hanser Publications, Cincinnati, OH.

135.   National Institute on Aging. (2007). *Why Population Aging Matters: A Global Perspective*, Publ. No. 07-6134. National Institute on Aging, Bethesda, MD.

136.   OECD. (1996). *Ageing in OECD Countries: A Critical Policy Challenge*. Organization for Economic Co-operation and Development, Paris.

137.   Beattie, W. (1998). Current challenges to providing personalized care in the long term care facility. *Int. J. Health Care Qual. Assur. Inc. Leadersh. Health Serv.*, 11(2–3): i–v.

138.   Ely, S. (2009). Personalized medicine: individual care of cancer patients. *Trans. Res.*, 154(6): 303–308

139.   Huang, S., Liu, P., Mokasdar, A., and Hou, L. (2013). Additive manufacturing and its societal impact. *Int. J. Adv. Manuf. Technol.*, 67: 1191–1203.

140.   He, Y., Ye, M., and Wang, C. (2006). A method in the design and fabrication of exact-fit customized implants based on sectional medical images and rapid prototyping technology. *Int. J. Adv. Manuf. Technol.*, 28: 504.

141.   Picariello, P. (2009). *Committee F42 on Additive Manufacturing Technologies*. ASTM International, West Conshohocken, PA, http://www.astm.org/COMMITTEE/F42.htm.

2 Additive Manufacturing Using Free Space Deposition in Metals
Experiment and Theory

Abinand Rangesh

CONTENTS

ABSTRACT

Additive manufacturing has the capability to build three-dimensional objects directly from a computer model to create complex internal or external geometries. The aim of this research was to design the foundations of a new additive manufacturing technology that could deposit shapes directly in free space without the use of any support structures and eliminate the limitation of

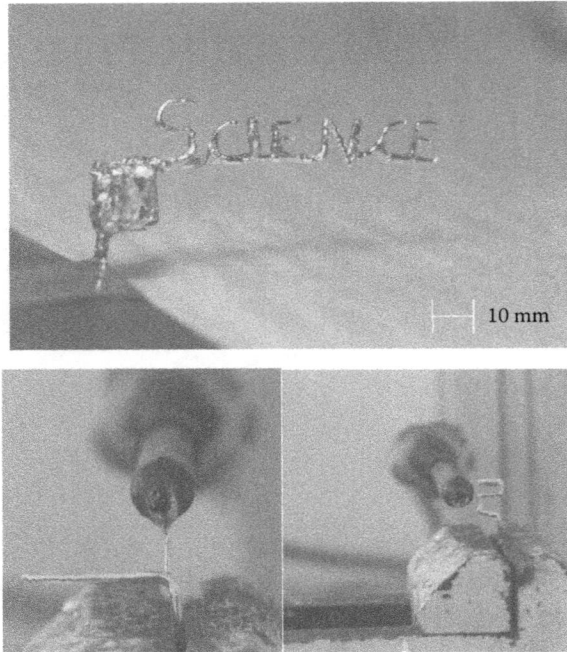

FIGURE 2.1 Photograph of track deposition in free space. (Top) Manual build of the letters "Science" built using free space deposition. (Bottom) Various shapes in the process of being deposited.

deposition in the *x-y* plane that most current technologies utilize. The contribution of this research is the use of a heated tool as a temporary moving support structure during material solidification to deposit tracks in free space. Notable results in free space track deposition were that the initial track diameter and volume affected the repeatability and quality of tracks. The amount of material fed to the soldering iron before commencing deposition affected the taper of tracks. At an initial volume of 7 mm^3 and an initial track diameter of 0.8 mm, none of the ten tracks deposited broke or showed taper > ~1°. The maximum deposition velocity for free space track deposition using lead-free solder was limited to 1.5 mm s^{-1}. Finite element modeling showed that the initial volume within the melt boundary, initial track radius, and distance between the soldering iron and the solidification front could be used to inform future experimental design. Selection of initial operating settings may be used to produce tracks profiles with standard deviations < 30% of initial track width.

2.1 INTRODUCTION

Additive manufacturing has the capability to build three-dimensional objects directly from a computer design by selectively depositing material. It allows the construction of parts with complete internal or external geometries. However, most conventional metal additive manufacturing technologies are limited in their ability to build overhanging structures. Kruth et al.[1] and Thomas and Bibb[2] found that selective laser melting (SLM) was limited in its ability to build overhanging structures with an angle less than 40° to 45° from the horizontal without building fixed support structures. Fixed support structures provide a means to support tracks and features during solidification and are usually removed during post-processing of parts. If the individual tracks are supported during solidification it may be possible to create overhanging tracks or tracks that start on a substrate but stretch into free space, as shown in Figure 2.1. Overhanging features can be built by layering individual overhanging tracks. After solidification it may not be necessary to continue supporting the track which would allow deposition in any plane or straight into free space.

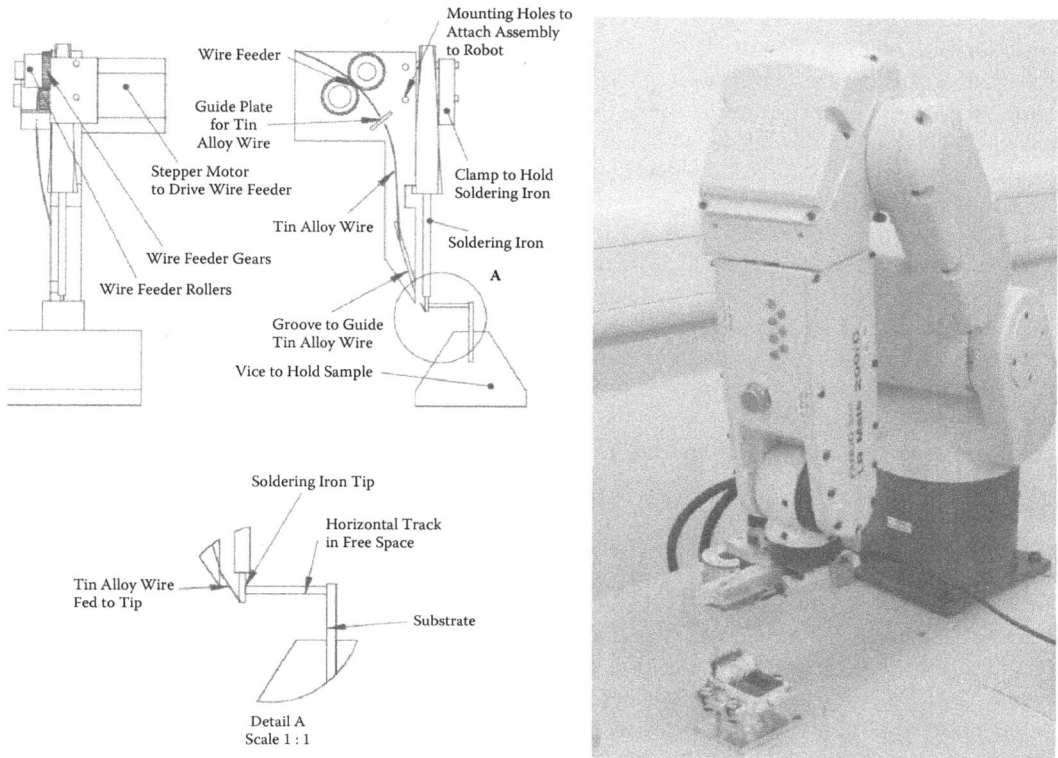

FIGURE 2.2 Photograph of experimental apparatus. The soldering iron and wire feeder are attached to a six-axis robotic arm and used to deposit lead-free solder tracks from a substrate mounted in the sample holder.

This research investigates the concept of a moving support structure during solidification using a soldering iron as the heat source and support structure and lead-free solder as the deposition material. The concept can then be extended to higher melting point materials such as stainless steel and titanium. Further information on the deposition of higher melting point materials can be found in Rangesh[3] and Rangesh and O'Neill.[4]

The research summary is divided into two sections: deposition of tracks in free space and the theory behind free space deposition. The first investigates the operating parameters that affect the repeatability and quality of free space tracks. The second section compares the experimental results from the first section with a computer model of the molten track boundary to predict the behavior of free space track deposition in a general case.

2.2 MODELING DEPOSITION OF TRACKS IN FREE SPACE

2.2.1 EXPERIMENTAL APPARATUS

Figure 2.2 shows the process setup. The apparatus is mounted on the robotic arm and can be rotated using the robot so the centerline of the soldering iron is horizontal or vertical. A soldering iron mounted on a FANUC LR Mate® 200iC (FANUC Robotics, Japan) robotic arm was used to perform the experiments. The wire feeder rollers were driven by a Trinamic PD1-013-32 stepper motor (Trinamic Motion Control GmbH, Germany). The soldering iron used was a 50-W Antex 660TC temperature-controlled soldering iron (Antex, United Kingdom) fitted with a modified Antex B110660 tip (2.3-mm diameter and 1.5-mm diameter at the tip. The lead-free solder used was Sn3.5Ag0.5Cu with a rosin flux core and a diameter of 1 mm unless otherwise specified. The material has a melting point of 217°C. This flux acts as a cleaning agent and improves wetting to the substrate or soldering iron by reducing the contact angle of the lead-free solder. Arenas and Acoff[5] found that, on a copper substrate, rosin-activated flux reduced the contact angle of Sn3.8Ag0.7Cu on a copper substrate from about 40° to 30° at 240°C. At 280°C, the contact angle with rosin-activated flux was reduced to 20° on a copper substrate.

The stepper motor for the wire feeder was controlled by sending commands from the computer's RS232 port, activating the stepper for a fixed time period at 200 steps per min. This equates to 1 mm³ of material fed to the soldering iron per second. A computer program was written to take information from the robot using a Velleman VM110 input/output card (Velleman, Inc., Fort Worth, TX) and use that information to control the stepper motor. The robot passed its speed and status—stationary or moving—and when to feed material to the computer. The computer program listened for a request from the robot for new material and then activated the stepper motor for a fixed time period set by the user. It then retracted the wire for 0.5 seconds so the solder wire was not permanently in contact with the soldering iron. The robot was programmed to pause for the same length of time as the feeding cycle after sending a request to the computer. Once material was fed, the robot could resume movement.

The robot was programmed with a sequence of relative coordinates, which means that the robot would follow a tool path that was offset from a starting position rather than as absolute movements from a fixed origin. The robot was controlled manually to the start position and the program was started. This would result in slight variations in starting position between experiments so the robot was programmed to return to the position it started from at the end of motion. The robot was fitted with FANUC's TCP speed option, which output the linear speed of the robot to analog port 1. The computer used its analog output to signal the robot's digital input port regarding the status of the feeding process. When the wire feeder finished feeding, it signaled the robot by setting the digital input so the robot could resume motion.

2.2.2 CLASSIFICATION OF RESULTS

The results can be classified quantitatively by measuring track width and qualitatively using the classification shown in Figure 2.3.

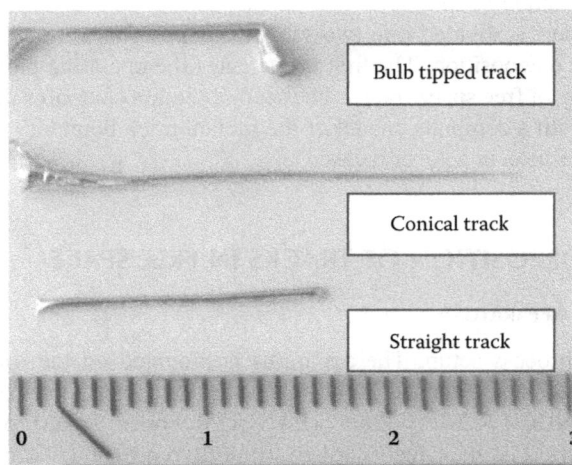

FIGURE 2.3 Qualitative classification of results. The scale is in millimeters. The tracks can be classified from top to bottom as a bulb tipped track, a conical track, and a straight track.

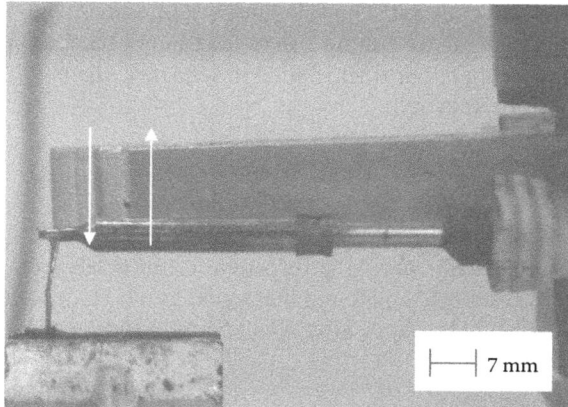

FIGURE 2.4 Depositing a vertical track of lead-free solder into free space from a wire mounted vertically in the sample holder. White arrows show the direction of movement. The robot moves down, melting a wire held in the sample holder, and then moves back up to create a track.

2.2.3 EXPERIMENTAL METHODOLOGY

2.2.3.1 Effect of Varying Material Feed Rate

Tracks of length of 10 mm were deposited using a temperature setting of 250°C and a velocity of 1 mm s^{-1}. A volume of 1 mm^3 of lead-free solder was fed to the soldering iron before commencing a track. Various feed settings were tested from 0.25 mm^3 s^{-1} to 2 mm^3 s^{-1} at 0.25-mm^3 s^{-1} intervals. This experiment was devised to determine the effect of feed rate and to determine the operating range within which straight tracks could be deposited. The results were classified qualitatively.

2.2.3.2 Effect of Varying Temperature

The effect of soldering iron temperature was investigated. The solder melts at ~217°C and the soldering iron has a maximum temperature of 450°C, so temperatures from 250 to 400°C at 25°C intervals were tested. The experiment was performed at a velocity setting of 1 mm s^{-1} after feeding 7 mm^3 before depositing tracks 10 mm in length. The purpose of this experiment was to determine the operating range for temperature.

2.2.3.3 Effect of Varying Initial Volume of Material Fed

The effect of the initial wetting diameter was tested by the previous experiment. It is likely that if the initial wetting diameter affects the precision and quality of tracks then other initial conditions may affect the quality of the track. The effects of the initial volume of material fed to the soldering iron on the track characteristics were tested (see Figure 2.4). Volumes between 5 mm^3 and 8 mm^3 at 0.5-mm^3 intervals were tested. The same methodology for deposition and measurement of tracks as described for the earlier experiments was used. Lead-free solder of 0.8-mm diameter was melted back to the required volume and tracks were deposited. Ten tracks at each volume were created unless more than four tracks failed in sequence. The mean track width, standard deviation, number of failed tracks, and number of tapered tracks were measured.

2.2.3.4 Effect of Varying Velocity

The effects of varying the velocity of the soldering iron while depositing tracks started from the 0.8-mm wire were investigated to determine the maximum deposition velocity. The same methodology for deposition and measurement of tracks as the previous experiment was used with five tracks deposited at each setting. The only parameter varied was the velocity. Temperature, volume, and initial wire diameter were held constant. The velocities tested were 0.25 to 2 mm s^{-1} at 0.25-mm s^{-1} intervals. Above 2 mm s^{-1}, intervals of 1 mm s^{-1} were tested until the maximum reliable deposition velocity was reached. If more than four tracks failed in sequence, the previous setting was deemed to be the maximum deposition velocity.

2.2.4   RESULTS

2.2.4.1   Effect of Material Feed Rate

Different feed settings were tested, varying at 0.25-mm^3 s^{-1} intervals from 0.25 to 2 mm^3 s^{-1}, but all settings gave unreliable results. It was found that tracks could be deposited using these settings but the tracks were susceptible to breakage after about 5 mm. Feeding material while the robot was stationary was considered.

2.2.4.2   Effect of Varying Soldering Iron Temperature

The shape of the track, tendency for track breakage, and level of oxidation were affected by the temperature setting. Qualitative observations are shown in Table 2.1.

2.2.4.3   Effect of Varying Initial Volume of Material Fed to Soldering Iron

Figures 2.5, 2.6, and 2.7 show the effect of varying the volume of material fed to the soldering iron. At volumes below 5 mm^3 all of the tracks failed. Between 5.5 mm^3 and 7.0 mm^3 tracks could be created, and fewer than two tracks were tapered, as can be seen by Figure 2.5. The best results were obtained at 7 mm^3, as there were no failures or tapered tracks. At volumes greater than 7 mm^3, the tracks were initially wider than the wire diameter and then began to taper, as can be seen by the 8-mm^3 volume tracks in Figure 2.7. The tracks are also asymmetrical and have one axis wider than the other. This asymmetry is likely to be due to the shape of the soldering iron tip. The soldering iron tip is a curved surface with a single axis of curvature. When the molten material wets such a curved surface, the projected shape of the molten material on the soldering iron is an ellipse rather than a circle. If the soldering iron were a flat plate or a sphere the projected wetting shape may be symmetrical.

TABLE 2.1

Effect of Temperature on Track Quality

Temperature (°C)	Observations
250	New material barely melted on contact with the soldering iron, and tracks had a visually dull surface. Most tracks were conical in shape.
275	New material melted on contact with the soldering iron, and the flux protected the molten material from oxidation. The molten material was easy to work with; tracks did not sag under their own weight, and straight tracks with no conical taper could be created if other settings were correct.
300	Similar working characteristics to 275°C.
325	Excess material was melted during feeding due to the residual heat. This molten material accumulated on the soldering iron tip as a large droplet and had a tendency to detach from the soldering iron surface because of its own weight. This excess material also increased the tendency for bulb tipped tracks.
350	Problems similar to the 325°C case. The flux began to break down, leaving burned flux residue on the substrate, and the lead-free solder would not wet the substrate surface so tracks could not be deposited.
375	Working conditions similar to the 350°C setting.
400	The molten solder no longer adhered to the surface of the soldering iron.

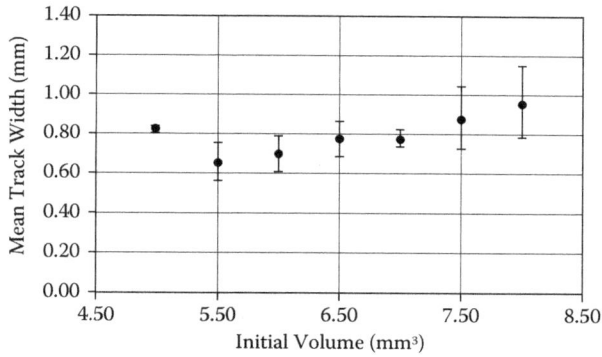

FIGURE 2.5   The effect of initial volume on track width. Tracks were created at 275°C at 0.5 mm s^{-1} velocity. Ten tracks at each volume were compared.

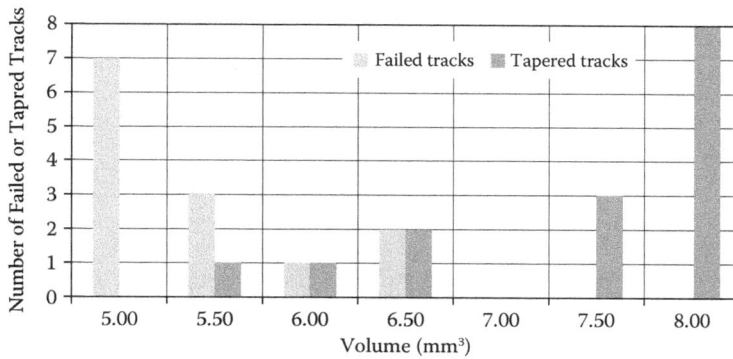

FIGURE 2.6   The effect of initial volume on the number of tapered and failed tracks.

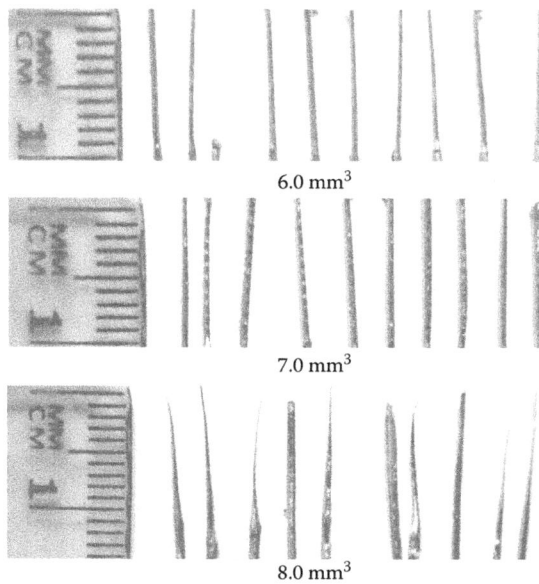

6.0 mm^3

7.0 mm^3

8.0 mm^3

FIGURE 2.7   Photographs of the tracks created with various volumes of material fed to the soldering iron tip. Experiments were conducted at a temperature of 275°C and velocity of 0.5 mm s^{-1}.

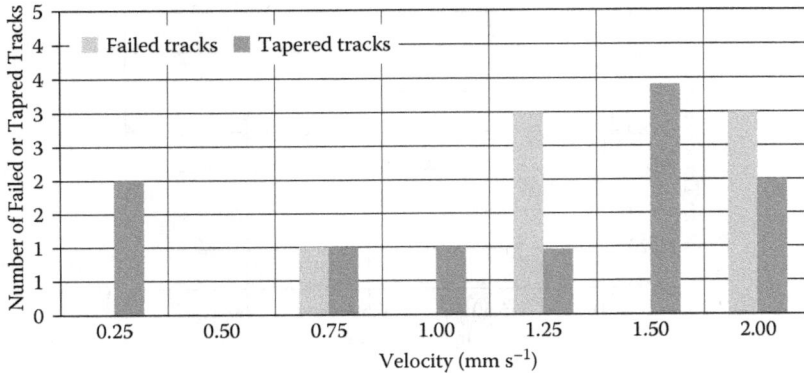

FIGURE 2.8   Number of failed or tapered tracks at each velocity setting.

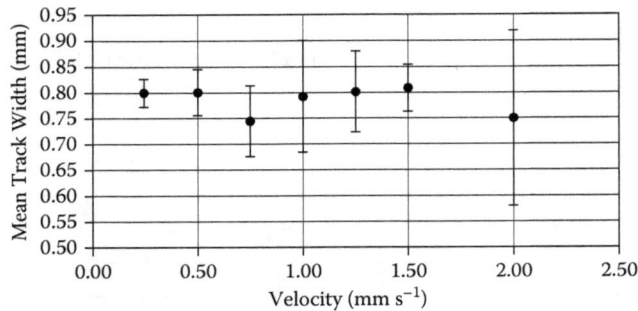

FIGURE 2.9   The effect of velocity on the mean track width. The velocity does not appear to affect the mean track width.

2.2.4.4 Effect of Varying Velocity

Figures 2.8 and 2.9 show the results for each velocity setting. At velocities greater than 2 mm s^{-1} it was not possible to deposit tracks, as all of the tracks broke at the start of the soldering iron motion. The reliability of the tracks above 1 mm s^{-1} is questionable, as there were three failures at the 1.25-mm s^{-1} and 2.00-mm s^{-1} settings. It can be seen that tapered tracks were found at all velocity settings except at 0.5 mm s^{-1}. Figure 2.8 shows the effect of velocity on the mean track width. All of the values appear to be similar to the wire diameter of 0.8 mm, and there appears to be no visible correlation between velocity and track width. The range was used instead of the 95% confidence interval in Figure 2.8 for the 1.25-mm s^{-1} and 2-mm s^{-1} settings due to the three track failures at these settings. The three failures at 2 mm s^{-1} and failure of all tracks at velocities greater than 2 mm s^{-1} suggest that the maximum usable velocity is 1.5 mm s^{-1} as there were no track failures. The velocity of 1.5 mm s^{-1} also has limitations, as four out of the five tracks were tapered. It appears that the most reliable setting is 0.5 mm s^{-1}, which demonstrated no track failures or taper.

2.2.5 Discussion

The series of experiments on free space deposition demonstrated the ability to deposit tracks starting from a substrate and extending into free space. They also indicate the potential to move away from layered building on the horizontal plane to depositing tracks and features in any plane. The operating range for temperature was found to be between 250°C and 310°C. This is above the

FIGURE 2.10 Wetting width on the electrode reduces as the volume of molten solder on the electrode reduces. The electrode has moved by 2 mm vertically between each of the three images.

217°C melting point and is the minimum temperature at which tracks could be deposited. Below 250°C, the metal freezes before it can be manipulated. If the temperature is greater than 325°C, the flux begins to break down and the molten material eventually stops adhering to the soldering iron surface. Within this temperature operating range, it was difficult to correlate the track width to temperature. Soldering irons do not have sophisticated temperature control mechanisms, and thermal imaging showed that the temperature varied by up to 25°C for a given temperature setting. Therefore, further analysis of the effects of temperature is not possible until a better method of controlling the tip temperature is found.

Varying the initial volume of material fed to the soldering iron had a marked effect on track characteristics. Some volumes were unstable; for example, the 5-mm^3 setting resulted in seven track failures. It is likely that the volume determines the melt boundary shape and some shapes are more stable than others. The results of varying the initial volume fed to the electrode showed that the 7-mm^3 setting produced the most repeatable results with no failures. Below this value of initial volume some tracks broke at the start of deposition, and at values greater than 7 mm^3 the tracks exhibited significant taper. A possible reason for the reliable results at the 7-mm^3 setting is that the tracks start with a stable melt boundary shape and do not neck or break during motion.

The maximum velocity was found to be 2 mm s^{-1}, but three out of the five tracks failed and two out of the three tracks deposited exhibited taper. The 1.5-mm s^{-1} setting showed no failures but resulted in four out of the five tracks having taper. Almost all of the tracks had a mean track width similar to the wire diameter. There appeared to be no correlation between track width and velocity. The most reliable velocity setting was 0.5 mm s^{-1}, as none of the tracks failed.

At velocities greater than 2 mm s^{-2}, track failures were inferred to be due to the initial acceleration of the soldering iron. The initial acceleration was set at the robot's default setting of 100%, so for a velocity of 2 mm s^{-1} the initial acceleration would be 2 mm s^{-2}. Testing track deposition at a velocity of 2.5 mm s^{-1} and an acceleration of 0.5 mm s^{-2} resulted in no tracks being deposited. This indicates that the track breakage may have to do with the stability of the melt boundary. Yadroitsev et al.[6] and Gusarov et al.[7] suggested that balling is a function of the length-to-width ratio of the tracks. They also found that using higher velocities on a substrate resulted in balling of tracks. Using a similar analogy, higher velocity settings may result in deposition of a longer segment of track that is molten; because this segment has a larger length-to-width ratio, it becomes unstable and breaks. After repeated observation, it appeared that the wetting width on the soldering iron reduced as material was deposited in cases where there was no track taper (see Figure 2.10). The reasons for the wetting width on the soldering iron reducing are not known at this stage, nor if there is any causality between the wetting width and necking.

To conclude, the experimental results suggest that initial conditions such as volume of molten material, starting diameter, and deposition velocity play significant roles in the stability of tracks deposited in free space. In the next section, finite element modeling is used to develop a melt boundary model and further investigate the reasons for track necking or taper.

2.3 MODELING FREE SPACE DEPOSITION

The melt boundary was modeled using Surface Evolver,[8] an interactive program for the study of surfaces shaped by surface tension and other energies.[9] It can be used to model a variety of surface tension problems, including the solder bridging phenomenon.[10] Surface Evolver attempts to solve for a surface by minimizing the energy of the surface.[8] The energy of the surface can be a result of surface tension, gravity, etc. or a result of user-specified functions. The problem to be solved is defined in a user data file containing information about the model geometry, volume, vertex and edge constraints, and user-defined functions and integrals. The model geometry can be given fixed volumes that act as constraints on the solution.

To study boundary shapes formed due to surface tension, Erle et al.[11] and Gillette and Dyson[12] considered the case of a soap film suspended between two parallel plates where the wetting diameter on each plate was identical. They considered the problem of the stability of this interface both theoretically and experimentally. The soap film was suspended between two plates of diameter d and distance l apart. The air inside the interface boundary was evacuated until the soap film burst. If the value of l/d increases, the onset of instability occurs at a higher volume. This suggests that the effects of volume and the stability ratio (the ratio of the separation between end faces to diameter, or l/d) may affect the shape of the modeled melt boundary. It also suggests that reducing volume will eventually lead to track taper and breakage. The critical stability ratio for a catenoid has been found to be 0.4718 to 0.6627, depending on the volume enclosed within.

Figure 2.11 shows the dimensions of the soldering iron tip (modified Antex B11060) and the variables for the solder surface in the x-z plane. $R(x)$ is defined as the radius of the solder surface at a distance x from the origin. The input variables for the model are the initial track radius (R_t), distance from the soldering iron to solidification front (l), and initial volume of solder within the melt boundary (V_0). Outputs from the model are the melt boundary shape $R(x)$, of which $R_1 = R(l + C)$, where $C = 0.95$ mm and signifies the track radius at the solidification front.

Brakke[13] modeled a liquid droplet wetting a cylinder by working in cylindrical coordinates, where a cylinder can be represented by a plane and then transforms the solved surface into Cartesian coordinates as shown in Figure 2.12. This approach was chosen as it provided a smoother edge at the

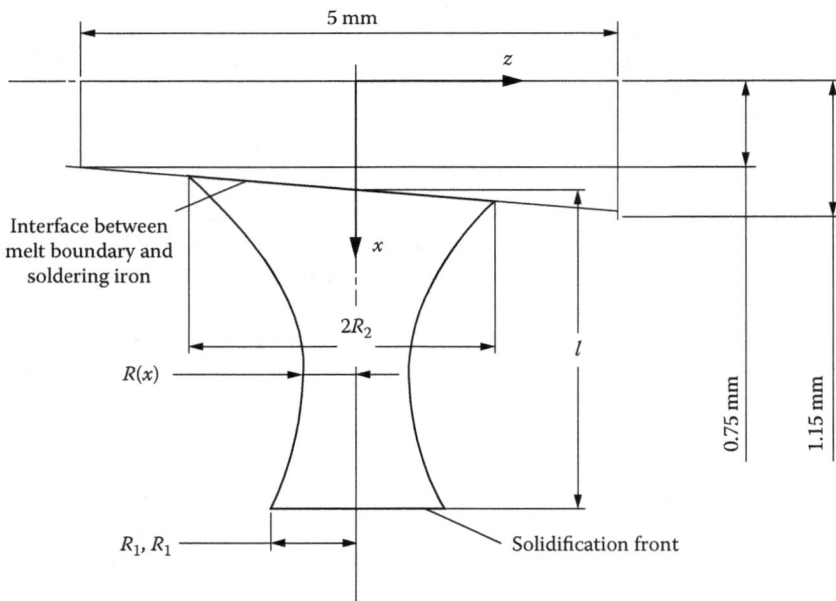

FIGURE 2.11 Geometry of liquid droplet between the solidification front and the soldering iron.

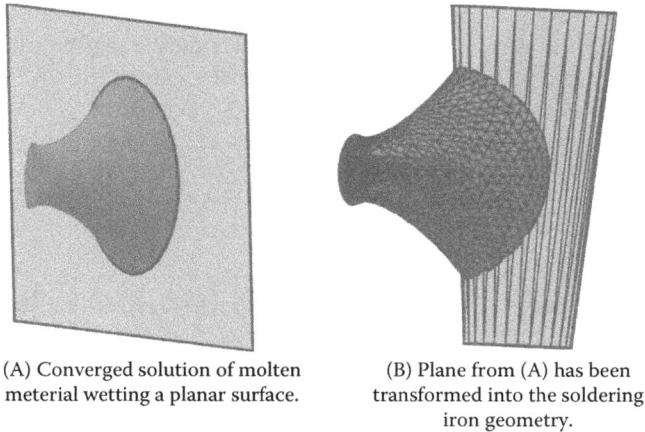

(A) Converged solution of molten
meterial wetting a planar surface.

(B) Plane from (A) has been
transformed into the soldering
iron geometry.

FIGURE 2.12 Sequence of convergence for solder wetting the electrode. The triangular mesh can be seen in the right-hand image.

interface between the cylinder and liquid, as opposed to directly modeling a liquid droplet wetting a cylinder. Other pertinent aspects of the model are the boundary constraints, in particular, to model the partial wetting between the lead-free solder and soldering iron.

The lead-free solder partially wets the surface of the soldering iron and makes a contact angle with its surface. To prevent over-constraining the problem, the software deals with a liquid partially wetting a solid by omitting the face in contact between the liquid and solid and using a line integral defining the energy for that surface. Using Green's theorem, the surface integral is converted into a line integral at the contact line between the solder and soldering iron given by[8,13]

$$\oiint \gamma_{la} \cos(\theta) \vec{k} . \vec{dS} = \oint \vec{w} . \vec{dl} \tag{2.1}$$

where

$$\vec{w} = \gamma_{la} \cos(\theta) z \vec{j} \tag{2.2}$$

Equations 2.1 and 2.2 show the energy integral calculation for the interface between the lead-free solder and the soldering iron where g_{la} is the liquid–air surface tension of the solder and q is the contact angle between the solder and soldering. The vectors are the Cartesian unit vectors. The composition of the lead-free solder modeled was Sn3.5Ag0.5Cu with rosin-activated flux and a surface tension value of 535 m Nm^{-1} at 275°C.[14] Arenas and Acoff[6] found that most of the commercially available solders with rosin-activated flux that they tested had a contact angle of between 20° and 22° with a copper substrate at 280°C. The contact angle used for the model was found by comparing the values of l and $2R_2$ measured from a magnified photograph (20× magnification) of a droplet of lead-free solder ($V = 5 \pm 1$ mm^3) wetting the soldering iron surface against the estimated values from the melt boundary model for contact angles between 15° and 30° in 5° intervals. It was found that a change of 15° in contact angle resulted in a change of ~15% in the value of $2R_2$ and ~25% in l. At a volume of 5.5 mm^3 and 20° contact angle, a percentage difference of less than 10% between the photograph and the model, with errors in measurement of ~2%, was obtained. To constrain the lead-free solder to the soldering iron surface, the boundary constraints

$$x = Az + C \tag{2.3}$$

and

$$x^2 + y^2 = (Az + C) \tag{2.4}$$

were added. These represent the boundary constraints before and after transformation to the soldering iron geometry, where $A = 0.083$ and $C = 0.95$ mm; these represent dimensional constants specific to the soldering iron tip used for the experiments described earlier (2.3 mm at the base and 1.5 mm at the tip, as shown in Figure 2.11). The value of C represents the radius of the soldering iron tip halfway between base and tip, and A is the gradient of the tip surface with respect to the tip's central axis. The model also assumed l to be constant, so the boundary constraint

$$x = l + C \tag{2.5}$$

was used at the solidification front. This assumption is valid while the soldering iron is at a constant temperature and the temperature of the melt boundary is in steady state. It can be used for comparison of the model against melt boundary photographs. This assumption has some limitations when it is used for estimating the future track width, as the distance of the solidification front from the soldering iron surface may change depending on the melt boundary shape, volume of molten material, and temperature of the soldering iron. This limitation may be overcome by estimating the melt boundary shape and track profile for a range of values for l. The second boundary constraint applied at the solidification front

$$R^2 = y^2 + z^2 \tag{2.6}$$

simulates the starting of tracks from a fixed wire diameter. To simulate track deposition, the volume of material within the melt boundary (V) was reduced in 0.25-mm³ intervals until 1 mm³ was remaining. The interval of 0.25 mm³ was chosen as it was the largest volume interval that did not produce jagged edges in the modeled melt boundary shape at $V < 4$ mm³. Reductions in volume below 1 mm³ again resulted in distortion and jagged edges on the modeled melt boundary. The effects of initial acceleration and gravity were neglected and may constitute a source of error. Gravity has been incorporated in the contact angle measurement from the 5.5-mm³ volume droplet of solder wetting the underside of the soldering iron. The experiments were performed at an initial acceleration of ~0.5 mm s⁻¹. The surface tension was assumed to be constant along the melt boundary. As shown by Selby,[15] surface tension varies with temperature above the melting point of the material and hence may vary along the melt boundary. However, for the purposes of obtaining an indication of the effect of operating settings and predictive factors for track taper these simplifying assumptions were made.

2.3.1 EXPERIMENTAL METHODOLOGY

2.3.1.1 Estimating the Track Profile

The melt boundary model was also used to estimate the track width at the settings $R_t = 0.4$ mm and V_0 at 5.5 mm³ and 7 mm³. For each parameter tested, the volume was reduced from V_0 to 1 mm³ in 0.25-mm³ intervals to simulate track deposition. The estimated values of R_1/R_t against V_0/V were compared to the experimental results obtained earlier. The values of R_1 were taken from the vertex data produced by the model. The mean of the values of $R(l + C)$ were used as the value of R_1 as there were four to six data points at $x = l + C$ where $C = 0.95$ mm, as previously mentioned. Volumes of 5.5 mm³ and 7 mm³ were chosen, as no more than one track created at these settings had exhibited taper greater than 1° over 10 mm. The width of the tracks in the x-y and x-z planes also varied by less than 10%, allowing the assumption that the tracks had a circular cross-section. This means that the volume can be estimated by multiplying the mean cross-sectional area by the distance traveled. These tracks are likely to have smaller values of uncertainty in volume estimation compared to the tracks that exhibited taper or had an elliptical cross-section, such as in the case of the tracks created at an initial volume setting of 8 mm³.

2.3.1.2 Effect of Initial Operating Settings on Track Taper

The melt boundary model was used to develop a more complete process map by varying initial volume, track radius, and separation to categorize tracks based on operating settings and to provide a method of predicting track taper. Erle et al.[11] and Meseguer et al.[16] showed that increasing l/d or reducing V leads to track instability. As track deposition occurs, the volume within the melt boundary reduces and may lead to track taper or breakage. A dimensionless initial track condition parameter I, where $I = l^4/(V_0R_t)$, was created as a proxy for l/d and V since the only parameters that can be controlled in the current experimental setup are V_0, R_t, and l. The standard deviation of R_1/R_t for each track profile in the x-y and x-z planes was plotted against the parameter I to determine if I could be used to predict the standard deviation. The standard deviation can be used as a predictor of track taper because the larger the standard deviation is the larger the variation in track profile (track width along the length). The parameters tested were V_0 from 4 mm^3 to 12 mm^3 in 2-mm^3 intervals and at 7 mm^3 (as this setting produced the smallest standard deviation in mean track width and lowest number of track failures in the experiments); R_t of 0.25 mm, 0.4 mm, and 0.6 mm; and l of 1.2 mm, 1.7 mm, and 2.2 mm. For each parameter tested, the volume was reduced from V_0 to 1 mm^3 in 0.25-mm^3 intervals, and the track data points in the x-y and x-z planes were recorded for each whole number value of volume. The process map was also used to suggest possible initial operating settings that lead to tracks with R_1/R_t values between 0.8 and 1.1.

2.3.2 Results

2.3.2.1 Estimating the Track Profile

Figure 2.13 shows dimensionless track radius predictions by the melt boundary model against the proportion of volume remaining within the melt boundary. As l depends on temperature, volume, and melt boundary shape, it may change as material is deposited, so a range of values for l has been tested in Figure 2.13. It can be seen that $V = 5.5$ mm^3 and $l = 1.5$ mm produces the closest fit, with a maximum difference between experiment and model of <15%. Other values of l at 5.5 mm^3 also produce better fits against experimental data compared to the 4.5-mm^3 or 6.5-mm^3 volumes. Each data point is the mean value measured from seven tracks for the 5.5-mm^3 volume and ten tracks for the 7-mm^3 volume. The error bars for R_1/R_t are calculated using the standard error calculated for the earlier experiments correct to the nearest 0.1 mm.

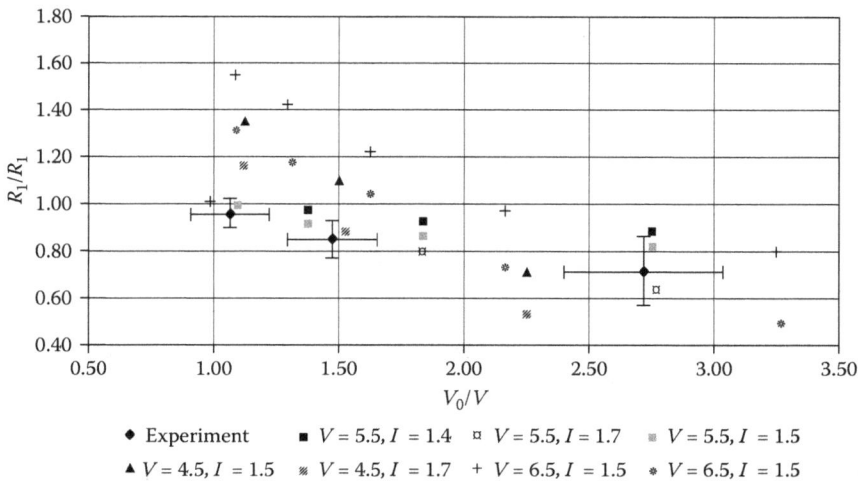

FIGURE 2.13 Track radius estimates compared to experimental results for $V = 5.5 \pm 1$ mm^3. The units of V are in mm^3 and l is in mm.

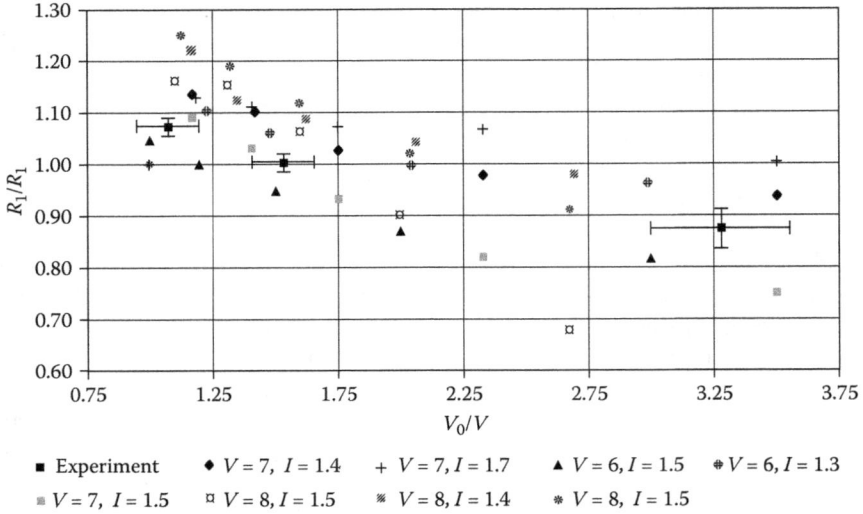

FIGURE 2.14 Track radius estimates compared to experimental results for $V = 7 \pm 1$ mm³. The units of V are in mm³ and l is in mm.

The error bars for V_0/V are equal to

$$\pm \frac{V_0}{V} \sqrt{\sum_{n=1}^{n=k} \frac{SE_n^2}{m_n^2}}$$

where SE is the standard error and m_n is the mean of each measured dimension from 1 to k. It can be seen that the error bars are larger as V_0/V increases. This is because the error is a larger proportion of V as the volume reduces. In Figure 2.14, the model estimates the track width with a maximum difference of ~15% between experiment and model for $V = 7$ mm³ and $l = 1.4$ mm. Other model data points have a larger difference of up to ~20% for $V = 8$ mm³ and $l = 1.7$ mm.

2.3.2.2 Effect of Initial Operating Conditions on Track Taper

Figures 2.15 and 2.16 show scatterplots of the standard deviation of R_1/R_t calculated for each value of V_0/V from 1 to V_0 against the dimensionless parameter I. Linear least squares regression was used to calculate the line of best fit. The correlation coefficient for the data points in Figure 2.15 is 0.62 and the coefficient of determination is 0.38, showing that 38% of the variability in the dependent variable is accounted for by the line of best fit. Other regression methods showed a poorer correlation between the two variables. Out of the 63 data points, 48 points can be found at $I \leq 6$ and have a standard deviation less than 0.3. It can also be seen from Figure 2.16 that 5 out of 9 data points at $I > 8$ have standard deviations greater than 0.3 compared with only 4 out of 54 data points for values of $I < 8$. To achieve standard deviations less than 0.2, choosing $I < 4$ would result in 35 out of 40 tracks being within limits. To achieve a standard deviation less than 0.1, choosing $I < 2$ would not suffice, as 11 out of 21 tracks would be outside the range selected, and choosing $1 < I < 2$ may not be reliable as there are fewer than 10 data points but that selection may provide a starting point for further experimentation.

It can be seen from Figure 2.16 that using a value of $1 < I < 2$ may produce tracks that have standard deviations less than 0.2, but to achieve standard deviations less than 0.1, a better selection method is necessary as there are only three data points below this value.

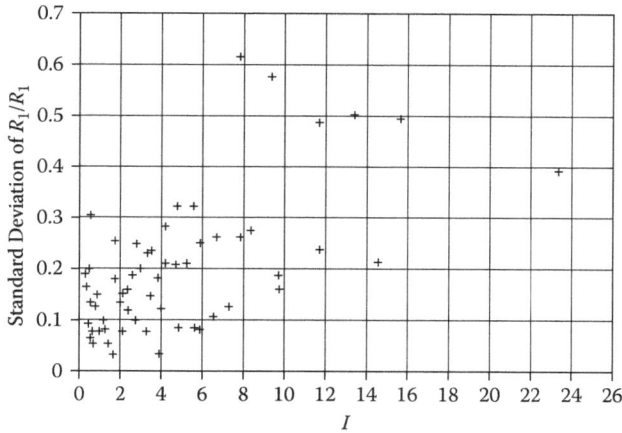

FIGURE 2.15 The effect of the dimensionless parameter I on the standard deviation of R_1/R_t in the x-z plane.

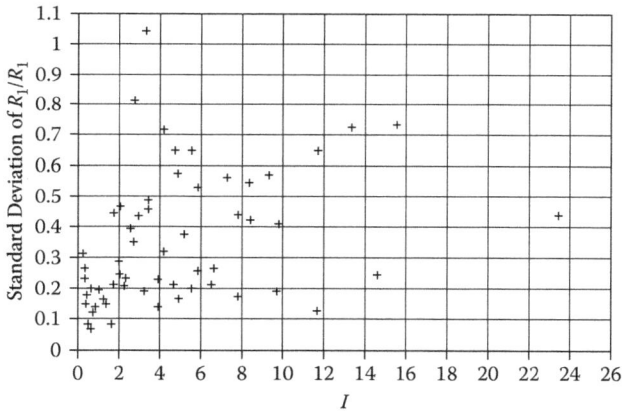

FIGURE 2.16 The effect of the dimensionless parameter I on the standard deviation of R_1/R_t in the x-y plane.

To categorize the track shapes, Figure 2.17 shows a line plot of a selection of I values in both the x-y and x-z planes. On the x-axis, the ratio of initial volume to molten volume is shown as the track is deposited, and on the y-axis the value of track diameter as a ratio to initial track diameter is plotted. It can be seen for $I = 1.02$, R_1/R_t is between 0.8 and 1.15. For $I = 1.4$, the value of R_1/R_t is between 0.9 and 1.5. This means that the track width lies between 0.7 mm and 1.2 mm. The results for $I = 1.02$ and 1.4 corresponds to $V = 7$ mm^3, $l = 1.3$ mm and 1.4 mm, and $R_t = 0.4$ mm. This may explain why the lowest standard deviation was obtained earlier for tracks with an initial volume of 7 mm^3 and a starting wire diameter of 0.8 mm. The difference in track shapes for $I = 1.02$ and 1.4 show that the tracks are sensitive to l.

The value of $I = 2.61$ corresponds to $V = 8$ mm^3, $l = 1.7$ mm, and $R_t = 0.4$ mm. It can be seen that if $R_1/R_t > 1$ and $V_0/V < 3$ in the x-y plane, then R_1/R_t tapers to less than 0.8. In the x-z plane, R_1/R_t initially rises to ~1.5 but then remains at ~1. The characteristics of the predicted track at $I = 2.61$ are similar in shape to the tracks observed for the 7.5-mm^3 and 8-mm^3 settings in the experiments. Some of the 8-mm^3 tracks had a track width of 1.4 mm, although the mean was 1.2 mm in both planes, making R_1/R_t ~ 1.6. The model predicts $R_1/R_t = 1.75$ in the x-z plane and $R_1/R_t = 1.3$ in the x-y plane. This discrepancy between model and experiment may be due to uncertainties in the value

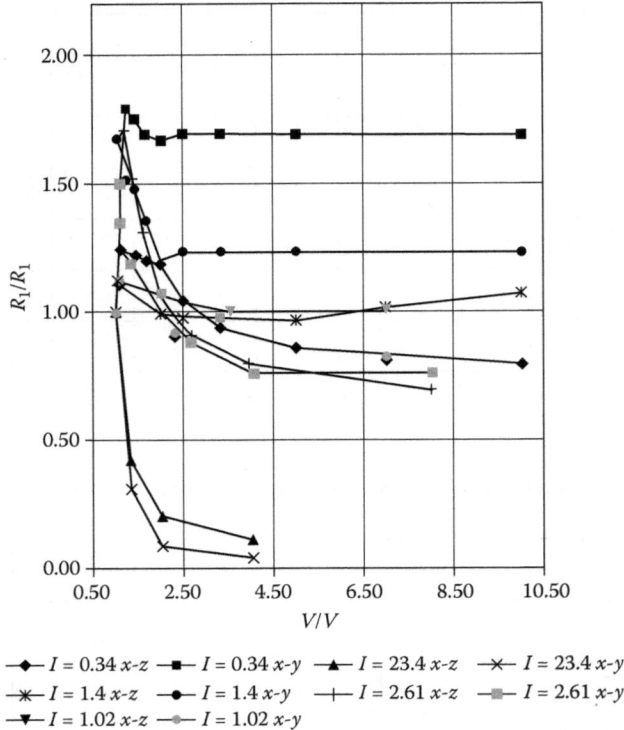

FIGURE 2.17 Effect of varying I on estimating track radius in the x-y and x-z planes.

of I or because the model is not predicting the shape at 8 mm^3 accurately. Although the initial melt boundary shape did not appear to be very different for volumes between 5.5 mm^3 and 7.5 mm^3 as seen in Figures 2.15 and 2.16, the earlier experimental results and Figure 2.17 show that the melt boundary is sensitive to small changes in initial conditions, particularly in the value of I.

2.3.3 DISCUSSION

The objectives of this section were to develop a melt boundary model, test the model results against photographs of the melt boundary, and test the ability of the model to estimate track width and provide some measure of predictive capability of track taper given a set of initial operating conditions. The melt boundary model has shown itself capable of estimating the boundary shape correct to <10% (except for one data point that had a difference of 18%) in three out of four cases and a track width correct to 15% in the two cases tested. The melt boundary model was also used to develop a process map and plot a dimensionless parameter representing initial operating settings against the standard deviation of R_1/R_t to estimate track width variation along its length. It was found that the parameter I could be used to exclude operating settings, reducing the chances of producing tracks with standard deviations greater than 0.2.

The melt boundary model can also be used to explain the experimental results found earlier which showed that starting the track from a fixed wire diameter rather than from a substrate increased the repeatability of tracks. Changing the initial track width causes a change in R_t that in turn affects the value of I so it is likely that track diameter affects the repeatability of tracks. In particular, increasing the value of R_t to greater than 0.6 mm will lead to values of $I < 1$. This will lead to tracks that are not axisymmetric, as can be seen by the values of R_1/R_t in the x-y and x-z planes in Figures 2.15

and 2.16. A possible reason for this may be the sensitivity of the melt boundary to l. The soldering iron does not have a temperature control system that keeps a constant temperature, and using a thermal camera showed that the temperature varied by as much as 25°C at a given temperature setting. Because $I \mu l^4$ and l is affected by changes in temperature, this could lead to poor repeatability of tracks as small changes in the value of l may cause large changes in the value of I. To improve control of the temperature, a possible solution is to use a pyrometer-based feedback control.

Considering values of $I > 2$, this is equivalent to $2VR_t < l^4$. If the value of I increases while l remains constant, this is akin to increasing l/d or a reduction in volume. The literature has already shown that this leads to necking and instability. From Figure 2.17, it can be seen that at $I = 23.4$ R_1/R_t has a steep negative gradient and approaches zero. In reality, such tracks are likely to break and may be the cause of the inability to deposit tracks at volumes less than 5 mm³. On the other end of the spectrum, where $I < 1$, this is equivalent to $VR_t > l^4$. This value of I corresponds to tracks deposited ~250°C. As can be seen from Figure 2.17, the tracks taper in the x-z plane.

The experiments also found that a velocity of 1.5 mm s⁻¹ resulted in track necking, and at velocities greater than 2 mm s⁻¹ all tracks broke. This could not be modeled using the current model but a possible reason for the track necking may be that the value of l increases due to the initial acceleration (as V and R_t are likely to remain constant), and increasing I to greater than two, leading to necking. A new model is needed to accurately depict the melt boundary to incorporate the effects of velocity and acceleration. Finally, the experiments also found that it was not possible to feed new material while the track was being deposited. This section has shown that track deposition is sensitive to volume. Unless the volume of material fed is carefully controlled such as by synchronously increasing the volume in line with the volume deposited, the track deposition may become unstable. The effect of the rate of volume deposition compared to the volume remaining within the melt boundary could not be investigated using the current model.

The use of the parameter I has some limitations. The first is that it does not take into account soldering iron geometry so it is related to this particular experimental apparatus. The soldering iron geometry affects R_2, but this was not included in I because its relationship to volume could not be quantified precisely as the value of I. The value of I as has been defined currently can be used for the purposes of exclusion of initial operating settings with the current experimental setup, but $1 < I < 2$ does not guarantee that the track will not be tapered or break.

The results in this section have shown that it is possible to model the melt boundary shape using the Surface Evolver software. The results also show that the melt boundary shape was not highly sensitive to volume between 5.5 mm³ and 7.5 mm³. However, when using the melt boundary to predict the future track radius, it was sensitive to initial conditions V, l, and R_t. A dimensionless parameter I was developed to categorize tracks and to determine a range for the value of I that would reduce the chances of necking. To reduce necking it was found the $I < 2$ may have a higher incidence of producing tracks with standard deviations that are less than 30% of initial track width. Some optimum values for I were 1 and 1.4, corresponding to an initial volume of 7 mm³ at 275°C and starting from a wire diameter of 0.8 mm, which are the same settings that produced the best results in the experiments.

2.4 CONCLUSIONS

A soldering iron was used to deposit tracks extending horizontally or vertically from a substrate into free space suspended between the substrate and the soldering iron. The operating temperature range was found to lie between 250°C and 310°C. The maximum reliable deposition velocity was 1.5 mm s⁻¹. Above this velocity, ~80% of tracks necked. It was also found that the volume of material on the soldering iron tip at the start of track deposition significantly affected the quality and characteristics of tracks produced. It was found that the best results were obtained at an initial volume of 7 mm³ while depositing tracks starting from a 0.8-mm diameter wire. Track failures occurred below this volume, and tapered tracks were created above this volume.

FIGURE 2.18 Overhanging walls built using tracks deposited in free space. Walls with a thickness of ~1 mm could be built, as shown by the wall farthest to the right.

The melt boundary was modeled using the Surface Evolver software to understand the reasons for track instability. The predictions for melt boundary shape compared favorably with the experimental results. Photographs of the melt boundary showed that the wetting width on the soldering iron appeared to reduce as the track was deposited. The model showed that the wetting width on the soldering iron reduces with the volume of material on the tip. As material is deposited, the volume of material on the soldering iron tip reduces, leading to the reduction in wetting diameter. The literature shows that with interfaces of revolution the stability of the shape is governed by the separation between end faces to end face diameter ratio. Reducing the volume reduces the value of the stability limit. The melt boundary model was used to estimate the track width; these showed a difference of less than 18% between the experimental results and the model. A non-dimensional parameter—$I = l^4/(V_0 R_t)$, where V_0 is the initial volume, l is the distance from the soldering iron to the solidification front, and R_t is the initial track radius—was used to make predictions for operating settings that are likely to produce track taper. The standard deviation of R_1/R_t over the length of the track for each operating parameter was plotted against I in both the x-y and x-z planes. It was found that tracks that had $I < 2$ had a lower incidence of having a standard deviation $< 30\%$ of the initial track width.

Extending the concept of free space deposition, it can be shown that solid objects such as walls can be built as shown in Figure 2.18. The last wall on the right had a thickness of ~1 mm. Solid or hollow shapes can also be built, as shown in Figure 2.19.

All of the experiments to date have been performed open loop and did not incorporate feedback control. The use of a feedback system by measuring the melt boundary profile could improve the precision of track deposition. The tracks deposited were straight line tracks, and no complex shapes were built. The feedback may make it possible to build complex wireframe shapes or solid objects. The primary advantage of this type of technology is the ability to deposit tracks in any direction. There are numerous other technologies such as SLM that can build solid or layered objects, so free space deposition is better reserved for creating complex wire frame shapes for both structural and

FIGURE 2.19 Three-dimensional objects built using free space deposition; manual build of a solid pyramid and a hollow pyramid.

aesthetic applications. Potential uses include building bone in growth areas for implants, developing art or architectural concepts, and strengthening or additively adding features to subtractive-built objects such as aerospace components.

Further research is needed in building multilayered structures. Figures 2.18 and 2.19 show that this is possible, but further refinement in the interface between the technology and software is necessary. Free space deposition techniques allow the use of tool paths that are not limited to a single plane; however, the scan strategy is likely to be more complex due to manipulating a track of molten material in free space. Therefore, consideration of particular strategies, such as for wall building, must be included in the tool path. Currently, many additive manufacturing machines import data from a CAD model in the stereolithography (STL) data format that uses triangular facets.[18,19] The formats that are ideal for layered building may not be optimal for free space deposition, so research into data formats that can represent the model and tool paths for free space deposition in an optimum way is needed. For example, Jacob et al.[18] suggested using formats that represent the part geometry utilizing a mathematically precise format using splines. Such a representation of the CAD model might be a better representation for free space deposition.

Another key area of future work is extending the research to the deposit of higher melting point materials such as stainless steel and titanium. Exploratory work has shown that it is possible to deposit higher melting point materials using an electrical arc to heat a tool in lieu of a soldering iron. Rangesh[3] reported that repeated tool tip failure due to oxidation indicated a need for further research to determine better tool materials and process parameters. The models described here suggest that free space deposition can be extended to other materials, as the shape of the molten material on the tool tip is the primary factor in determining track shape. However, practical considerations such as tool tip oxidation, choice of heat source, and material choice are likely to affect part quality and shape.

To conclude, the key contribution of this research is the development of the concept of a moving support structure to allow free space construction in metals. From the survey of the literature, as far as the author is aware, this is the first time free space metal deposition has been reported, but it offers some unique capabilities for three-dimensional building and additive manufacturing.

Writing metal in free space allows wire frame objects to be built. No other additive manufacturing technology currently has this capability. Fine features such as jewelry, bone structures, and other intricate shapes can be built using the concept of a moving support structure. Overhanging features can be built without support structures, as shown in Figure 2.18. This means that closed features such as a hollow object could be built as one piece, as shown in Figure 2.19, and parts could be built in any orientation. The method of using a heated tool to deposit and shape molten material is independent of heat source. It can be designed to work with other heat sources such as a laser or a resistively heated tool. Multiple-material deposition such as metal deposition onto polymers is also potentially feasible. Although the concept has been proved experimentally and analyzed theoretically at a basic level, there are significant opportunities for further research and development in this area.

ACKNOWLEDGMENTS

This research was made possible by the support of mentors, colleagues, friends, and family. I am grateful for the guidance and access to research funding provided by my supervisor, Dr. William O'Neill. I also thank my colleagues at the University of Cambridge, in particular Dr. Ali Khan, Dr. Andrew Cockburn, and Prof. Steven Barrett, for their intellectual support and for reviewing my research. Part of the experimental analysis has previously appeared in the *Journal of Materials Processing Technology*[17] and has been reproduced here with the kind permission of Elsevier. Finally, I thank the institutions that have funded this research: the Engineering and Physical Sciences Research Council and the Department of Engineering at the University of Cambridge.

REFERENCES

1. Kruth, J.P., Mercelis, P., Van Vaerenbergh, J., Froyen, L., and Rombouts, M. (2005). Binding mechanisms in selective laser sintering and selective laser melting. *Rapid Prototyping J.*, 11: 26–36.
2. Thomas, D. and Bibb, R. (2008). Identifying the geometric constraints and process specific challenges of selective laser melting. In: *Proceedings of Time Compression Technologies Rapid Manufacturing Conference 2008* [CD-ROM]. Rapid News Publications, Coventry, U.K.
3. Rangesh, A. (2010). Free Space Deposition in Metals, doctoral dissertation, University of Cambridge.
4. Rangesh, A. and O'Neill, W. (2011). Rapid prototyping by consolidation of stainless steel powder using an electrical arc. *Rapid Prototyping J.*, 17: 280–287.
5. Arenas, M.F. and Acoff, V.L. (2004). Contact angle measurements of Sn-Ag and Sn-Cu lead-free solders on copper substrates. *J. Electron. Mater.*, 33: 1452–1458.
6. Yadroitsev, I., Gusarov, A., Yadroitsava, I., and Smurov, I. (2010). Single track formation in selective laser melting of metal powders. *J. Mater. Process. Technol.*, 201: 1624–1631.
7. Gusarov, A.V., Yadroitsev, I., Bertrand, P.H., and Smurov, I. (2009). Model of radiation and heat transfer in laser-powder interaction zone at selective laser melting. *J. Heat Transfer*, 131: 72–101.
8. Brakke, K. (1991). *Surface Evolver Manual*, Research Report GCG-31. The Geometry Center, Minneapolis, MN.
9. Brakke, K. (1992). The Surface Evolver. *Exp. Math.*, 1: 141–165.
10. Brakke, K. and Singler, T. (1996). Computer simulation of solder bridging phenomena. *Trans. ASME*, 118: 122–126.
11. Erle, M.A., Gillette, R.D., and Dyson, D.C. (1970). Stability of interfaces of revolution with constant surface tension: the case of the catenoid. *Chem. Eng. J.*, 1: 96–109.
12. Gillette, R.D. and Dyson, D.C. (1971). Stability of fluid interfaces between equal solid circular plates. *Chem. Eng. J.*, 2: 44.
13. Brakke, K. (2004). *Surface Evolver Workshop*, http://www.susqu.edu/brakke/evolver/workshop/workshop.htm.
14. Kaban, I., Mhiaoui, S., Hoyer, W., and Gasser, J.G. (2005). Surface tension and density of binary lead and lead-free Sn-based solders. *J. Phys.: Condens. Matter*, 17: 7867–7873.
15. Selby, A.L. (1890). Variation of surface-tension with temperature. *Proc. Phys. Soc. Lond.*, 11: 119–122.
16. Meseguer, J., Slobozhanin, L.A., and Perales, J.M. (1995). A review on the stability of liquid bridges. *Adv. Space Res.*, 16: 5–14.
17. Rangesh, A. and O'Neill, W. (2012). The foundations of a new approach to additive manufacturing: characteristics of free space metal deposition. *J. Mater. Process. Technol.*, 212: 203–210.
18. Jacob, G.G.K., Kai, C.C.K., and Mei, T. (1999). Development of a new rapid prototyping interface. *Comput. Indust.*, 39: 61–71.
19. Chiu, Y.Y., Liao, Y.S., and Lee, S.C. (2004). Slicing strategies to obtain accuracy of feature relation in rapidly prototyped parts. *Int. J. Mach. Tools Manuf.*, 44: 797–806.

3 Additive Manufacturing of Metals via Selective Laser Melting
Process Aspects and Material Developments

Jean-Pierre Kruth, Sasan Dadbakhsh, Bey Vrancken,
Karolien Kempen, Jef Vleugels, and Jan Van Humbeeck

CONTENTS

ABSTRACT

This chapter provides a basic insight into the selective laser melting (SLM) process. It reports the state of the art in SLM of metals and describes the latest developments enabled by the efforts of researchers at the University of Leuven (KU Leuven). After an introduction, the SLM machine components (including the basic units and process layout) as well as hardware additions (such as a preheating stage and monitoring systems) are explained. The melting/solidification conditions and the corresponding quality issues are briefly reviewed as well. Important SLM parameters and additional considerations such as geometrical aspects or laser remelting/erosion abilities are described. After this, a survey of various SLM materials, including a number of steels, aluminum alloys, and titanium alloys, in addition to functional and advanced materials such as refractory tantalum and shape-memory NiTi, is given. The microstructural and mechanical properties of SLM materials are discussed and compared with those produced by conventional methods. Possible methods to mitigate SLM problems such as cracking and residual stresses are also suggested. Finally, this chapter explains different post-processing treatments and illustrates that conventional post-processing techniques might not be efficient or suitable for some SLM materials.

3.1   INTRODUCTION

Selective laser melting (SLM) is an additive manufacturing (AM) technique to produce complex three-dimensional parts through solidifying successive layers of powder materials on the basis of a CAD model. SLM is associated with complete melting of the powder material rather than sintering or partial melting of the powder particles which is the dominant mechanism in the selective laser sintering (SLS) process. Figure 3.1 shows a schematic layout of the SLM process. In this process (carried out under a protective atmosphere), a metal base plate is used to anchor the part during the building process. A layer of powder (most commonly from metals) is spread on top of the base plate and is subsequently melted by a laser beam projected from above. The laser scans the powder bed according to the shape defined in a CAD file (that has been sliced into many different layers). After each layer has been scanned, the powder bed is moved down over a distance of one layer thickness, followed by an automated leveling system that distributes a new layer of powder. The laser then melts a new cross-section. The process is repeated to form the desired solid metal part (comprised of hundreds or possibly thousands of thin layers).[1–6]

Many advantages are associated with the SLM process, including (1) high density and strength of the parts; (2) negligible waste of material (unused powders can be recycled); (3) possibility of producing complicated shapes (e.g., a steel mold with curved internal cooling channels, which is common to other AM methods); (4) ability to process a wide variety of metals and their mixtures (due to the powder-based nature of SLM); and (5) no need for any distinct binders or melt phases, so

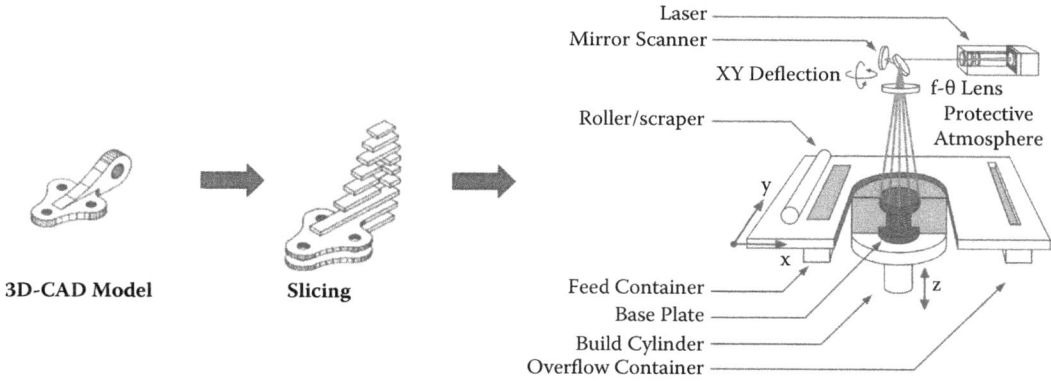

FIGURE 3.1 Schematic diagram of the SLM process.

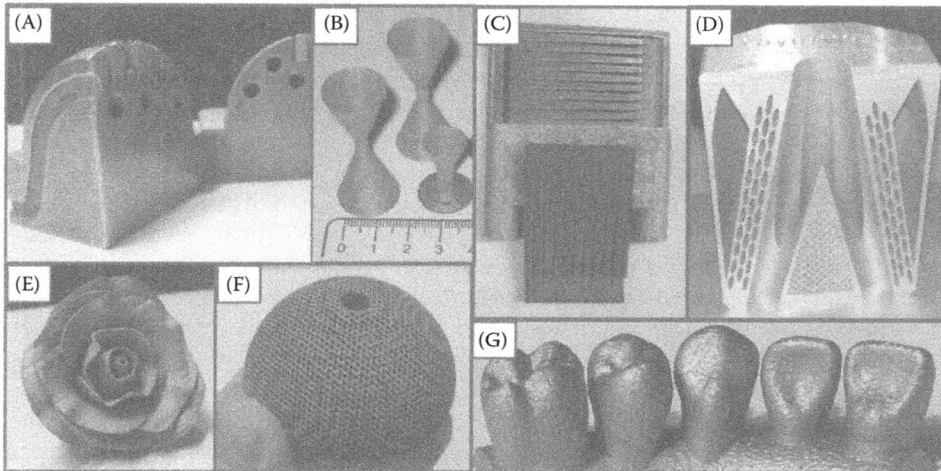

FIGURE 3.2 Some complex SLM parts: (A) stainless steel mold with conformal cooling to enhance the productivity in injection molding, (B) Ti6Al4V thin-wall structures, (C) 316L stainless steel heating plate for aerospace industry, (D) advanced nozzle with internal cooling system from AlSi10Mg alloy, (E) stainless steel artistic flower, (F) Ti6Al4V biomedical acetabular cup with advanced cellular porosity for improved biocompatibility, and (G) CoCr dental parts.

the process can directly produce single material parts (e.g., steel, Ti, or Al alloys), rather than first producing a composite green part that requires such secondary processing steps as debinding* and furnace sintering (as done with some other AM methods).

Customized medical parts, thin-wall structures, tooling inserts with conformal cooling channels, and functional components with high geometrical complexity are common examples of the SLM applications.[7–9] Some of these SLM applications, produced at the University of Leuven (KU Leuven), are shown in Figure 3.2. The great liberty afforded by SLM in design and geometry (undercuts, overhangs, free forms, and lattice structures, as well as elementary shapes) is an important advantage. From a material perspective, specific CoCr, stainless steel, tool steel, titanium, aluminum, and nickel alloys are now commercially available, although scientific and research and development institutes are dedicated to expanding this small range of commercially used SLM materials.

---

* Debinding is a secondary step to remove the binding additives from the green compacts.

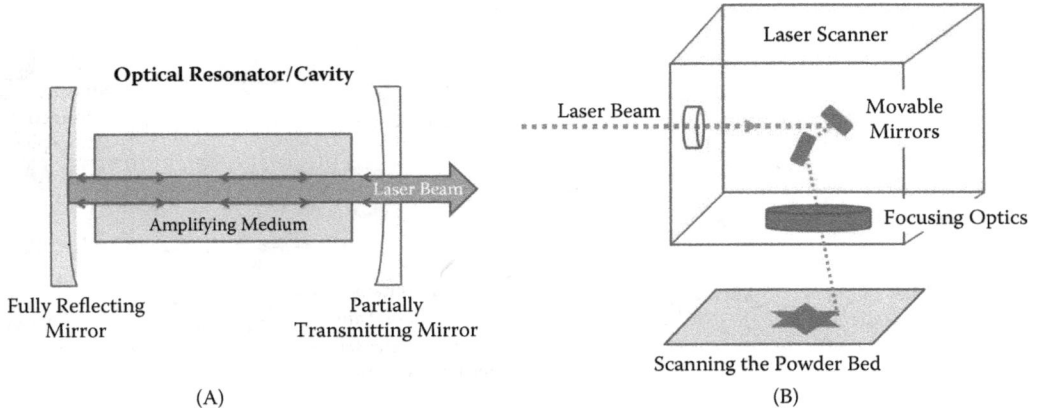

FIGURE 3.3   Simplified schematic of a typical (A) laser and (B) scanner system.

Selective laser melting is a newly developed process faced with various challenges. For example, SLM suffers from melt pool instabilities leading to imperfections such as low-quality down-facing surfaces, greater upper-surface roughness, and the risk of internal pores. The resulting coarse and grainy surface finish may require a secondary machining or polishing process. Additionally, high temperature gradients in SLM increase the risk of delamination and distortion due to large thermal or residual stresses. From an economical aspect, the high cost of a high-power laser source, long processing times, and a small palette of available materials are the main obstacles. Despite these challenges, SLM is increasingly becoming a competitive manufacturing alternative.

Several research institutes have made major contributions to the development of the SLM technology.[1–2,5–6] The additive manufacturing research group at KU Leuven has particularly accomplished several developments in terms of production parameters, machine tool innovations, materials, post-processing, etc. This chapter reports the state of the art in SLM of metals and describes the latest developments enabled by the efforts of researchers at KU Leuven. This review aims at developing an understanding of the SLM principles and parameters, machine functions, materials specifications, and post-processing in order to form a solid background for this technology. Also discussed are the available and under development machines at KU Leuven, in-house innovations in terms of *in situ* process monitoring, and complementary enhancements such as laser remelting and erosion techniques. The reviewed materials consist of a number of steels, aluminum alloys, and titanium alloys, in addition to functional and advanced materials such as refractory tantalum (Ta) and shape-memory NiTi.

3.2   SLM MACHINE AND EQUIPMENT

3.2.1   BASIC UNITS AND PROCESS LAYOUT

An SLM machine consists of three main units: (1) laser and scanner system, (2) controller system, and (3) build chamber. The laser (Figure 3.3A) consists of an elongated cavity with two mirrors at its ends in which light oscillates back and forth between the mirrors.[10] Between the mirrors, there is a light-amplifying medium, where stimulated emission occurs. The mirrors guide the light back into the amplifying medium repeatedly for continued growth of the developing beam.[11] After amplification, the emissions pass through a partially transmitting mirror. The resultant beam is then guided to enter a scanning system. Most laser scanner systems use a galvano scanner* to steer the laser

---

* In a galvano scanner, two electromechanical actuators rotate two mirrors (deflecting the laser beam in the *x* and *y* directions), in response to electric current flowing through their coils in a magnetic field.

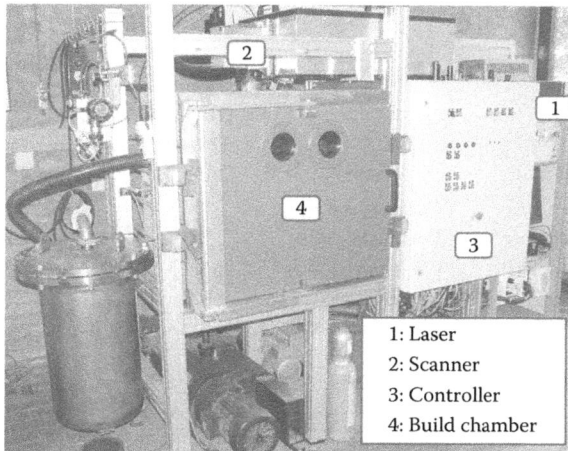

1: Laser
2: Scanner
3: Controller
4: Build chamber

FIGURE 3.4 In-house SLM machine developed at KU Leuven (PMA division).

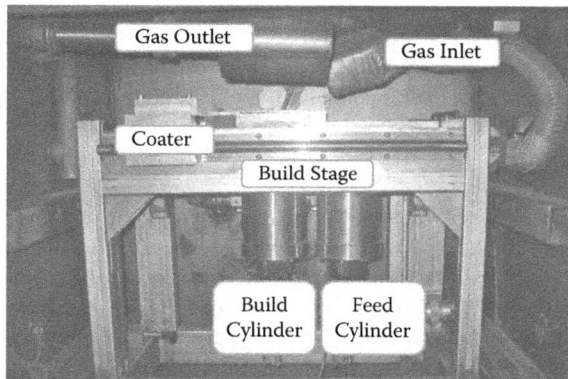

FIGURE 3.5 An example of a SLM machine build chamber at KU Leuven.

beam, as shown in Figure 3.3B. The scanner guides the beam through focusing optics (to further focus the beam) to scan any desired geometry on the powder bed (Figure 3.3B). In addition to the machine, the scanning is automated and controlled via a dedicated controller and computer unit.

Scanning the powder bed is carried out in a secured container called the build chamber (see Figure 3.4). The build chamber is designed to provide an inert/protective atmosphere (using mostly N_2 or Ar gas circulation) and contains a build stage (Figure 3.5). The build stage is a platform to hold the build cylinder (in which the base plate is mounted), the feed cylinder (holding the powder), the elevator systems (to bring the powder level up/down), and the coating system (to uniformly distribute a thin powder layer onto the base plate).

Figure 3.4 shows an in-house SLM machine developed at KU Leuven. This unique SLM machine is not commercially available; it has been designed, developed, and progressively improved by KU Leuven researchers over the years. The current machine is equipped with a 300-W Yb–YAG fiber laser (wavelength around 1070 nm), a galvano scanner, a vacuum chamber, in-house-developed controller software, a monitoring system, etc. The unique monitoring system includes a near-infrared, high-speed complementary metal-oxide semiconductor (CMOS) camera able to capture images at up to 20,000 FPS and a photodiode for monitoring the melt pool. This allows real-time monitoring and feedback control of the SLM process.[12]

TABLE 3.1

Metal SLM Machines Available at KU Leuven

Manufacturer	Laser	Mode	Output Power (W)	Spot Size Diameter[a] (μm)	Preheating Option	Monitoring Option
In-house-developed	Yb fiber	Continuous wave	300	80	Up to 400°C	Yes
Concept Laser Mlab	Yb fiber	Continuous wave	100	40	No	No
Concept Laser M1	Yb fiber	Continuous wave	200	150	No	Yes
Concept Laser M3 linear	Nd:YAG	Continuous wave/pulsed	100	180	Up to 250°C	No
EOS M250	CO_2	Continuous wave	200	300	No	No

[a] The reported values may differ in various laser powers.

In addition to the SLM machine developed in-house, KU Leuven possesses four commercial selective laser sintering (SLS)/SLM machines for metals. The specifications of these machines are summarized in Table 3.1. The type of laser has a significant influence on the consolidation of powder particles for several reasons: (1) the laser absorption of materials depends on the laser wavelength, (2) different lasers may provide a different range of energies, and (3) the laser mode (e.g., continuous wave, pulsed) has a great influence on the consolidation.[2]

3.2.2 PREHEATING STAGE

To increase the efficiency of the SLM process or to improve the quality of manufactured parts (in case of specific alloys), a preheating system may be incorporated in the SLM machine. This can be done by heating the base plate in order to mitigate the temperature gradient between the base plate and the top layers. Figure 3.6 shows an overview of a heating module that was designed and installed on one of the SLM machines at KU Leuven (Concept Laser M3). The heating element itself (labeled 2 in Figure 3.6) is installed underneath the building platform and enclosed by insulation material. The temperature of the base plate can be monitored by a thermocouple probe. A proportional and integral control loop (PI-controller) controls the power to the heating element to achieve the desired temperature on the base plate within a range of ±2°C.

3.2.3 PROCESS MONITORING AND CONTROL

Apart from the equipment necessary to perform SLM (e.g., laser, building platform), complementary systems can be incorporated in the machine to enhance the process quality. One very useful system monitors and controls the SLM process. At KU Leuven, Mercelis and other researchers have

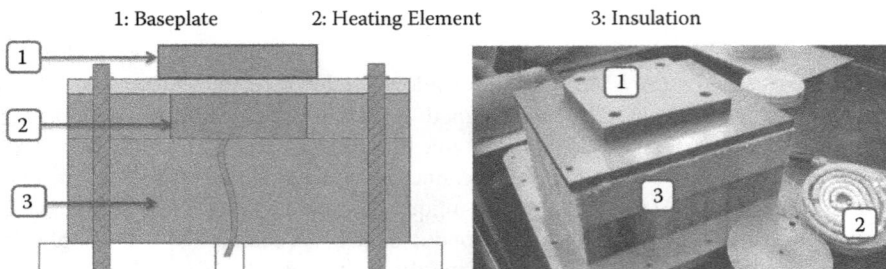

FIGURE 3.6   A customized preheating system developed by KU Leuven.[13]

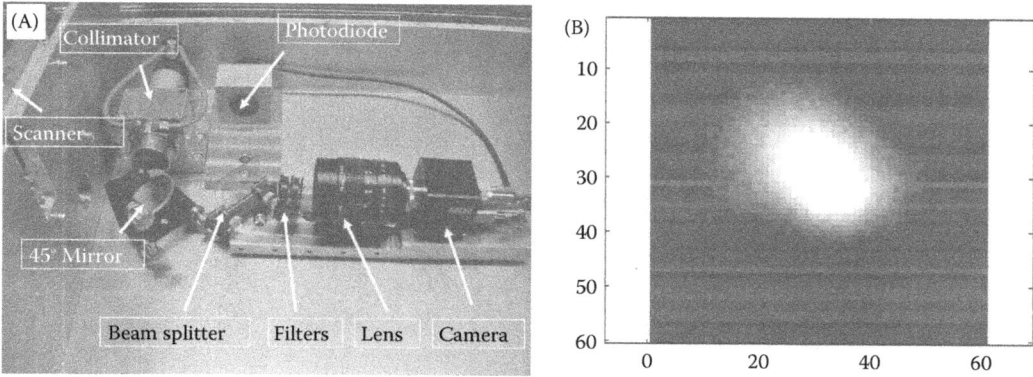

FIGURE 3.7   (A) Unique monitoring system installed on a SLM machine developed in-house, and (B) typical melt pool image during processing.[16]

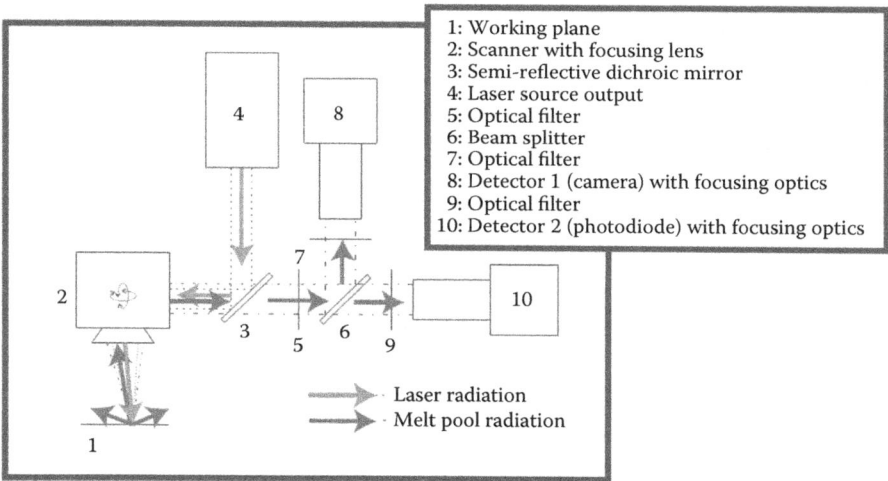

FIGURE 3.8   Schematic overview of the setup of the monitoring system.

developed a unique technique to monitor the melt pool during actual laser processing.[12,14–16] Within this system, a high-speed CMOS camera takes images of the melt pool (demonstrating melt pool size and its emission; see Figure 3.7). A schematic setup of this system is shown in Figure 3.8.

As seen in Figure 3.8, the laser source (4) is deflected by means of a semireflective dichroic mirror (3) toward the galvano scanner with focusing lens or F-theta lens (2). Part of the radiation from the melt pool is transmitted back to the scanner and passes through the dichroic mirror toward a beam splitter (6). The split beam is directed toward a near-infrared photodiode (8) and a high-speed near-infrared CMOS camera (10).[15] After interpreting the radiations received by the high-speed camera (Figure 3.7B), melt pool geometry can be monitored and mapped in real time.[16,17]

3.3   SLM PRINCIPLES AND QUALITY ISSUES

The final properties of laser-processed materials are affected by the laser–material interaction. The typical issues, such as porosity, balling, cracks, and residual stress, originate from this stage. This section briefly reviews melting pool characteristics, rapid solidification effects, and quality issues in laser-melted material.

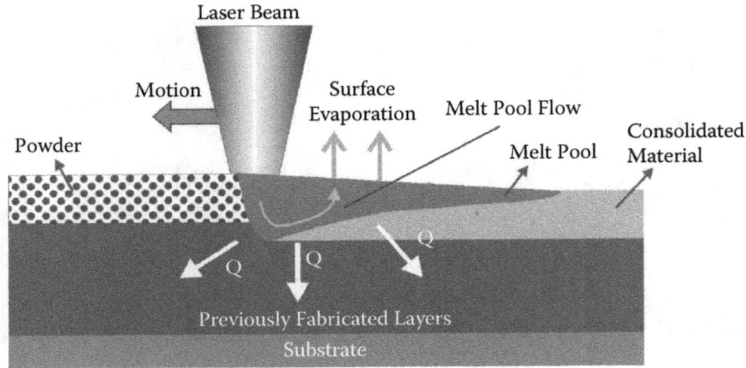

FIGURE 3.9   A schematic view of the laser–material interactions leading to different zones, flow of the melt, and evaporation.[20]

3.3.1   MELT POOL AND RAPID SOLIDIFICATION

During SLM, the short interaction of the powder bed and the heat source (due to the high scanning speed of the laser beam) leads to rapid heating and melting followed by rapid solidification. The resultant heat transfer and fluid flow (i.e., Marangoni flow) affect the size and shape of the melt pool, the cooling rate, and the transformation reactions in the melt pool and heat-affected zone (see Figure 3.9). In turn, the melt pool geometry influences the grain growth and the microstructure of the part.[2,18,19] Radial temperature gradients on the order of 10^2 to 10^4 K/mm develop between the center of the melt pool and the cooler solid–melt interface. These temperature gradients stimulate a fluid flow from the center of the melt pool toward the cooler solid–melt interface. Under certain conditions, Marangoni flow can also appear in the opposite direction of the flow. The generated Marangoni flow may even become a dominant convection mechanism in the laser melt pool.[20,21] This may lead to an unsteady motion of the solid–liquid interface, even when the scanning velocity is steady.[22] The rapid extraction of thermal energy that occurs during rapid solidification causes a large deviation from equilibrium (the laser melting cooling rates are estimated to be over 10^4 K/s). This offers some important advantages, including (1) a refined microstructure with a very small grain size and fine precipitates; (2) extension of solid solubility, even by orders of magnitude; (3) increased chemical homogeneity (e.g., reduced segregation, uniform distribution of secondary phase particles); and (4) formation of metastable (crystalline, quasicrystalline, amorphous) phases.[23,24]

3.3.2   POTENTIAL DEFECTS

3.3.2.1   Porosity

Porosity is a common defect in SLM products because the powder consolidation process is driven only by temperature changes, gravity, and capillary forces, without the application of external pressure. Porosity can be found as large irregular pores due to a lack of melting, shrinkage micropores due to a lack of feeding within interdendritic zones,[25] spherical pores caused by trapped gas, etc. Adjustments of the laser processing parameters can improve the densification above 99% of the theoretical full density values, even well above 99.9% for Ti6Al4V (see Figure 3.24 later in the chapter).[2,18,26]

3.3.2.2   Balling

Balling occurs when the molten material fails to wet the underlying substrate (due to the surface tension), spheroidizing the liquid. This results in a rough and bead-shaped scan track, increasing the surface roughness and increasing the porosity.[1,27,28] Generally, both material properties and processing variables can influence wettability and consequently balling.[2] Because liquid metals do not wet surface oxide films in the absence of a chemical reaction, it is very important to avoid oxidation and contamination. Another possibility to improve wetting is the addition of certain alloying elements, such as phosphor in selective laser melting of iron-based powder.[29]

3.3.2.3 Residual Stress

Laser-based processes (e.g., laser welding, laser cladding, SLM) are known to introduce large amounts of residual stresses due to the large thermal gradients that intrinsically exist in the process The residual stress originates partly from the cooling and shrinkage of the newly molten layer and partly from strain-induced stresses in the solid layers on the substrate underneath the newly applied layer (see Section 3.5.4 for more details). This imposes tensile stresses on the newly deposited layer and creates compressive stresses at the bottom.[29,30] The residual stresses will be partially relieved after cutting the part from the base plate, causing some deformation. The stress is usually tensile at the top or bottom and compressive in the center of the part. Depending on the commonly unfavorable effects of residual stress, (1) a subdivision of the scanned area to smaller scanning sections (known as chessboard or island scanning; see Figure 3.12 later in the chapter), (2) heating the substrate, or (3) post-processing (e.g., stress-relieving heat treatment) can be employed to mitigate the residual stress.[29–31]

3.3.2.4 Cracks

Cracks commonly originate from the high temperature gradient between the melt pool and surrounding solids leading to excessive thermal stress and rupture. Compositional segregations in some alloys may intensify the cracking. Alloys that are prone to hot cracking and solidification cracking have been proven difficult to process by SLM.

3.4 PROCESSING ASPECTS

3.4.1 STARTING POWDER

The apparent density of the powder, which influences the final density of the SLM parts, depends on the powder size, shape, and size distribution (an example is shown in Figure 3.10). Generally, packing of spheres leads to a higher density than other shapes. The spherical particles with smooth surfaces may also improve the powder flowability and deposition. Moreover, finer powders may result in a higher apparent density (to some extent), indicating a higher final density and mechanical

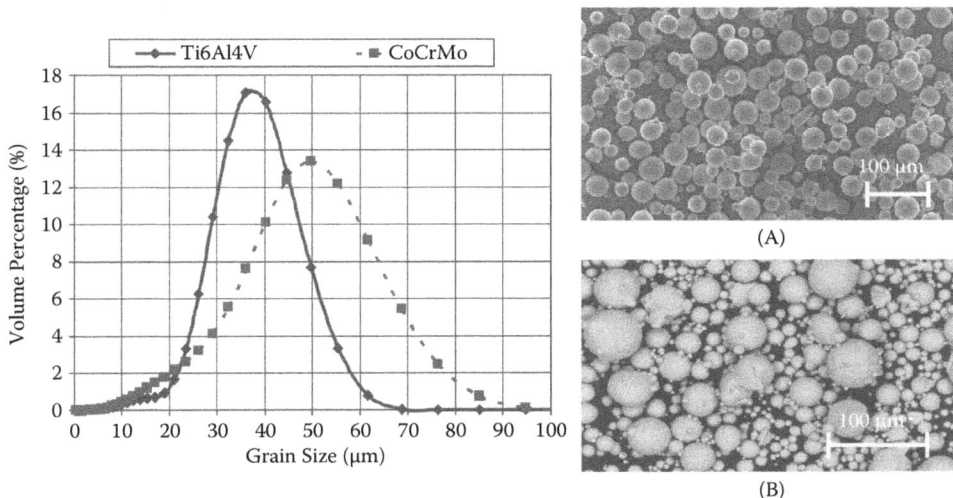

FIGURE 3.10 Powder size distribution and micrographs of (A) titanium and (B) cobalt-chromium powder.[8]

properties.[32–34] For mono-sized spheres, the highest obtainable packing density is theoretically 74%. However, the apparent powder density can be increased by mixing different powder sizes (see Figure 3.10B). In fact, finer particles can fill the voids between the larger powders and subsequently increase the powder density.[34] It is worth mentioning that higher packing density may also increase the cooling/solidification rate (by increasing the thermal conductivity), resulting in finer microstructural features. Higher mechanical properties may be achieved from these finer microstructural features.[35]

3.4.2 INPUT PARAMETERS (LASER POWER, SCANNING SPEED, LAYER THICKNESS, SCAN LINE SPACING)

The SLM process is set by adjusting various parameters. An optimal combination of laser power, scanning speed, powder layer thickness, and scan line spacing (also known as hatch spacing) is required to minimize the potential defects (by achieving optimal melt pools) and to produce high-quality parts. In order to improve the precision, an offset (about the radius of the laser beam) can be applied. A fill contour can also be used for a higher precision or an improved outer boundary quality (Figure 3.11). The effect of the process parameters can be combined and presented as the *energy density*, which is an engineering parameter representing the energy delivered to a unit volume of powder material. This is achieved by combining the laser power, scanning speed, scan line spacing (or hatch spacing), and layer thickness as shown below:

$$E_\rho = P/vst \ (\text{J/mm}^3) \tag{3.1}$$

or

$$E_\rho = P/vs \ (\text{J/mm}^2) \tag{3.2}$$

where E_ρ = energy density (J/mm^3 or J/mm^2), P = laser power (W), v = scanning speed (mm/s), s = scan line spacing (mm), and t = layer thickness.[36–38] Accordingly, increasing the laser power and decreasing the scanning speed, scan line spacing, or layer thickness increases the laser energy density.

3.4.3 SCANNING STRATEGIES

Different scanning strategies (Figure 3.12) affect the thermal history during the SLM process and consequently alter the material properties, including density, thermal and residual stress, and microstructure. It has been reported, for example, that an alternating scanning strategy (rotating the

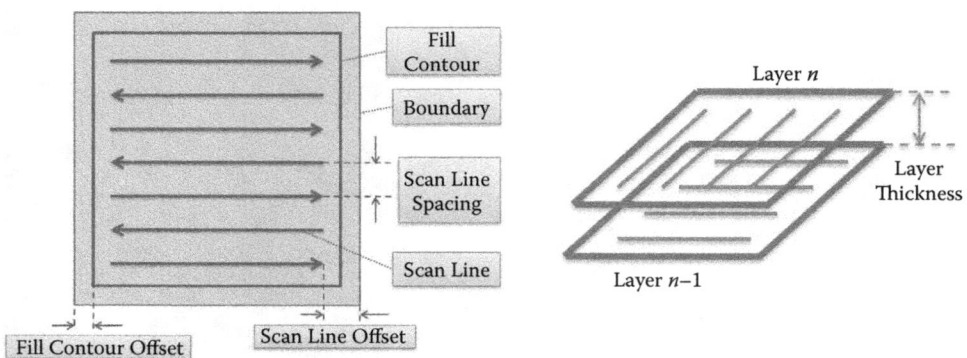

FIGURE 3.11 Schematic representation of scanning parameters that can be altered to improve quality. Boundary parameters (e.g., scan line offset, fill contour) increase precision and the quality of borders.

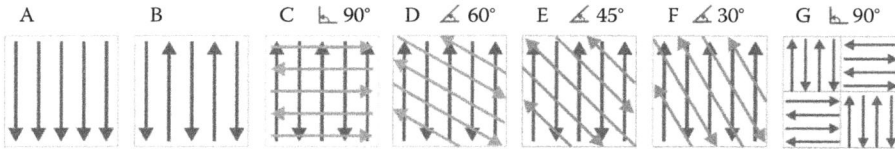

FIGURE 3.12 Schematic examples of possible scanning strategies: (A) unidirectional, (B) bidirectional, (C–F) alternating bidirectional (with different angles of 90°, 60°, 45°, and 30°), and (G) chessboard scanning strategy.[40]

scanning direction in each layer) improves the density of Ti6Al4V parts.[39] The scanning strategy also influences the residual stresses and can be modified to mitigate quality issues such as warpage, cracks, and delamination.

3.4.4 GEOMETRICAL CONSIDERATIONS

3.4.4.1 Overhang

As described, the ability to produce arbitrary complex parts is the main advantage of the SLM technology; however, challenges exist to produce overhanging surfaces. An overhanging structure is a down-facing surface of a component that is not supported by previously solidified material or a solid substrate. The melt pool is only supported by the powder material that cannot properly hold the melt on top. This results in dross formation at the down-facing surfaces.[41] The most common methods to resolve this issue are optimizing the orientation of the part (to minimize the overhanging surfaces) and the use of support structures (to carry the down-facing surfaces).[42,43] An alternative method (practiced at KU Leuven) is to manipulate the scanning parameters to mitigate the problem of overhanging surfaces.[41,44–46] In this technique, a number of additive layers of the overhanging surfaces are selected to be scanned using different laser energies (Figure 3.13A), where the laser scans to sinter the powder particles instead of fully melt them. This provides a smooth surface at the bottom by avoiding agglomerated particles, balling, and warpage. An example of a successful SLM production of a Ti6Al4V part with various overhanging surfaces is shown in Figure 3.13B.

3.4.4.2 Biomedical Scaffolds

One of the very successful applications of the SLM process is producing biomedical scaffolds for bone replacement applications (Figure 3.14). Bone scaffolds should possess a combination of good biocompatibility, appropriate mechanical properties, and high interconnected porosity. The scaffolds are manufactured with a controlled pore size, pore shape, and permeability to provide the desired biomedical properties;[47] however, there is always a geometrical mismatch between the scaffold design

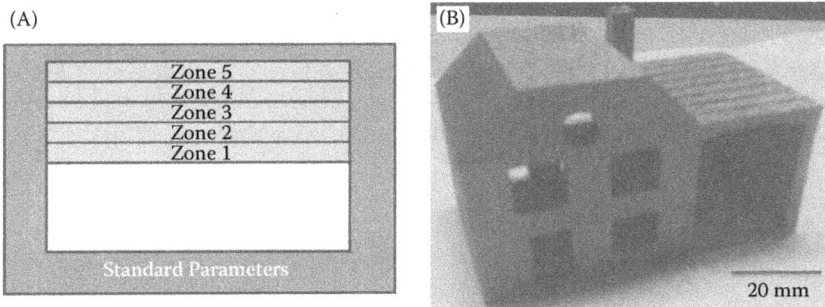

FIGURE 3.13 Production of overhanging surfaces using adjusted scanning parameters: (A) dividing the down-facing surface into different scanning zones[44] and (B) SLM of a Ti6Al4V part without use of support.[45]

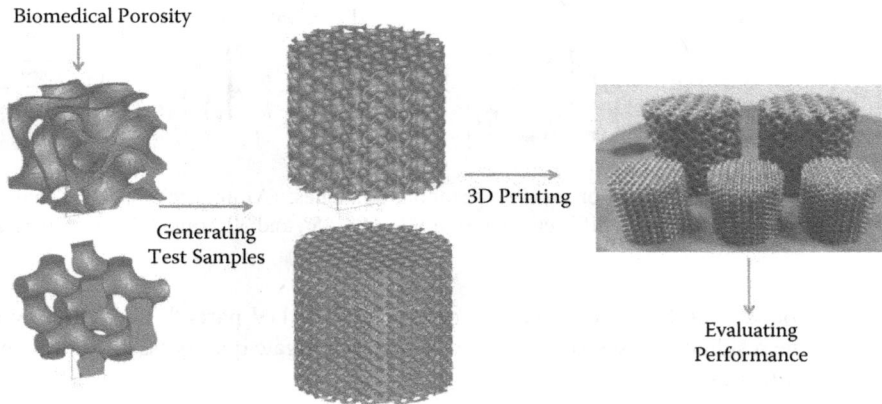

FIGURE 3.14   Production of biomedical scaffolds, from design to manufacture.

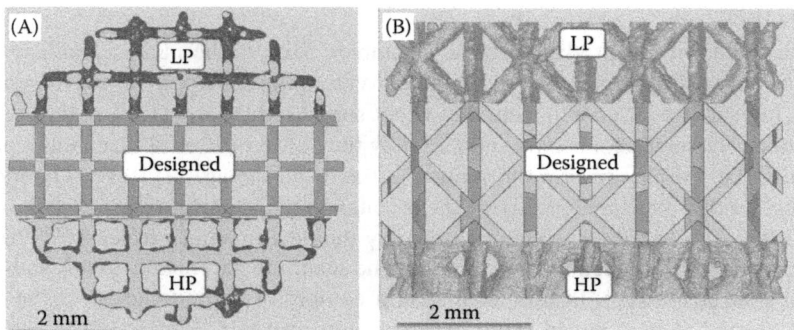

FIGURE 3.15   Comparison of nickel–titanium scaffolds made by LP parameters (low speed, low power: $P = 40$ W, $v = 160$ mm/s) and HP parameters (high speed, high power: $P = 250$ W, $v = 1100$ mm/s), with as-designed CAD file (180 µm strut thickness): (A) top view (x–y section) and (B) side view (x–z section).[49]

and the obtained products.[48] This mismatch originates from the finite laser beam diameter (a smaller laser spot allows higher precision), complex hatching of additive sections, and delays in the time to reach the set laser speed (the scanner requires a very short time until it reaches the desired speed).

In the case of continuous laser scanning, the geometrical mismatch between scaffold design and outcome structure depends greatly on the scanning parameters. In fact, when a higher scanning speed is set, the mismatch will increase due to the longer delay (for the laser to reach its speed). This is evident in Figure 3.15, where the scaffolds made by a higher scanning speed and adjusted to a higher power (see HP zone) show a larger mismatch than that of scaffolds made with lower speed and power (see LP zone) parameters but with a comparable laser energy density.[49] Because the laser does not reach the equilibrium speed during fast scanning, the effective energy density will be higher, resulting in a larger melt pool and larger geometrical mismatches.

3.4.5   LASER REMELTING AND LASER EROSION

In SLM, the surface roughness is an important parameter of the overall quality of the part.[8,50] A number of surface modification technologies are commonly applied after the SLM process, including mechanical methods (abrasive sandblasting and machining), chemical processes (acid etching and oxidation), and thermal processes (plasma spraying).[51] Because these methods are applied after SLM, they require removing the part from the building platform. This increases the

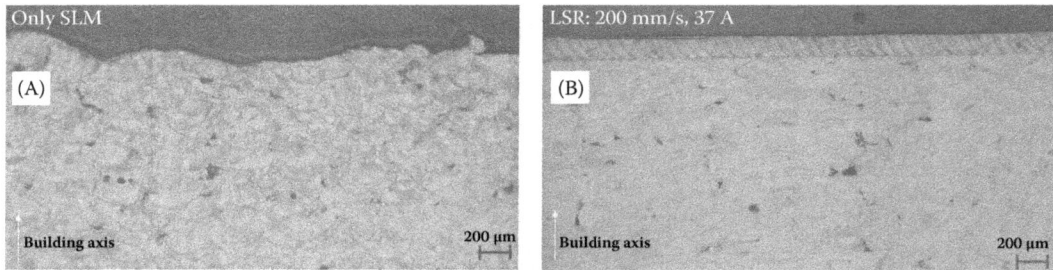

FIGURE 3.16 Surface quality enhancement of 316L stainless steel with laser remelting: (A) before laser remelting and (B) after laser remelting with 200-mm/s scanning speed and 95-W laser power.[39]

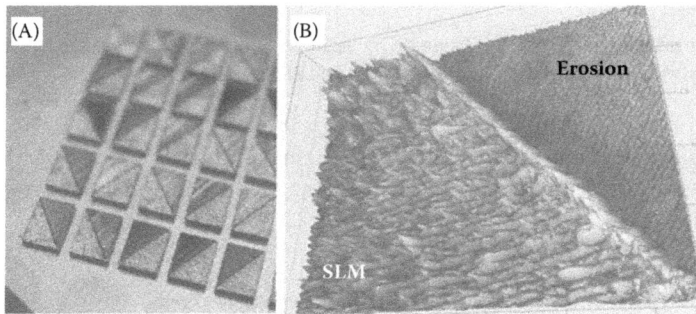

FIGURE 3.17 The surface quality before and after laser erosion of the SLM surface.[54]

overall production time, precision errors, and final cost; however, SLM itself can be directly used to improve the surface quality through a controlled laser remelting of the surface. For example, Figure 3.16 shows a successful laser remelting experiment to smoothen the surface without the use of any secondary treatment.[39,52] This surface improvement can be attributed to the *surface shallow melting* mechanism. In this mechanism, melting a thin material layer fills valleys in between the peaks and removes aggregated spheres of particles.[53]

Another technique to directly manipulate the surface of SLM products (using the SLM machine) is to evaporate the material by using an incident laser beam (pulse laser is usually needed). This is called *selective laser erosion* (SLE) and is employed to improve the surface (Figure 3.17) or even to engrave a desired geometry on additive layers. In this method, the laser beam can be narrowed to a dozen micrometers, allowing for very small internal radii and fine details, thus giving SLE the capability of micromachining. The main challenge is to achieve an appropriate adjustment when many parameters are involved.[16] The parameters may include scanning speed, pulse frequency, laser power, scan spacing, and number of eroded layers.[54] Figure 3.18 demonstrates an application of combining SLM and SLE to produce small internal and external features. Although SLM alone is limited to achieving holes or engravings below 400 to 500 µm, SLE of additive layers allows very precise evaporation of material to make features as small as 50 µm (Figure 3.18). Therefore, these techniques can be combined, such that dense additive layers are first made using SLM and then the hollow features can be eroded in each layer, to achieve the narrowest holes and engravings.[52]

3.5   SLM MATERIALS ASPECTS

In addition to machine equipment and processing aspects, material also strongly affects the obtained quality in SLM. This section is based primarily on the material investigations performed at KU Leuven and briefly explains the relation of the processing parameters with the material

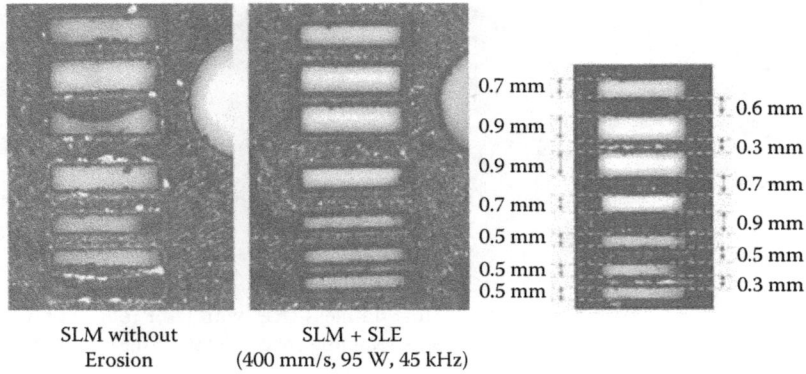

	0.7 mm		0.6 mm
	0.9 mm		0.3 mm
	0.9 mm		0.7 mm
	0.7 mm		0.9 mm
	0.5 mm		0.5 mm
	0.5 mm		0.3 mm
	0.5 mm		

SLM without   SLM + SLE
Erosion   (400 mm/s, 95 W, 45 kHz)

FIGURE 3.18 Comparison of thin slits/ribs made by SLM without erosion and made by SLM combined with SLE (after every layer) with nominal widths of the slits and ribs on the right.[52]

properties such as density, microstructure, and mechanical properties of the SLM products. It also suggests some possible post-treatments in order to enhance mechanical performance and surface quality of the products.

3.5.1 Overview of Metals Investigated at KU Leuven

Within KU Leuven, different metal alloys have been developed and processed by SLM. Table 3.2 gives an overview of these materials, their type, and their intended applications. These materials are the basis of the upcoming discussions.

TABLE 3.2
Overview of SLM Metals Utilized at KU Leuven

Material	General Properties	Applications	Type/Alloy
Titanium alloys	Lightweight; high strength; good corrosion resistance; biocompatible	Aerospace, automotive, biomedical, etc.	Ti6Al4V, Ti grade 2, NiTi
Aluminum alloys	Lightweight; medium/high strength; good corrosion resistance; high thermal/electrical conductivity	Aerospace, automotive, heat exchangers, electrical components, prototyping, etc.	AlSi10Mg, Al6061, Al7075
CoCrMo alloys	Biocompatible; good corrosion resistance; high strength and wear resistance	Turbines, dental and orthopedic implants, etc.	CoCr F75
Maraging steel	High hardness and strength	Tooling, molds and dies	18Ni300
Stainless steel	Excellent corrosion resistance	Functional components, chemical equipment, medical parts, etc.	316 L
Tool steels	High hardness at high temperature	Tooling	M2 HSS, H13
Refractory metals	High melting point, high density	High temperatures, biomedical implants (tantalum), chemical industry (tantalum)	Tantalum, tungsten
New alloys	To be explored	To be explored	Ti6Al4V+Mo, Al7075+Si

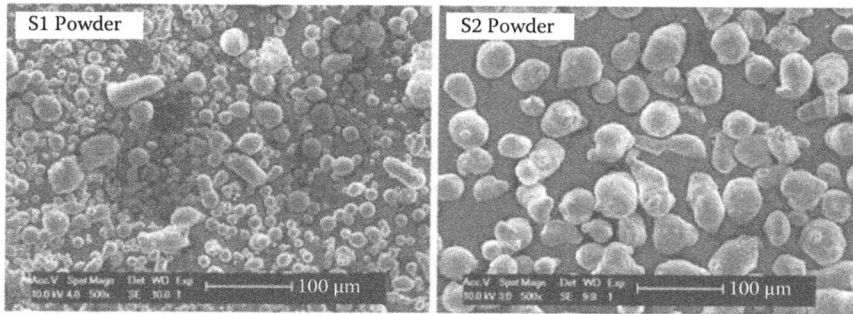

FIGURE 3.19 Comparison of AlSi10Mg powders by two different suppliers. For S1 powder (left image) the average particle size is 16.3 µm; for the S2 powder (right image), the average size is 48.4 µm. The S1 powder also contains a larger variety of particle sizes than does the S2 powder.[55]

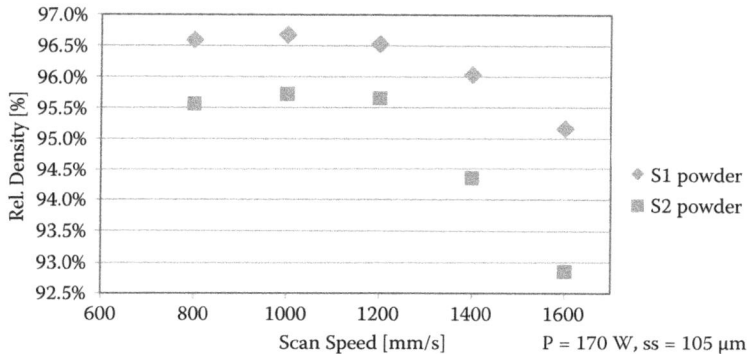

FIGURE 3.20 Relative density results for parts produced with identical scanning parameters but with the different powders shown in Figure 3.19.[55]

3.5.2 DENSITY

3.5.2.1 Powder Particle Size

As mentioned, powder particle size distribution is an important factor influencing the deposition and the SLM part density. For example, for the average particle size and the particle size distributions of AlSi10Mg shown in Figure 3.19, S1 powder with a finer size and larger variety of sizes leads to an improved density of the parts made by the same SLM parameters (Figure 3.20).[55]

3.5.2.2 Single Track Scans

Single tracks (Figure 3.21) can be used to explore the shape of the melt pool. The generated knowledge is a shortcut to assessing the processing windows (i.e., the successful range of scanning speeds and laser powers) for SLM of a specific material. The following factors are considered to evaluate the quality of a single track:[56,57]

1. Being uninterrupted (see Figure 3.21A,B for examples of good and bad tracks), preventing pores and irregularities
2. Controlled penetration (into the layer below), demonstrating good wetting and bonding to the previous layer (Figure 3.21C)
3. Controlled height for building up layers of good height (Figure 3.21C)

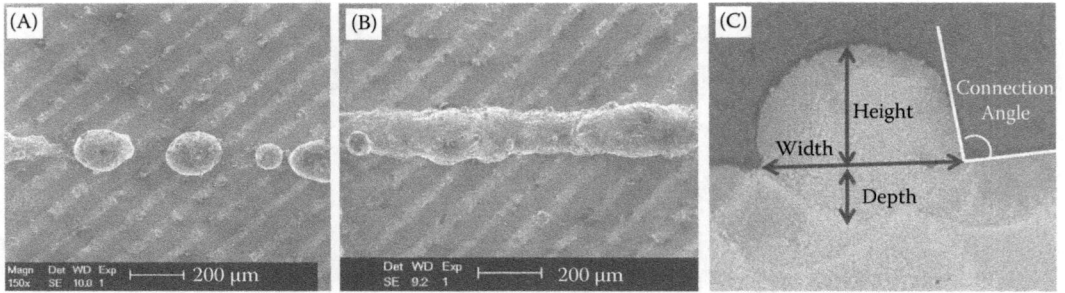

FIGURE 3.21 Important factors in single track experiments: (A) bad/interrupted track from top view, (B) good/continuous track from top view, and (C) side view factors demonstrated on a good track.[57]

   4. Optimum connection angle of about 90° between the scan track and the layer below (Figure 3.21C), which enhances the dimensional accuracy and density and minimizes the required overlap between adjacent scan tracks

Figure 3.22 shows the results of single track experiments on AlSi10Mg.[57] The parameters in the bottom right zone (low power/high speed) lead to poor wetting and droplet formation (spheroidization). This is due to the low laser energy density (for this range of parameters) which leads to porous parts and a weak bonding of the tracks to the layer below. In contrast, when the energy density is excessively high (top left corner of the figure), the tracks will be very short (due to strong penetration into the previous layers and partial evaporation). Moderate energy inputs, reached with 170- to 200-W laser power and 800- to 1400-mm/s scanning speed, can be considered as an appropriate process window to manufacture fully dense AlSi10Mg parts. This is the area that leads to continuous tracks, good depth/height of track, and almost 90° connection angles.[57]

   It is worth mentioning that at excessively high laser energy densities, such as can be seen in the top left scenarios in Figure 3.22, vaporization increases and the melt pool penetrates deeply into the layer. This may lead to the formation of a *keyhole*, as it is similarly known in laser welding.[58] One

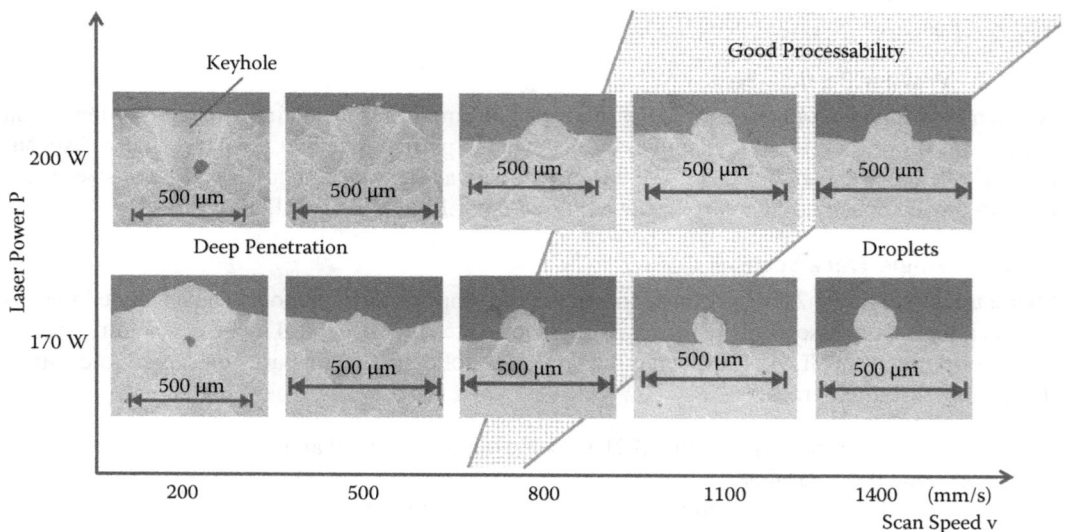

FIGURE 3.22 Process parameter map for single track scans of AlSi10Mg. The process window is indicated by hatch lines.[57]

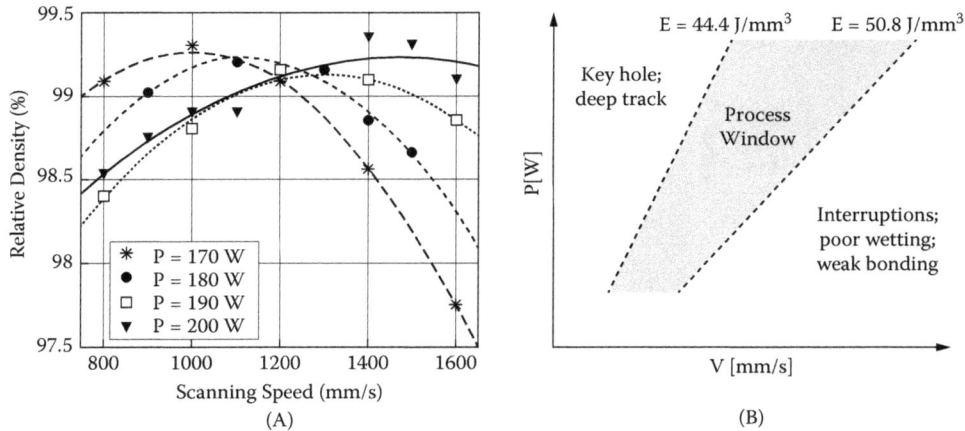

FIGURE 3.23 (A) Relative density of AlSi10Mg parts for different combinations of laser power and scanning speed (the scan spacing and layer thickness are set at 105 μm and 30 μm, respectively), and (B) the corresponding process window.[57]

of these so-called keyhole pores can be seen in the cross-section for a scan track produced at 200 W and 200 mm/s. Under these conditions, vapor cavities can be formed inside the melt pool. High Marangoni flow[56] can also contribute to the formation of entrapped gas pores.

3.5.2.3 Density Optimization

The results achieved from single track scans can be used to maximize the density of the SLM parts; for example, dense AlSi10Mg parts can be produced in the range of the process window developed in Figure 3.22 (as shown in Figure 3.23). In this case, the scan spacing was chosen to be 105 μm, 70% of the melt pool width (i.e., 30% overlap between adjacent scan tracks). The maximum density is reached when the laser power is adjusted to the scanning speed, generating an energy density of 44.4 to 50.8 J/mm^3 according to Equation 3.1. Density optimization has also been carried out for other metallic alloys (such as Ti6Al4V, M2 high-speed steel, and maraging steel).[8,13,59] Figure 3.24 shows the cross-sectional micrographs of Ti6Al4V SLM parts for visually assessing the porosity. The porosity is significantly reduced at higher energy densities of about 250 to 293 J/mm^3 when the part reaches above 99.8% of its theoretical density. It should be noted that, although the higher energy density may contribute to higher density of the products (Figure 3.24),[8,26] excessively high energies may lead to high temperature gradients, leading to internal stresses or part distortion,[30] and they may increase the risk of balling and dross formation in the melt pool, resulting in a bad surface finish.[2] Optimum energy densities to reach dense parts can even be generated using different sets of parameters. For example, both high power combined with a high scanning speed and low power combined with a low scanning speed are able to produce dense NiTi parts.[60]

3.5.2.4 Laser Remelting to Improve Density

As described in Section 3.4.5, laser remelting is capable of reducing the roughness of up-facing outer surfaces. Remelting can also be employed to increase the density of the SLM parts. In fact, rescanning (thus, remelting) each layer (or every few layers) can smoothen the previously melted layers which in turn can increase the density of the part in three dimensions. This technique has been successfully applied to enhance the density of different SLM materials such as Ti6Al4V, 316L stainless steel, 18Ni300 maraging steel, M2 high-speed steel, and AlSi10Mg alloy.[13,59,61] Figure 3.25, for example, demonstrates the effectiveness of laser remelting in the density improvement of 316L stainless steel. The SLM standard process parameters optimized for maximum density (scanning speed of 380 mm/s, laser power of 105 W, and scan spacing of 125 μm at a spot diameter of 200 μm)

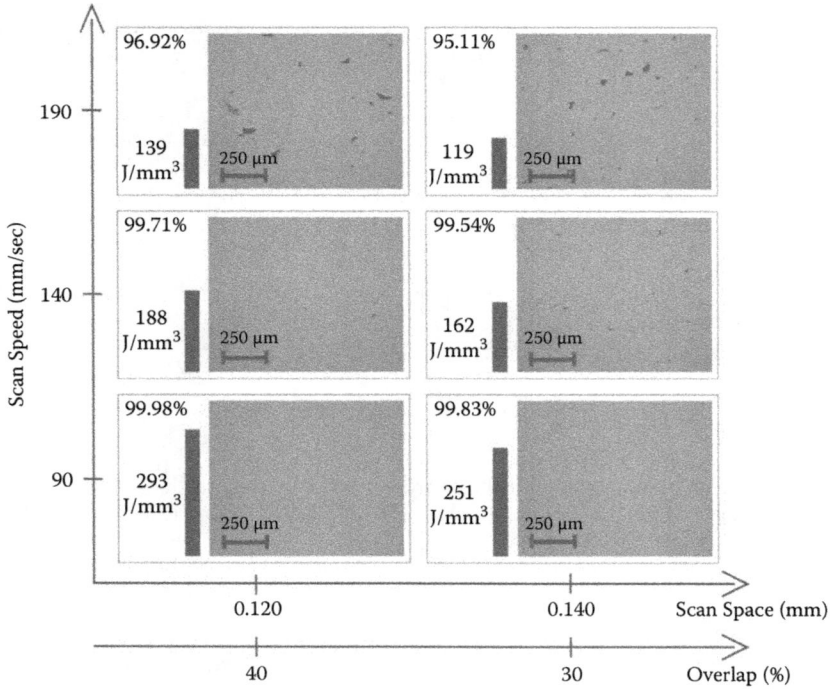

FIGURE 3.24   Results of parameter study for Ti6Al4V (laser power, 95 W).[8]

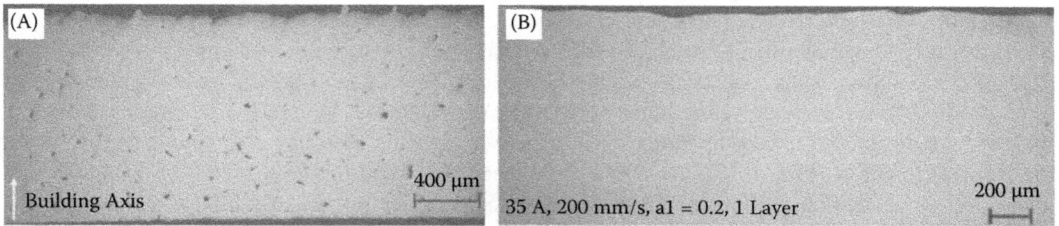

FIGURE 3.25   Cross-section of a 316L stainless steel parts made using optimum SLM parameters: (A) with no remelting and (B) after successful remelting.[61]

still lead to an evidently porous part (Figure 3.25A). However, applying laser remelting (using a scanning speed of 200 mm/s, laser power of 85 W corresponding to a pump current of 35 A, and scan spacing of 200 μm at a spot diameter of 200 μm) reduces the porosity by a factor of about 20. This may lead to an evidently dense part, as seen in Figure 3.25B.[61]

3.5.3   MICROSTRUCTURE, TEXTURE, AND MECHANICAL PROPERTIES

3.5.3.1   Microstructure

The microstructure of metals produced via SLM can be categorized into two different types of solidification structures. The first category is that of materials that have large and columnar grains after SLM, as shown in Figure 3.26A. During solidification of the first layers, a multitude of grains nucleate and grow preferentially along the direction of the maximum thermal gradients. The easy-growth direction (e.g., <100> for cubic metals[62]) of some of these grains will be aligned to this maximum gradient and hence their rapid growth will dominate the texture. The maximum gradient,

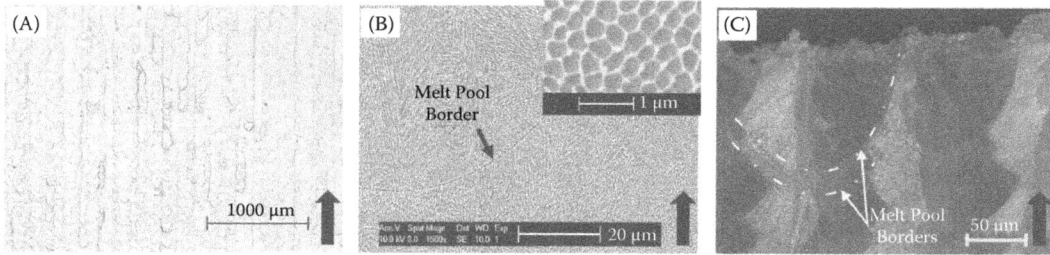

FIGURE 3.26 Side view (upward black arrows indicate the building direction) of the microstructure after SLM: (A) columnar Ta;[63] (B) cellular 18Ni300 maraging steel, where the insert shows the cellular structure;[65] and (C) AlSi10Mg grains oriented toward the center of each melt pool.[66] Notice the difference in scale.

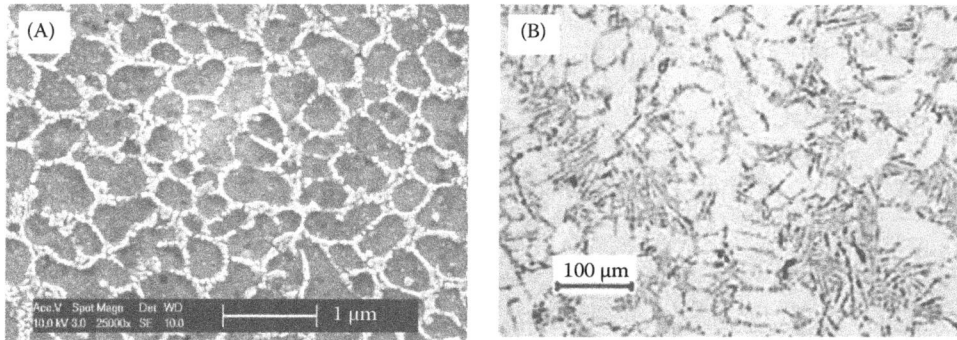

FIGURE 3.27 Comparison of the microstructure of AlSi10Mg after (A) SLM[65] and (B) casting.[67] Notice the difference in scale.

on the other hand, is generally oriented along the building direction, but it can be slightly altered by modifying the scanning strategy. During deposition of successive layers, the underlying material is remelted, allowing new solid material to form with the same crystallographic orientation of the grains below. By this mechanism, which is called *epitaxial growth*, the grains can grow over several layers, forming large and columnar grains. This type of solidification requires the stability of the planar solidification front, which is generally the case for pure and low-alloy metals. Therefore, pure metals (such as tantalum, titanium, and tungsten), as well as alloys with stable planar fronts (such as Ti6Al4V), show this type of morphology.[63-65]

The second category includes medium-/high-alloy metals for which the planar solidification front is unstable. In these alloys, local perturbations of the front lead to the development of a fine cellular structure in which the grains, consisting of multiple cells, grow toward the center of the melt pool. Examples include 18Ni300 maraging steel and AlSi10Mg, shown in Figure 3.26B and Figure 3.26C, respectively. The intercellular spacing is on the order of 0.5 μm for most materials.

In Figure 3.27A, dark aluminum cells in AlSi10Mg produced via SLM are decorated by a fine eutectic, in which the white particles are the Si phase. For comparison, the microstructure of as-cast material is shown in Figure 3.27B, where the eutectic containing relatively large Si particles/plates (with a darker color) is dispersed between the Al dendrites. Accordingly, the finer microstructure of AlSi10Mg produced via SLM leads to a higher strength than that of material produced by conventional casting methods.

The rapid solidification in SLM can be controlled using different combinations of parameters, such as a high laser power combined with a high scanning speed (HP parameters) or a low laser power combined with a low scanning speed (LP parameters). The former leads to faster heating and cooling, and the latter results in a slower heating and cooling. This can be applied to manipulate

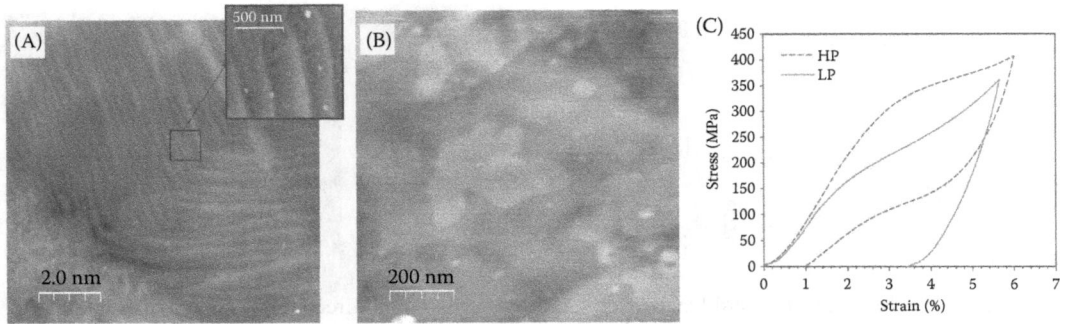

FIGURE 3.28 AFM microstructural details of as-SLM NiTi parts (at room temperature) made by using (A) LP parameters (exhibiting internally twinned martensitic structure) and (B) HP parameters (showing some fine flake-shaped grain structure). (C) Typical loading and unloading behavior of as-SLM NiTi parts showing a strong pseudoelasticity for the HP sample.[60]

the microstructure and properties of the SLM products. For example, in the case of shape-memory* NiTi alloy, faster heating and cooling can reduce the reversible martensite to austenite transformation temperatures by decreasing the grain size, suppressing precipitation, etc. This can be used to implement martensite stability using the LP parameters (Figure 3.28A), in contrast to more stable austenite for SLM parts made using the HP parameters (Figure 3.28B). The martensitic structure will lead to a pronounced thermal memory† in comparison with the austenitic structure, which will result in pseudoelasticity (Figure 3.28C).[60]

3.5.3.2 Texture

As a result of the directional solidification, metals produced by SLM develop a texture. Whereas the maximum thermal gradients are commonly oriented along the building direction, changing the scanning strategy can somehow influence these orientations. Consequently, this can alter the solidification direction and the texture. This is illustrated in Figure 3.29, in which the <100>, <110>, and <111> pole figures of AlSi10Mg are shown for three different scanning strategies.[66] In Figure 3.29B, the bidirectional scanning strategy, in which each layer is scanned using a zigzag pattern with a 90° rotation between layers, results in an expected <100> texture (<100> is the direction of maximum thermal gradients for cubic metals[62]). Chessboard (island) scanning also produces the same texture but it is weaker to some extent (Figure 3.29C).

In comparison, unidirectional scanning (without rotation between layers) leads to the strongest texture but it also tilts the texture perpendicular to the scan vectors (Figure 3.29A). Because AlSi10Mg falls within the second solidification category (see Section 3.5.3.1), the grains in AlSi10Mg grow toward the center of the melt pool (Figure 3.29C), orienting the <100> directions in the same direction. Partial remelting of the adjacent track (in the same layer) leads to a sideways heat conduction that tilts the texture to the left; the tilt is generated in all scanning strategies, but rotations between layers randomize it in Figure 3.29B,C. Moreover, the overlap remelts the grains on one side of the melt pool, leaving only the grains on the other side. The combination of these effects results in an intensified and tilted texture perpendicular to the scan vectors.

For pure and low-alloy metals (the first category) that solidify in large columnar grains, the dependence of texture on the scanning strategy is different. For these materials (such as Ta[63] and Ti6Al4V[40]), applying a rotation between the layers intensifies the fiber-like texture rather than

---

* Shape-memory alloy is a type of material that can regain its pre-deformed shape via heating above specific temperatures after deformation.
† Thermal memory is the ability of a shape memory alloy to recover its primary shape at a higher temperature after deformation at a lower temperature.

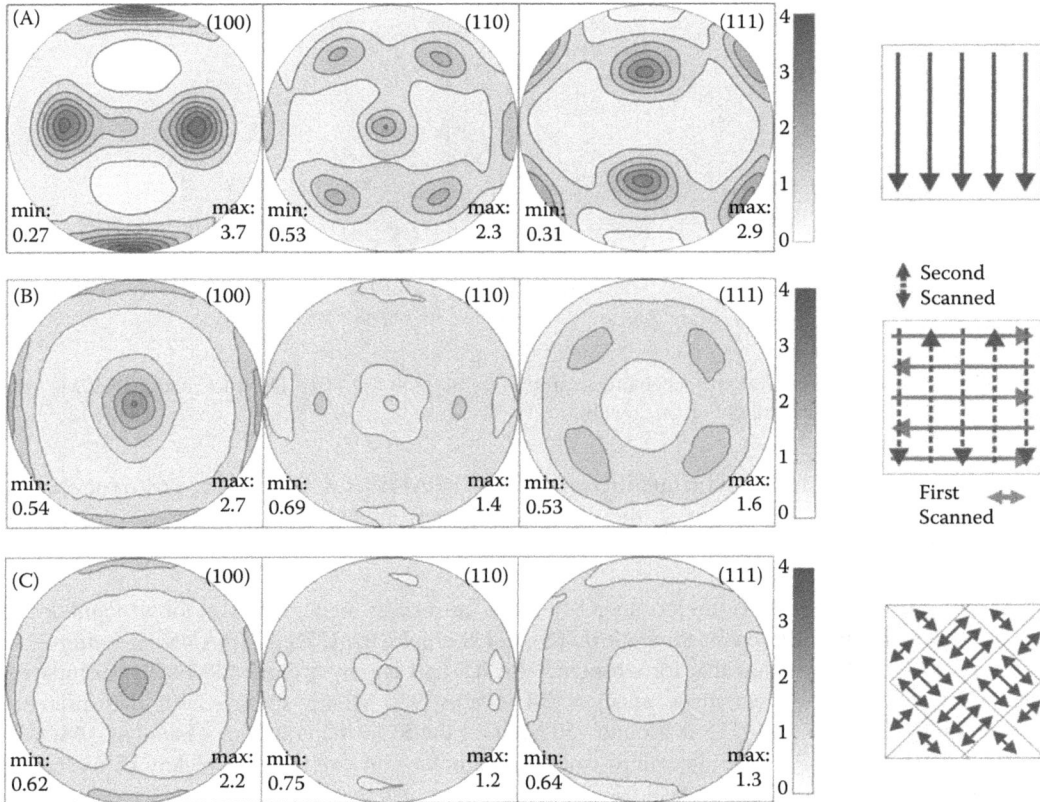

FIGURE 3.29 Pole figures of AlSi10Mg produced via SLM, for different scanning strategies: (A) Unidirectional scanning, no rotation between layers; (B) bidirectional scanning, 90° rotation between layers; and (C) chessboard (island) scanning. Each subsection is scanned using bidirectional scanning. (Adapted from Thijs, L. et al., *Acta Mater.*, 61(5), 1809–1819, 2013.)

weakening it. Because rotation directs the maximum thermal gradient toward the building direction, grains will have a better chance to continue their growth on the grains below. This more stringent selection of grains ultimately leads to a stronger texture.

3.5.3.3 Mechanical Properties

By now, it has been established that the high cooling rates in SLM lead to a fine and textured microstructure compared to conventionally produced material. This translates into different mechanical properties. In Figure 3.30, the yield stress and ductility of SLM-produced 18Ni300 maraging steel, Ti6Al4V-ELI,* and AlSi10Mg are compared to those of cast or rolled† (wrought) material.[68] As seen, the finer microstructure of materials produced via SLM leads to a stronger material (Figure 3.30A). However, in the case of Ti6Al4V(-ELI), this strength is offset by a much lower ductility of about 8%, whereas the conventional material reaches 18%. ASTM Standard F2924[69] for Ti6Al4V produced by powder bed fusion (SLM or EBM) calls for a minimum ductility of 10% in the highest

---

* Ti6Al4V-ELI (Grade 23) is similar to Ti6Al4V (Grade 5), except that ELI contains reduced levels of oxygen, nitrogen, carbon and iron. ELI is short for "extra low interstitials." The lower concentrations of interstitials are to improve ductility and fracture toughness.
† Ti6Al4V is after mill-annealing. This is a specific heat treatment, applied after rolling. Generally, it is used for stress relief, but it is also for slight microstructural modifications of the material deformed at high temperatures.

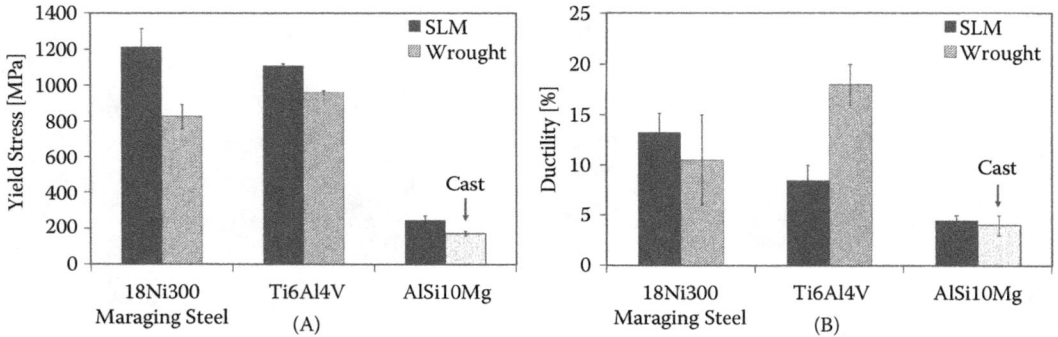

FIGURE 3.30 Comparison of the mechanical properties of 18Ni300, Ti6Al4V-ELI, and AlSi10Mg produced via SLM and in the wrought or cast condition.[59,68,70]

quality classification. Because the ductility of as-built Ti6Al4V is not sufficient, post-process heat treatments are suggested to increase the ductility. Apart from Ti6Al4V, other metals and alloys processed by SLM possess a ductility that is at least as good as conventional material, as illustrated by Figure 3.30B.[59,70]

As the yield stress of metals produced via SLM is in most cases higher than that for wrought or cast material, so is the ultimate tensile strength (UTS). For example, the UTS of Ti6Al4V-ELI after SLM is around 1250 MPa, whereas that for wrought Ti6Al4V-ELI is only around 1000 MPa. Comparing these UTS values to the yield stress values (1100 MPa vs. 950 MPa in Figure 3.30), the difference between the yield stress and UTS is around 150 MPa for the SLM material, but it is only 50 MPa for the wrought material. Because the microstructures of both materials are completely different, the strain hardening behavior during plastic deformation is also different, leading to this discrepancy.

3.5.4 Residual Stresses

As mentioned in Section 3.3.2, a disadvantage of the high cooling rates during the SLM process is the occurrence of thermally induced residual stresses. This is a major factor that has kept the SLM processable materials limited up to now. These stresses cause defects such as deformation, warpage/delamination, dimensional inaccuracy, and macro- or micro-cracks such as those shown in Figure 3.31 for Hastelloy C276.[65] The cracks can greatly reduce the mechanical performance, and residual tensile stresses at the surface harm the fatigue properties.

The residual stresses have different origins. One origin is strain-induced stress in solid substrate underneath the layer being melted; when the laser scans above that substrate, thermal-induced compressive stresses are induced in the substrate, which is not free to expand. Those compressive stresses

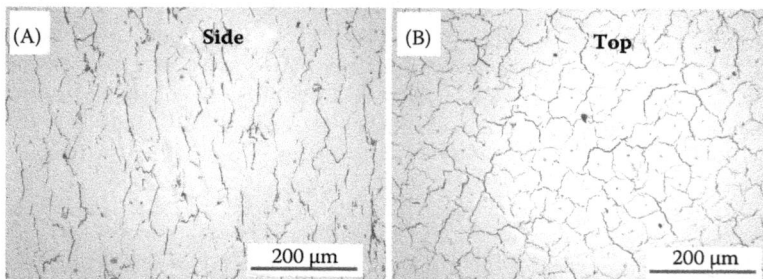

FIGURE 3.31 Cracks in SLM-produced Hastelloy C276.[65]

No preheating   T = 90°C   T = 150°C   T = 200°C

FIGURE 3.32 Cracks/delaminations formed in M2 high-speed steel parts without preheating during SLM; they are gradually reduced and eliminated by preheating the base plate during the process (preheating temperatures are shown below the parts).[13]

can reach the yield strength, inducing plastic deformation. When cooling down, the substrate tends to shrink but this shrinkage is also constrained. Therefore, the compressive stresses become tensile stresses that may even reach the UTS and hence may propagate cracks in the material. Those strain-induced stresses are complemented by stresses due to the solidification and cooling of the molten top layer. As the molten layer cools, the solidified material will shrink, but the shrinkage is again restricted by the adjacent cooler layers or the solidified tracks. This also leads to tensile stresses at the top surface and compressive stresses in the center of the part. Those tensile stresses cause the part to bend upward at the sides, which is again restricted by the attachment to the base plate causing large vertical tensile stresses at the sides. This is the major cause of cracks, which usually occur at a corner or other stress concentration zones near the base plate. If the part has a complex shape, additional support structures might be needed to restrict the curling of the part upward.[30] Mercelis and Kruth[30] have modeled induced stresses for different part thicknesses. The results indicate that, for medium thick parts, the final stress profile (after removal from the base plate) might be parabolic, with tensile stresses at the top and bottom of the part and compressive stresses in between.

Although the explanation above theoretically explains the origin of residual stresses in SLM using its layer-wise nature, it disregards the fact that each layer is made up of several individual scan vectors. Because of the elongated shape of the melt pool, the shrinkage is different along the vector compared to perpendicular to the vector. This leads to anisotropy in the local residual stresses, which are on average two times greater in the longitudinal direction. Therefore, shorter scan vectors can be used to mitigate the residual stress issue.[30,31,71,72] An alternative method to lower the residual stress is to increase the processing temperature. This can be done by preheating the base plate. This lowers the thermal gradients and thus the residual thermal stresses, as demonstrated by the reduction in crack density in M2 high-speed steel SLM parts shown in Figure 3.32.

Figure 3.33 illustrates the effect of the internal residual stresses on the mechanical behavior. Compact tension specimens, normally used for fracture toughness testing, were cut in half and the stresses were measured via the contour method.* The results indicate that the stresses in the building direction are two to three times larger than the stresses in the horizontal plane. Furthermore, the residual stress distribution affects the shape of the crack and consequently the crack growth rate.[73]

3.5.5 POST-PROCESSING

3.5.5.1 Post-Process Heat Treatment

Although the static mechanical properties of SLM-produced metallic components are commonly equal or superior to those of conventionally produced material, a post-process heat treatment may still be required for several purposes. First, it may still improve the general mechanical properties such as the hardness or ductility such that the material is superior to the best available conventional

---

* In the contour method, the surfaces of a symmetric cut are analyzed using a coordinate measurement machine (a device to measure the geometry and shape of an object). The measurements of the surface deformations are translated into residual stresses through an FEM model. The stress maps show stresses perpendicular to the measured plane.

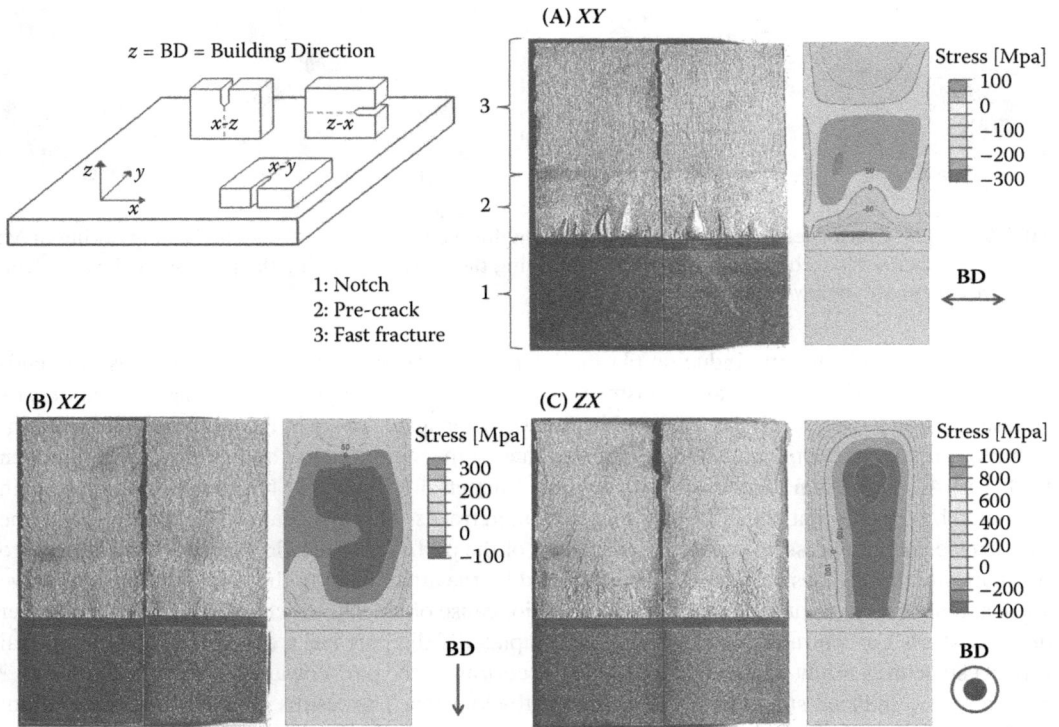

FIGURE 3.33 Residual stresses in compact tension specimens of SLM-produced Ti6Al4V. As indicated by the fracture surfaces, the residual stresses influence the crack propagation and therefore the behavior of the material under loading conditions. Notice the large stresses in the z,x specimen compared to the x,z and x,y specimens. The stresses in the z,x specimen are parallel to the building direction.[73]

material or achieves the recommended standard. Second, it can tune the microstructure to fit other needs, such as good fatigue performance or high creep resistance. Third, it relieves residual stresses. Ti6Al4V is one of the few materials for which a heat treatment is recommended to improve the mechanical properties and to comply with the international standards.[69] As shown in Figure 3.34, the martensitic α' plates coarsen into a lamellar $\alpha+\beta$ structure upon heat treatment above 600°C. The coarsening will be even more severe when the temperature is near the β transus around 1000°C (Figure 3.34C). Although this coarsening leads to softening of the material, it significantly improves the ductility, so all mechanical properties will meet the standard specifications via an optimal treatment between 850°C and 950°C (Figure 3.35).[68]

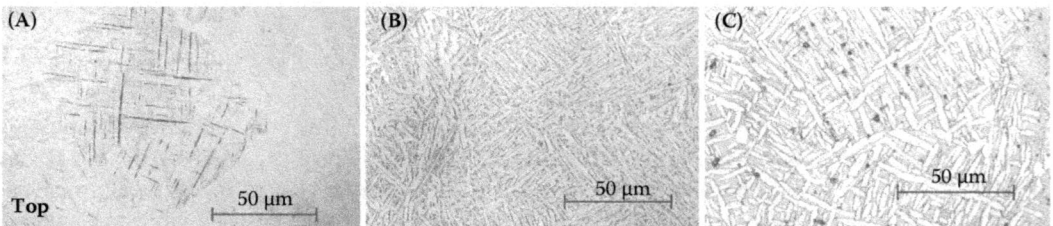

FIGURE 3.34 Microstructural coarsening of SLM-made Ti6Al4V after different heat treatments: (A) as built; (B) heat treated at 850°C for 2 hours, followed by air cooling; and (C) heat treated at 950°C for 2 hours, followed by air cooling.[68]

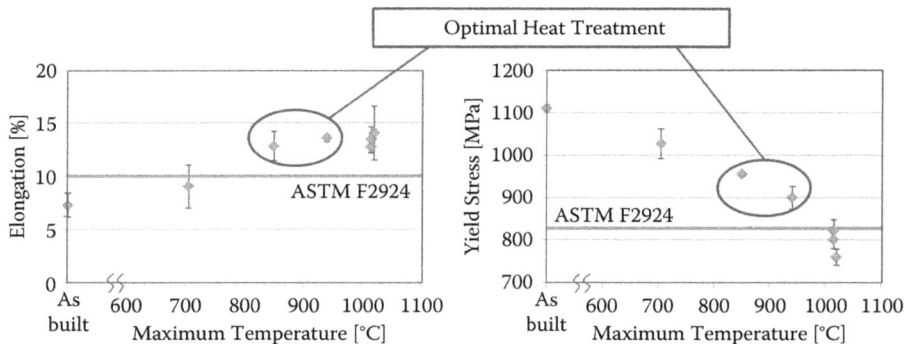

FIGURE 3.35 Ductility and yield stress of SLM-produced Ti6Al4V as a function of the maximum temperature of heat treatment (for 2 hours). Heat treatment at 850°C to 950°C produces ASTM standard ductility, while the standard strength is still obtainable. (Adapted from Vrancken, B. et al., *J. Alloys Compd.*, 541, 177–185, 2012.)

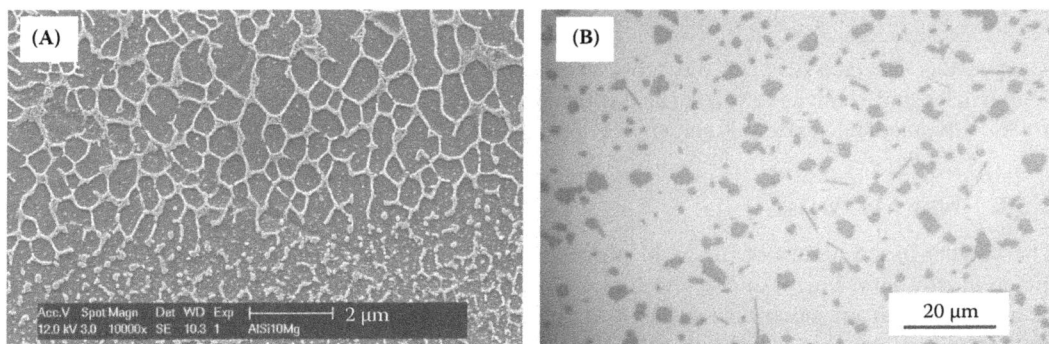

FIGURE 3.36 Microstructure of AlSi10Mg after SLM: (A) as built, and (B) after solution heat treatment at 540°C for 8 hours.[74]

As opposed to the beneficial post-heat treatment for Ti6Al4V, the post-heat treatment on SLM-produced AlSi10Mg should be avoided. In fact, this product has a hardness of 135 to 150 HV, which is equal to high-pressure die-cast and aged material (HPDC-T6, 130 to 135 HV). However, subsequent aging treatment (using various combinations of time/temperature) provides no significant improvement. Applying solution annealing prior to aging (being conventionally common for this material) will even reduce the hardness. This is because the high-temperature solution treatment changes the fine grain structure of the SLM material, as can be seen in Figure 3.36. Subsequent aging fails to elevate the hardness back to the hardness prior to the heat treatment—that is, to the as-built condition.[74]

3.5.5.2 Surface Cleaning of Biomedical Scaffolds

In addition to the microstructure, residual stress, and mechanical performance, the surface finish is another important concern in SLM that necessitates post-processing. This is particularly important for biomedical scaffolds (with open porosity), where numerous powder grains are heterogeneously attached to the strut surfaces (Figure 3.37A). These particles can be released and harm the living body. Therefore, the surface of biomedical scaffolds should be cleaned of the powder grains prior to the application. Surface cleaning of the biomedical scaffolds is a complicated process that requires uniform and successful cleaning throughout the open porous structure. This can be done by chemical or electrochemical treatment, which works by penetration of acid-based solutions into the porous structures through the interconnected pores. During these treatments, chemical reactions

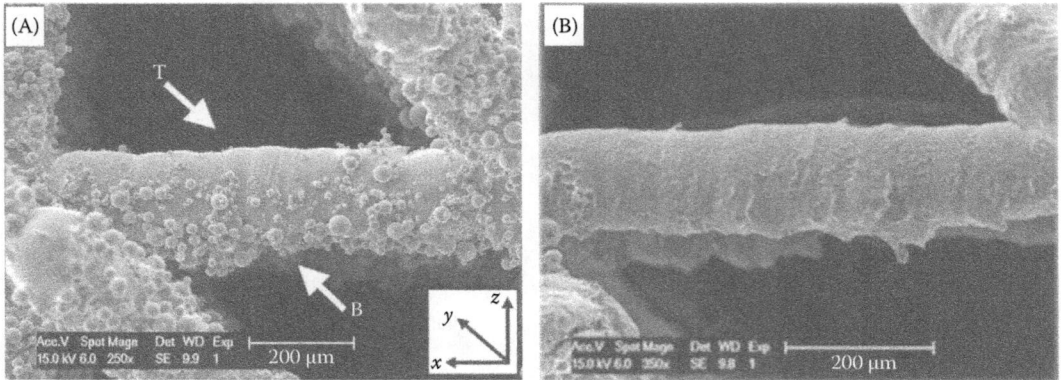

FIGURE 3.37 A typical Ti6Al4V strut (A) after SLM where non-melted powder grains are attached to the top and bottom of the strut, and (B) after chemical etching (with a solution of 0.5 mL HF + 50 g H_2O), and the powder grains have been successfully removed.[75]

occur at the interface between the substrate and the chemical solution, leading to separation of the particles and changes in the surface roughness. A successful chemical etching treatment of a Ti6Al4V biomedical scaffold is shown in Figure 3.37B.[51,75,76]

3.5.6 NEW MATERIALS

The powder-based nature of the SLM process allows easy and quick processing of mixtures of different powders. Furthermore, the high cooling rates may allow the successful combination of materials that are otherwise difficult to achieve. Finally, due to the limited size of the melt pool, any particles that remain unmelted during the process will be distributed homogeneously throughout the part, something that is nearly impossible when casting complex shapes. Figure 3.38 illustrates the significant influence of additional alloying elements on the mechanical properties, comparing the stress–strain curves of Ti6Al4V-ELI combined with 10 wt% Mo to only Ti6Al4V-ELI. The elongation is more than doubled, to 20%, while the material still has a yield stress of 850 MPa after the addition of Mo. Furthermore, due to severe solute distribution during

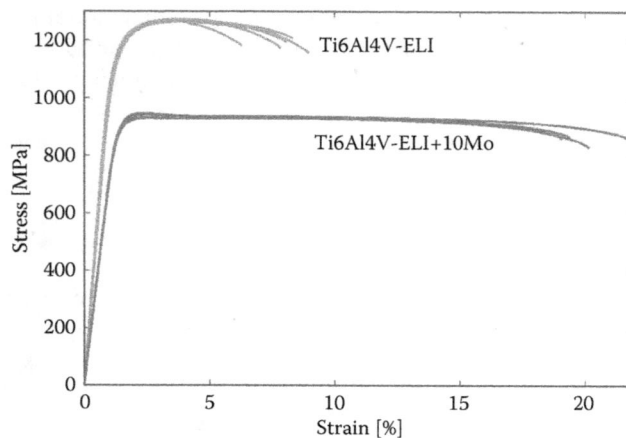

FIGURE 3.38 Comparison of the mechanical properties of Ti6Al4V-ELI and Ti6Al4V-ELI mixed with 10 wt% Mo.[77]

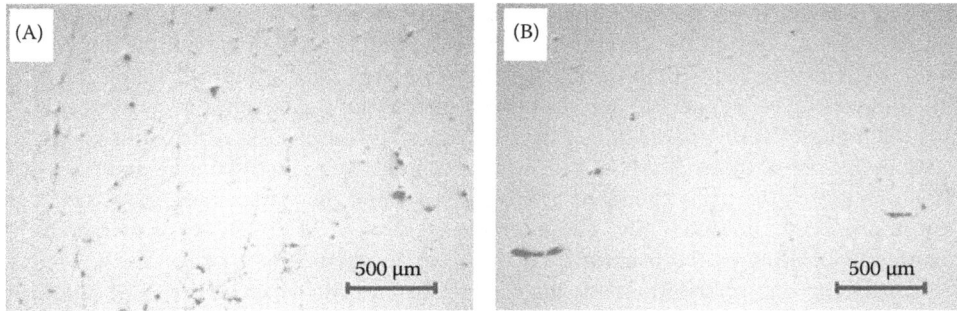

FIGURE 3.39 Side views of (A) Al7075 and (B) Al7075 + 4 wt% Si produced via SLM. The cracks are eliminated by the addition of Si.

solidification, cellular solidification (the second solidification category) appears instead of the planar solidification front, eliminating the large columnar grains (see SLM solidification of low-alloy metals, first mode, as described in Section 3.5.3.1). The manipulated microstructure consists of a β phase instead of α′ martensite (Figure 3.34A), as Mo is a β stabilizing element. This is also evident from the slope of the elastic region in Figure 3.38 (α′ has a higher elastic modulus than the β phase).[77]

The addition of elemental particles to the main powder for SLM is pursued for different purposes. First, it can be used to manipulate current alloys or to develop new alloys. This allows improving the properties or achieving advanced products that are almost impossible to produce using other production techniques. This also includes the production of metal–matrix composites in which ceramic particles are homogeneously dispersed in a metal matrix.[78,79] Second, additional alloying elements may facilitate the production of materials with low processability. The poor processability of Al7075 is a good example that can be improved by additional alloying elements. In fact, this alloy is very sensitive to cracking after SLM (Figure 3.39), but the addition of only 4 wt% of Si (a common alloying element in cast Al alloys) eliminates these cracks and enables the SLM production of a modified version of Al7075. Perhaps the addition of Si changes the thermal properties, lowers the melt pool viscosity, and reduces the dendritic sizes (and hence the compositional segregations), which in turn improves the processability.

3.6 SUMMARY AND CONCLUSIONS

This chapter has given a basic overview of the SLM process, which is a versatile powder-based additive manufacturing method to produce complicated metallic components. A typical SLM machine is composed of various units such as a laser/scanner system, controller, and build chamber. This layout can be upgraded with additional units (e.g., dedicated monitoring systems, heating stage) to enhance the SLM efficiency, as has been shown for an enhanced machine built by KU Leuven. The suitable sets of SLM parameters (e.g., laser power, scanning speed, scan line spacing, layer thickness, scanning strategy) need to be individually developed for each new alloy. This is essential to manufacture high-quality components by avoiding SLM-related weaknesses such as porosity, balling, residual stress, or cracks. In addition to these issues, extra attention must be paid to other factors such as the geometry of the part (e.g., down-facing surfaces, small features, deviation of the scaffolds from the original design) and the desired surface quality. To improve the part quality (e.g., density, roughness) or to achieve higher precision (in particular, to manufacture micro features), complementary SLM techniques such as laser remelting and laser erosion can be implemented. Developing new materials is another growing field within the SLM technology enabling the manufacture of superior components.

The SLM parameters are normally first optimized to reach the highest density. This is achieved with an optimum laser energy density for which full melting is reached but balling, excessive thermal gradients, enhanced melt pool turbulence and instability, gas entrapments, keyholes, etc. are avoided. Single track experiments can be a shortcut to determine such optimal laser energies. After reaching the highest density, performing materials characterization is essential. This is particularly important due to the extreme SLM solidification conditions (e.g., rapid heating and rapid solidification) which makes the SLM solidified microstructures and their associated properties different from those produced conventionally. The microstructure of SLM products is generally very fine and textured (depending on the material). The resulting mechanical properties can be comparable or even superior to conventionally made materials (when a suitable set of the SLM parameters is used). SLM-produced parts may require post-processing prior to their implementation into the final application. The post-processing can be pursued to manipulate the microstructure, residual stresses, and mechanical performance or to improve the surface finish. The post-treatment is particularly important for specific applications such as biomedical scaffolds. However, it should be emphasized that post-process heat treatments applied to conventionally made products may not be suitable for SLM components, which necessitates performing individual studies for each individual material.

ACKNOWLEDGMENTS

The authors acknowledge the support of KU Leuven through the European projects Marie Curie ITN BioTiNet (Grant No. 264635), ZeDAM, and HI-MICRO (Grant No. 314055), as well as the national Belgian projects GOA/2002/06, GOA/2010/12, IWT-SBO DiRaMaP, IWT-SBO eSHM-AM, and SIM-IWT-ICON EXPAMET. The authors also thank all current and past researchers at KU Leuven whose input enabled the provision of this article: S. Clijsters, L. Thijs, M. Speirs, S. Buls, R. Mertens, E. Yasa, P. Mercelis, T. Craeghs, M. Rombouts, and J. Van Vaerenbergh.

REFERENCES

1. Abe, F., Osakada, K., Shiomi, M., Uematsu, K., and Matsumoto, M. (2001). The manufacturing of hard tools from metallic powders by selective laser melting. *J. Mater. Process. Technol.*, 111(1–3): 210–213.
2. Kruth, J.-P., Levy, G., Klocke, F., and Childs, T.H.C. (2007). Consolidation phenomena in laser and powder-bed based layered manufacturing. *CIRP Ann. Manuf. Technol.*, 56(2): 730–759.
3. Levy, G.N., Schindel, R., and Kruth, J.-P. (2003). Rapid manufacturing and rapid tooling with layer manufacturing (LM) technologies, state of the art and future perspectives. *CIRP Ann. Manuf. Technol.*, 52(2): 589–609.
4. Li, R., Shi, Y., Liu, J., Yao, H., and Zhang, W. (2009). Effects of processing parameters on the temperature field of selective laser melting metal powder. *Powder Metall. Met. Ceram.*, 48(3–4): 186–195.
5. Osakada, K. and Shiomi, M. (2006). Flexible manufacturing of metallic products by selective laser melting of powder. *Int. J. Mach. Tools Manuf.*, 46(11): 1188–1193.
6. Meiners, W., Wissenbach, K., and Gasser, A. (2001). Selective Laser Sintering at Melting Temperature, U.S. Patent 6215093.
7. Santos, E.C., Shiomi, M., Osakada, K., and Laoui, T. (2006). Rapid manufacturing of metal components by laser forming. *Int. J. Mach. Tools Manuf.*, 46(12–13): 1459–1468.
8. Vandenbroucke, B. and Kruth, J.-P. (2007). Selective laser melting of biocompatible metals for rapid manufacturing of medical parts. *Rapid Prototyping J.*, 13(4): 196–203.
9. Petrovic, V., Vicente Haro Gonzalez, J., Jordá Ferrando, O., Delgado Gordillo, J., Ramón Blasco Puchades, J., and Portolés Griñan, L. (2010). Additive layered manufacturing: sectors of industrial application shown through case studies. *Int. J. Prod. Res.*, 49(4): 1061–1079.
10. Steen, W.M. and Mazumder, J. (2010). *Laser Material Processing*, 4th ed. London: Springer.
11. Silfvast, W.T. (1996). *Laser Fundamentals*. Cambridge: Cambridge University Press.
12. Kruth, J.-P., Mercelis, P., Van Vaerenbergh, J., and Craeghs, T. (2007). Feedback Control of Selective Laser Melting, paper presented at 15th International Symposium on Electromachining, Pittsburgh, PA, April 23–27.

13. Kempen, K., Thijs, L., Vrancken, B., Buls, S., Van Humbeeck, J., and Kruth, J.-P. (2013). Lowering Thermal Gradients in Selective Laser Melting by Pre-Heating the Baseplate, paper presented at 24th Solid Freeform Fabrication Symposium, Austin, TX, August 12–13.

14. Bechmann, F., Berumen, S., Craeghs, T., Herzog, F., and Kruth, J.-P. (2012). Method for Producing a Three-Dimensional Component, Google Patent US20130168902 A1.

15. Craeghs, T., Bechmann, F., Berumen, S., and Kruth, J.-P. (2010). Feedback control of layerwise laser melting using optical sensors. *Phys. Proc.*, 5(B): 505–514.

16. Craeghs, T., Clijsters, S., Yasa, E., Bechmann, F., Berumen, S., and Kruth, J.-P. (2011. Determination of geometrical factors in layerwise laser melting using optical process monitoring. *Optics Lasers Eng.*, 49(12): 1440–1446.

17. Clijsters, S., Craeghs, T., Buls, S., Kempen, K., and Kruth, J.-P. (2014). *In situ* quality control of the selective laser melting process using a high-speed, real-time melt pool monitoring system. *Int. J. Adv. Manuf. Technol.*, 75(5–8): 1089–1101.

18. Childs, T.H.C., Hauser, C., and Badrossamay, M. (2005). Selective laser sintering (melting) of stainless and tool steel powders: experiments and modelling. *Proc. I. Mech. E. Part B J. Eng. Manuf.*, 219(4): 339–358.

19. Das, S. (2003). Physical aspects of process control in selective laser sintering of metals. *Adv. Eng. Mater.*, 5(10): 701–711.

20. Rombouts, M. (2006). Selective Laser Sintering/Melting of Iron-Based Powders, doctoral thesis, KU Leuven, Leuven.

21. Ion, J.C. (2005). Surface melting. In: *Laser Processing of Engineering Materials* (Ion, J.C., Ed.), Chap. 11. Butterworth-Heinemann, Oxford.

22. Mohanty, P.S. and Mazumder, J. (1998). Solidification behavior and microstructural evolution during laser beam–material interaction. *Metall. Mater. Trans. B*, 29(6): 1269–1279.

23. Birol, Y. (1996). Microstructural characterization of a rapidly solidified Al-12 wt% Si alloy. *J. Mater. Sci.*, 31(8): 2139–2143.

24. Lavernia, E. and Srivatsan, T.S. (2010). The rapid solidification processing of materials: science, principles, technology, advances, and applications. *J. Mater. Sci.*, 45(2): 287–325.

25. Campbell, J. (2003). *Castings*. Butterworth-Heinemann, Oxford.

26. Vandenbroucke, B. and Kruth, J.-P. (2008). Direct digital manufacturing of complex dental prostheses. In: *Bio-Materials and Prototyping Applications in Medicine* (Bártolo, P.J. and Bidanda, B., Eds.), Chap. 7. Springer, New York.

27. Li, R., Liu, J., Shi, Y., Wang, L., and Jiang, W. (2012). Balling behavior of stainless steel and nickel powder during selective laser melting process. *Int. J. Adv. Manuf. Technol.*, 59(9–12): 1025–1035.

28. Toyserkani, E., Khajepour, A., and Corbin, S. (2005). *Laser Cladding*. CRC Press, Boca Raton, FL.

29. Kruth, J.-P., Froyen, L., Van Vaerenbergh, J., Mercelis, P., Rombouts, M., and Lauwers, B. (2004). Selective laser melting of iron-based powder. *J. Mater. Process. Technol.*, 149(1–3): 616–622.

30. Mercelis, P. and Kruth, J.-P. (2006). Residual stresses in selective laser sintering and selective laser melting. *Rapid Prototyping J.*, 12(5): 254–265.

31. Shiomi, M., Osakada, K., Nakamura, K., Yamashita, T., and Abe, F. (2004). Residual stress within metallic model made by selective laser melting process. *CIRP Ann. Manuf. Technol.*, 53(1): 195–198.

32. Kruth, J.-P., Van der Schueren, B., Bonse, J.E., and Morren, B. (1996). Basic powder metallurgical aspects in selective metal powder sintering. *CIRP Ann. Manuf. Technol.*, 45(1): 183–186.

33. Simchi, A. (2006). Direct laser sintering of metal powders: mechanism, kinetics and microstructural features. *Mater. Sci. Eng. A*, 428(1–2): 148–158.

34. Zhu, H.H., Fuh, J.Y.H., and Lu, L. (2007). The influence of powder apparent density on the density in direct laser-sintered metallic parts. *Int. J. Mach. Tools Manuf.*, 47(2): 294–298.

35. Rajabi, M., Simchi, A., Vahidi, M., and Davami, P. (2008). Effect of particle size on the microstructure of rapidly solidified Al-20Si-5Fe-2X (X = Cu, Ni, Cr) powder. *J. Alloys Compd.*, 466(1–2): 111–118.

36. Gu, D. and Shen, Y. (2006). Processing and microstructure of submicron WC-Co particulate reinforced Cu matrix composites prepared by direct laser sintering. *Mater. Sci. Eng. A*, 435–436: 54–61.

37. Gu, D. and Shen, Y. (2009). Effects of processing parameters on consolidation and microstructure of W–Cu components by DMLS. *J. Alloys Compd.*, 473(1–2): 107–115.

38. Olakanmi, E.O. (2013). Selective laser sintering/melting(SLS/SLM) of pure Al, Al–Mg, and Al–Si powders: effect of processing conditions and powder properties. *J. Mater. Process. Technol.*, 213(8): 1387–1405.

39. Kruth, J.-P., Badrossamay, M., Yasa, E., Deckers, J., Thijs, L., and Van Humbeeck, J. (2010). Part and Material Properties in Selective Laser Melting of Metals, paper presented at 16th International Symposium on Electromachining (ISEM XVI), Shanghai, China, April 19–23.

40. Thijs, L., Vrancken, B., Kruth, J.-P., and Van Humbeeck, J. (2013). The Influence of Process Parameters and Scanning Strategy on the Texture in Ti6Al4V Parts Produced by Selective Laser Melting, paper presented at Materials Science and Technology (MS&T) 2013, Montreal, Canada, October 27–31.

41. Van Elsen, M., Corbeels, P., vandezande, H., and Kruth, J.-P. (2006). Production of Overhanging Structures in Stainless Steel with Selective Laser Melting, paper presented at 7th Annual International Conference on Rapid Prototyping Technologies for Competitive Tooling, Cape Town, South Africa, November 1–3.

42. Hussein, A., Hao, L., Yan, C., Everson, R., and Young, P. (2013. Advanced lattice support structures for metal additive manufacturing. *J. Mater. Process. Technol.*, 213(7): 1019–1026.

43. Strano, G., Hao, L., Everson, R.M., and Evans, K.E. (2013). A new approach to the design and optimisation of support structures in additive manufacturing. *Int. J. Adv. Manuf. Technol.*, 66(9–12): 1247–1254.

44. Mertens, R., Kempen, K., Clijsters, S., and Kruth, J.-P. (2014). Production of AlSi10Mg Parts with Downfacing Areas by Selective Laser Melting, paper presented at International Conference on Polymers and Moulds Innovations (PMI2014), Guimarães, Portugal, September 10–12.

45. Clijsters, S., Craeghs, T., and Kruth, J.-P. (2011). *A Priori* Process Parameter Adjustment for SLM Process Optimization, paper presented at 5th International Conference on Advanced Research in Virtual and Rapid Prototyping, Leiria, Portugal, September 28–October 1.

46. Mertens, R., Clijsters, S., Kempen, K., and Kruth, J.-P. (2014). Optimization of scan strategies in selective laser melting of aluminum parts with downfacing areas. *J. Manuf. Sci. Eng.*, 136(6): 061012–061018.

47. Van Bael, S., Chai, Y.C., Truscello, S., Moesen, M., Kerckhofs, G., Van Oosterwyck, H., Kruth, J.-P., and Schrooten, J. (2012). The effect of pore geometry on the *in vitro* biological behavior of human periosteum-derived cells seeded on selective laser-melted Ti6Al4V bone scaffolds. *Acta Biomater.*, 8(7): 2824–2834.

48. Speirs, M., Van Humbeeck, J., Schrooten, J., Luyten, J., and Kruth, J.-P. (2013). The effect of pore geometry on the mechanical properties of selective laser melted Ti-13Nb-13Zr scaffolds. *Proc. CIRP*, 5: 79–82.

49. Speirs, M., Dadbakhsh, S., Buls, S., Kruth, J.-P., Van Humbeeck, J., Schrooten, J., and Luyten, J. (2013). The Effect of SLM Parameters on Geometrical Characteristics of Open Porous NiTi Scaffolds, paper presented at 6th International Conference on Advanced Research in Virtual and Rapid Prototyping, Leiria, Portugal, October 1–5.

50. Kruth, J.-P., Vandenbroucke, B., Van Vaerenbergh, J., and Mercelis, P. (2005). Benchmarking of Different SLS/SLM Processes as Rapid Manufacturing Techniques, paper presented at International Conference on Polymers and Moulds Innovations (PMI2005), Gent, Belgium, April 20–23.

51. Liu, X., Chu, P.K., and Ding, C. (2004). Surface modification of titanium, titanium alloys, and related materials for biomedical applications. *Mater. Sci. Eng. R Rep.*, 47(3–4): 49–121.

52. Yasa, E., Kruth, J.-P., and Deckers, J. (2011). Manufacturing by combining selective laser melting and selective laser erosion/laser re-melting. *CIRP Ann. Manuf. Technol.*, 60(1): 263–266.

53. Dadbakhsh, S., Hao, L., and Kong, C.Y. (2010). Surface finish improvement of LMD samples using laser polishing. *Virtual Phys. Prototyping*, 5(4): 215–221.

54. Kruth, J.-P., Yasa, E., and Deckers, J. (2008). Roughness Improvement in Selective Laser Melting, paper presented at International Conference on Polymers and Moulds Innovations (PMI2008), Gent, Belgium, September 17–19.

55. Kempen, K., Thijs, L., Yasa, E., Badrossamay, M., Verheecke, W., and Kruth, J.-P. (2011). Process Optimization and Microstructural Analysis for Selective Laser Melting of AlSi10Mg, paper presented at International Solid Freeform Fabrication Symposium, Austin, TX, August 6–8.

56. Yadroitsev, I., Gusarov, A., Yadroitsava, I., and Smurov, I. (2010). Single track formation in selective laser melting of metal powders. *J. Mater. Process. Technol.*, 210(12): 1624–1631.

57. Kempen, K., Thijs, L., Van Humbeeck, J., and Kruth, J.-P. (2014). Processing AlSi10Mg by selective laser melting: parameter optimization and material characterization. *Mater. Sci. Technol.*, 31(8): 917–923.

58. Ion, J.C. (2005). Keyhole welding. In: *Laser Processing of Engineering Materials* (Ion, J.C., Ed.), Chap. 16. Butterworth-Heinemann, Oxford.

59. Kempen, K., Yasa, E., Thijs, L., Kruth, J.-P., and Van Humbeeck, J. (2011). Microstructure and mechanical properties of selective laser melted 18Ni-300 steel. *Phys. Proc.*, 12(A): 255–263.

60. Dadbakhsh, S., Speirs, M., Kruth, J.-P., Schrooten, J., Luyten, J., and Van Humbeeck, J. (2014). Effect of SLM parameters on transformation temperatures of shape memory nickel titanium parts. *Adv. Eng. Mater.*, 16(9): 1140–1146.

61. Yasa, E., Deckers, J., and Kruth, J.-P. (2011). The investigation of the influence of laser re-melting on density, surface quality and microstructure of selective laser melting parts. *Rapid Prototyping J.*, 17(5): 312–327.
62. Messler, R.W. (2008). *Principles of Welding: Processes, Physics, Chemistry, and Metallurgy.* Wiley-VCH, New York.
63. Thijs, L., Montero Sistiaga, M.L., Wauthle, R., Xie, Q., Kruth, J.-P., and Van Humbeeck, J. (2013). Strong morphological and crystallographic texture and resulting yield strength anisotropy in selective laser melted tantalum. *Acta Mater.*, 61(12): 4657–4668.
64. Thijs, L., Verhaeghe, F., Craeghs, T., Van Humbeeck, J., and Kruth, J.-P. (2010). A study of the micro structural evolution during selective laser melting of Ti-6Al-4V. *Acta Mater.*, 58(9): 3303–3312.
65. Vrancken, B., Wauthlé, R., Kruth, J.-P., and Van Humbeeck, J. (2013). Study of the Influence of Material Properties on Residual Stress in Selective Laser Melting, paper presented at Solid Freeform Fabrication Symposium, Austin, TX, August 12–14.
66. Thijs, L., Kempen, K., Kruth, J.-P., and Van Humbeeck, J. (2013). Fine-structured aluminium products with controllable texture by selective laser melting of pre-alloyed AlSi10Mg powder. *Acta Mater.*, 61(5): 1809–1819.
67. Bassani, P., Previtali, B., Tuissi, A., Vedani, B.M., Vimercati, G., and Arnaboldi, S. (2005). Solidification behaviour and microstructure of A360–SIC P cast composites. *Metall. Sci. Technol.*, 23(2): 3–10.
68. Vrancken, B., Thijs, L., Kruth, J.-P., and Van Humbeeck, J. (2012). Heat treatment of Ti6Al4V produced by selective laser melting: microstructure and mechanical properties. *J. Alloys Compd.*, 541: 177–185.
69. ASTM. (2014). *Standard Specification for Additive Manufacturing Titanium-6 Aluminum-4 Vanadium ELI (Extra Low Insterstitial) with Powder Bed Fusion*, ASTM F2924. ASTM International, West Conshohocken, PA.
70. Kempen, K., Thijs, L., Van Humbeeck, J., and Kruth, J.-P. (2012. Mechanical properties of AlSi10Mg produced by selective laser melting. *Phys. Proc.*, 39: 439–446.
71. Klingbeil, N.W., Beuth, J., Chin, R., and Amon, C. (1998). Measurement and Modeling of Residual Stress-Induced Warpage in Direct Metal Deposition Processes, paper presented at Solid Freeform Fabrication Symposium, Austin, TX, August 10–12.
72. Kruth, J.-P., Deckers, J., Yasa, E., and Wauthle, R. (2012). Assessing and comparing influencing factors of residual stresses in selective laser melting using a novel analysis method. *Proc. Inst. Mech. Eng. B J. Eng. Manuf.*, 226(B6): 980–991.
73. Vrancken, B., Cain, V., Knutsen, R., and Van Humbeeck, J. (2014). Residual stress via the contour method in compact tension specimens produced via selective laser melting. *Scripta Mater.*, 87: 29–32.
74. Vrancken, B., Kempen, K., Thijs, L., Kruth, J.-P., and Van Humbeeck, J. (2014). Adapted Heat Treatment of Selective Laser Melted Materials, paper presented at Euro PM2014 International Congress & Exhibition, Salzburg, Austria, September 21–24.
75. Pyka, G., Burakowski, A., Kerckhofs, G., Moesen, M., Van Bael, S., Schrooten, J., and Wevers, M. (2012). Surface modification of Ti6Al4V open porous structures produced by additive manufacturing. *Adv. Eng. Mater.*, 14(6): 363–370.
76. Pyka, G., Kerckhofs, G., Papantoniou, I., Speirs, M., Schrooten, J., and Wevers, M. (2013). Surface roughness and morphology customization of additive manufactured open porous Ti6Al4V structures. *Materials*, 6(10): 4737–4757.
77. Vrancken, B., Thijs, L., Kruth, J.-P., and Van Humbeeck, J. (2014). Microstructure and mechanical properties of a novel β titanium metallic composite by selective laser melting. *Acta Mater.*, 68: 150–158.
78. Dadbakhsh, S. and Hao, L. (2012). *In situ* formation of particle reinforced Al matrix composite by selective laser melting of Al/Fe$_2$O$_3$ powder mixture. *Adv. Eng. Mater.*, 14(1–2): 45–48.
79. Dadbakhsh, S., Hao, L., Jerrard, P.G.E., and Zhang, D.Z. (2012). Experimental investigation on selective laser melting behaviour and processing windows of *in situ* reacted Al/Fe$_2$O$_3$ powder mixture. *Powder Technol.*, 231: 112–121.

4 Projection Microstereolithography as a Micro-Additive Manufacturing Technology
Processes, Materials, and Applications

Jae-Won Choi, Yanfeng Lu, and Ryan B. Wicker

CONTENTS

ABSTRACT

Since the advent of microstereolithography (μSL) in 1993, much research on this technology has contributed to overcoming the resolution limitation in macro-additive manufacturing. Microstereolithography has been widely utilized as a micro-additive manufacturing technology to produce complex and delicate three-dimensional (3D) microstructures. Various applications such as tissue engineering, microfluidics, microactuators, metamaterials, and biomedical microdevices have been explored using this technology. Even though this technology evolved from conventional stereolithography (SL), it is no longer a means of prototyping but instead a novel manufacturing process to produce 3D functional microstructures that cannot be achieved in stereolithography. Over the last two decades, a variety of microstereolithography techniques have been developed and many materials have been used. In addition to the use of existing materials, trials to evaluate new materials and various applications have been conducted. We have reviewed more than 200 archival publications, including conference proceedings and journal articles, and have concluded that this technology holds much promise with regard to creating functional 3D microstructures for advanced applications. A wide range of systems, materials, and applications is presented and further advancement of applications is discussed.

4.1 INTRODUCTION

A microstereolithography (usually abbreviated as μSL, MSL, or MSTL) technology is referred to as a process of microvat photopolymerization to produce complex three-dimensional (3D) microstructures with the resolution of a few microns (possibly down to submicron depending on the diffraction limit) and minimum feature size of a few tens of microns depending on the curing characteristics of the material used. The first demonstration was presented by two groups[1,2] almost simultaneously. More than 20 years have passed since development of the first microstereolithography technique, a technology that has grown steadily. It is attractive and promising because of its manufacturing capabilities, including complex 3D, high-aspect-ratio fabrications, as well as multi-material and multi-scale fabrications. In addition to development of microstereolithography systems themselves, many commercial materials and laboratory-based synthesized or blended materials are being utilized by this technology, allowing researchers to explore a broad range of applications. In particular, specific applications involving highly intricate inner architectures, such as tissue engineering with biodegradable/biocompatible materials and periodic microstructures with ceramic-loaded materials, have been engineered.

Advancements made in the areas of electronics, optics, and materials, among many others, have contributed to the rapid development of microstereolithography. Many researchers have been working in this area, and several types of systems have been developed, including scanning and projection microstereolithography and two-photon micro-/nanostereolithography processes that are classified by the method used to solidify a layer of the material. The first microstereolithography approach adopted a scanning method with a focused laser spot or broadband lamp as the energy source.[1,2] Nakamoto and Yamaguchi[3] developed a mask-based process by irradiating a uniform ultraviolet (UV) laser beam through a patterned mask. Bertsch et al.[4] developed the first dynamic mask generator using a liquid crystal display (LCD) device. The LCD device creates a cross-sectional image transferred from a computer which makes a light pattern to cure a photopolymer. This technology is regarded as a faster process due to only a single exposure being required to solidify a layer, compared to the scanning type. Further advancements have been made utilizing a digital micromirror device (DMD) as a dynamic mask generator in a projection type of process.[5,6] The introduction of DMDs improved the resolution and fabrication speed dramatically compared to LCD-based projection processes.

The authors have reviewed more than 200 publications, including journal articles and proceedings papers, to analyze trends in the research on microstereolithography. We assumed that there is a correlation between the number of publications and the number of research activities. It should be noted that there might be overlaps of research topics among some journal articles and proceedings papers. Also, one would expect a time delay between the accomplishment of the actual research

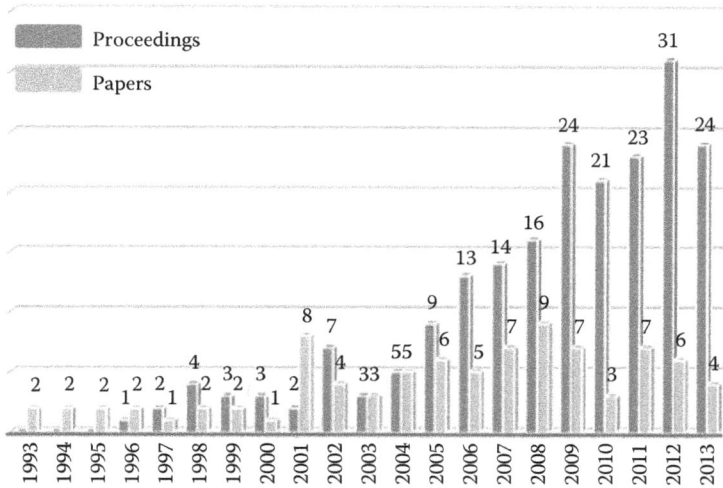

FIGURE 4.1   Number of publications by years.

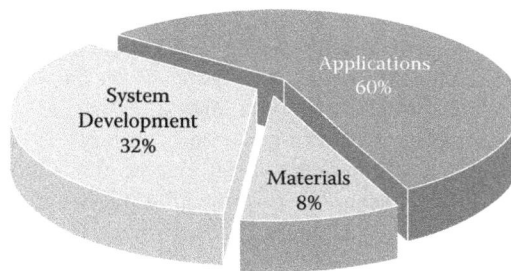

FIGURE 4.2   Distribution of major topics in publications.

work and its publication, which is not reflected in the following figures. The number of publications has gradually increased, most remarkably in the last decade (Figure 4.1). Figure 4.2 shows the distribution of these publications within such major topic areas as system development, material development, and applications. This distribution indicates that more than half of the publications focus mainly on applications, suggesting that this technology offers unique advantages over other manufacturing technologies; however, the distribution in Figure 4.2 does not reflect multiple topics appearing in one publication, so this distribution could change if multiple topics are considered. System development is still attracting considerable research focusing on multi-material fabrication processes,[7,8] improved resolution and accuracy,[9,10] and process optimization.[11,12] In addition to system development, the use and synthesis of new materials have also been intensively studied. The following sections describe the technology, materials, applications, and prospects in more detail.

4.2   WHAT IS MICROSTEREOLITHOGRAPHY?

4.2.1   MICROSTEREOLITHOGRAPHY

Evolved from stereolithography technology, microstereolithography offers higher resolutions and is generally described as stacking layers of a solidified photopolymer in series. The resolutions in microstereolithography are submicron to a few tens of microns, whereas those of stereolithography are usually greater than 50 μm. Due to the higher resolutions, microstereolithography can build very intricate and complex 3D microstructures with virtually any photopolymers.

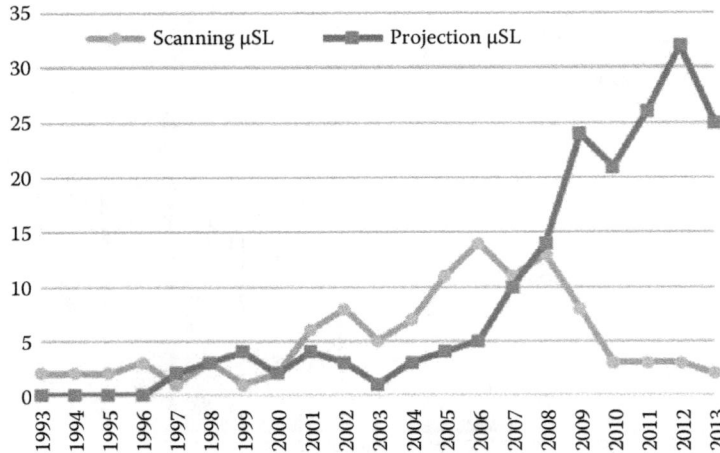

FIGURE 4.3   Research trends of projection and scanning microstereolithography by years.

Since the first microstereolithography system was developed in 1993, a number of research groups have been working on system development.[4,13–22] Generally, microstereolithography can be classified as either a scanning or a projection process, based on the means used to create a layer. In the scanning process, each layer is cured by scanning a focused UV lamp or laser on the photopolymer. In the projection process, a pattern generator is adopted to create an integral image of a layer on the surface of the photopolymer; as a result, one layer is solidified at a time. Figure 4.3 clearly illustrates the trend in using scanning or projection microstereolithography over the past several decades. More research activities regarding the projection type have been accomplished than for the scanning type. The projection type has been applied at a greater rate since 2006, whereas use of the scanning type has gradually decreased. The reasons for these trends can be found in the remarkable fabrication capabilities of projection microstereolithography such as fast production time, higher resolution, and lower setup costs due to the relatively low-cost light source compared to scanning microstereolithography.

The energy source is an important factor in microstereolithography in terms of the fabrication resolution, speed, and cost. Various types of light sources have been employed. The light sources can be classified as one of two types: laser or broadband lamp, as shown in Table 4.1. Various wavelengths of both types have been used. Comparing these two light sources, it appears that broadband lamps, usually used in projection microstereolithography, are more attractive due to their higher efficiency and lower cost. The following section describes one of the most popular versions of projection microstereolithography; discussions regarding the other microstereolithography systems can be found in References 1 to 4, 17, and 23 to 29.

TABLE 4.1
Light Sources Used in the Microstereolithography Process

Light Type	Source
Laser	Helium–cadmium (HeCd),[1,19,29–32] argon,[13,23,25–27,33–45] and neodymium-doped yttrium aluminum garnet (Nd:YAG)[46–49]
Broadband lamp	Xenon[2,50–52] and mercury[6,8,53–71]
Semiconductor	Light-emitting diode (LED) using commercial projectors[72–75]

4.2.2 DMD-BASED PROJECTION MICROSTEREOLITHOGRAPHY

The emergence of DMDs has led to the development of advanced manufacturing technologies using dynamic mask projection in both two-dimensional (2D) and 3D micro/macromanufacturing fields. In particular, 3D micromanufacturing using DMDs is regarded as a promising process to provide functional microstructures. Most DMD-based 3D micromanufacturing systems, which are referred to as *dynamic mask projection microstereolithography*, employ liquid photopolymers to produce fully 3D microstructures in a layer-by-layer fashion. This system has x- and y-resolutions of a few microns and a z-resolution of a few tens of microns, which is limited by cure depth of the chosen material.[5-6,8,58,65,76]

A DMD-based projection microstereolithography system is generally composed of a light source, DMD, beam delivery optics, Z stage, and several mechanical parts such as optical plates, blocks, material container, and a platform, as shown in Figure 4.4. The system in the figure has as its light source a mercury lamp (OmniCure™ S2000; Lumen Dynamics Co., Canada) with output of 200 W. The filtered light (bandpass filter with a center wavelength of 365 nm) was delivered through an optical fiber (Lumen Dynamics). The DMD (~786,000 micromirrors, size of each mirror = 13.68 μm × 13.68 μm) generates a binary pattern (black and white image), which is identical to a sliced cross-section of the model to be fabricated. Each micromirror is tilted at ±12° by an electrostatic force, and a group of tilted mirrors forms the desired pattern. The patterned image is delivered through a tube lens with a focal length of 120 mm and a diameter of 40 mm (Acromat doublet lens; Melles Griot, Albuquerque, NM), and a reflecting mirror with a diameter of 50.8 mm (Newport Corp., Irvine, CA). The image is finally focused on the photopolymer by objective lenses (CFI Plan Fluor 4× and 10×; Nikon, Japan). During the lamp exposure, the light pattern from the DMD is focused on the photopolymer surface, and an individual layer is created. The positions of each component were initially determined through the optical design[76] and then finally adjusted by a 2D patterning experiment.

To determine the resolution of the system, a 2D patterning experiment was performed using 1-pixel and 2-pixel grid patterns with an exposure time of 1 sec, which is equivalent to the exposure energy of 26.4 mJ/cm². Figure 4.5 shows the fabricated patterns on a quartz substrate. The resolution of the system was determined to be 1.67 μm/pixel for a reduction ratio of 0.122 (1.67 μm/13.68 μm). The working area (i.e., field of view) was determined to be ~1.71 mm × 1.28 mm, which corresponds to a total DMD pixel count of 1024 × 768. This resolution and working area were

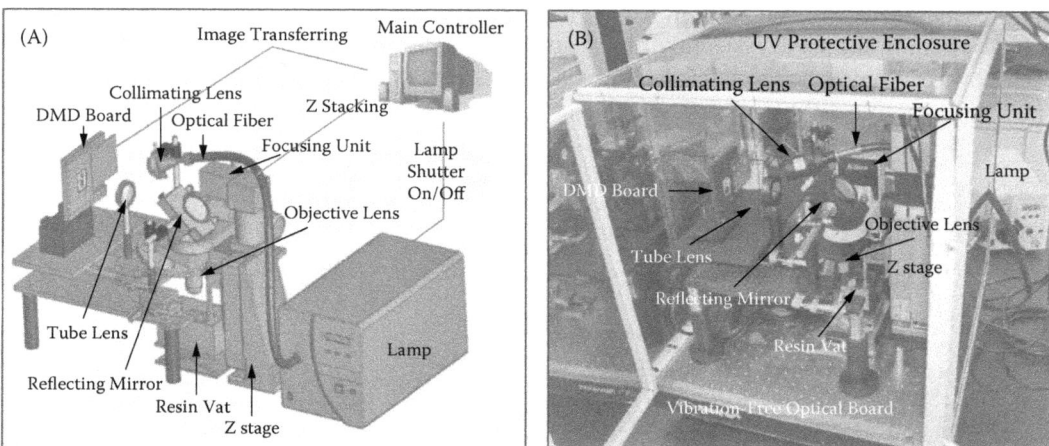

FIGURE 4.4 Developed microstereolithography system: (A) schematic diagram, and (B) photograph.

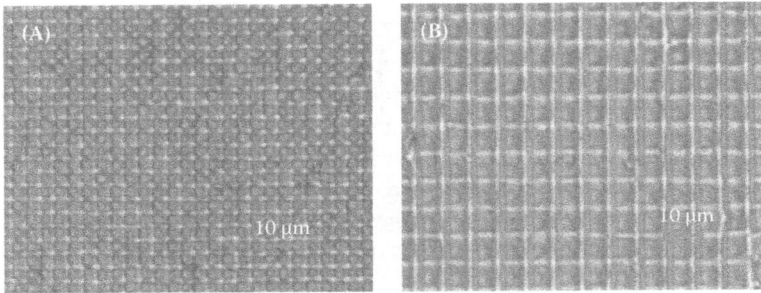

FIGURE 4.5 Fabricated 2D patterns using (A) 1-pixel and (B) 2-pixel grid patterns.

determined using an objective lens with numerical aperture (N.A.) of 0.30, and the current system can employ another objective lens with N.A. of 0.13, which results in a lower resolution of 4.50 μm/pixel (reduction ratio = 0.329) and a broader working area of 4.61 mm × 3.46 mm. In addition to the resolution in 2D, the vertical resolution along the z-axis also needs to be considered. To stack the individual layers, a Z stage (ATS100-050; Aerotech, Pittsburgh, PA) with a resolution of 1 μm was chosen with a custom-made stainless steel platform. The resolution in the z-axis is also dependent on the cure depth, which directly affects micromanufacturing capability. Cleaning and post-curing processes are needed to obtain the final microstructures. A more detailed description of the developed system can be found in References 76 to 78.

4.2.3 FABRICATION EXAMPLES

Using the developed system described above, several 3D microstructures were fabricated. The photopolymers used here were SI 40 provided by 3D Systems (Rock Hills, SC)[77] and WaterShed 11120, ProtoTherm 12120, and Somos® 14120 White provided by DSM Somos® (New Castle, DE).[8] Photoinitiator 2,2-dimethoxy-2-phenylacetophenone (DMPA; Fisher Scientific, Waltham, MA)[58,65] was chosen to alter curing characteristics. Figure 4.6 shows micro-chess piece sets using two different objective lenses with a 0.3 N.A. (Figure 4.6A) and a 0.13 N.A. (Figure 4.6B). These fabrication examples illustrate the different resolutions and volumes fabricated using different objective lenses even in the same system. The reduction ratio can be selectively used for desired models. Figure 4.7 shows scanning electron microscopy (SEM) images of various 3D microstructures fabricated using a N.A. of 0.3 in the same system. All microstructures have overhanging features, and cure depth control was accomplished by adjusting the concentration of the light absorber in the materials. The

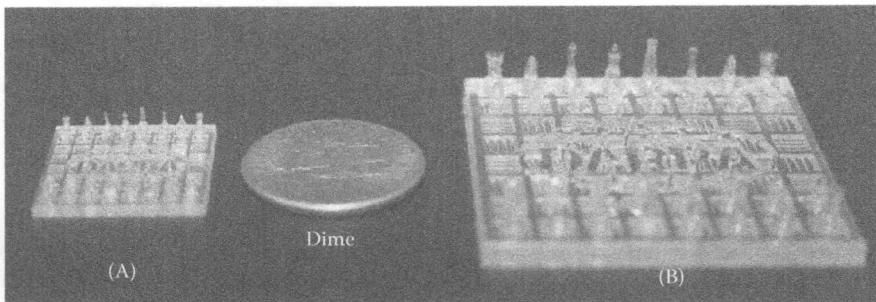

FIGURE 4.6 Fabricated micro-chess piece sets using N.A. of (A) 0.3 and (B) 0.13. The coin is a dime, and the chess boards were manufactured using the Viper si² SLA® system. (From Choi, J.W. et al., *Int. J. Adv. Manuf. Technol.*, 49(5–8), 543–551, 2010.)

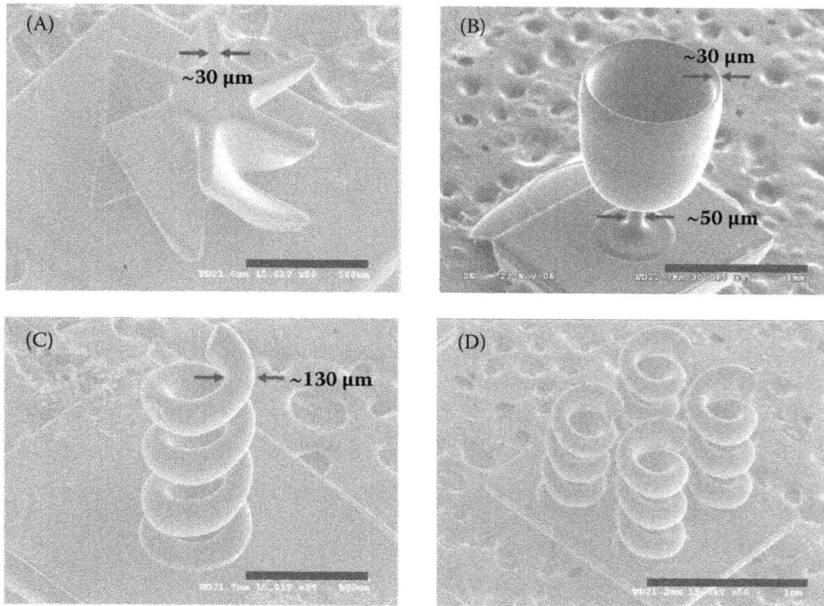

FIGURE 4.7 Fabricated microstructures: (A) microfan (scale bar: 500 µm); (B) microcup (scale bar: 1 mm); (C) microspring (scale bar: 500 µm); and (D) microsprings (scale bar: 1 mm). (Adapted from Choi, J.W. et al., *Rapid Prototyping J.*, 15(1), 59–70, 2009.)

total numbers of layers for the microfan, microcup, microspring, and microspring array are 50, 300, 298, and 298 with layer thicknesses of 10 µm, 4 µm, 4 µm, and 4 µm, respectively. The minimum feature achievable is ~30 µm. These fabrication examples definitively demonstrate the 3D micro-manufacturing capability of the current system and materials.

4.3 ADVANCED MICROSTEREOLITHOGRAPHY

4.3.1 MULTI-MATERIAL MICROSTEREOLITHOGRAPHY

Multi-material manufacturing, especially in a micromanufacturing process, is one of the challenges in additive manufacturing. Choi et al.[8] developed a multi-material microstereolithography system, introducing a syringe pump to dispense different kinds of materials into a removable small vat. In a regular microstereolithography process, after one layer structure is obtained, the vat moves down with the Z stage to make a new layer of the material to cover the built layer. During the multi-material process, however, the syringe pump is used to dispense a prescribed amount of the material, which makes a given layer thickness, into the vat. Once the fabrication of one material is finished, a manual rinsing step is implemented to wash out the residual material completely, followed by the dispensing of the second material, in the same fashion as described above. The material changeover process is repeated until the fabrication is complete. A microstructure of the king chess piece was fabricated in order to verify the dispensing process (Figure 4.8), and multi-material microstructures were then fabricated (Figure 4.9) to demonstrate the entire manufacturing process. Figure 4.9 shows a rook chess piece consisting of three materials used in different layers and a microspring and post consisting of two different materials used in the same layers.[8]

The fabrication examples with two or three kinds of material indicate that multi-material microstereolithography is a feasible approach to fabricating multi-material microstructures. The promising fabrication process can be widely used in various microfabrication fields, such as microtransistors, microsensors, and microactuators. There are some issues, however, in multi-material fabrication

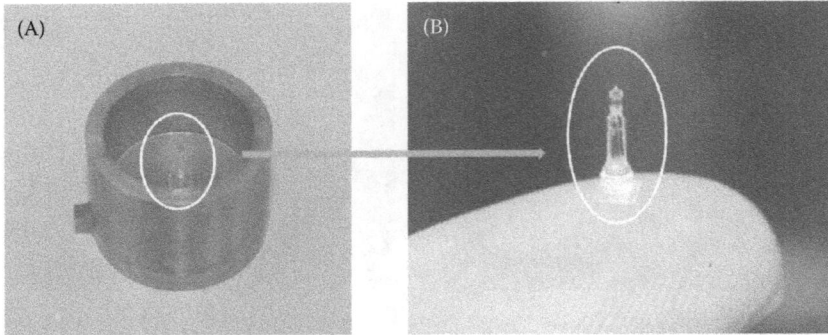

FIGURE 4.8 Micromanufacturing in microstereolithography using a syringe pump: fabricated king chess piece (A) in the vat and (B) on the index finger. (From Choi, J.W. et al., *Int. J. Adv. Manuf. Technol.*, 49(5–8), 543–551, 2010.)

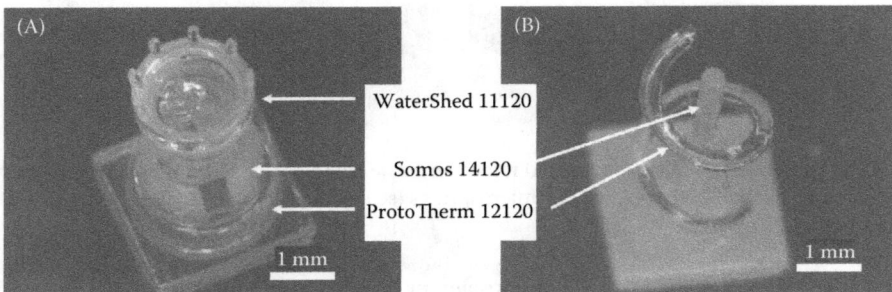

FIGURE 4.9 Multi-material microstructures: (A) rook micro-chess piece, and (B) microspring and post (From Choi, J.W. et al., *Int. J. Adv. Manuf. Technol.*, 49(5–8), 543–551, 2010.)

using the microstereolithography process, such as material contaminants, difficulty in precise control of the material amount to be supplied, and weak bonding between different parts. More investigations are needed to improve the fabrication ability of multi-material microstereolithography.

Han et al.[69,79] also developed another type of microstereolithography system: digital micromirror device projection printing (DMD-PP), which can switch materials in process to fabricate a heterogeneous microstructure. Jo et al.[7] developed a material switching system consisting of a material supply module to provide a photopolymer and control module to level the photopolymer to fabricate a multi-material biomicrostructure. Maruo et al.[29] also developed a multi-material system to fabricate optical waveguides with two different photopolymers having different refractive indexes.

4.3.2 LARGE-AREA MICROSTEREOLITHOGRAPHY

Several groups have investigated the development of new systems or new methods to enable large-area microstereolithography fabrication. Ikuta et al.[15] developed a mass production system using an optical-fiber multibeam scanning method. A single light bundle is divided into several light bundles with fibers. The divided individual light is focused on the photopolymer surface; thus, this process can produce as many structures as possible by increasing the number of optical fibers.

In projection microstereolithography, the working area along the *x-y* plane is confined by the reduction ratio described in Section 4.2.2 and is incompatible with large-area manufacturing. To accomplish large-area micromanufacturing for mass production or production of a macro-part with microfeatures using the projection system, an XY stage was installed under the

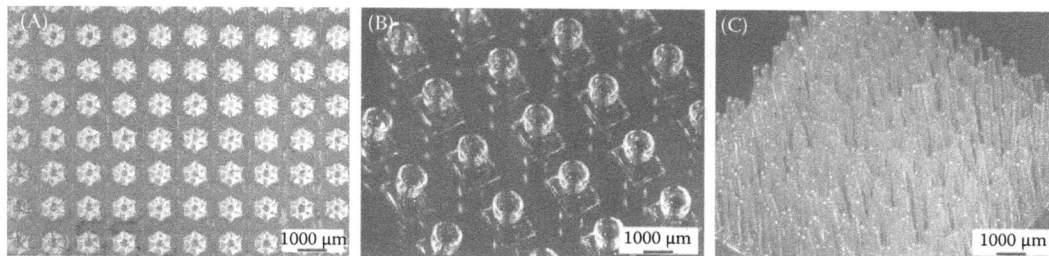

FIGURE 4.10   Array-type microstructures: (A) microfan array, (B) magnified image of a microfan, and (C) microcup.

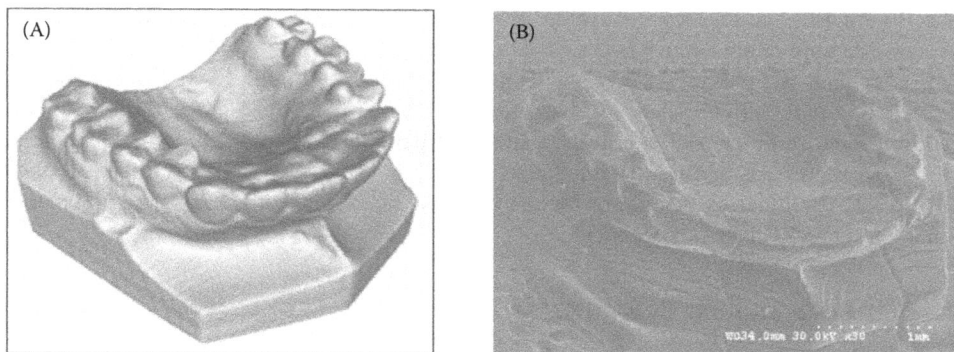

FIGURE 4.11   Fabricated tooth model demonstrating use of large-area micromanufacturing: (A) 3D model, and (B) fabricated teeth.

microstereolithography system.[60,62,77] To demonstrate the capability of large-area micromanufacturing, microstructure arrays were fabricated (Figure 4.10).[77] In Figure 4.10A,B, the microfan and microglass are identical to those in Figure 4.7, and Figure 4.10C shows the micropillar array with 150 layers and a layer thickness of 20 μm. Working areas for the microfan, microglass, and micropillar arrays were 11.5 mm × 11.5 mm, 10 mm × 10 mm, and 9.5 mm × 9.5 mm for 100, 25, and 576 microstructures, respectively. Total fabrication times were 5.5 hr, 8.3 hr, and 16 hr, respectively. The XY stage had a resolution of 1 μm in the x- and y-directions and was mechanically stable against the load of the microstereolithography system. In addition to the array-type microstructures, Figure 4.11 shows a fabricated tooth model created using large-area micromanufacturing; for fabrication, the working area was divided into several regions.[77] In particular, this application can be potentially extended to multi-scale micromanufacturing.

4.4   MATERIALS

4.4.1   Photopolymers

Photocrosslinkable liquid materials called photopolymers are necessary for microstereolithography because a solidification process from liquid to solid has to occur during the fabrication. Principally, microstereolithography systems can use virtually any commercially available photocrosslinkable resins. Materials with a high viscosity, however, are very difficult to use in microstereolithography systems because, generally, no recoater is adopted and specific times are required for the material to be refreshed. To reduce the viscosity, various diluents, depending on the matrix material properties, have been selected. Choi[77] employed SI 40 from 3D Systems (Rock Hills, SC), and WaterShed®

11120, ProtoTherm 12120™, and Somos® 14120 White from Somos®–DSM (New Castle, DE).[8] Because these photopolymers are relatively too viscous to be used in microstereolithography, acrylate- or diacrylate-based monomers, which are photocurable, were added as diluents. In general, a viscosity of less than ~200 cP is recommended for microstereolithography.[76–77] As diluents, several low-viscosity monomers such as propoxylated (2) neopentyl glycol diacrylate (PNGD) provided by Sartomer, Inc. (Warrington, PA);[8,78] 1,6-hexanediol diacrylate (HDDA); and isobornyl acrylate (IBXA)[65,77] have been used. In addition to diluted commercial photopolymers, monomers, oligomers, and polymers can be used by adding a photoinitiator such as 2,2-dimethoxy-2-phenylacetophenone (DMPA; Fisher Scientific, Waltham, MA).[65,77] Table 4.2 provides more information on monomers/oligomers, photoinitiators, nanoparticles for composite resins, and photoabsorber/photoinhibitors to stabilize photocrosslinking.

Two important issues associated with micromanufacturing are overcure and cure depth, as illustrated in Figure 4.12, which shows the microstructures for the same 3D model used in Figure 4.7. It can be seen that cure depth control is key in 3D micromanufacturing. To aid in cure depth control, a light absorber such as Tinuvin® 327[65,77] and Tinuvin® 400 (Ciba, Duluth, GA)[8,78] (see Table 4.2) can be introduced to reduce light penetration depth with successful results. Photocrosslinkable materials are characterized by critical energy (E_c) and light penetration depth (D_p), which can be calculated by linear regression of cured-depth data according to dosed energies. A number of materials, including diluted commercial resins and blended reactive monomers, oligomers, and polymers, have been characterized by curing experiments.[65]

4.4.2 Photocrosslinkable Biomaterials

Microstereolithography has gained increasing attention in the areas of biomedical and tissue engineering due to its unique advantages. Photocrosslinkable biomaterials have been rapidly developed in response to this promising technology. Choi et al. fabricated microneedles[77] and scaffolds[76] using poly(propylene fumarate) (PPF) and nerve guidance conduits for peripheral nerve regeneration using poly(ethylene glycol) (PEG) diacrylate. Diethyl fumarate (DEF) is commonly used as a diluent to decrease the viscosity of PPF. The structure has a higher elastic modulus and fracture strength fabricated using PPF with the appropriate amount of DEF.[130]

Kang et al.[100] utilized aminoethyl methacrylated hyaluronic acid (HAAEMA) and trimethylene carbonate/trimethylolpropane (TMC/TMP) to fabricate 3D scaffolds with Irgacure® 819 as a photoinitiator, and TMC/TMP proved to be a biocompatible and biodegradable polymer. Gauvin et al.[96] synthesized collagen-based gelatin methacrylate (GelMA) hydrogels with Irgacure® 2959 as a photoinitiator and 2-hydroxy-4-methoxy-benzphenone-5-sulfonic acid (HMBS) as a photoabsorber to fabricate 3D scaffolds for cell culture. Applications of the use of biomaterials are discussed in detail in the following section. Liska et al.[131] developed biophotopolymers based on commercially available reactive diluents and modified gelatin to fabricate an artificial cellular bone. Baudis et al.[132] developed a series of acrylate-based photoelastomers for artificial vascular grafts.

4.4.3 Photocrosslinkable Composites

Composite materials mixed with nano- or microscale particles, such as ceramic or magnetic particles with photopolymers, have been adopted in microstereolithography to achieve specific mechanical properties or functions. Zhang et al.[25] developed an aqueous UV-curable ceramic suspension mixed with deionized (DI) water, monomers (acrylamide and methylenebissacrylamide), fine alumina powders, dispersant (DARVAN® C), and photoinitiator (Irgacure® 2959) to obtain non-aqueous and aqueous UV-curable ceramic suspensions. This photopolymer composite was successfully used to fabricate microstructures, such as microgears, microtubes, and micro convex cone structures. Chen et al.[133] dispersed 3Y-ZrO$_2$ (1-μm average particle size; Tosho Co., Tokyo, Japan) and Al$_2$O$_3$ (170-nm average particle size; Taimei Chemical Co., Nagoya, Japan) powders at a 4:1 weight ratio in a

TABLE 4.2

Materials Used in Microstereolithography

Material Type	Material Used	Refs.
Monomers/ oligomers	1,6-Hexanediol diacrylate (HDDA)	Zhang et al.,[25] Jiang et al.,[35] Zissi et al.,[80] Provin and Monneret,[81] Baker et al.,[82] Goswami et al.,[83] Ibrahim et al.,[84] Gandhi and Bhole[85]
	Poly(ethylene glycol) diacrylate	Leigh et al.,[86] Lee and Fang,[87] Lee et al.,[88,89] Bail et al.[129]
	Dipentaerythritol penta-/hexa acrylate	Leigh et al.[86,90]
	Poly(propylene fumarate)/diethyl fumarate (PPF/ DEF)[a]	Lee et al.,[41,118–120,122–124] Lan et al.[121]
	Hyaluronic acid[a]	Kang et al.[100]
	Gelatin methacrylate[a]	Gauvin et al.[96]
Photoinitiator	Benzoin ethyl ether (BEE)	Sun et al.,[6] Choi et al.,[8,67] Goswami et al.,[83] Sun and Zhang,[91] Thomas et al.[92]
	2,2-Dimethoxy-2-phenylacetophenone (DMPA)	Lu and Chen,[75] Zissi et al.,[80] Bertsch et al.,[93] Park et al.,[94] Jariwala et al.[95]
	Irgacure® series (2959,[a] 784, 819[a])	Stampfl et al.,[48] Heller et al.,[49] Lu et al.,[72] Provin and Monneret,[81] Baker et al.,[82] Gauvin et al.,[96] Wu et al.,[97] Monneret et al.,[98,99] Kang et al.,[100] Zhou et al.[101]
	Lucirin® TPO	Ronca et al.[128]
Micro/ nanoparticles	Alumina	Zhang et al.,[25] Provin and Monneret,[81] Bertsch et al.,[93] Monneret et al.,[98] Sun and Zhang,[102] Kirihara,[103] Noritoshi et al.,[104] Kirihara et al.,[105–107] Goswami et al.[108]
	Titania	Kirihara et al.[109–111]
	Zirconate titanate	Jiang et al.,[35] Sun and Zhang[102]
	Zirconium oxide	Kirihara and Tasaki[112]
	$B_2O_3 \cdot Bi_2O_3$	Kirihara and Nakano[113]
	Nickel oxide	Komori et al.[114]
	$Fe_{72}B_{14.4}Si_{9.6}Nb_4$	Kirihara and Nakano[113]
	Ferrite particles FA-700	Yasui et al.[115]
	Hydroxyapatite[a]	Lee et al.,[120] Seol et al.,[125–127] Ronca et al.[128]
	Nano HAP powder	Ronca et al.[128]
Photosenstizer/ absorber	Sudan I	Baker et al.,[82] Lee and Fang,[87] Khairu et al.[116]
	Tinuvin® 324	Sun and Zhang[91]
	Tinuvin® 327	Choi et al.[65]
	Tinuvin® 400	Choi et al.[117]
	2-Hydroxy-4 methoxy-benzphenone-5-sulfonic acid[a]	Gauvin et al.[96]
	Tocopherol	Ronca et al.[128]

[a] These materials have been used for bioapplications.

photocurable acrylate resin (JL2129; D-MEC Co., Tokyo, Japan) at 40 vol% loading in order to fabricate a photonic crystal with a diamond structure for electromagnetic wave control. Seol et al.[127] used a slurry mixture of hydroxyapatite (HAp, $Ca_{10}(PO_4)_6(OH)_2$) and tricalcium phosphate (TCP, $Ca_3(PO_4)_2$) powder mixed with a photocurable resin to create ceramic scaffolds. Ronca et al.[134] also embedded nanosized hydroxyapatite (Nano HAP) particles into synthesized poly(D,L-lactide) oligomers to fabricate porous structures. Yasui et al.[115] mixed ferrite particles with a photocurable resin (D-MEC SCR 770; density, 1.17 g/cm³) to fabricate magnetic nanoactuators. Komori et al.[114] used

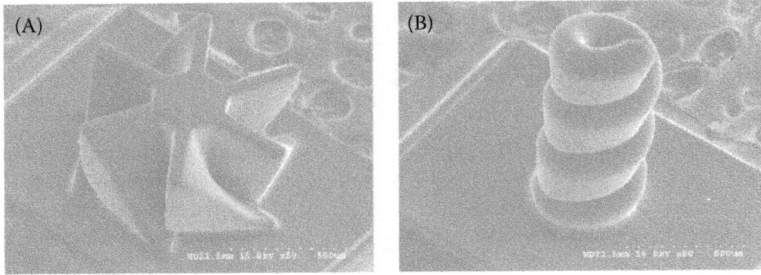

FIGURE 4.12 Microstructures with overcure: (A) microfan, and (B) microspring. (From Choi, J.W. et al., *Rapid Prototyping J.*, 15(1), 59–70, 2009.)

zirconia and nickel oxide mixed with a photosensitive acrylic resin to fabricate dendritic electrodes for solid oxide fuel cells. In this research, a mechanical knife edge was used to spread the slurry paste uniformly at a 5-μm layer thickness. Microstereolithography of many other particles, including alpha-alumina,[81] zirconate titanate,[35] titania,[109] $Fe_{72}B_{14.4}Si_{9.6}Nb_4$,[113] and $B_2O_3 \cdot Bi_2O_3$,[113] has been investigated as a means to introduce the desired material properties to 3D microstructures.

4.5 APPLICATIONS

Microstereolithography has been used in a wide range of applications, which can be classified into several different categories including biomedical engineering, optical engineering, actuators and sensors, fluidics, and many others. Figure 4.13 shows the distribution based on several major fields of applications based on the references. It should be noted that more than 50% of publications have been focused on the biomedical field, due to the fast development of advanced materials. Microstereolithography is also an ideal choice in microelectromechanical systems (MEMS), where parts generally require a small size, complex geometry, and high resolution.

4.5.1 MICROSTEREOLITHOGRAPHY IN BIOMEDICAL ENGINEERING

As shown in Figure 4.13, biomedical engineering is the most popular application of microstereolithography technology because of the superior manufacturing capabilities it offers, such as high-resolution 3D microfabrication with high-speed production. Ikuta et al.[135] carried out the first trial to fabricate biomedical microdevices in 1995 and then a series of investigations followed based on biochemical IC chips.[136–140] A growing number of researchers have undertaken research into other biochemical

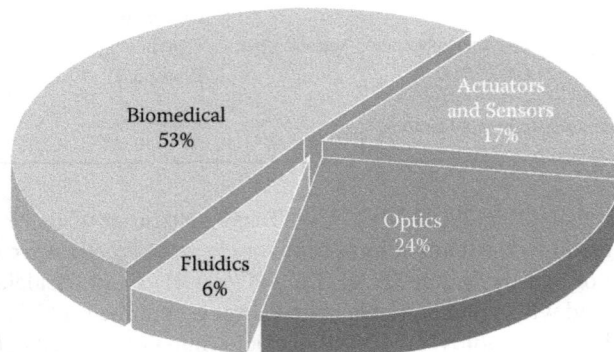

FIGURE 4.13 The distribution of microstereolithography applications.

applications. Kwon and Matsuda[51] fabricated microarchitectural constructs that included a 3D micropillar array and multi-microtunnels using acrylated trimethylene carbonate-based liquid prepolymers. Nagakura et al.[141] constructed an osmotic valve to control insulin injection. Ortega et al.[142] fabricated an artificial limbus with micropockets for delivering cells to the cornea. Other research groups have focused on applications in tissue engineering, another emerging area for 3D printing and microstereolithography. In general, research activities on tissue engineering include 3D scaffold fabrication for bone tissue regeneration,[41,43–45,64,76,120,143–147,173] scaffolds for general cell culture,[63,72,86,96,118,119,121–124,148–164] and cell culture wells or sacrificial moldings for scaffold construction.[165–167] The following sections describe several representative biomedical applications in more detail.

4.5.1.1 Scaffolds

If human tissues are damaged and need to be replaced, the patient's own cells can be cultured and transplanted.[76] This has the advantage of preventing the immune rejection response, which often occurs when transplanting tissues from donors. The tissues cultured in a template using the patient's cells are biologically similar to the originals.[77] For the regeneration of tissues, a mechanical structure known as a scaffold is required. This scaffold has to be porous and have a desired shape. The scaffold plays an important role in sustaining the cultured cells until tissues are regenerated and additionally must undergo degradation; therefore, the scaffold has to be biocompatible and biodegradable. To produce controlled porous scaffolds, microstereolithography has been widely used due to the novel 3D manufacturability.[76] Microstereolithography is potentially capable of producing microscaffolds with photocrosslinkable biomaterials. As mentioned in Section 4.2, poly(propylene fumarate) (PPF), which is biocompatible and biodegradable, diluted with diethyl fumarate (DEF) is one such suitable material in this process. Prior to producing biocompatible/biodegradable microscaffolds, several microscaffolds (Figure 4.14) were produced using diluted commercial resins such as SI 40 and IBXA.[77] Then, microscaffolds (Figure 4.15) were produced using the synthesized PPF/DEF.[76] The fabricated PPF/DEF lattice and kidney microscaffolds have micropores of ~100 μm. The PPF/DEF microscaffolds (Figure 4.15) are comparable to those shown in Figure 4.14, thus the PPF/DEF biomaterial system is promising with regard to producing tissue engineering scaffolds.

4.5.1.2 Microneedles

Microneedles have been regarded as an alternative drug delivery device for those with a low pain threshold or for time-release applications.[77] Microstereolithography could be an effective means to mass produce microneedles. Many researchers have produced microneedles using microelectromechanical systems and lithography, electroforming, and molding technologies. These microneedles, however, have been limited in their geometric complexity due to the nature of the processes. For successful insertion and drug delivery, the geometry of the microneedle should be optimally designed to provide high mechanical strength, low insertion force, and proper drug dosage.

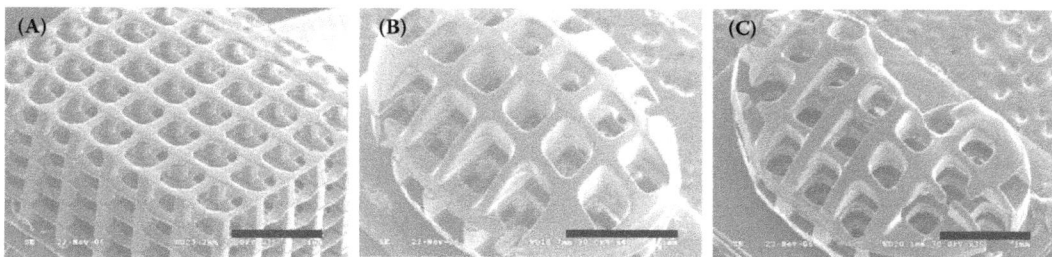

FIGURE 4.14 Fabricated microscaffolds using the diluted SI 40: (A) lattice, (B) oval, and (C) kidney microscaffolds (all scale bars are 1 mm). (From Choi, J.W., Development of Projection-Based Microstereolithography Apparatus Adapted to Large Surface and Microstructure Fabrication for Human Body Application, doctoral dissertation, Pusan National University, South Korea, 2007.)

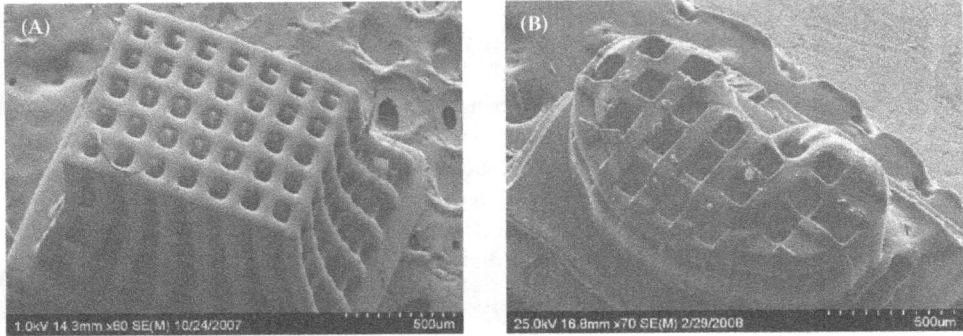

FIGURE 4.15 Fabricated microscaffolds using PPF/DEF: (A) lattice, and (B) kidney microscaffolds. (From Choi, J.W. et al., *J. Mater. Process. Technol.*, 209, 5494–5503, 2009.)

Using diluted SI 40, the microneedle array shown in Figure 4.16A was fabricated. Figure 4.16B shows four fabricated microneedles and a 26-gauge hyperdermic needle for comparison. The diameters of the microneedle's tip and base are about 25 μm and 180 μm, respectively. The height of the microneedle is ~900 μm, and the microneedles are spaced ~900 μm apart. The large-area microstereolithography system was used to fabricate a microneedle array with 400 individual microneedles within a range of 10 mm × 10 mm (Figure 4.16C). The number and distance of the microneedles in the array can be adjusted according to the amount of the drug, mechanical strength, and many other factors.

One important issue associated with microneedle insertion is hyperdermal retention, which would necessitate a biocompatible/biodegradable structure. PPF/DEF has the potential to avoid this issue due to its biocompatibility, biodegradability, and good mechanical strength. In addition, some drugs can be dissolved in the PPF or PPF/DEF solution. Microneedle arrays containing drugs dissolved in the PPF/DEF prepolymer blended with *bis*(2,4,6-trimethylbenzoyl) phenylphosphine oxide (BAPO) as a photoinitiator were fabricated using microstereolithography. PPF with an average molecular weight (M_n) of 1510 Da was synthesized and then blended with DEF at the ratio of 70:30 (w/w).[124] DEF can reduce the viscosity of PPF and enhance the mechanical strength of the fabricated parts. A protein model drug, bovine serum albumin (BSA), was dissolved in dimethyl sulfoxide (DMSO) and autoclaved water to obtain the drug solution, which was fully mixed with the PPF/DEF solution. Two microneedle arrays without (Figure 4.17A) and with (Figure 4.17B) the 2.5 wt% drug loading were fabricated. Each microneedle had a circular base and cone top. The microneedle had an overall height of 1000 μm (base height of 700 μm and cone height of 300 μm) and a tip diameter of 20 μm; the base diameter was 200 μm. Each array contained 25 microneedles. This example of microstereolithography of PPF/DEF microneedles shows the potential capabilities for building drug delivery devices for which the shapes must be spatially controlled at micron resolutions.

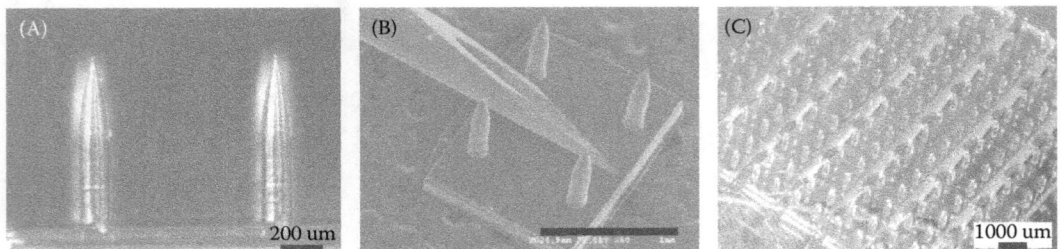

FIGURE 4.16 Fabricated microneedles using diluted SI 40: (A) microscope image; (B) SEM image with a 26-gauge hyperdermic needle (scale bar: 1 mm); and (C) a microneedle array.

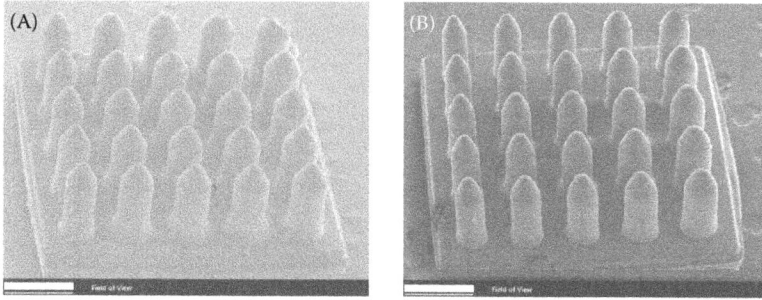

FIGURE 4.17   SEM images of microneedle arrays: (A) 0 wt%, and (B) 2.5 wt% drug loading. (Both scale bars are 500 μm.)

In addition to the research on PPF/DEF microneedles, several investigations have addressed similar topics. Gittard et al.[168] developed a technique for scalable production of microneedle devices by visible light dynamic mask microstereolithography. Boehm et al.[169] produced biodegradable acid anhydride copolymer microneedles containing quantum dots. Miller et al.[170] fabricated a hollow microneedle array with carbon fiber electrodes incorporated to attach cells.

4.5.1.3   Stents

Vascular scaffolds are widely used in the treatment of vascular diseases such as coronary artery disease and peripheral vascular disease, which are the major causes of mortality in the world.[171] A stent, which is inserted into the vessel by a balloon catheter, occasionally with drug eluting, would seem to be a perfect tool to treat these diseases.[172] The microstereolithography process could be a potential approach to fabricating a stent with high resolution and complex features, which are required for the unique functions of a stent in the vessel. HDDA, an acrylate-based monomer, with 2 wt% of DMPA as a photoinitiator and 1 wt% of Tinuvin® 327 as a photoabsorber, were used to demonstrate the capability of the microstereolithography technology for stent fabrication.[77] Figure 4.18 shows the design and the fabricated stent obtained from the microstereolithography system. Some medicines to prevent the artery from being blocked can be mixed with the matrix material instead of coating them on the surface so drugs can be continuously released over a long term.[173,174]

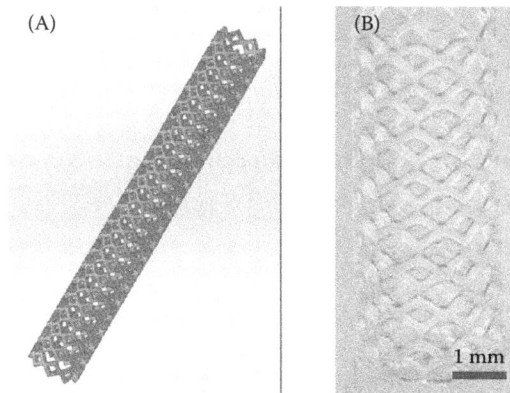

FIGURE 4.18   Stent structure fabricated by microstereolithography: (A) design of the stent, and (B) microscope image of the fabricated stent.

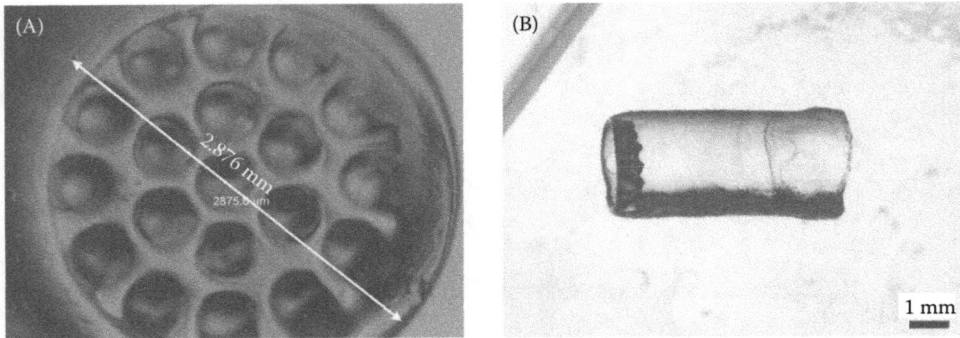

FIGURE 4.19 Fabricated NGCs: (A) top view with multi-lumen, and (B) side view. (From Choi, J.W. et al., *Proceedings of TERMIS 2nd World* Congress, 2009.)

4.5.1.4 Nerve Guidance Conduits

Nerve guidance conduits (NGCs), used for the regeneration of peripheral nerves, require multiple lumens of micron dimensions to increase surface area.[175] Microstereolithography was used to fabricate multi-lumen, poly(ethylene glycol) (PEG)-based NGCs. PEG solutions were prepared by dissolving 20% (w/w) of PEG (MW 3400) in water. To make the solution photoreactive, 50 μL of N-vinyl-2-pyrrolidone (NVP) solution containing 15 mg of 2,2-dimethoxy-2-phenylacetophenone (DMPA) were added in 5 g of the PEG solution. Multi-lumen NGCs with an outer diameter of 3 mm, inner diameter of 2 mm, and height of 8 mm were fabricated. The inner diameter included 19 lumens of 300-μm diameter. Figure 4.19 shows the fabricated NGC. This bioactive NGC can be potentially used in an *in vivo* peripheral nerve regeneration study using a rat sciatic nerve defect model.[175]

4.5.2 Microstereolithography in Optics

Microstereolithography provides a suitable means to fabricate precise optical microdevices. Maruo et al.[29] developed a new type of microstereolithography system using multiple kinds of photocurable polymers. This system was used to fabricate optical waveguides utilizing two different kinds of polymers having different refractive indexes of 1.51 and 1.55. Several groups have performed research on microlens arrays and optical crystals using microstereolithography.

4.5.2.1 Microlens Arrays

A microlens can be used for collimating light in an optoelectronic sensor, an optical communication device, and confocal microscopes.[176] Given the superior capability of 3D micromanufacturing by microstereolithography, a microlens array is an interesting application.[75] Figure 4.20 shows fabricated elliptical microlens with a major axis of ~140 μm and minor axis of ~100 μm. Each microlens consists of 10 layers with a layer thickness of 1 μm. Large-area microstereolithography was used to fabricate a microlens array (Figure 4.20B), for which the microlens count was 18,000 in an area of 20 mm × 20 mm.[176]

4.5.2.2 Photonic Crystals

Photonic crystals were widely used to control microwave waves in a terahertz frequency range. Microstereolithography offers an excellent route to produce these photonic crystals. Kirihara et al.[103,105,107,110,177–188] have been working intensively on using microstereolithography to fabricate 3D photonic crystals applied in terahertz resonators. Photonic crystals with a lattice diamond structure were fabricated using photopolymers with various types of powders such as alumina or ceramic. Chartier et al.[189] developed alumina and zirconia formulations with powder loading of more than 50 vol% to fabricate complex microdevices in millimeter and submillimeter wavelength domains.

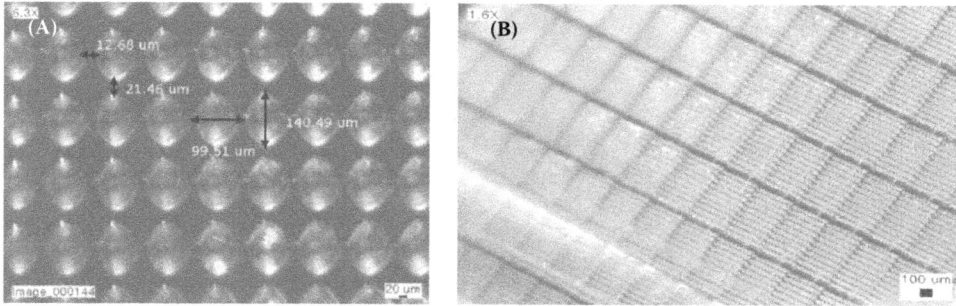

FIGURE 4.20 Fabricated microlens: (A) microlens with dimensions, and (B) microlens array. (From Park, I.B. et al., *J. Korean Soc. Precision Eng.*, 25(2), 123–130, 2008.)

4.5.3 MICROSTEREOLITHOGRAPHY IN SENSORS AND ACTUATORS

The conventional method to fabricate microscale sensors and actuators is MEMS technology, which has numerous limitations, including time-consuming and high-cost processes. Microstereolithography offers superior fabrication capabilities. Reports of applications in sensors and actuators over the past two decades can be found in References 88, 92, and 190 to 207. Kang et al.[208,209] utilized a UV laser scanning-based microstereolithography technology to make a bellows-shaped microactuator with a commercial photopolymer (SL5180). In their research, two different microactuators (e.g., grippers) with different internal structure were fabricated. Baker et al.[82] fabricated a 3D polymeric microspring structure to harvest vibration energy that consisted of a two-by-two array of 3D constant-pitch helical coils and used HDDA and Irgacure® 819 as a photoinitiator. Vatani et al.[215] developed a flexible tactile sensor using a hybrid fabrication process with microstereolithography and direct print. Figure 4.21 shows the fabricated flexible sensor, for which the skin structure was built using a microstereolithography machine retrofitted from a commercial beam projector and sensing elements (black strips) were fabricated using a direct-print process.

4.5.4 MICROSTEREOLITHOGRAPHY IN MICROFLUIDICS

The fabrication of microfluidics is an intriguing application area for microstereolithography, which can meet the requirements of microfluidics for complex internal structures, precise sizes, and certain mechanical properties. The first microstereolithography apparatus was used to build microfluidics such as a micro-integrated fluidic system in 1994.[210] Bertsch et al.[211] fabricated two

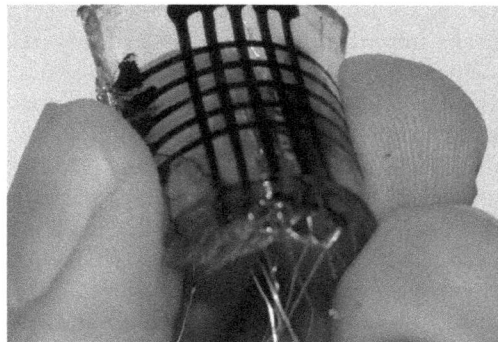

FIGURE 4.21 Flexible tactile sensor fabricated by microstereolithography and direct print. (From Vatani, M. et al., Fabrication and Characterization of 3D Printed Compliant Tactile Sensors, paper presented at Rapid Additive Manufacturing Solutions, Pittsburgh, PA, June 10–13, 2013.)

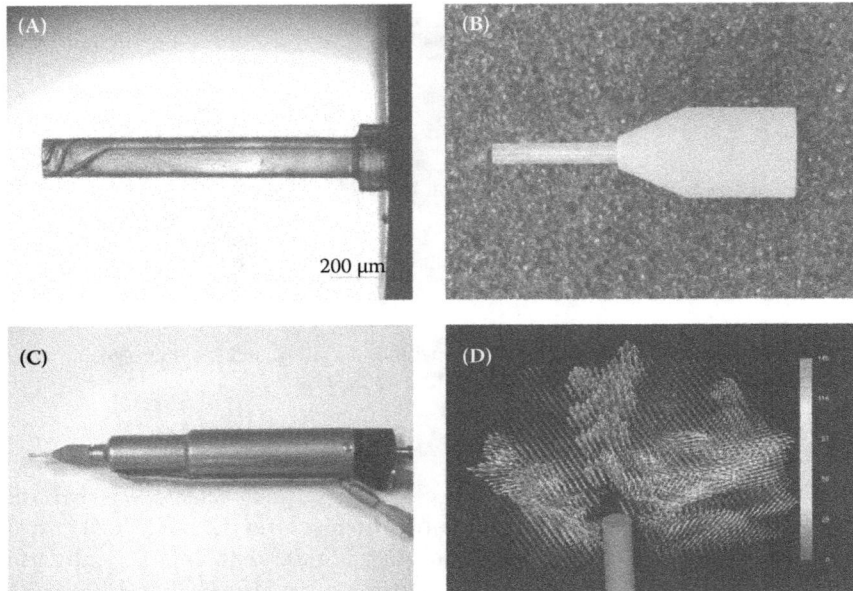

FIGURE 4.22 Fabricated sleeve with 3D microvane: (A) 3D microvane within the sleeve; (B) combined sleeve with the threaded fitting; (C) commercial phacoemulsifier with the fabricated sleeve; and (D) particle image velocimetry. (From Choi, J.W. et al., *Biomed. Microdevices*, 12(5), 875–886, 2010; Choi, J.W. et al., in *Proceedings of the 20th Annual International Solid Freeform Fabrication Symposium*, 2009, pp. 553–568.)

types of micromixers with complex internal structures, microchannels, and helix elements. Kang et al.[38,212] developed a blood typing system having a passive valve for controlling the flow of blood and antibodies. Monneret et al.[99] developed fluidic microchambers with the complex shapes of a well, channel, and walls that were put inside a system to control the contact of beads and living cells. Hasegawa et al.[213] designed and fabricated a multidirectional microswitching valve chip for an air-pressure-driven microdispenser system. Snowden et al.[214] produced small-scale microfluidic components for an electrochemical flow detector.

Practical applications of microstereolithography also include cataract surgery, for which phacoemulsification has been widely used to remove cataracts.[68,78] An ultrasonic irrigation and aspiration handpiece emulsifies and removes the internal lens of the eye through incision sites, and an artificial lens is inserted. During this operation, the irrigation solution flowing through the sleeve of a phacoemulsifier may damage the corneal endothelial cells in the inner eye due to impinging flow. This damage commonly results from shear stress, so reduced flow velocities are required.[68,78] A recent effort to reduce this velocity involved fabrication of a sleeve with 3D microvanes to generate swirl using microstereolithography. The sleeve with the microvanes was fabricated by microstereolithography with diluted ProtoTherm™ 12120. A threaded fitting to attach the fabricated sleeve to the commercial phacoemulsifier was fabricated using an Eden 333 (Stratasys, Inc.; Eden Prairie, MN), a commercial additive manufacturing machine, and was integrated as shown in Figure 4.22. A 3D microvane with a thickness of ~150 µm and width of ~220 µm was fabricated within the sleeve (Figure 4.22A). Both the sleeve and threaded fitting (Figure 4.22B) were successfully fabricated and assembled. Figure 4.22C exhibits a commercial phacoemulsifier equipped with the fabricated sleeve and threaded fitting. With a new tip fitted to the phacoemulsifier, the velocity distribution within 0.5-mm intervals was measured using particle image velocimetry (PIV) by submerging the tip in water. The measurement plane was normal to the axis of the tip, and the mean velocity vectors are shown in Figure 4.22D. This technique verified that water irrigated from the head of the tip, and the irrigation gushed in three radial directions when the 3D microvane was applied.

4.6 DISCUSSION

The projection microstereolithography technology has been advanced and widely utilized in various applications across mechanical, electrical, and biomedical research fields in conjunction with functional processible materials. Among the several reasons cited for the popularity of this technology are its ability to offer spatially high resolution, real three dimensions, a high aspect ratio, and complex manufacturing capabilities. Many structures described in the literature could not otherwise be easily achieved without the aid of microstereolithography. In addition to these exceptional capabilities, microstereolithography systems can be easily implemented for the rapid production of fully customized low-cost structures. Furthermore, the greater availability of numerous photopolymers has contributed to the advancement of microstereolithography.

Despite the many advantages over other technologies, several limitations and challenges still must be addressed. These include seamless multimaterial fabrication across layers or within a single layer, which requires a process to avoid material contamination between material changeovers; multi-scale fabrications ranging from microns to a few centimeters, a problem that can be overcome with continuous projection; and the higher resolutions (e.g., submicron) necessary for further optical applications. In addition, more research on material development must be performed to enhance manufacturability and create more applications.

Additional applications are expected to arise due to the development of transforming new materials. Many novel materials are being used in various structures, sensors, and actuators, as well as in bioengineering and optical engineering applications. Microstereolithography of these materials can be difficult, but it is expected that future research on microstereolithography will allow nonprocessable materials to become microstereolithography-processable materials, which would require further system advancement with regard to new light sources, better optical devices, and more reliable stacking mechanisms.

4.7 SUMMARY

More than 200 publications were reviewed to identify research trends in microstereolithography. We found that even though microstereolithography was first demonstrated almost 20 years ago it is still regarded as a promising technology for creating complex 3D microstructures and providing practical solutions in the biomedical, optical, mechanical, and other engineering and science fields. During the last two decades, a number of researchers have focused on the development of cutting-edge systems, and the hardware technology seems mature. In particular, projection microstereolithography has become more popular due to its ease of implementation and low cost. With existing and newly synthesized materials, many practical applications have been introduced, including tissue engineering scaffolds, biomedical devices, drug-delivery devices, sensors and actuators, fluidics, and optical devices. It is believed that research activities involving microstereolithography will continue to be carried out due to the superior manufacturing capabilities offered by the technology. In addition, new materials will provide more opportunities for the use of microstereolithography and create new applications.

REFERENCES

1. Takagi, T. and Nakajima, N. (1993). Photoforming applied to fine machining. In: *Proceedings of the 4th International Symposium on Micro Machine and Human Science (MHS)*, pp. 173–178.
2. Ikuta, K. and Kirowatari, K. (1993). Real three dimensional micro fabrication using stereo lithography and metal molding. In: *Proceedings of the 6th IEEE Workshop on Micro Electro Mechanical Systems (MEMS'93)*, pp. 42–47.
3. Nakamoto, T. and Yamaguchi, K. (1996). Consideration on the producing of high aspect ratio micro parts using UV sensitive photopolymer. In: *Proceedings of the Seventh International Symposium on Micro Machine and Human Science*, pp. 53–58.

4. Bertsch, A., Jézéquel, J.Y., and André, J.C. (1997). Study of the spatial resolution of a new 3D micro-fabrication process: the microstereophotolithography using a dynamic mask-generator technique. *J. Photochem. Photobiol. A Chem.*, 107(1–3): 275–281.

5. Bertsch, A., Bernhard, P., and Renaud, P. (2001). *Microstereolithography: Concepts and Applications.* IEEE, New York, pp. 289–298.

6. Sun, C., Fang, N., Wu, D.M., and Zhang, X. (2005). Projection micro-stereolithography using digital micro-mirror dynamic mask. *Sens. Actuators A*, 121: 113–120.

7. Jo, K.H., Park, I.B., Ha, Y.M., and Lee, S.H. (2013). Material switching system for multi-material bio-structure in projection micro-stereolithography. In: *Proceedings of the 37th International MATADOR Conference*, pp. 289–302.

8. Choi, J.W., MacDonald, E., and Wicker, R.B. (2010). Multi-material microstereolithography. *Int. J. Adv. Manuf. Technol.*, 49(5–8): 543–551.

9. Zhou, C. and Chen, Y. (2012). Additive manufacturing based on optimized mask video projection for improved accuracy and resolution. *J. Manuf. Proc.*, 14: 107–118.

10. Minev, E., Popov, K., Minev, R., Dimov, S., and Gagov, V. (2011). Grid method for accuracy study of micro parts manufacturing. *Micro Nanosys.*, 3(3): 263–267.

11. Xu, G.S., Pan, H., Ma, X.M., Lou, S., and Qiu, R.H. (2010). Investigation of UV light intensity distribution of integral micro-stereolithography system. *Adv. Mater. Res.*, 97–101: 3985–3988.

12. Dean D., Mott, E., Luo, X., Busso, M., Wang, M.O., Vorwald, C., Siblani, A., and Fisher, J.P. (2014). Multiple initiators and dyes for continuous digital light processing (cDLP) additive manufacture of resorbable bone tissue engineering scaffolds. *Virtual Phys. Prototyping*, 9(1): 3–9.

13. Takagi, T. and Nakajima, N. (1994). Architecture combination by micro photoforming process. In: *Proceedings of the 7th IEEE Workshop on Micro Electro Mechanical Systems (MEMS '94)*, pp. 211–216.

14. Yamaguchi, K. and Nakamoto, T. (1995). Consideration on the optimum conditions to produce micro-mechanical parts by photo polymerization using direct focused beam. In: *Proceedings of the Sixth International Symposium on Micro Machine and Human Science*, pp. 71–76.

15. Ikuta, K., Ogata, T., Tsubio, M., and Kojima, S. (1996). Development of mass productive micro stereo lithography (Mass-IH process). In: *Proceedings of the 9th IEEE Workshop on Micro Electro Mechanical Systems (MEMS '96)*, pp. 301–306.

16. Maruo, S. and Kawata, S. (1997). Two-photon-absorbed photopolymerization for three-dimensional microfabrication. In: *Proceedings of the 10th IEEE Workshop on Micro Electro Mechanical Systems (MEMS '97)*, pp. 169–174.

17. Bertsch, A., Zissi, S., Jézéquel, J.Y., Corbel, S., and André, J.C. (1997). Microstereophotolithography using a liquid crystal display as dynamic mask-generator. *Microsys. Technol.*, 3(2): 42–47.

18. Bertsch, A., Lorenz, H., and Renaud, P. (1998). Combining microstereolithography and thick resist UV lithography for 3D microfabrication. In: *Proceedings of the 11th IEEE Workshop on Micro Electro Mechanical Systems (MEMS '98)*, pp. 18–23.

19. Maruo, S. and Ikuta, K. (1998). New microstereolithography (Super-IH process) to create 3D freely movable micromechanism without sacrificial layer technique. In: *Proceedings of the 1998 International Symposium on Micromechatronics and Human Science (MHS '98)*, pp. 115–120.

20. Maruo, S., Ikuta, K., and Korogi, H. (2001). Remote light-driven micromachines fabricated by 200 nm microstereolithography. In: *Proceedings of the 1st IEEE Conference on Nanotechnology*, pp. 507–512.

21. Deshmukh, S. and Gandhi, P.S. (2009). Optomechanical scanning systems for microstereolithography (MSL): analysis and experimental verification. *J. Mater. Process. Technol.*, 209: 1275–1285.

22. Luo, N., Gao, Y., Zhang, Z., Xiao, M., and Wu, H. (2012). Three-dimensional microstructures of photo-resist formed by gradual gray-scale lithography approach. *Opt. Appl.*, 42(4): 853–864.

23. Chatwin, C., Farsari, M., Huang, S., Heywood, M., Birch, P., Young, R., and Richardson, J. (1998). UV microstereolithography system that uses spatial light modulator technology. *Appl. Opt.*, 37: 7514–7522.

24. Monneret, S., Loubère, V., and Corbel, S. (1999). Microstereolithography using a dynamic mask generator and a non-coherent visible light source. In: *Proceedings of SPIE 3680: Design, Test, and Microfabrication of MEMS and MOEMS*, pp. 553–560.

25. Zhang, X., Jiang, X.N., and Sun, C. (1999). Micro-stereolithography of polymeric and ceramic micro-structures. *Sens. Actuators*, 77: 149–156.

26. Chatwin, C.R., Farsari, M., Huang, S., Heywood, M.I., Young, R.C.D. et al. (1999). Characterisation of epoxy resins for microstereolithographic rapid prototyping. *Int. J. Adv. Manuf. Technol.*, 15: 281–286.

27. Farsari, M., Claret-Tournier, F., Huang, S., Chatwin, C.R., Budgett, D.M., Birch, P.M., Young, R.C.D., and Richardson, J.D. (2000). A novel high-accuracy microstereolithography method employing an adaptive electro-optic mask. *J. Mater. Proc. Technol.*, 107: 167–172.

28. Bertsch, A., Bernhard, P., Vogt, C., and Renaud, P. (2006). Rapid prototyping of small size objects. *Rapid Prototyping J.*, 6: 259–266.
29. Maruo, S., Ikuta K., and Ninagawa T. (2001). Multi-polymer microstereolithography for hybrid opto-MEMS. In: *Proceedings of the 14th IEEE Workshop on Micro Electro Mechanical Systems (MEMS '01)*, pp. 151–154.
30. Burmberger, G. and Färber, G. (2000). Integration of COTS components with a DSP coprocessor for microstereolithography machine control. In: *Proceedings of the 2000 IEEE, International Conference on Control Applications*, pp. 588–593.
31. Varadan, V.K. and Xie, J. (2002). Three dimensional MEMS with functionalized carbon nanotubes. In: *Proceedings of SPIE 4700: Smart Structures and Materials: Smart Electronics, MEMS, and Nanotechnology*, pp. 1–10.
32. Maruo, S. and Ikuta, K. (2002). Submicron stereolithography for the production of freely movable mechanisms by using single-photon polymerization. *Sens. Actuators A*, 100: 70–76.
33. Farsari, M., Huang, S., Young, R.C.D. et al. (1998). Four-wave mixing studies of UV curable resins for microstereolithography. *J. Photochem. Photobiol. A Chem.*, 115: 81–87.
34. Huang, S., Heywood, M.I., Young, R.C.D., Farsari, M., and Chatwin, C.R. (1998). Systems control for a micro-stereolithography prototype. *Microproc. Microsys.*, 22: 67–77.
35. Jiang, X.N., Sun, C., Zhang, X., Xu, B., and Ye, Y.H. (2000). Microstereolithography of lead zirconate titanate thick film on silicon substrate. *Sens. Actuators*, 87: 72–77.
36. Chung, S., Park, S., Lee, I., Jeong, H., and Cho, D. (2004). A study on microreplication of real 3D-shape structures using elastomeric mold: from pure epoxy to composite based on epoxy. *Int. J. Mach. Tools Manuf.*, 44: 147–154.
37. Lee, I.H. and Cho, D.W. (2004). An investigation on photopolymer solidification considering laser irradiation energy in micro-stereolithography. *Microsys. Technol.*, 10(8–9): 592–598.
38. Kang, H.W., Lee, I. H., and Cho, D.W. (2004). Development of an assembly-free process based on virtual environment for fabricating 3D microfluidic systems using microstereolithography technology. *J. Manuf. Sci. Eng.*, 126: 766–771.
39. Cho, Y.H., Lee, I.H., and Cho, D.W. (2005). Laser scanning path generation considering photopolymer solidification in micro-stereolithography. *Microsys. Technol.*, 11(2–3): 158–167.
40. Lee, J.W., Lee, I.H., and Cho, D.W. (2006). Development of micro-stereolithography technology using metal powder. *Microelectron. Eng.*, 83: 1253–1256.
41. Lee, J.W., Lan, P.W., Kim, B., Lim, G., and Cho, D.W. (2007). 3D scaffold fabrication with PPF/DEF using micro-stereolithography. *Microelectron. Eng.*, 84(5–8): 1702–1705.
42. Kim, J.Y., Lee, J.W., Lee, S.J., Park, E.K., Cho, D.W., and Kim, S.Y. (2006). Development of bone scaffold using HA nano powder and MSTL technology. In: *Proceedings of the 32nd International Conference on Micro- and Nano-Engineering*, pp. 339–340.
43. Lee, S.J., Kim, B., Lee, J.S., Lim, G., Kim, S.W., Rhie, J.W., and Cho, D.W. (2006). Design and fabrication of 3D scaffolds using micro-stereolithography. In: *Proceedings of the Regenerate World Congress on Tissue Engineering and Regenerative Medicine*, p. 330.
44. Lee, S.J., Kang, H.W., Kang, T.Y., Kim, B., Lim, G., Rhie, J.W., and Cho, D.W. (2007). Development of a scaffold fabrication system using an axiomatic approach: development of a scaffold fabrication system using an axiomatic approach. *J. Micromech. Microeng.*, 17: 147–153.
45. Lee, S.J., Kang, H.W., Park, J.K., Rhie, J.W., Hahn, S.K., and Cho, D.W. (2008). Application of microstereolithography in the development of three-dimensional cartilage regeneration scaffolds. *Biomed. Microdevices*, 10: 233–241.
46. Devaux, F., Mosset, A., Lantz, E., Monneret, S., and Gall, H.L. (2001). Image upconversion from the visible to the UV domain: application to dynamic UV microstereolithography. *Appl. Opt.*, 40: 4953–4957.
47. Mapili, G., Lu, Y., Chen, S., and Roy, K. (2005). Laser-layered microfabrication of spatially patterned functionalized tissue-engineering scaffolds. *J. Biomed. Mater. Res.*, 75B: 414–424.
48. Stampfl, J., Baudis, S., Heller, C., Liska, R., Neumeister, A., Kling, R., Ostendorf, A., and Spizbart, M. (2008). Photopolymers with tunable mechanical properties processed by laser-based high-resolution stereolithography. *J. Micromech. Microeng.*, 18: 1–9.
49. Heller, C., Schwentenwein, M., Russmueller, G., Varga, F., Stampfl, J., and Liska, R. (2009). Vinyl ester: low cytoxicity monomers for the fabrication of biocompatible 3D scaffolds by lithography based additive manufacturing. *J. Polym. Sci. A Polym. Chem.*, 47: 6941–6954.
50. Oda, G., Miyoshi, T., Takaya, Y., Ha, T., and Kimura, K. (2004). Microfabrication of overhanging shape using LCD microstereolithography. In: *Proceedings of SPIE 5662: Fifth International Symposium on Laser Precision Microfabrication*, pp. 649–654.

51. Kwon, I.K. and Matsuda, T. (2005). Photo-polymerized microarchitectural constructs prepared by microstereolithography (μSL) using liquid acrylate-end-capped trimethylene carbonate-based prepolymers. *Biomaterials*, 26: 1675–1684.

52. Lee, D., Miyoshi, T., Takaya, Y., and Ha, T. (2006). 3D microfabrication of photosensitive resin reinforced with ceramic nanoparticles using LCD microstereolithography. *J. Laser Micro/Nanoeng.*, 1: 142–148.

53. Cheng, Y.L., Li, M.L., Lin, J.H., Lai, J.H., Ke, C.T., and Huang, Y.C. (2005). Development of dynamic mask photolithography system. In: *Proceedings of the IEEE International Conference on Mechatronics (ICM '05)*, pp. 467–471.

54. Choi, J.W., Ha, Y.M., Kim, H.S., Won, M.H., Choi, K.H., and Lee, S.H. (2005). An implementation of RP-based micro fabrication apparatus for micro structures. In: *Proceedings of High Aspect Ratio Micro Structure Technology Workshop (HARMST 05)*, pp. 190–191.

55. Choi, J.W., Ha, Y.M., Won, M.H., Choi, K.H., and Lee, S.H. (2005). Fabrication of 3-dimensional microstructures using dynamic image projection. In: *Proceedings of Asian Symposium for Precision Engineering and Nanotechnology (ASPEN 2005)*, pp. 472–476.

56. Choi, J.W., Ha, Y.M., Choi, K.H., and Lee, S.H. (2005). Curing characteristics of 3-dimensional microstructures using dynamic pattern projection. In: *Proceedings of SPIE 6050: Optomechatronic Micro/Nano Devices and Components*, pp. 1–5.

57. Choi, J.W., Park, I.B., Ha, Y.M., Jung, M.G., Lee, S.D., and Lee, S.H. (2006). Insertion force estimation of various microneedle array-type structures fabricated by a microstereolithography apparatus. In: *Proceedings of SICE-ICASE 2006 International Joint Conference*, pp. 3678–3681.

58. Choi, J.W., Ha, Y.M., Lee, S.H., and Choi, K.H. (2006). Design of microstereolithography system based on dynamic image projection for fabrication of three-dimensional microstructures. *J. Mech. Sci. Technol.*, 20(12): 2094–2104.

59. Choi, J.W., Park, I.B., Kim, H.C., and Lee, S.H. (2007). Fabrication of high-aspect ratio microneedles using microstereolithography. In: *Proceedings of High Aspect Ratio Micro Structure Technology Workshop (HARMST 07)*.

60. Ha, Y.M., Choi, J.W., Lee, S.H., and Kim, H.C. (2007). Fabrication of 3D micro-structure on large surface using projection type micro-stereolithography. In: *Proceedings of Asian Symposium for Precision Engineering and Nanotechnology (ASPEN 2007)*, pp. 492–495.

61. Han, L.S., Mapili, G., Chen, S., and Roy, K. (2008). Projection microfabrication of three-dimensional scaffolds for tissue engineering. *J. Manuf. Sci. Eng.*, 130: 1–4.

62. Ha, Y.M., Choi, J.W., and Lee, S.H. (2008). Mass production of 3-D microstructures using projection microstereolithography. *J. Mech. Sci. Technol.*, 22(3): 514–521.

63. Choi, J.W., Lee, S.H., Choi, K.H., Jung, I., Ha, C.S., and Wicker, R.B. (2008). 3D PPF micro-scaffold fabrication using DMD-based maskless projection microstereolithography. In: *Proceedings of the 24th Southern Biomedical Engineering Conference (SBEC 2008)*, pp. 205–206.

64. Choi, J.W., Park, I.B., Wicker, R.B., Lee, S.H., and Kim, H.C. (2008). Fabrication of complex 3D microscale scaffolds and drug delivery devices using dynamic mask projection microstereolithography. In: *Proceedings of the 19th Annual International Solid Freeform Fabrication Symposium*, pp. 652–675.

65. Choi, J.W., Wicker, R.B., Cho, S.H., Ha, C.S., and Lee, S.H. (2009). Cure depth control for complex 3D microstructure fabrication in dynamic mask projection microstereolithography. *Rapid Prototyping J.*, 15(1): 59–70.

66. Park, B., Choi, J.W., and Lee, S.H. (2009). Sacrificial layer fabrication method for enhancement of dimensional accuracy of fabricated microstructures in microstereolithography. *Int. J. Precision Eng. Manuf.*, 10: 91–98.

67. Choi, J.W., Irwin, M.D., and Wicker, R.B. (2010). DMD-based 3D micro-manufacturing. In: *Proceeding of SPIE 7596: Emerging Digital Micromirror Device Based Systems and Application II*, pp. 1–11.

68. Choi, J.W., Yamashita, M., Sakakibara, J., Kaji, Y., Oshika, T., and Wicker, R. (2010). Combined micro and macro additive manufacturing of a swirling flow coaxial phacoemulsifier sleeve with internal microvanes. *Biomed. Microdevices*, 12(5): 875–886.

69. Han, L.H., Suri, S., Schmidt, C.E., and Chen, S. (2010). Fabrication of three-dimensional scaffolds for heterogeneous tissue engineering. *Biomed. Microdevices*, 12(4): 721–725.

70. Park, I.B., Ha, Y.M., and Lee, S.H. (2010). Cross-section segmentation for improving the shape accuracy of microstructure array in projection microstereolithography. *Int. J. Adv. Manuf. Technol.*, 46: 151–161.

71. Park, I.B., Ha, Y.H., Kim, M.S., and Lee, S.H. (2010). Fabrication of a micro-lens array with a non-layered method in projection microstereolithography. *Int. J. Precision Eng. Manuf.*, 11: 1–8.

72. Lu, Y., Mapili, G., Suhali, G., Chen, S., and Roy, K. (2006). A digital micro-mirror device-based system for the microfabrication of complex, spatially patterned tissue engineering scaffolds. *J. Biomed. Mater. Res.*, 77A: 396–405.

73. Zhao, X. and Zhang, C. (2013). Digital manufacturing system design for large area microstructure based on DLP projector. *J. Theor. Appl. Inform. Technol.*, 48(1): 490–495.

74. Suwandi, D., Whulanza, Y., and Istiyanto, J. (2014). Visible light maskless photolithography for bioma-chining application. *Appl. Mech. Mater.*, 493: 552–557.

75. Lu, Y. and Chen, S. (2008). Direct write of microlens array using digital projection photopolymerization. *Appl. Phys. Lett.*, 92: 041109.

76. Choi, J.W., Wicker, R.B., Lee, S.H., Choi, K.H., Ha, C.S., and Chung. I. (2009). Fabrication of 3D biocompatible/biodegradable micro-scaffolds using dynamic mask projection microstereolithography. *J. Mater. Process. Technol.*, 209: 5494–5503.

77. Choi, J.W. (2007). Development of Projection-Based Microstereolithography Apparatus Adapted to Large Surface and Microstructure Fabrication for Human Body Application, doctoral dissertation, Pusan National University, South Korea.

78. Choi, J.W., Yamashita, M., Sakakibara, J., Kaji, Y., Oshika, T., and Wicker, R.B. (2009). Functional micro/macro fabrication combining multiple additive fabrication technologies: design and development of an improved micro-vane phacoemulsifier used in cataract surgery. In: *Proceedings of the 20th Annual International Solid Freeform Fabrication Symposium*, pp. 553–568.

79. Wu, S., Han, L.H., and Chen, S. (2009). Three-dimensional selective growth of nanoparticles on a poly-mer microstructure. *Nanotechnology*, 20: 1–4.

80. Zissi, S., Bertsch, A., Jézéquel, J.Y., Corbel, S., Lougnot, D.J., and André. J.C. (1996). Stereolithography and microtechniques. *Microsys. Technol.*, 2: 97–102.

81. Provin, C. and Monneret, S. (2001). Complex ceramic-polymer composite microparts made by micro-stereolithography. In: *Proceedings of SPIE 4408: Design, Test, Integration, and Packaging of MEMS/ MOEMS*, pp. 535–542.

82. Baker, E., Reissman, T., Zhou, F., Wang, C., Lynch, K., and Sun, C. (2012). Microstereolithography of three-dimensional polymeric springs for vibration energy harvesting. *Smart Mater. Res.*, 2012: 1–9.

83. Goswami, A., Umarji, A.M., and Madras, G. (2012). Thermomechanical and fractographic behavior of poly (HDDA-co-MMA): a study for its application in microcantilever sensors. *Polym. Adv. Technol.*, 23: 1604–1611.

84. Ibrahim, R., Raman, I., Ramlee, M.H.H. et al. (2012). Evaluation on the photoabsorber composition effect in projection microstereolithography. *Appl. Mech. Mater.*, 159: 109–114.

85. Gandhi, P. and Bhole, K. (2013). Characterization of "bulk lithography" process for fabrication of three-dimensional microstructures. *J. Micro Nano-Manuf.*, 1(4): 041002.

86. Leigh, S.J., Gilbert, H.T., Barker, I.A., Becker, J.M., Richardson, M. et al. (2013). Fabrication of 3-dimen-sional cellular constructs via microstereolithography using a simple, three-component, poly(ethylene glycol) acrylate-based system. *Biomacromolecules*, 14(1): 186–192.

87. Lee, H. and Fang, N.X. (2012). Micro 3D printing using a digital projector and its application in the study of soft materials. *J. Vis. Exp.*, 69: e4457.

88. Lee, H., Zhang, J., Jiang, H., and Fang, N.X. (2012). Prescribed pattern transformation in swelling gel tubes by elastic instability. *Phys. Rev. Lett.*, 108: 214304.

89. Lee, H., Lu, J., Georgiadis, J., and Fang, N. (2013). A study of the concentration dependent water dif-fusivity in polymer using magnetic resonance imaging. In: *Proceedings of American Physical Society March Meeting*, abstract no. W32.007.

90. Leigh, S.J., Purssell, C.P., Bowen, J., Hutchins, D.A., Covington, J.A., and Billson, D.R. (2011). A min-iature flow sensor fabricated by micro-stereolithography employing a magnetite/acrylic nanocomposite resin. *Sens. Actuators A*, 168: 66–71.

91. Sun, C. and Zhang, X. (2002). Experimental and numerical investigations on microstereolithography of ceramics. *J. Appl. Phys.*, 92, 4796–4802.

92. Thomas, K.A., Singh, A., and Natarajan, V. (2009). Fabrication of polymeric microcantilevers. *Defence Sci. J.*, 59: 616–621.

93. Bertsch, A., Jiguet, S., and Renaud, P. (2004). Microfabrication of ceramic components by microstereo-lithography. *J. Micromech. Microeng.*, 14: 197–203.

94. Park, I.B., Ha, Y.M., and Lee, S.H. (2011). Still motion process for improving the accuracy of latticed microstructures in projection microstereolithography. *Sens. Actuators A*, 167: 117–129.

95. Jariwala, A.S., Ding, F., Boddapati, A., Breedveld, V., Grover, M.A., Henderson, C.L., and Rosen, D.W. (2011). Modeling effects of oxygen inhibition in mask-based stereolithography. *Rapid Prototyping J.*, 17(3): 168–175.

96. Gauvin, R., Chen, Y.C., Lee, J.W., Soman, P., Zorlutuna, P., Nichol, W.J., Bae, H., Chen, S., and Khademhosseini, A. (2012). Microfabrication of complex porous tissue engineering scaffolds using 3D projection stereolithography. *Biomaterials*, 33: 3824–3834.

97. Wu, D., Fang, N., Sun, C., and Zhang, X. (2002). Fabrication and characterization of THz plasmonic filter. In: *Proceedings of the 2nd IEEE Conference on Nanotechnology*, pp. 229–231.

98. Monneret, S., Provin, C., Gall, H.L., and Corbel, S. (2002). Microfabrication of freedom and articulated alumina-based components. *Microsys. Technol.*, 8: 368–374.

99. Monneret, S., Belloni, F., and Soppera, O. (2007). Combining fluidic reservoirs and optical tweezers to control beads/living cells contacts. *Microfluid. Nanofluid.*, 3: 645–652.

100. Kang, T., Park, J.K., Yeom, J., Kang, H.W., Hahn, S.K., and Cho, D.W. (2009). Fabrication of 3D scaffolds using microstereolithography with HA-AEMA and TMC-TMP, paper presented at the 3rd International Conference on Mechanics of Biomaterials & Tissues, Clearwater Beach, FL, December 13–17.

101. Zhou, F., Bao, Y., Cao, W., Stuart, C.T., Gu, J., Zhang, W., and Sun, C. (2011). Hiding a realistic object using a broadband terahertz invisibility cloak. *Sci. Rep.*, 78: 10.1038/srep00078.

102. Sun, C. and Zhang, X. (2002). The influences of the material properties on ceramic micro-stereolithography. *Sens. Actuators A*, 101: 364–370.

103. Kirihara, S. (2009). Structural joining of ceramics nanoparticles: development of photonic crystals for terahertz wave control by using micro stereolithography. *KONA Powder Part. J.*, 27: 107–118.

104. Noritoshi, O., Toshiki, N., and Soshu, K. (2009). Visualizations of terahertz frequency amplifications in water cells. *Trans. JWRI*, 39(2): 151–153.

105. Kirihara, S., Niki, T., and Kaneko, M. (2010). Teraherts wave properties of ceramic photonic crystals with graded structure fabricated by using micro-stereolithography. *Mater. Sci. Forum*, 631–632: 299–304.

106. Kirihara, S., Ohta, N., Niki, T., Uehara, Y., and Tasaki, S. (2011). Development of photonic and thermo-dynamic crystals conforming to sustainability conscious materials tectonics. *WIT Trans. Ecol. Environ.*, 154: 103–114.

107. Kirihara, S., Ohta, N., Takinami, Y., and Tasakai, S. (2012). Smart processing of micro photonic crystals for terahertz wave control-freeform fabrication by stereolithographic technique. *Mater. Sci. Forum*, 706–709: 1925–1930.

108. Goswami, A., Ankit, K., Balashanmugam, N., Umarjia, A.M., and Madras, G. (2014). Optimization of rheological properties of photopolymerizable alumina suspensions for ceramic microstereolithography. *Ceram. Int.*, 40: 3655–3665.

109. Kirihara, S., Komori, N., Nakano, M., Ohta, N., and Niki, T. (2010). Terahertz wave properties of micro patterned titania and metallic glass particles in hexagonal tablets fabricated using microstereolithography. In: *Characterization and Control of Interfaces for High Quality Advanced Materials III* (Ewsuk, K. et al., Eds.), Chap. 10. John Wiley & Sons, Hoboken, NJ.

110. Kirihara, S., Nikil, T., and Kaneko, M. (2010). Terahertz wave harmonization in geometrically patterned dielectric ceramics through spatially structural joining. In: *Advances in Multifunctional Materials and Systems* (Akedo J., Ohsato H., Shimada T., and Singh, M., Eds.), Chap. 12. John Wiley & Sons, Hoboken, NJ.

111. Kirihara, S., Tasaki, S., and Itakura, Y. (2013). Creation of titania artificial interfaces with geometric patterns by using microstereolithography and aqueous solution techniques. *Int. J. Appl. Ceram. Technol.*, 10(3): 468–473.

112. Kirihara, S. and Tasaki, S. (2010). Fabrication of zirconium oxide solid electrolytes with ordered porous structures by using micro stereolithography. *Trans. JWRI*, 39(2): 279–280.

113. Kirihara, S. and Nakano, M. (2013). Freeform fabrication of magnetophotonic crystals with diamond lattices of oxide and metallic glasses for terahertz wave control by micro patterning stereolithography and low temperature sintering. *Micromachines*, 4: 149–156.

114. Komori, N., Tasaki, S., and Kirihara, S. (2012). Fabrication of dendritic electrodes for solid oxide fuel cells by using micro stereolithography. In: *Advanced Processing and Manufacturing Technologies for Structural and Multifunctional Materials VI* (Ohji, T. et al., Eds.), Chap. 13. John Wiley & Sons, Hoboken, NJ.

115. Yasui, M., Ikeuchi, M., and Ikuta, K. (2013). Density controllable photocurable polymer for three-dimensional magnetic microstructures with neutral buoyancy. *Appl. Phys. Lett.*, 103: 201901.

116. Khairu, K., Ibrahim, M., Hassan, S., Hehsan, H., Kasmin, A., and Fazial, E. (2013). Study on layer fabrication for 3D structure of photoreactive polymer using DLP projector. *Appl. Mech. Mater.*, 911: 465–466.

117. Choi, J.W., MacDonald, E., and Wicker, R. (2009). Multiple material microstereolithography. In: *Proceedings of the 20th Annual International Solid Freeform Fabrication Symposium*, pp. 781–792.
118. Lee, J.W., Lan, P.X., Seol, Y.J., and Cho, D.W. (2007). PPF/DEF-HA Composite Scaffold Using Micro-Stereolithography, paper presented at the European Chapter Meeting of the Tissue Engineering and Regenerative Medicine International Society (TERMIS-EU 2007), London, UK, September 4–7.
119. Lee, J.W., Jung, J.H., Kim, J.Y., Lim, G., and Cho, D.W. (2009). Estimating cell proliferation by various peptide coatings at the PPF/DEF 3D scaffold. *Microelectron. Eng.*, 86(4–6): 1451–1454.
120. Lee, J.W., Ahn, G.S., and Cho, D.W. (2008). Development of nano- and micro-scale composite 3D scaffolds using PPF/DEF-hydroxyapatite. In: *Proceedings of the 34th International Conference on Micro- and Nano-Engineering*, p. 332.
121. Lan, P.X., Lee, J.W., Seol, Y.J., and Cho, D.W. (2009). Development of 3D PPF/DEF scaffolds using micro-stereolithography and surface modification. *J. Mater. Sci. Mater. Med.*, 20: 271–279.
122. Lee, J.W., Lee, S.H., Kim, J.Y., Lee, B.K., and Cho, D.W. (2009). *In vivo* bone formation of a scaffold that releases BMP-2 fabricated using solid freeform fabrication. In: *Proceedings of TERMIS 2nd World Congress*, p. S29.
123. Lee, J.W., Jung, J.H., Kim, D.S., Lim, G., and Cho, D.W. (2009). Estimation of cell proliferation by various peptide coating at the PPF/DEF 3D scaffold. *Microelectron. Eng.*, 86: 1451–1454.
124. Lee, J.W., Ahn, G., Kim, D.S., and Cho, D.W. (2009). Development of nano- and microscale composite 3D scaffolds using PPF/DEF-HA and micro-stereolithography. *Microelectron. Eng.*, 86: 1465–1467.
125. Seol, Y.J., Kang, H.W., Kim, S.W., Rhie, J.W., Yun, W.S., and Cho, D.W. (2009). Development of 3-Dimensional Hydroxyapatite Structures for Tissue Engineering, paper presented at the 3rd International Conference on Mechanics of Biomaterials & Tissues, Clearwater Beach, FL, December 13–17.
126. Seol, Y.J., Kim, J.Y., Park., E.K., Kim, S.Y., and Cho, D.W. (2009). Fabrication of a hydroxyapatite scaffold for bone tissue regeneration using microstereolithography and molding technology. *Microelectron. Eng.*, 86: 1443–1446.
127. Seol, Y.J., Park, D.Y., Park, J.Y., Kim, S.W., Park, S.J., and Cho, D.W. (2013). A new method of fabricating robust freeform 3D ceramic scaffolds for bone tissue regeneration. *Biotechnol. Bioeng.*, 110(5): 1444–1455.
128. Ronca, A., Ambrosio, L., and Grijpma, D.W. (2013). Preparation of designed poly(D,L-lactide)/nanosized hydroxyapatite composite structures by stereolithography. *Acta Biomater.*, 9, 5989–5996.
129. Bail, R., Patel, A., Yang, H., Rogers, C.M., Rose, F.R.A.J., Segal, J.I., and Ratchev, S.M. (2013). The effect of a type I photoinitiator on cure kinetics and cell toxicity in projection-microstereolithography. *Proc. CIRP*, 5: 222–225.
130. Fisher, J., Dean, D., and Mikos, G.A. (2002). Photocrosslinking characteristics and mechanical properties of diethyl fumarate/poly(propylene fumarate) biomaterials. *Biomaterials*, 23: 4333–4343.
131. Liska, R., Schuster, M., Inführ, R., Turecek, C., Fritscher, C. et al. (2007). Photopolymers for rapid prototyping. *J. Coat. Technol. Res.*, 4: 505–510.
132. Baudis, S., Heller, C., Liska, R., Stampfl, J., Bergmeister, H., and Weigel, G. (2009). (Meth)acrylate-based photoelastomers as tailored biomaterials for artificial vascular grafts. *J. Polym. Sci. A Polym. Chem.*, 47: 2664–2676.
133. Chen, W., Kirihara, S., and Miyamoto, Y. (2008). Fabrication and characterization of three-dimensional ZrO_2-toughened Al_2O_3 ceramic microdevices. *Int. J. Appl. Ceram. Technol.*, 5: 353–359.
134. Ronca, A., Ambrosio, L., and Grijpma, D.W. (2013). Preparation of designed poly (D,L-lactide)/nanosized hydroxyapatite composite structures by stereolithography. *Acta Biomater.*, 9: 5989–5996.
135. Ikuta, K. (1995). Biomedical micro device fabricated by micro stereo lithography (IH process). In: *Proceedings of the Sixth International Symposium on Micro Machine and Human Science*, pp. 67–70.
136. Ikuta, K., Maruo, S., Hasegawa, T., and Adachi, T. (2001). Micro-stereolithography and its application to biochemical IC chip. In: *Proceedings of SPIE 4274: Laser Applications in Microelectronic and Optoelectronic Manufacturing*, pp. 360–374.
137. Ikuta, K., Takahashi, A., and Maruo, S. (2001). In-chip cell-free protein synthesis from DNA by using biochemical IC chips. In: *Proceedings of the 14th IEEE Workshop on Micro Electro Mechanical Systems (MEMS '01)*, pp. 455–458.
138. Ikuta, K., Maruo, S., Hasegawa, T., Adachi, T., Takahashi, A., and Ikeda, K. (2002). Biochemical IC chips fabricated by hybrid microstereolithography. *Mater. Res. Soc. Symp. Proc.*, 758: 193–204.
139. Ikuta, K., Sasaki, Y., Maegawa, H., and Maruo, S. (2003). Biochemical IC chip for pretreatment in biochemical experiments. In: *Proceedings of the 16th IEEE Workshop on Micro Electro Mechanical Systems (MEMS '03)*, pp. 343–346.

140. Ikuta, K., Maruo, S., Hasegawa, T., Itho, S., Korogi, H., and Takahashi, A. (2004). Light-drive biomedical micro tools and biochemical IC chips fabricated by 3D micro/nano stereolithography. In: *Proceedings of SPIE 5604: Optomechatronic Micro/Nano Components, Devices, and Systems*, pp. 52–66.

141. Nagakura, T., Inada, K., Susuki, Y., Yoshida, N., Yamada, A., Ikeuchi, M., and Ikuta, K. (2010). The study of micro liter insulin injection system by osmotic pressure for diabetes therapy. *IFMBE Proc.*, 25(8): 382–383.

142. Ortega, I., Deshpande, P., Gill, A.A., MacNeil, S., and Claeyssens, F. (2013). Development of a micro-fabricated artificial limbus with micropockets for cell delivery to the cornea. *Biofabrication*, 5: 025008.

143. Lee, S.J., Kim, B., Lee, J.W., Kim, J.Y., Kang, H.W. et al. (2006). Assessment of microstereolithography system for development of various types of three-dimensional scaffolds in cartilage and bone tissue regeneration. In: *Proceedings of the European Chapter Meeting of the Tissue Engineering and Regenerative Medicine International Society (TERMIS-EU 2006)*, p. 227.

144. Kim, J.Y., Lee, J.W., Lee, S.J., Park, E.K., Kim, S.Y., and Cho, D.W. (2007). Development of a bone scaffold using HA nanopowder and micro-stereolithography technology. *Microelectron. Eng.*, 84(5–8): 1762–1765.

145. Lee, J.W., Ahn, G., Kim, J.Y., and Cho, K.W. (2010). Evaluating cell proliferation based on internal pore size and 3D scaffold architecture fabricated using solid freeform fabrication technology. *J. Mater. Sci. Mater. Med.*, 21: 3195–3205.

146. Naitoh, M., Kinoshita, H., Gotoh, K., and Ariji, E. (2012). Cone-beam computed tomography images of phantoms with simulating trabecular bone structure fabricated using micro stereolithography. *Okajimas Folia Anat. Jpn.*, 89(2): 27–33.

147. Lee, S.H., Lee, J.W., Lee, B.K., Kang, K.S., Kim, J.Y., and Cho, D.W. (2011). Bone regeneration using a microstereolithography-produced customized poly(propylene fumarate)/diethyl fumarate photopolymer 3D scaffold incorporating BMP-2 loaded PLGA microspheres. *Biomaterials*, 32(3), 744–752.

148. Lee, S.J., Kim, B., Lee, J.S., Kim, S.W., Kim, M.S., Kim, J.S., Lim, G., and Cho, D.W. (2006). Three-dimensional microfabrication system for scaffold in tissue engineering. *Key Eng. Mater.*, 326–328: 723–726.

149. Lee, J.W., Kim, B., Lim, G., and Cho, D.W. (2007). Scaffold fabrication with biodegradable poly(propylene fumarate) using microstereolithography. *Key Eng. Mater.*, 342–343: 141–144.

150. Lee, S.J., Kim, B., Lim, G., Kim, S.W., Rhie, J.W., and Cho, D.W. (2007). Effect of three-dimensional scaffold geometry on chondrocyte adhesion. *Key Eng. Mater.*, 342–343: 97–100.

151. Lee, J.W., Lan, P.X., Kim, B., Lim, G., and Cho, D.W. (2008). Fabrication and characteristic analysis of a poly(propylene fumarate) scaffold using micro-stereolithography technology. *J. Biomed. Mater. Res. Part B Appl. Biomater.*, 87B: 1–9.

152. Kang, H.W., Rhie, J.W., Kim, S.W., and Cho, D.W. (2008). Three-dimensional scaffold fabrication by indirect micro-stereolithography using various biomaterials. In: *Proceedings of the TERMIS North America Annual Conference & Exposition*, abstract no. 76.

153. Kang, H.W. and Cho, D.W. (2007). Indirect solid freeform fabrication (SFF) using microstereolithography technology. In: *Proceedings of 2007 Annual Conference of Tissue Engineering and Regenerative Medicine International Society—Asian Pacific Region (TERMIS-AP)*, p. 98.

154. Lee, S.J., Kang, T., Rhie, J.W., and Cho, D.W. (2007). Development of three-dimensional hybrid scaffold using chondrocyte-encapsulated alginate hydrogel. *Sens. Mater.*, 19(8): 445–451.

155. Lee, S.J., Kim, B., Lim, G., Rhie, J.W., Kang, H.W., and Cho, D.W. (2007). Development of three-dimensional alginate encapsulated chondrocyte hybrid scaffolds using microstereolithography. In: *Proceedings of ASME 2007 International Manufacturing Science and Engineering Conference*, pp. 91–95.

156. Lee, S.J., Kim, B., Lim, G., Rhie, J.W., Kim, D.S., and Cho, D.W. (2007). Development of Three-Dimensional Hybrid Scaffold Using Chondrocytes Encapsulated Alginate Hydrogel, paper presented at the Society for Biomaterials 2007 Annual Meeting, Chicago, IL, April 18–21.

157. Seol, Y.J., Kim, J.Y., Park, E.K., Kim, S.Y., and Cho, D.W. (2007). Indirect solid free-form fabrication of HA scaffold using MSTL system. In: *Proceedings of 2007 Annual Conference of Tissue Engineering and Regenerative Medicine International Society—Asian Pacific Region (TERMIS-AP)*, p. 134.

158. Kang, H.W., Kang, T., and Cho, D.W. (2008). Development of a bi-pore scaffold using indirect solid free-form fabrication based on microstereolithography technology. In: *Proceedings of the 34th International Conference on Micro and Nano Engineering*, p. 241.

159. Lee, J.W., Anh, N.T., Kang, K.S., Seol, Y.J., and Cho, D.W. (2008). Development of growth factor-embedded scaffold with controllable pore size and distribution using micro-stereolithography. In: *Proceedings of the European Chapter Meeting of the Tissue Engineering and Regenerative Medicine International Society (TERMIS-EU 2008)*, p. 835.

160. Lee, S.J., Park, J.H., Seol, Y.J., Lee, I.H., and Cho, D.W. (2008). Development of hybrid scaffold and bioreactor for cartilage regeneration. In: *Proceedings of the 34th International Conference on Micro- and Nano-Engineering*, p. 527.

161. Kang, H.W., Seol, Y.J., and Cho, D.W. (2009). Development of an indirect solid freeform fabrication process based on microstereolithography for 3D porous scaffolds. *J. Micromech. Microeng.*, 19: 1–8.

162. Kang, H.W., Jin, C.Z., Rhie, J.W., and Cho, D.W. (2009). Chitosan-alginate scaffold based on SFF technology. In: *Proceedings of TERMIS 2nd World Congress*, p. S217.

163. Xia, C. and Fang, N.X. (2009). 3D microfabricated bioreactor with capillaries. *Biomed. Microdevices*, 11: 1309–1315.

164. Barker, I.A., Ablett, M.P., Gilbert, H.T.J., Leigh, S.J., Covington, J.A., Hoyland, J.A., Richardson, S.M., and Dove, A.P. (2014). A microstereolithography resin based on thiol-ene chemistry: towards biocompatible 3D extracellular constructs for tissue engineering. *Biomater. Sci.*, 2: 472–475.

165. Inoue1, Y. and Ikuta, K. (2012). Cell culture biochemical IC chip with cell-level biocompatibility. In: *Proceedings of the 25th IEEE Workshop on Micro Electro Mechanical Systems (MEMS '12)*, pp. 788–791.

166. Kang, H.W. and Cho, D.W. (2012). Development of an indirect stereolithography technology for scaffold fabrication with a wide range of biomaterial selectivity. *Tissue Eng. Part C Methods*, 18(9): 719–729.

167. Inoue, Y. and Ikuta, K. (2013). Detoxification of the photocurable polymer by heat treatment for microstereolithography. *Proc. CIRP*, 5: 115–118.

168. Gittard, S.D., Miller, P.R., Jin, C., Martin, T.N., Boehm, R.D. et al. (2011). Deposition of antimicrobial coatings on microstereolithography-fabricated microneedles. *J. Miner. Met. Mater. Soc.*, 63(6): 59–68.

169. Boehm, R.D., Miller, P.R., Hayes, S.L., Monteiro-Riviere, N.A., and Narayan, R.J. (2011). Modification of microneedles using inkjet printing. *Adv. Phys.*, 1(2): 022139.

170. Miller, P.R., Gittard, S.D., Edwards, T.L., Lopez, D.M., Xiao, X. et al. (2011). Integrated carbon fiber electrodes within hollow polymer microneedles for transdermal electrochemical sensing. *Biomicrofluidics*, 5(1): 013415.

171. Serruys, P.W., Onuma, Y., and Dudek, D. (2011). Evaluation of the second generation of a bioresorbable everolimus-eluting vascular scaffold for the treatment of *de novo* coronary artery stenosis: 12-month clinical and imaging outcomes. *J. Am. Coll. Cardiol.*, 58(15): 1578–1588.

172. Kuniyal, H. and Vakil, H. (2010). A simple and effective technique to fabricate diagnostic and surgical stent. *Int. J. Oral Implantol. Clin. Res.*, 1(3): 173–175.

173. Su, S., Chao, R.Y., Landau, C.L., Nelson, K.D., Timmons, R.B., Meidell, R.S., and Eberhart, R.C. (2003). Expandable bioresorbable endovascular stent. I. Fabrication and properties. *Ann. Biomed. Eng.*, 31: 667–677.

174. Yu, X., Wan, C., and Chen, H. (2008). Preparation and endothelialization of decellularised vascular scaffold for tissue-engineered blood vessel. *J. Mater. Sci. Mater. Med.*, 19: 319–326.

175. Choi, J.W., Mann, B., MacDonald, E., and Wicker, R.B. (2009). Fabrication of multi-material, multi-lumen, poly (ethylene glycol)-based nerve guidance conduits using microstereolithography. In: *Proceedings of TERMIS 2nd World Congress*.

176. Park, I.B., Lee, S.D., Kwon, T.W., Choi, J.W., and Lee, S.H. (2008). Fabrication of elliptical micro-lens array with large surface using μSL. *J. Korean Soc. Precision Eng.*, 25(2): 123–130.

177. Kaneko, M. and Kirihara, S. (2010). Millimeter wave control using TiO_2 photonic crystal with diamond structure fabricated by micro-stereolithography. *Mater. Sci. Forum*, 631–632: 293–298.

178. Chen, W., Kirihara, S., and Miyamoto, Y. (2007). Three-dimensional microphotonic crystals of ZrO_2 toughened Al_2O_3 for terahertz wave applications. *Appl. Phys. Lett.*, 91: 153507.

179. Chen, W., Kirihara, S., and Miyamoto, Y. (2007). Fabrication and measurement of micro three-dimensional photonic crystals of SiO_2 ceramic for terahertz wave applications. *J. Am. Ceram. Soc.*, 90: 2078–2081.

180. Miyamoto, Y., Kanaoka, H., and Kirihara, S. (2008). Terahertz wave localization at a three-dimensional ceramic fractal cavity in photonic crystals. *J. Appl. Phys.*, 103: 103106.

181. Chen, W., Kirihara, S., and Miyamoto, Y. (2008). Microfabrication of three-dimensional photonic crystals of SiO_2–Al_2O_3 ceramics and their terahertz wave properties. *Int. J. Appl. Ceram. Technol.*, 5: 228–233.

182. Sano, D. and Kirihara, S. (2010). Fabrication of metal photonic crystals with graded lattice spacing by using micro-stereolithography. *Mater. Sci. Forum*, 631–632: 287–292.

183. Kanaoka, H., Kirihara, S., and Miyamoto, Y. (2008). Terahertz wave properties of alumina microphotonic crystals with a diamond structure. *J. Mater. Res.*, 23: 1036–1041.

184. Sano, D. and Kirihara, S. (2009). Microwave emission from metal photonic crystals fabricated by using stereolithography. *Ferroelectrics*, 388: 23–30.

185. Kirihara, S., Kaneko, M., and Niki, T. (2009). Terahertz wave control using ceramic photonic crystals with a diamond structure including plane defects fabricated by microstereolithography. *Int. J. Appl. Ceram. Technol.*, 6: 41–44.

186. Ohta, T., Niki, T., and Kirihara, S. (2011). Fabrication of terahertz wave resonators with alumina diamond photonic crystals for frequency amplification in water solvents. *IOP Conf. Ser. Mater. Sci. Eng.*, 18: 072015.

187. Kirihara, S., Ohta, N., Niki, T., and Tasaki, S. (2012). Fabrications of terahertz wave resonators in micro liquid cells introduced into alumina photonic crystals with diamond structures. *ISRN Mater. Sci.*, 2011: 897235.

188. Kirihara, S. and Nakano, M. (2012). Terahertz wave properties of magnetophotonic crystals fabricated by micro stereolithography. *Trans. JWRI*, 41(2): 19–22.

189. Chartier, T., Duterte, C., Delhote, N., Baillargeat, D., Verdeyme, S., Delage, C., and Chaput, C. (2008). Fabrication of millimeter wave components via ceramic stereo- and microstereolithography processes. *J. Am. Ceram. Soc.*, 91: 2469–2474.

190. Beluze, L., Bertsch, A., and Renaud, P. (1999). Microstereolithography: a new process to build complex 3D objects. In: *Proceedings of Design, Test, and Microfabrication of MEMS and MOEMS*, pp. 808–817.

191. Bertsch, A., Lorenz, H., and Renaud, P. (1999). 3D microfabrication by combining microstereolithography and thick resist UV lithography. *Sens. Actuators*, 73: 14–23.

192. Kawata, S., Sun, H.B., Tanaka, T., and Takada, K. (2001). Finer features for functional microdevices. *Nature*, 412: 697–698.

193. Ji, T.S., Vinoy K.J., and Varadan, V.K. (2002). RF MEMS phase shifter by microstereolithography on silicon. In: *Proceedings of SPIE 4700: Smart Structures and Materials: Smart Electronics, MEMS, and Nanotechnology*, pp. 58–65.

194. Sun, H.B. and Kawata, S. (2003). Two-photon laser precision microfabrication and its applications to micro-nano devices and systems. *J. Lightwave Technol.*, 21: 624–633.

195. Maruo, S., Ikuta, K., and Ogawa, M. (2004). Laser-driven multi-degrees-of-freedom nanomanipulators produced by two-photon microstereolithography. In: *Proceedings of SPIE 5662: Fifth International Symposium on Laser Precision Microfabrication*, pp. 89–94.

196. Kim, D.S., Lee, I.H., Kwon, T.H., and Cho, D.W. (2004). A barrier embedded Kenics micromixer. *J. Micromech. Microeng.*, 14: 1294–1301.

197. Kobayashi, K. and Ikuta, K. (2008). Three-dimensional magnetic microstructures fabricated by microstereolithography. *Appl. Phys. Lett.*, 92: 262505.

198. Hutchins, D.A., Billson, D.R., Bradley, R.J., and Ho, K.S. (2011). Structural health monitoring using polymer-based capacitive micromachined ultrasonic transducers (CMUTs). *Ultrasonics*, 51: 870–877.

199. Goswami, A., Phani, A., Krisna, A., Balashanmugam, N., Madras, G., and Umarji, A.M. (2011). Poly-HDDA Microstructure Fabrication Using Microstereolithography for Microcantilever-Based Sensor Technology, paper presented at SPIE 7926: Micromachining and Microfabrication Process Technology XVI, San Francisco, CA, January 22.

200. Snowden, M.E., Unwin, P.R., and Macpherson, J.V. (2011). Single walled carbon nanotube channel flow electrode: hydrodynamic voltammetry at the nanomolar level. *Electrochem. Comm.*, 13: 186–189.

201. Nakamoto, S., Kobayashi, K., Ikeuchi, M., and Ikuta, K. (2011). Mobile micro screw pump with flow sensing capability for on-site flow control in microchannel device. In: *Proceedings of the 24th IEEE Workshop on Micro Electro Mechanical Systems (MEMS '11)*, pp. 1166–1169

202. Lee, D., Hiroshima, H., Zhang, Y., Itoh, T., and Maeda, R. (2011). Cylindrical projection lithography for microcoil structures. *Microelectron. Eng.*, 88: 2625–2628.

203. Vidyaa, V. (2012). Hybrid design electrothermal polymeric microgripper with integrated force sensor. In: *Proceedings of the 8th Annual COMSOL Conference in Bangalore*.

204. Leigh, S.J., Bowen, J., Purssell, C.P., Covington, J.A., Billson, D.R., and Hutchins, D.A. (2012). Rapid manufacture of monolithic micro-actuated forceps inspired by echinoderm pedicellariae. *Bioinspir. Biomim.*, 7(4): 044001.

205. Monri, K. and Maruo, S. (2013). Three-dimensional ceramic molding based on microstereolithography for the production of piezoelectric energy harvesters. *Sens. Actuators A*, 200: 31–36.

206. Baker, E., Reissman, T., Zhou, F., and Sun, C. (2012). Material property manipulation of photopolymer vibration energy harvesters. In: *Proceedings of the 6th International Conference on Micro- and Nanosystems*, pp. 263–272.

207. Sugiyama, K., Monri, K., and Maruo, S. (2013). Three-dimensional vibration energy harvester using a spiral piezoelectric element. In: *Proceedings of the 2013 International Symposium on Micro-NanoMechatronics and Human Science (MHS)*, pp. 1–6.

208. Kang, H.W., Lee, I.H., and Cho, D.W. (2006). Development of a micro-bellows actuator using micro-stereolithography technology. *Microelectron. Eng.*, 83: 1201–1204.
209. Kang, H.W., Lee, I.H., Cho, D.W., and Yun, W.S. (2005). Development of micro-bellows actuator using micro-stereolithography technology. In: *Proceedings of the 35th International Conference on Micro- and Nano-Engineering*.
210. Ikuta, K., Hirowatari, K., and Ogata, T. (1994). Three dimensional micro integrated fluid system (MIFS) fabricated by stereo lithography. In: *Proceedings of the 7th IEEE Workshop on Micro Electro Mechanical Systems (MEMS '94)*, pp. 1–6.
211. Bertsch, A., Heimgartner, S., Cousseau, P., and Renaud, P. (2001). 3D micromixers—downscaling large scale industrial static mixers. In: *Proceedings of the 14th IEEE Workshop on Micro Electro Mechanical Systems (MEMS '01)*, pp. 507–510.
212. Kang, T., Lee, S.J., Kim, Y., Lee, G.W., and Cho, D.W. (2010). Intelligent micro blood typing system using a fuzzy algorithm. *J. Micromech. Microeng.*, 20: 015024.
213. Hasegawa, T., Nakashima, K., Omatsu, F., and Ikuta, K. (2008). Multi-directional micro-switching valve chip with rotary mechanism. *Sens. Actuators A*, 143: 390–398.
214. Snowden, M.E., King, P.H., Covington, J.A., Macpherson, J.V., and Unwin, P.R. (2010). Fabrication of versatile channel flow cells for quantitative electroanalysis using prototyping. *Analyt. Chem.*, 82: 3124–3131.
215. Vatani, M., Lu, Y., Engeberg, E.D., and Choi, J.W. (2013). Fabrication and Characterization of 3D Printed Compliant Tactile Sensors, paper presented at Rapid Additive Manufacturing Solutions, Pittsburgh, PA, June 10–13.

5 Printed and Hybrid Electronics Enabled by Digital Additive Manufacturing Technologies

Pooran C. Joshi, Teja Kuruganti, and Chad E. Duty

CONTENTS

ABSTRACT

Additive manufacturing and printed electronics technologies have the potential to enable new products and markets. The capabilities, versatility, and reliability of additive manufacturing and direct-write techniques are growing at a rapid pace to meet the cost and performance demands of future manufacturing technologies. This chapter provides an overview of the emerging additive manufacturing concepts and techniques for the processing and two-dimensional (2D) and three-dimensional (3D) integration of functional electronic components and devices on microscale structures. Key enabling processing technologies for flexible electronic applications and their impact on material and device integration are also discussed. Additive manufacturing technology in combination with printed electronics has the potential to define the path toward hybrid technology integration of sensors and electronics on engineered 3D geometries. Recent developments in flexible hybrid electronics exploiting additive manufacturing and direct-write printing technologies are also highlighted.

5.1 INTRODUCTION

The three-dimensional (3D) additive manufacturing (AM) technologies enabling design-driven product development have advanced at a rapid pace since their introduction more than 20 years ago. AM is on the verge of reaching maturity for industrial manufacturing, and the market size is expected to reach \$5.2 billion by 2020. The design freedom offered by AM systems is being exploited for a variety of innovative applications ranging from conventional prototyping and rapid tooling to more advanced applications such as medical and dental, building construction, aerospace and automotive manufacturing, 3D electronic devices, consumer products, and microsystems.[1–10] AM technologies employing metal, dielectric, polymer, and ceramic materials are creating a world of possibilities surpassing limitations of conventional technologies and decoupling product

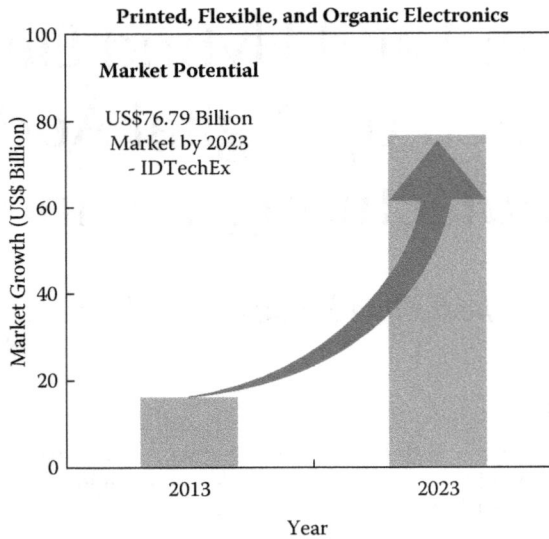

FIGURE 5.1   Market share projections for printed, flexible, and organic electronics.

development time and complexity of design.[11–15] In addition to offering the ability to manufacture highly complex parts at high resolution without adding cost, additive manufacturing can lead to substantial reductions in material waste and product development time compared to conventional subtractive manufacturing processes.

Printed electronics are becoming a more and more viable technology every day with rapidly expanding material and device sets and a wide application range.[16–18] Although the semiconductor industry is continuously pushing toward higher density electronic circuits to meet cost and performance demands, the printed electronics technology is aiming to exploit large-area roll-to-roll manufacturing techniques to meet future manufacturing technology demands, with applications extending beyond the traditional sphere to military, automotive, aerospace, medical, and consumer applications.[19–25] The emerging industry of large-area manufacturing and organic and printed electronics is bringing about new opportunities for the realization of sensors on unconventional substrates. The total market for printed, flexible, and organic electronics, as shown in Figure 5.1, is projected to grow from $16.04 billion in 2013 to $76.79 billion in 2023 with huge growth potential in the stretchable electronics, logic, memory, and thin-film sensors segments as they emerge from research and development (R&D).[26] Over 3000 organizations are pursuing printed, flexible, and organic electronics, including printing, electronics, materials, and packaging companies. The combination of cost-effective additive manufacturing processes and digital printing techniques offers a roll-to-roll manufacturing platform for the development of conformal, flexible electronics.[27–30] Inkjet-based contact-free printing technologies are attracting interest for low-cost, large-area production of high-performance electronic devices and systems that meet cost, functionality, and environmental demands.[31–35] Unlike complementary metal-oxide semiconductor (CMOS) device processing or printed circuit board (PCB) manufacturing technology, which are subtractive in nature, inkjet printing allows maskless, non-contact integration of functional components, circuits, and devices. The possibility of manufacturing semiconductor devices on flexible substrates opens up a wide expanse of new applications that were not practical with traditional electronics. Current state-of-the-art inkjet printers are capable of printing functional electronic circuits with line widths as small as 10 μm. Benefiting from the rapid growth in the nanomaterials processing techniques, low-temperature flexible substrates, advances in ink technology, and roll-to-roll-compatible low-temperature curing techniques, inkjet printing technology will play an increasingly dominant role in shaping printed electronics.

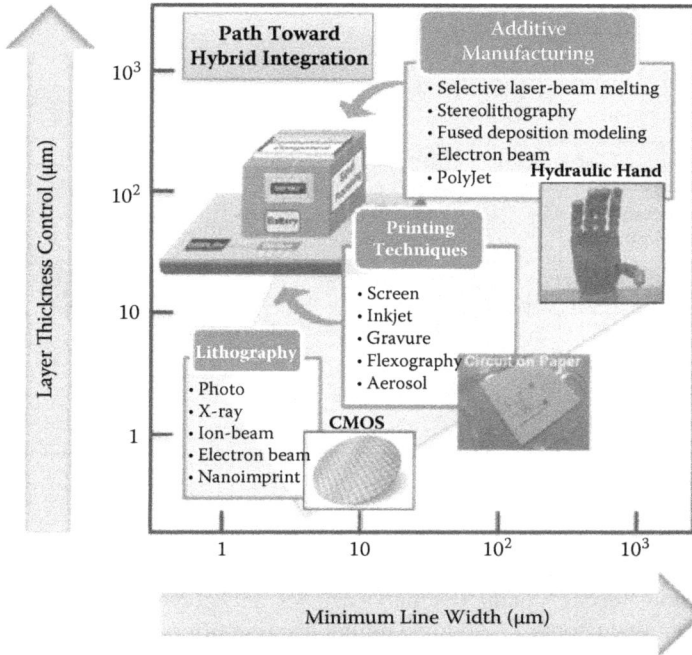

FIGURE 5.2 Path toward hybrid technology development combining flexible electronics and printed 3D structures.

The capabilities of 3D printing are continuously growing, and the path toward hybrid technology integration is becoming clearer as shown in Figure 5.2.[36-40] The additive printed electronics technology, in combination with additive manufacturing technologies, is opening up new possibilities for conformal electronic circuits and systems embedded in advanced 3D structures. Not only does additive manufacturing offer a high degree of design freedom, but highly complex geometries can also be fabricated more quickly while consuming less material and using less energy. From an engineering and design standpoint, AM technologies are becoming more accurate, with feature size control extending below 100 μm. Direct digital printing allows the development of conformal electronics on diverse surfaces. Several multifunctional devices that include sensing, communication, computation, and signal processing abilities can be integrated into a flexible platform using printing techniques. This provides an opportunity to develop innovative products combining multifunctional printed circuits and additively manufactured 3D structures. Overall, the combination of 3D microscale structures and additive printed electronics is going to have a significant impact on the functionality and component design, manufacturing processes, supply chains, and business models.

This chapter focuses on the key developments in the flexible printed electronics and additive manufacturing technology that are defining the path toward a hybrid technology for potential applications in aerospace, defense, health, and electronics sectors. The core enabling components, as shown in Figure 5.3, and their impact on the materials and device technologies are highlighted in this chapter.

5.2 INKJET PRINTING TECHNOLOGY

Over the past decade, rapid advances in additive printing technologies have been opening up an entirely new world of electronic components, devices, and systems. Growing interest in all printed electronics technology is aimed at exploiting its key advantages of low cost, high throughput, hybrid integration, and green manufacturing. The range of materials for printed electronics produced

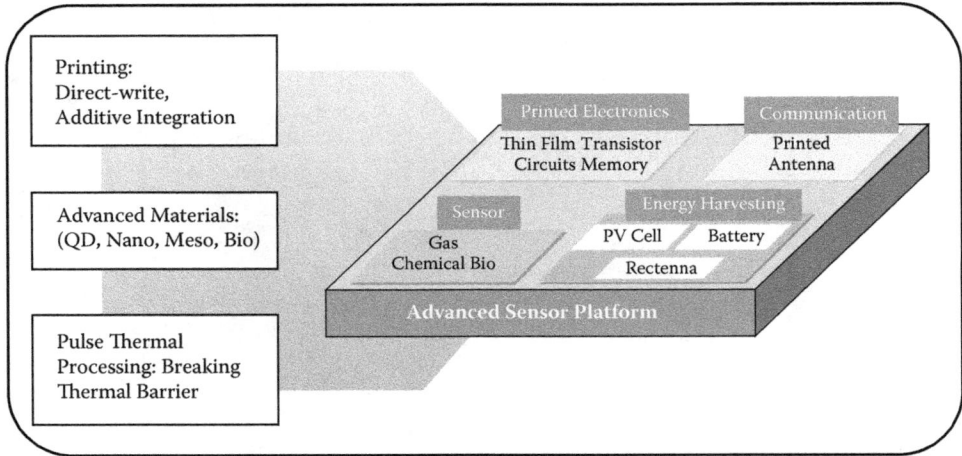

FIGURE 5.3   Core components to realize a multifunctional sensor platform.

by direct-writing techniques is growing rapidly, and the material list includes organic/inorganic semiconductors, biomaterials, conductive nanoparticles and polymers, dielectrics, ferromagnetic materials, and superconductors.[41–50] Additive digital printing offers new possibilities for conformal integration of active and passive devices meeting the manufacturing technology demands of higher functionality and improved energy efficiency over existing technologies. Printed electronics show promise for use in a wide range of discrete and integrated electronic device applications, such as thin-film transistors, organic light-emitting diode (OLED) lighting, solar cells, radiofrequency identification (RFID), antennas, inductors, capacitors, interconnects, sensors, and displays. All kinds of hybrid applications can also be envisioned as a result of cointegration with conventional technology.

Among the ink-based printing techniques, screen printing has been the one most extensively developed for printed electronic applications.[51–53] The screen printing technique is suitable for low-cost, high-volume production of printed electronics products on a wide variety of planar and curved substrates, including glass, plastic, ceramics, and fabric. High print resolution (<50 µm), design flexibility, and low infrastructure costs of the screen-printing technology have been well exploited for printed circuit boards (PCBs), photovoltaic (PV) technology, batteries, sensors, and display applications using a wide range of available inks. Even though the versatility of screen printing has led to the development of diverse high-density electronic products, the transition of the technology toward a roll-to-roll production platform is limited due to major disadvantages related to limited line resolution and thickness control, low throughput, poor line definition on flexible substrates, and high demands on ink viscosity control.

Although screen printing has dominated most of the printed electronics applications to date, the focus is shifting to inkjet printing and traditional high-speed printing methods to meet the cost, functionality, throughput, and scalability demands of roll-to-roll manufacturing technology.[54–58] Figure 5.4 highlights some of the emerging roll-to-roll printing technologies. The gravure additive printing technique is attractive for the mass production of flexible electronic devices at high throughput and high resolution on a variety of suitable substrate materials. In gravure printing, an entire print pattern is transferred to the substrate in a single pass by means of an impression cylinder that applies pressure to allow the substrate to come in contact with the ink-filled recessed cells of the engraved cylinder. The main challenge for the gravure printed electronics technology is the limit of achievable resolution considering the ink viscosity, ink–substrate interaction, and mechanical design constraints. Flexography (or flexo printing) is the fastest growing analog printing process and has the potential to achieve the necessary speed, resolution (<20 µm), and precision for high-volume printing on a wide variety of substrates. Flexo and gravure printing techniques differ in the

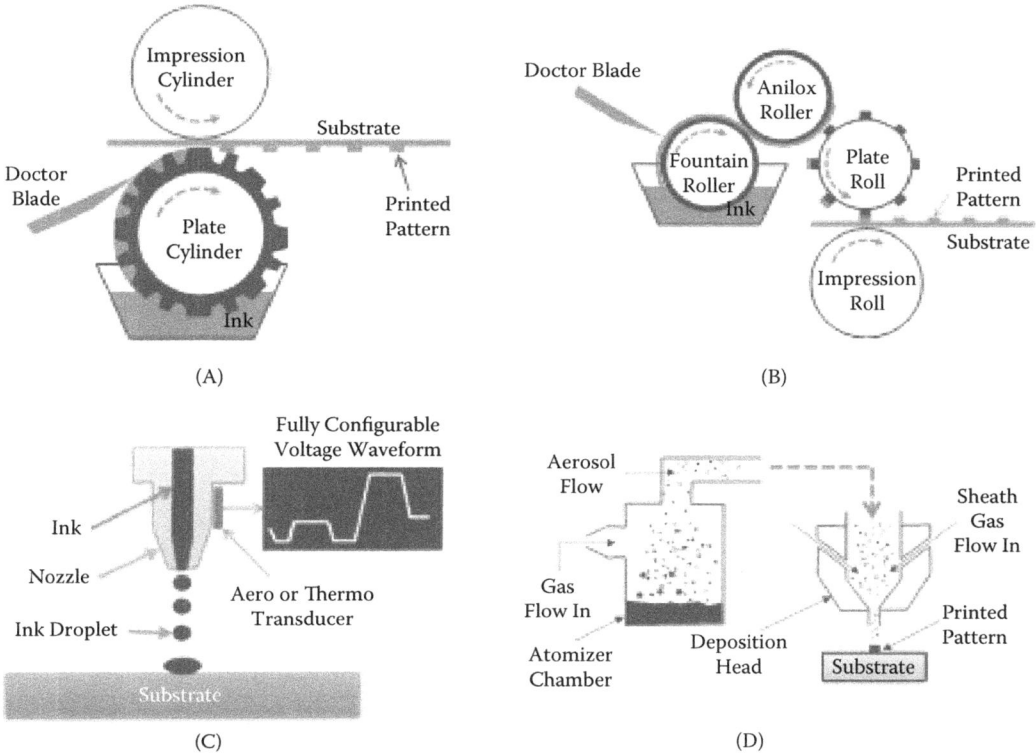

FIGURE 5.4   Emerging printing techniques for flexible electronics: (A) gravure, (B) flexographic, (C) inkjet, and (D) aerosol jet.

application of inks and imprint surfaces. As shown in Figure 5.4, flexographic printing incorporates different rolls—fountain, anilox, and plate cylinder—to achieve high-speed, continuous roll-to-roll processing. Flexo is the most flexible printing process, as the image transfer is facilitated by means of a soft flexible plate that can conform to diverse substrates. Although flexography is promising for a broad range of electronic and packaging applications, it faces the challenges of mechanical design and ink formulations impacting print resolution and flexo-plate integrity.

Inkjet printing is rapidly becoming the most promising direct-write technology for printed electronics. The inkjet printing approach combines material deposition and patterning in a single step, resulting in a significant reduction of device processing steps as required in a standard semiconductor cleanroom setup. As shown in Figure 5.5, conventional integrated circuit processing requires lithography techniques and masks for pattern generation, vacuum techniques for thin-film deposition and etching, coaters and developers for photoresist processing, and sizable financial support for maintenance. The printing technique eliminates the masking and etching steps, which alone results in a substantial reduction in material waste, cross-contamination, and energy consumption, as well as the processing time and steps required. Additionally, the non-contact, digital inkjet printing techniques offer significant advantages with regard to conformal integration, vacuum-free processing, and green infrastructure for roll-to-roll manufacturing of flexible electronics.

In general, inkjet printing techniques can be divided into two main categories depending on the method of drop generation: (1) continuous inkjet (CIJ) and (2) drop-on-demand (DOD).[59-61] CIJ, one of the oldest inkjet technologies, has been used extensively for a wide range of printing applications (e.g., automobile parts, medical diagnostics, electronic components, cosmetics) on structured and uneven surfaces. In single- or multi-nozzle CIJ systems, a continuous stream of ink is ultrasonically

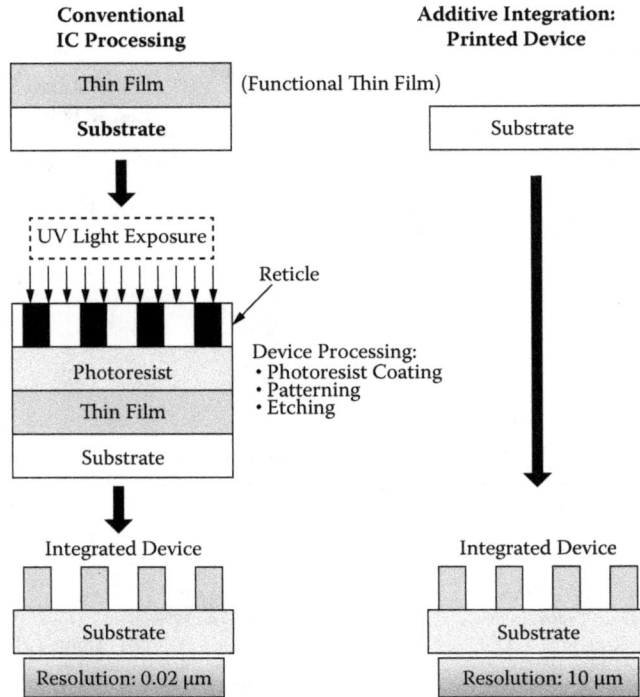

FIGURE 5.5 A comparison of subtractive integrated circuit processing and the additive printing approach for thin film and device integration.

stimulated to break down into small drops. Droplet size control below 10 μm has been reported for fine-line printing. The droplets are generated at a high rate (60 kHz to 1 MHz) and electrostatically steered for printing onto the substrate.[57] The unused droplets are recycled to repeat the process. Even though the flexibility and rapid prototyping capabilities of CIJ have been well exploited for numerous applications, the issues of droplet size and velocity control, fluid recycling efficiency, achievable resolution for micromanufacturing, and throughput on complex geometries are limiting its widespread use for functional electronic applications.

In a DOD printing system, ink drops are generated on demand by the application of a transient pressure pulse to the ink chamber behind a nozzle array. The pressure pulse is generated by a piezo-electric or thermal element. The drop size can be as small as 15 μm, while drop generation rates as high as 30 kHz have been reported, which are suitable for high-resolution printing applications. A fundamental difference between CIJ and DOD is in material throughput, as the DOD mode requires higher energy to produce a droplet as compared to the CIJ mode. The drop spacing in CIJ is about 2.3 times the drop diameter, resulting in a continuous process. In DOD, the minimum spacing can be 20 times the drop diameter, thus significantly impacting the printing throughput;[33] however, no ink recycling is required in the DOD printer system which eases demands on ink formulation and viscosity control. Thermal-mode DOD printers are far behind piezoelectric-mode printers, as the thermal mode makes the fluid an integral part of the energy input device, putting limitations on ink–solvent selection and fluid property control. Overall, the piezoelectric DOD inkjet technology is the most efficient of the inkjet technologies for high-resolution, high-speed printing of complex electronic components on a wide variety of printing surfaces.

The aerosol jet technology offers 3D conformal printing of a wide range of materials (metals, insulators, ferrites, polymers, adhesives, biological materials) onto almost any substrate with line-width control over three orders of magnitude from 5 μm to 5 mm.[62–64] The aerosol jet system can handle inks with viscosities in the range of 1 to 2500 cP, and its ability to print high-solids-content inks (≥60 wt%) exceeds that of conventional inkjet printers. The unique ability of the aerosol jet system to print on non-planar surfaces makes it an ideal solution for hybrid sensor technology development in combination with 3D additive manufacturing techniques. The printing process utilizes a dense aerosol of microdroplets that is aerodynamically focused using a flow guidance deposition head. The material beam is focused to as small as a tenth the size of the nozzle orifice (typically, 100 μm). Aerosol jet systems can precisely deposit materials on both planar and non-planar substrates. Patterning is accomplished by attaching the substrate to a computer-controlled platen or by translating the flow guidance head while the substrate position remains fixed. The relatively large (>5 mm) standoff distance from the deposition head to the substrate allows accurate material deposition on non-planar substrates, over existing structures, and into channels. Aerosol jet systems can locally process high-quality thin film (as fine as 10 nm) with excellent edge definition and near-bulk properties on low-temperature substrates using a laser treatment process. Current aerosol jet systems can write at speeds up to 200 mm/s with high dynamic accuracy (±6 μm). Aerosol jet systems can be equipped with multi-nozzle deposition heads to meet the demands of high-volume roll-to-roll manufacturing. The aerosol jet technology has been exploited for a wide range of applications, including flexible displays, electromagnetic shielding, solder-free electronics, high-efficiency solar cells, and embedded components, including sensors, resistors, and antennas. Although the aerosol jet technology shows promise for conformal 3D printing of electronic components and devices, solving problems related to its process complexity, overspray control, and resolution control below 20 μm will dictate the path taken toward manufacturing technology integration.

5.3 LOW-THERMAL-BUDGET ANNEALING

Novel low-temperature, low-thermal-budget annealing techniques are required to realize highly functional materials and devices on flexible substrates. Material and device characteristics are strongly influenced by the thermal treatments, whereas the device integration platform defines the acceptable limit of processing temperature. The development of high-performance quantum dots, thin films, thick films, and bulk ceramics is strongly influenced by the sintering step. Thermal treatments are designed to activate dopants, control grain growth, influence crystallinity, improve material density, impact the material–substrate interface, minimize bulk or interfacial defects, and promote substrate surface reaction kinetics.[65–68] Furnace annealing is used extensively for bulk ceramics and thick-film processing where high temperatures (>800°C) and long annealing cycles (>1 hour) are typically required to form high-quality materials. It is a simple, scalable, and relatively inexpensive technique suitable for low-throughput applications. However, the high thermal budget for furnace curing does not meet the large-area processing demands of roll-to-roll manufacturing technology. At the same time, the low heating and cooling rates of conventional furnaces limit process throughput. Increasingly, furnace annealing techniques are being replaced by rapid thermal annealing (RTA) methods to meet the demands of submicron CMOS technology, nanomaterials processing, and low-temperature flexible electronics. In general, RTA techniques offer the significant advantages of short cycle time, reduced thermal exposure, and lot size flexibility as compared to conventional furnace annealing. Figure 5.6 shows the promising low thermal budget technologies for advanced material and device development.

Each annealing technique offers a different thermal budget, energy density and distribution, and heat coupling efficiency to control the bulk and interfacial properties of advanced materials and devices. Spike annealing using conventional tungsten–halogen lamps allows material processing on a time scale of a few seconds. CMOS electronics have effectively exploited the spike annealing technique for shallow junction formation; however, the heating and cooling rates are limited by the thermal mass of the lamp, which limits its applicability for deep submicron CMOS technology. The slow ramp rates result in a material thermal state similar to the heating source, resulting in a uniform heating of the entire structure that is not desirable for low-temperature flexible electronics. Laser sintering techniques offer the high energy density necessary to overcome the

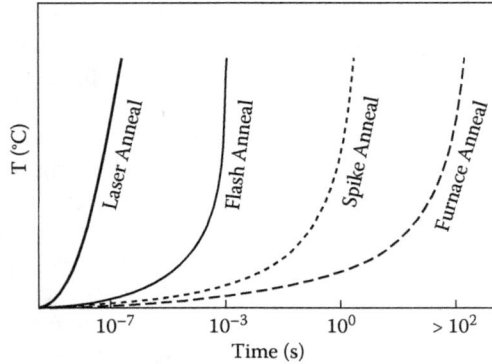

FIGURE 5.6   Key low thermal budget annealing technologies for advanced material and device development.

thermal activation barrier and low thermal budget for ultra-shallow junction formation in CMOS devices. Laser processing techniques are also a promising alternative for low-temperature material and device integration on glass or plastic substrates. The ultrafast (nanoseconds) and low thermal budget laser curing allows for surface annealing without impacting the underlying fragile layers and devices. However, the cost and throughput are major obstacles to integrating laser processing tools into large-area roll-to-roll manufacturing platforms.

5.4   PULSE THERMAL PROCESSING TECHNOLOGY

The pulse thermal processing (PTP) technique is being explored for the development of cost-effective, high-yield, and high-quality integrated thin films and devices on low-temperature substrates. The PTP technique is a new alternative to flash annealing techniques. The moderate power levels of flash annealing systems are not suitable for low-temperature materials processing considering short pulse lengths. Additionally, it is a batch process with a low processing rate (~1 wafer/minute). The PTP technology offers the ability to expose large areas of material to an extremely high energy flux during a very short period of time, as shown in Figure 5.7, thus meeting the demands of roll-to-roll manufacturing technology.[69,70] One of the key advantages of PTP processing as compared to laser annealing is that the entire substrate is exposed to the light. Laser annealing requires a complex scanning scheme and a sample translation stage to accomplish large-area processing as the laser spot size is typically smaller than 50 μm, limiting process throughput, and expensive optics are

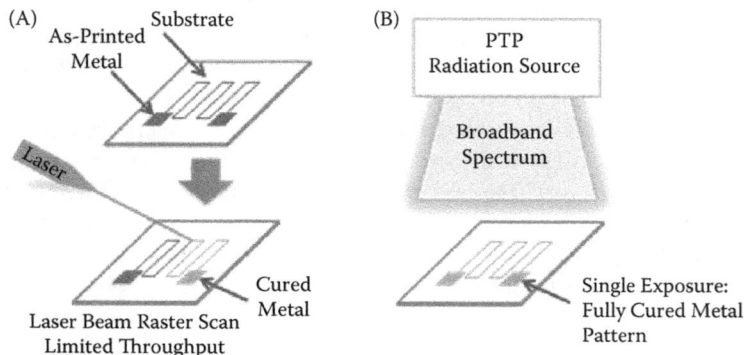

FIGURE 5.7   Impact on process throughput of (A) laser annealing, and (B) pulse thermal processing.

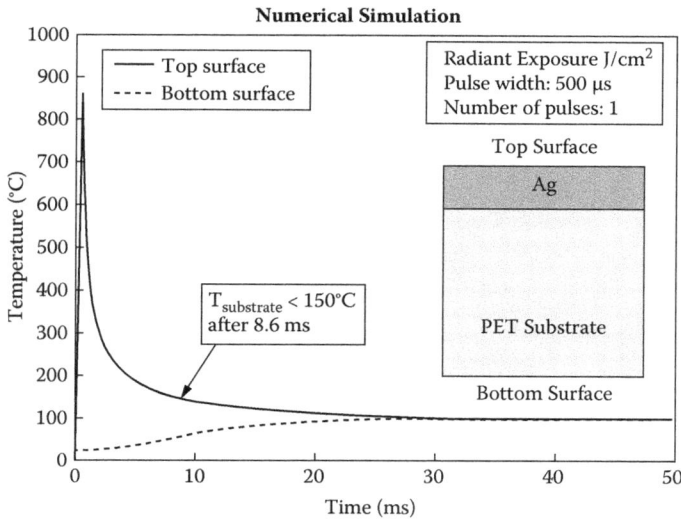

FIGURE 5.8   Simulated temperature profile on a silver (1 µm) and PET (125 µm) structure.

required for large-area beam shaping. The power efficiency of laser systems of interest for thin-film electronics is typically below 2%, whereas the PTP system uses high-intensity flash lamps with an efficiency of about 25%.

The general benefits of pulse thermal processing with a plasma arc lamp include low thermal budget, high throughput, rapid heating and cooling rates, low-temperature substrate integration, and nanometer-scale diffusion control. The PTP technique has the ability to thermally process thin layers of material at several hundred degrees Celsius while restricting the thermal exposure of underlying materials to a small percentage of that temperature.[71] As an example, Figure 5.8 shows the temperature profile of a 1-µm-thick silver layer printed on a 125-µm-thick polyethylene terephthalate (PET) substrate. PET has a maximum working temperature of about 150°C. A 500-µm-wide PTP pulse of light is able to heat the silver (Ag) film to over 800°C while the substrate remains unaffected, as the heating time is shorter than the thermal equilibrium time in the substrate. The high-conductivity Ag line is formed after a processing time more than three orders lower than the processing time of a few minutes typically required in an oven. The low processing time is significant for the processing of metals such as copper that show high reactivity toward oxygen and require inert or reducing atmospheres during thermal treatment to prevent nonconductive oxide layer growth.

The PTP system based on a high-density plasma arc lamp (PulseForge® 3300; NovaCentrix, Austin, TX) is capable of producing an extremely high power density (up to 20 kW/cm²), which meets the processing demands of diverse materials and device technologies such as displays, batteries, and photovoltaics. The broadband lamp spectrum can be exploited for processing both thermal and ultraviolet (UV)-processed material sets. The PTP technique is capable of delivering extremely high power in very short pulses (as low as 30 µs), thus enabling functional device integration on low-temperature substrates such as glass, plastic, paper, and fabric. The PTP platform is suitable for small-scale and pilot production, as well as full-speed production (>300 m/min) when needed. The low-temperature, low-thermal-budget PTP technique has been explored for the development of diverse functional materials ranging from quantum dots to bulk ceramics, as shown in Figure 5.9. The fluence and radiation spectrum of PTP have been successfully exploited for diverse applications, including ultra-shallow junction control in CMOS, quantum efficiency enhancement for solid-state lighting, cathode material crystallization for batteries, glass strengthening for structural applications, fabricating metal sheets from nanoparticles, flexible solar cells, and structured materials for automobiles, among others.

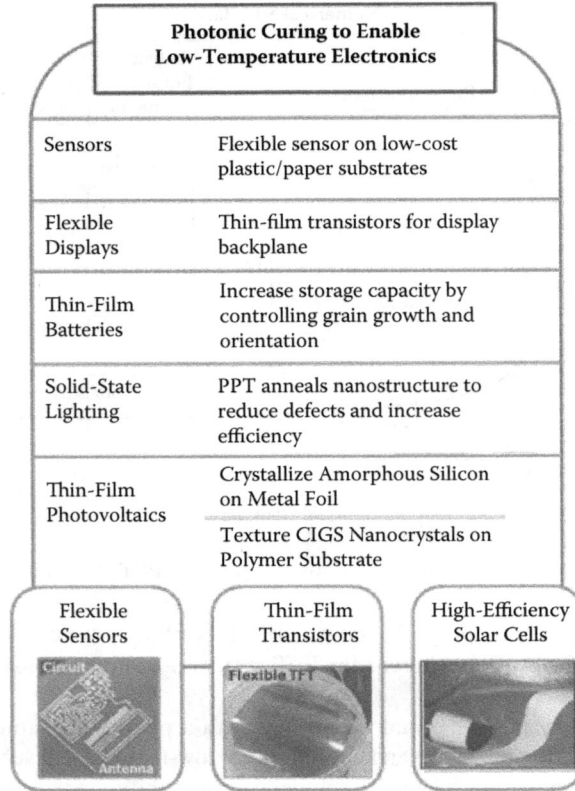

FIGURE 5.9 Low-temperature electronics enabled by pulse thermal processing technology.

5.5 INK DEVELOPMENT AND DEVICE APPLICATIONS

High-quality inks with controlled rheological properties are required to exploit the structural, mechanical, optical, and electrical properties of nanomaterials for functional device development. Over the years, extensive efforts have been directed toward high-performance ink development, and the range of functional materials that can be printed (metals, semiconductors, ceramics, biomaterials, quantum dots, and polymers) is rapidly expanding. Ink technology is continuously improving to keep pace with the evolving printing systems, substrate materials of interest, and potential markets.[72–76] For optimal performance and practical applications, the ink technology has to meet the critical demands of high-resolution printing, long shelf life, low-temperature curing, and green formulation. In addition to functional inks, the substrate selection is critical for the development of a successful device technology meeting the cost and performance demands of the manufacturing technology. Substrate selection is strongly affected by cost, weight, electrical and mechanical properties, ink–substrate compatibility, chemical and thermal resistance, and achievable printing resolution. A wide range of substrate materials such as glass, silicon, ceramic, paper, plastic, and fabric have been used for the development of high-resolution printed electronics.[77–81] With the available ink formulations and substrate materials, the printing resolution of drop-on-demand inkjet technology has already surpassed that of conventional screen printing technology, and the line width values are fast approaching 10 μm.

Inkjet printing of conductive metal nanoparticle inks as electrical contacts and interconnects in printed electronic circuits and devices is probably the most extensively researched topic. For practical electronic applications, metal-based inks have to meet the critical requirements of high

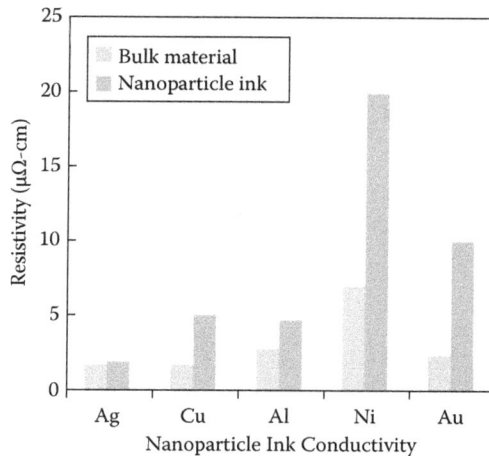

FIGURE 5.10 Nanoparticle metal ink performance in comparison to bulk material.

electrical conductivity of the printed patterns at line widths below 100 μm and a curing temperature compatible with low-temperature substrates such as PET and paper. Metal nanoparticle-based
conductive inks (e.g., silver, copper, gold, nickel) are being extensively investigated to achieve low
curing temperatures and exploit the high surface-to-volume ratios of nanomaterials. As shown in
Figure 5.10, metal ink performance has improved significantly over the years, and printed metal
conductivity values are quickly approaching bulk values with advances being made in nanomaterial
technology and low thermal budget annealing techniques.[82–85]

Printing techniques in combination with low-cost substrates have been exploited for a wide
spectrum of electronic devices (e.g., photovoltaic cells, transistors, OLEDs, photocells, antennas,
disposable diagnostic devices, batteries) and systems.[86–95] Some typical device configurations are
highlighted in Figure 5.11. Developments in printing techniques are continuously introducing new
ways to sense, detect, and communicate at high throughput, good resolution, and low production
cost. Inkjet printing can be used to fabricate single-use, cost-competitive, and small sensors with fast
response times and low detection limits. Inkjet printing of conducting polymers and bioenzymes has
been explored for glucose sensors. Graphene-based inkjet printing of flexible bioelectronic circuits
and sensors holds promise for use with nanomaterial-based sensor technology. For chemical and gas
sensing applications, a number of colorimetric sensor arrays, optoelectronic noses, chemresistors,
and surface-enhanced Raman spectroscopy (SERS) sensors have been developed. Nanoparticle and
polymer inks have been engineered to print Schottky diodes, OLEDs, photoconductors, and photodiode devices operating in the visible and infrared part of the optical spectrum. Reliable inkjet-
printed silver interconnects have been developed for foil-type lithium-ion batteries. Inkjet-printed
strain gauges and active pressure sensors will have a significant impact on structural condition
monitoring in the fields of aerospace engineering, automobiles, and medical science. Inkjet-printed
actuators have been demonstrated for microfluidic lab-on-a-chip (LOC) systems. Inkjet printing
is also being explored as a powerful microfabrication tool for the manufacture of ceramic components.[50] Bulk and thick-film ceramic sensor technology is well established, and inkjet printing
techniques can be further exploited for advanced hybrid sensor technology development.

Printing techniques are promising for the ultra-low-cost realization of flexible antennas and
radiofrequency (RF) electronics.[96–100] Radiofrequency identification (RFID) is a rapidly developing wireless technology for remote object identification and data collection, and inkjet printing
has the potential to evolve as one of the most promising RFID mass-production techniques. Direct
inkjet printing of antennas on paper and plastic substrates with the use of electrically conductive
inks is a promising approach to the production of low-cost RFID tags. Inkjet-printed systems on

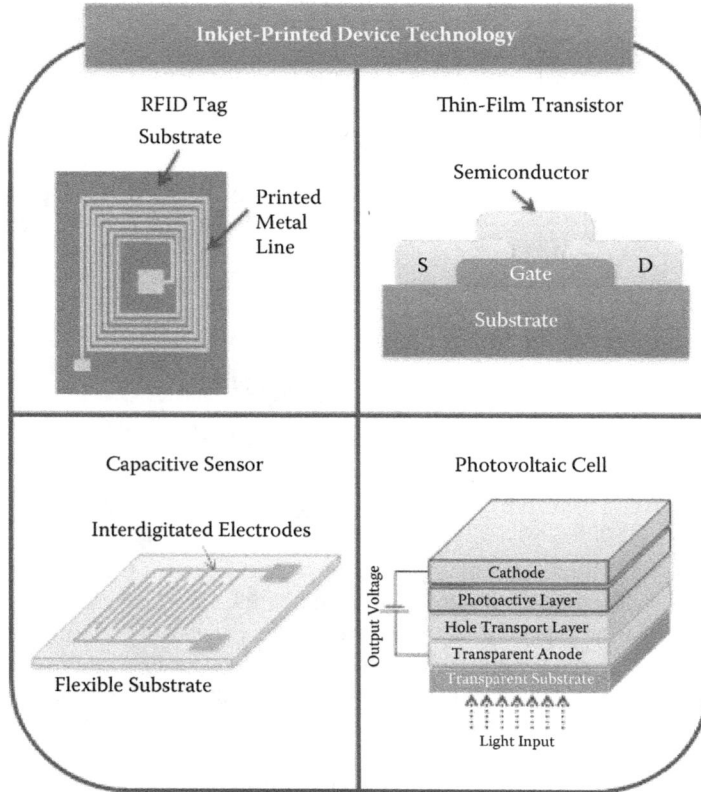

FIGURE 5.11   Select printed device technologies for low-cost electronics.

paper or plastic substrates could set the foundation for conformal antennas and packages that are wearable. Inkjet-printed flexible antennas are key to implementing low-cost wireless sensor nodes for target sensor data collection, precision localization, and tracking applications. Ultra-wideband (UWB) antennas operating at high frequencies up to 10 GHz have been reported.[101] Inkjet-printed, graphene-based thin films are showing promise for wideband antenna applications.[102] Inkjet printing has been used to create high-quality passive devices such as spiral inductors, interconnects, and parallel plate capacitors for an RFID front end.

Inkjet techniques are being intensively investigated for the development of inexpensive renewable energy sources to achieve grid parity. Combining the inkjet printing process with the use of flexible substrates has the potential to significantly reduce manufacturing costs by eliminating the patterning process and allowing for a mass production technique, such as roll-to-roll fabrication. Metallization is a key processing step that impacts the performance of solar cells optically and electrically. The use of direct-write metal contacts has been demonstrated for diverse solar cell technologies, including silicon solar cells, dye-sensitized solar cells (DSCs), and copper–indium–gallium–selenide (CIGS) solar cells.[103–105] Additive integration of multiple metal layers by printing techniques has been shown to improve the contacting process and cell performance for silicon (Si) photovoltaics. It is possible to achieve conductivities close to the bulk metals in 30- to 40-μm wide metal lines while minimizing material waste. Direct writing enables quick prototyping of different contact schemes and geometries at a uniformity level not possible with screen printing techniques. Inorganic chalcopyrite $CuIn_xGa_{1-x}Se_2$ thin-film solar cells exhibiting an efficiency of 5.04% have been fabricated by an inkjet printing technique offering potential cost advantages over conventional fabrication processes involving deposition, patterning, and etching of selected materials. The initial cell performance is suitable for self-powered sensor platforms, and further improvements in efficiency beyond the 10% level will make the printed PV technology attractive for stand-alone applications.

Solar cells based on organic semiconductors have become a promising alternative to inorganic photovoltaics due to their low weight, low cost, easy processing, and flexibility when compared to traditional silicon solar cells. Among the organic photovoltaic fabrication techniques, inkjet printing is emerging as a key process for organic photovoltaic (OPV) development due to its key advantages of low cost, large-area roll-to-roll manufacturing capability, high registration accuracy, and non-contact conformal thin-film processing. Cost-effective polymer solar cells prepared with inkjet-printed active layers have demonstrated the best solar cell performance with a power conversion efficiency (PCE) of 3.7% and an incident photon-to-current efficiency (IPCE) of 55%. In general, OPV efficiencies have reached levels of 12.0%, and devices demonstrate superior performance under high-temperature and low-light conditions as compared to traditional crystalline Si and thin-film PV.[106] As the printed OPV cells fill the performance gap, the path toward roll-to-roll manufacturing will be clearly defined by exploiting additive printing techniques. Inkjet-printed non-metallic contacts such as PEDOT:PSS, graphene, single-walled carbon nanotubes (SWCNTs), and indium–zinc–tin oxide (IZTO) with high transparency and high electrical conductivity are emerging as potential replacements for indium–tin oxide (ITO) contacts currently used in diverse solar cell technologies.[107–109] The new electrode materials have the potential to meet the challenges of high cost, high processing temperature, and material reliability with ITO contacts currently used in diverse organic and inorganic thin-film solar cells. Overall, the combination of inkjet printing and emerging high-efficiency solar cell materials promises to make solar power competitive with electricity.

Printing techniques have been extensively investigated for the development of organic and inorganic thin-film transistors (TFTs) for display, optoelectronic, biomedical, radiofrequency identification, and logic and data processing applications.[110–113] The rapid advances in ink technology and nanomaterial processing are paving the way for ubiquitous integration of low-cost, flexible TFT devices on low-temperature glass, paper, fabric, and plastic substrates. Printed transistors cannot compete with traditional CMOS transistors in terms of operating speed and electrical performance; however, the additive integration approaches offer significantly reduced device processing steps, reduced material wastage, and a higher cost/performance advantage for practical applications. Metal nanoparticle inks and ink blends are suitable for source, drain, and gate contact development at process temperatures below 150°C. Diverse organic and semiconductor materials have been explored for the development of high-performance TFTs. There are many reports of organic thin-film transistors (OTFTs) exhibiting mobilities greater than 1 cm^2/Vs and switching performances comparable to those of conventionally fabricated polymer TFTs. The printed inorganic TFTs have shown high mobilities on the order of 20.6 cm^2/Vs.[114] The SWCNTs are a promising alternative to polymers and semiconducting oxides for inkjet-printing-based fabrication, and high-performance TFTs with mobilities exceeding 30 cm^2/Vs have been demonstrated. Inkjet printing has also been explored for the large-area fabrication of graphene devices with mobilities of up to ~95 cm^2/Vs, paving the way to all-printed, flexible, and transparent graphene devices on arbitrary substrates.[113] Inkjet printing offers design flexibility in implementing complex device geometries without adding process complexity that would be required in a standard CMOS process flow. Selectively printed high-κ/low-κ double-layer dielectric structures have demonstrated effective tuning of threshold voltage and corresponding change from depletion to enhancement operation modes in TFTs. Hybrid complementary TFTs involving organic and inorganic semiconductors have proven to be superior to an all-organic approach in terms of speed, gain, and compatibility to fabricate complex circuits. Ambipolar TFTs based on organic semiconductors have been experimentally realized with different active-layer configurations. A novel inkjet printing process is being utilized to demonstrate all inkjet-printed and fully self-aligned transistors. A wide range of functional devices and circuits have been developed including transistor arrays, complementary thin-film transistors, inverters, ring oscillators, differential amplifiers, and digital-to-analog converters (DACs). The performance

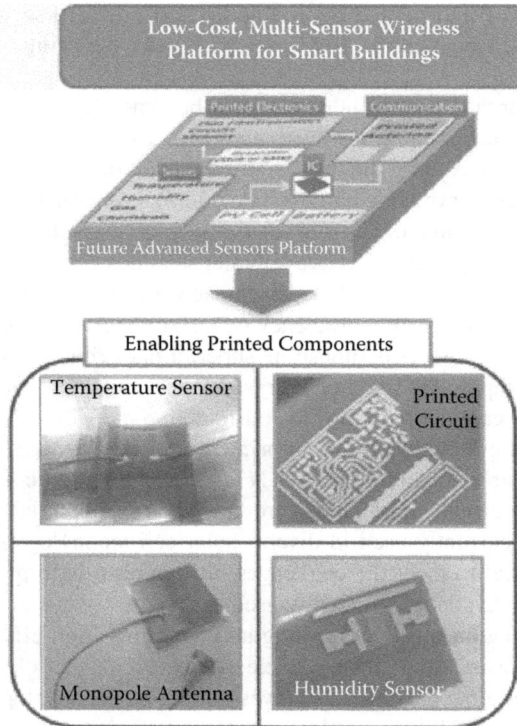

FIGURE 5.12 Exploiting conductive silver inks to print sensor, communication, and signal processing components.

of the current state-of-the-art printed TFTs and additive device integration exploiting sheet-to-sheet and roll-to-roll processes over large areas is promising for the development of a mass-scale manufacturing technology.

Flexible and printed sensors based on printed conductive patterns have the potential to significantly impact smart building technology. Buildings consume up to 40% of energy produced in the United States (www.eia.gov). Advanced sensors and controls have the potential to reduce the energy consumption of buildings by 20 to 40%. Currently, installation and wiring costs for sensors are significantly high and prohibitive for building managers to deploy large quantities of advanced sensors. Printable electronics represent a technology that has the potential to generate fully printable wireless sensors at cost. Printed electronics can establish a high degree of coordination among complexity, functionality, innovation, and expected benefits of the advanced sensor platform shown in Figure 5.12. Low-thermal-budget pulse thermal processing and drop-on-demand inkjet printing of silver conductive lines below 100-µm resolution have been well demonstrated. Environmental sensors for temperature, humidity, and gas sensing can be easily printed on low-cost flexible substrates to realize "peel-and-stick" wireless sensors.[115,116] Printed antennas have shown a return loss below −10 dB in the frequency range of 433 MHz to 5.5 GHz that encompasses many industrial, scientific, and medical (ISM) radio bands for low-power, short-range communication platforms. The printed temperature and humidity sensors on plastic and paper substrates show high sensitivity and selectivity for additive integration of environmental sensors. Evolving thin-film printing techniques and high-performance flexible devices such as thin-film transistors, solar cells, thin-film batteries, and antennas offer a system-level solution to realize a multifunctional sensor platform meeting the cost and energy-efficiency demands of future smart buildings.

5.6 HYBRID ELECTRONICS: MERGING PRINTED ELECTRONICS AND ADDITIVE MANUFACTURING

Additive manufacturing techniques and processes offer significant cost and time savings in comparison to conventional manufacturing technologies while enabling precise dimension control and unique geometrical complexity in 2D and 3D structures. Since their introduction more than 20 years ago, the versatility and flexibility of AM technologies have been exploited for a variety of applications ranging from conventional prototyping and rapid tooling to more advanced applications such as aerospace and defense, health care, buildings technology, automotive, consumer products, and space technology.[117–121] Rapid advances in AM technologies are changing the face of manufacturing processes with their ability to produce parts for every phase of a product's life cycle (from pre-development to prototyping to manufacturing). The representative AM processes currently being extensively investigated to build intricate structures and prototypes include stereolithography (SL), selective laser sintering (SLS), electron beam melting (EBM), fused deposition modeling (FDM), ultrasonic additive manufacturing (UAM), and 3D printing (3DP). The geometrical accuracy of these AM techniques is now well below the level of 100 μm.

The rapid advances in 3D additive manufacturing techniques and high-resolution digital printing techniques are paving the path toward achieving hybrid technologies offering unique functionalities and market opportunities. Hybrid AM blurs the boundary line between component development and system integration aided by the design freedom of AM techniques. Advances in silicon manufacturing have reduced the physical dimensions and power requirements while improving the processing capabilities of various devices. Several multifunctional devices with communication, computation, and signal-processing capabilities are currently available in the commercial market. These devices can be integrated into hybrid sensor platforms combining the flexibility of printed circuits and the processing power of commercial silicon manufacturing along with 3D integration of metal, dielectric, and semiconducting materials exploiting AM techniques. Hybrid sensor technology with conformal integration of advanced sensors and electrical interconnects on 3D structures with arbitrary and complex forms will lead to advanced electronic systems while addressing the cost and performance demands of the manufacturing technology. Figure 5.13 shows hybrid sensor platforms with printed, embedded, or bonded sensors, actuators, microprocessors, communication components, energy harvester devices, and signal-processing circuitry that are enabling new applications with capabilities extending beyond the scope of conventional materials and integration approaches.[122–133]

Embedded sensors in 3D custom-shaped structures and packaging configurations enabled by additive manufacturing and printed electronics offer continuous monitoring, significantly impacting the productivity, reliability, and safety aspects critical for technology integration. Passive conductive structures including circuit layouts and interconnect structures can be printed conformally on the substrate of interest and post-processed to achieve the desired functionality. This facilitates hybrid circuit manufacturing to embed electronics in various systems to address practical constraints. Currently, flexible circuits manufactured using conventional non-additive techniques have demonstrated numerous advantages in packaging, integration, and mobility in various applications. Printing provides the ability to utilize flexible circuits along with the ability to integrate circuits on existing systems (e.g., printing circuits on enclosures to reduce the volume of a device). Layered two-dimensional structures demonstrate how three-dimensional functional circuits can dramatically improve packaging requirements. Embedded piezoresistive and capacitive sensors processed in a single build process show rapid prototyping of multifunctional sensors. 3D printing for rapid prototyping of structural electronics has been demonstrated for flexible circuit boards for mobile applications; tactile sensors for robotic hands; electronic die with integrated microprocessors, accelerometers, and LEDs; lab-on-a-chip devices for *in situ* analysis; mechatronic components; custom microbatteries in individual or array-type geometries integrated directly onto a device; electroactive polymer actuators

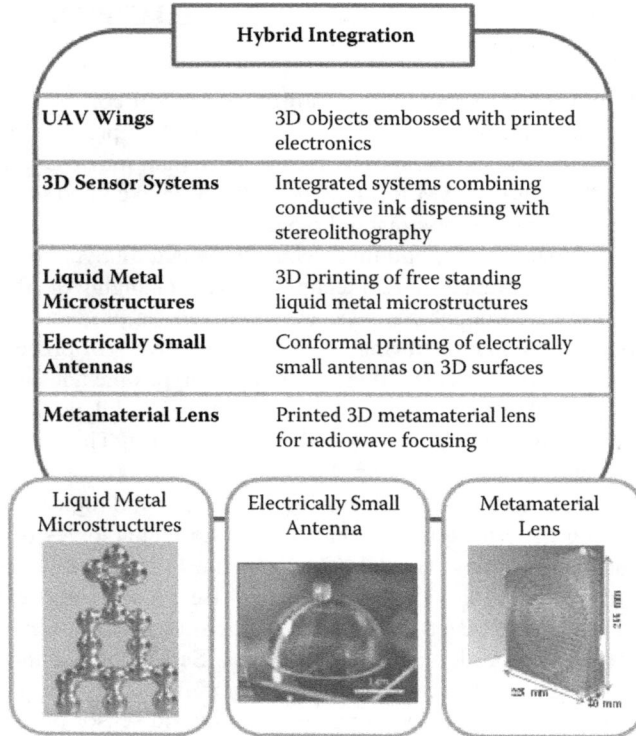

FIGURE 5.13 Combining additive manufacturing and direct-write techniques to define a path toward hybrid integration. (From Ladd, C. et al., *Adv. Mater.*, 25(36), 5081–5085, 2013; Adams, J.J. et al., *Adv. Mater.*, 23, 1335–1340, 2011; Ehrenberg, I.M. et al., *J. Appl. Phys.*, 112(7), 073114, 2012. With permission.)

for soft robotics; curved plastic parts with printed conductive tracks and components; highly integrated RF system-in-package (SiP); and electrically small antennas for embedded wireless sensor nodes. The emerging hybrid manufacturing concept is evident in the high-resolution (line width of ~10 μm) printing of a conformal sensor, antenna, and power and signal circuitry directly onto the FDM-printed wing of an unmanned aerial vehicle (UAV) model. AM technology is finding increasing use in tissue engineering and biomedical sensors. 3D-printed bionic ears with electromagnetic responses beyond normal audible frequencies show the potential of a 3D biosensor technology offering structural and electronic performances far exceeding those of normal human sensory organs.[129]

Advanced printing techniques offer an entirely new way of integrating sensors into structures. The ability of retroactively depositing these sensors on existing structures provides unique opportunities to improve the reliability of legacy systems as well as the low-cost manufacturing of next-generation intelligent systems. For example, depositing low-cost strain gauges on automotive components provides the unique ability to understand real-time stress fatigue of various high-value components to enhance preventive maintenance and improve efficiency. These sensors can be printed on flexible substrates that can be easily bonded to various structures or in some cases directly on the structures of interest. This allows unique process visibility into system performance, which is typically inferred from sparse measurements or simulation. Additionally, three-dimensional additive manufacturing provides opportunities for directly embedding sensors in various structures for *in situ* measurements with ease compared to traditional manufacturing. Direct integration of strain- and temperature-sensing fibers into metallic structural units and nylon components shows the viability of monitoring structures in ways not possible with conventional design and manufacturing.

Frequency-agile RF metamaterials are a natural fit for realizing multimode and wideband small printed antennas for commercial and defense communication applications through 2D/3D design considerations. Antenna structures for various frequencies provide the ability to integrate antennas into enclosures. Passive radiofrequency devices, including radiofrequency identification systems and resonators, can be manufactured and integrated into products to provide end-to-end visibility of device location and avoid counterfeiting. Significant R&D efforts worldwide are focused on exploiting the material design-enabled functionalities that extend beyond the scope of conventional materials. The unprecedented control of electromagnetic properties afforded by metamaterials opens the door to diverse applications such as superlenses, antennas, invisibility cloaks, perfect absorbers, energy harvesters, and sensors. The continued improvements in digital manufacturing techniques in terms of resolution and material handling capabilities coupled with the inherent advantages of high-throughput, large-area processing and low cost are leading toward the development of metamaterials-based disruptive frequency agile technology.[125,135] Various metamaterials can be additively manufactured to enhance the communication component performance. 3D metamaterial implementations of gradient index (GRIN) lenses, metallic fishnet structures, and RF concave lenses have already shown practical possibilities for microwave communication applications. Additive manufacturing techniques offer the opportunity to implement novel metamaterial structures through 3D printing of diverse dielectric and metallic periodic structures with a controlled fill factor.

5.7 CONCLUSIONS

Driven by increasing global competition and a shortage of resources, additive manufacturing processes and systems are being extensively investigated to establish a mainstream manufacturing technology for diverse technology areas and industries. Additive manufacturing and printed electronics technologies offer new opportunities for advanced device integration platforms and new markets. Design-driven manufacturing and rapid prototyping capabilities of AM systems are being exploited for a variety of innovative applications ranging from unique or specialized functional components to more advanced large-scale applications such as medical and dental, building construction, aerospace and automotive manufacturing, 3D electronic devices, consumer products, and microsystems. Advances in nanomaterials processing, direct-write printing techniques, and low-thermal-budget curing technology are paving the path toward multifunctional sensor platforms incorporating sensing elements, communication components, signal processing circuitry, and energy harvesters. Printed flexible electronics are the backbone of emerging wearable device technology that has been projected to thrive by 2025. Low-cost, multifunctional sensors and controls will enable new applications and business opportunities for wearable devices combined with the Internet of Things (IoT). The combination of additive manufacturing techniques and direct-write printing techniques is resulting in the development of hybrid electronics technologies based on the cointegration of 3D structures with printed components and devices. Rapid advances in innovative additive manufacturing processes, systems, and materials coupled with high-resolution digital printing techniques show promise for a disruptive hybrid electronics manufacturing technology development.

REFERENCES

1. Berman, B. (2012). 3-D printing: the new industrial revolution. *Bus. Horizons*, 55(2): 155–162.
2. Berger, U. (2012). Rapid manufacturing of mechatronic components—applications of stereolithography. In: *Proceedings of IEEE 2012 9th France–Japan & 7th Europe–Asia Congress on Mechatronics (MECATRONICS)/13th International Workshop on Research and Education in Mechatronics (REM)*, pp. 128–134.
3. Melchels, F.P.W., Domingos, M.A.N., Klein, T.J., Malda, J., Bartolo, P.J., and Hutmacher, D.W. (2012). Additive manufacturing of tissues and organs. *Prog. Polym. Sci.*, 37(8): 1079–1104.

4. Strano, G., Hao, L., Everson, R.M., and Evans, K.E. (2013). A new approach to the design and optimization of support structures in additive manufacturing. *Int. J. Adv. Manuf. Tech.*, 66(9–12): 1247–1254.
5. Lim, S., Buswell, R.A., Le, T.T., Austin, S.A., Gibb, A.G.F., and Thorpe, T. (2012). Developments in construction-scale additive manufacturing processes. *Automat. Constr.*, 21: 262–268.
6. Becker, B. (2014). Additive changes to advanced ceramics. *Ceram. Ind. Mag.*, 164(4): 21–24.
7. Hlavin, M. (2014). 3-D printing: the next industrial revolution. *Appliance Des.*, March 21, pp. 22–23.
8. Rasmussen, R. (2013). The way I see it: 3D printed electronics: the future is now. *SMT Mag.*, June, pp. 8–11.
9. Behrendt, U. (2009). Laser-sintering in the aerospace industry is on the rise. *Laser Tech. J.*, 6(5): 44–47.
10. Xue, L., Li, Y., and Wang, S. (2011). Direct manufacturing of net-shape functional components/testpieces for aerospace, automotive, and other applications. *J. Laser Appl.*, 23(4): 042004.
11. Kelbassa, I., Wohlers, T., and Caffrey, T. (2012). *Quo vadis*, laser additive manufacturing? *J. Laser Appl.*, 24(5): 050101.
12. Monzon, M.D., Diaz, N., Benitez, A.N., Marrero, M.D., and Hernandez, P.M. (2010). Advantages of fused deposition modeling for making electrically conductive plastic patterns. In: *Proceedings of 2010 International Conference on Manufacturing Automation (ICMA)*, pp. 37–43.
13. Chartier, T. and Badev, A. (2013). Rapid prototyping of ceramics. In: *Handbook of Advanced Ceramics: Materials, Applications, Processing, and Properties* (Somiya, S., Ed.), pp. 489–526. Academic Press, Waltham, MA.
14. Rossiter, J., Walters, P., and Stoimenov, B. (2009). Printing 3D dielectric elastomer actuators for soft robotics. In: *Proceedings of SPIE 7287: Electroactive Polymer Actuators and Devices (EAPAD)*, pp. 1–10.
15. Zhu, F., Macdonald, N.P., Cooper, J.M., and Wlodkowic, D. (2013). Additive manufacturing of lab-on-a-chip devices: promises and challenges. In: *Proceedings of SPIE 8923: Micro/Nano Materials, Devices, and Systems*, pp. 1–14.
16. Sekitani, T. and Someya, T. (2012). Stretchable organic integrated circuits for large-area electronic skin surfaces. *MRS Bull.*, 37(03): 236–245.
17. Parashkov, R., Becker, E., Riedl, T., Johannes, H.H., and Kowalsky, W. (2005). Large area electronics using printing methods. *Proc. IEEE*, 93: 1321–1329.
18. Kamyshny, A. and Magdassi, S. (2014). Conductive nanomaterials for printed electronics. *Small*, 10(17): 3515–3535.
19. Subramanian, V., Chang, J.B., Vornbrock, A., Huang, D.C. et al. (2008). Printed electronics for low-cost electronic systems: technology status and application development. In: *Proceedings of the 38th European Solid-State Device Research Conference (ESSDERC 2008)*, pp. 17–24.
20. Tarapata, G., Paczesny, D., and Kawecki, K. (2013). Characterization of inkjet-printing HF and UHF antennas for RFID applications. In: *Proceedings of SPIE 8903: Photonics Applications in Astronomy, Communications, Industry, and High-Energy Physics Experiments*, pp. 1–7.
21. Rida, A., Yang, L., Vyas, R., and Tentzeris, M.M. (2009). Conductive inkjet-printed antennas on flexible low-cost paper-based substrates for RFID and WSN applications. *Antennas Propagat. Mag.*, 51(3): 13–23.
22. Andò, B. and Baglio, S. (2011). Inkjet-printed sensors: a useful approach for low cost, rapid prototyping. *Instrum. Meas. Mag.*, 14(5): 36–40.
23. Halonen, E., Kaija, K., Mäntysalo, M., Kemppainen, A., Österbacka, R., and Björklund, N. (2009). Evaluation of printed electronics manufacturing line with sensor platform application. In: *Proceedings of the European Microelectronics and Packaging Conference (EMPC 2009)*, pp. 1–8.
24. Mantysalo, M., Pekkanen, V., Kaija, K., Niittynen, J., Koskinen, S., Halonen, E., Mansikkamaki, P., and Hameenoja, O. (2009). Capability of inkjet technology in electronics manufacturing. In: *Proceedings of the 59th Electronic Components and Technology Conference (ECTC 2009)*, pp. 1330–1336.
25. Chalamala, B.R. and Temple, D. (2005). Big and bendable (flexible plastic-based circuits). *Spectrum*, 42(9): 50–56.
26. Das, R. and Harrop, P. (2015). *Printed, Organic & Flexible Electronics: Forecasts, Players & Opportunities 2015–2025*. IDTechEx, Cambridge, MA.
27. Briand, D., Oprea, A., Courbat, J., and Bârsan, N. (2011). Making environmental sensors on plastic foil. *Mater. Today*, 14(9): 416–423.
28. Hua, B., Chen, W., and Zhou, J. (2013). High performance flexible sensor based on inorganic nanomaterials. *Sens. Actuators B Chem.*, 176: 522–533.
29. Koo, J.H., Seo, J., and Lee, T. (2012). Nanomaterials on flexible substrates to explore innovative functions: from energy harvesting to bio-integrated electronics. *Thin Solid Films*, 524: 1–19.

30. Brosseau, C. (2011). Emerging technologies of plastic carbon nanoelectronics: a review. *Surf. Coat. Technol.*, 206(4): 753–758.
31. Briand, D., Molina-Lopez, F., Quintero, A.V., Ataman, C., Courbat, J., and de Rooij, N.F. (2011). Why going towards plastic and flexible sensors? *Proc. Eng.*, 25: 8–15.
32. Kim, N.S. and Han, K.N. (2010). Future direction of direct writing. *J. Appl. Phys.*, 108(10): 102801.
33 Hon, K.K.B., Li, L., and Hutchings, I.M. (2008). Direct writing technology—advances and developments. *CIRP Ann. Manuf. Technol.*, 57(2): 601–620.
34 Futera, K., Jakubowska, M., Koziol, G., Arazna, A., and Janeczek, K. (2012). Low cost inkjet printing system for organic electronic applications. In: *Mechatronics: Recent Technological and Scientific Advances* (Jablonski, R. and Brezina, T., Eds.), pp. 713–721. Springer, Heidelberg.
35 Teunissen, P., Rubingh, E., van Lammeren, T., Abbel, R., van de Geijn, S., and Groen, P. (2013). Towards high speed inkjet printed electronics—technology transfer from S2S to R2R production. In: *Proceedings of the 29th International Conference on Digital Printing Techniques (NIP 2013) and Digital Fabrication*, pp. 484–488.
36. Koiva, R., Zenker, M., Schurmann, C., Haschke, R., and Ritter, H.J. (2013). A highly sensitive 3D-shaped tactile sensor. In: *Proceedings of the 2013 IEEE/ASME International Conference on Advanced Intelligent Mechatronics (AIM)*, pp. 1084–1089.
37. Karioja, P., Mäkinen, J.T., Keränen, K., Aikio, J., Alajoki, T., Jaakola, T., Koponen, M. et al. (2012). Printed hybrid systems. In: *Proceedings of SPIE 8344: Nanosensors, Biosensors, and Info-Tech Sensors and Systems*, pp. 1–13.
38. Petrovic, V., Gonzalez, J.V.H., Ferrando, O.J., Gordillo, J.D., Puchades, J.R.B., and Grinan, L.P. (2011). Additive layered manufacturing: sectors of industrial application shown though case studies. *Int. J. Prod. Res.*, 49(4): 1061–1079.
39. Kawahara, Y., Hodges, S., Cook, B.S., Zhang, C., and Abowd, G.D. (2013). Instant inkjet circuits: lab-based inkjet printing to support rapid prototyping of ubicomp devices. In: *Proceedings of the 2013 ACM International Joint Conference on Pervasive and Ubiquitous Computing (UbiComp 2013)*, pp. 363–372.
40. Hutchings, I.M. and Martin., G.D. (2013). Introduction to inkjet printing for manufacturing. In: *Inkjet Technology for Digital Fabrication* (Hutchings, I.M. and Martin, G.D., Eds.), pp. 1–20. John Wiley & Sons, West Sussex.
41. Kaija, K., Pekkanen, V., Mäntysalo, M., Koskinen, S., Niittynen, J., Halonen, E., and Mansikkamäki, P. (2010). Inkjetting dielectric layer for electronic applications. *Microelectron. Eng.*, 87(10): 1984–1991.
42. Mäntysalo, M., Pekkanen, V., Kaija, K., Niittynen, J., Koskinen, S., Halonen, E., Mansikkamäki, P., and Hämeenoja. O. (2009). Capability of inkjet technology in electronics manufacturing. In: *Proceedings of the 59th Electronic Components and Technology Conference*, pp. 1330–1336.
43. Hayes, D.J., Grove, M.E., Wallace, D.B., Chen, T., and Cox, W.R. (2002). Inkjet printing in the manufacture of electronics, photonics, and displays. In: *Proceedings of SPIE 4809: Nanoscale Optics and Applications*, pp. 1–6.
44. Imai, H., Mizuno, S., Makabe, A., Sakurada, K., and Wada, K. (2006). Application of inkjet printing technology to electro packaging. In: *Proceedings of the 39th International Symposium on Microelectronics*, pp. 484–490.
45. Suganuma, K. (2014). *Introduction to Printed Electronics*. Springer, New York, pp. 49–74.
46. Sanchez-Romaguera, V., Madec, M.B., and Yeates, S.G. (2008). Inkjet printing of 3D metal–insulator–metal crossovers. *React. Funct. Polym.*, 68(6): 1052–1058.
47. Teichler, A., Perelaera, J., and Schubert, U.S. (2013). Inkjet printing of organic electronics–comparison of deposition techniques and state-of-the-art developments. *J. Mater. Chem. C*, 1(10): 1910–1925.
48. Secor, E.B., Prabhumirashi, P.L., Puntambekar, K., Geier, M.L., and Hersam, M.C. (2013). Inkjet printing of high conductivity, flexible graphene patterns. *J. Phys. Chem. Lett.*, 4(8): 1347–1351.
49. Hwang, M.S., Jeong, B.Y., Moon, J., Chun, S.K., and Kim, J. (2011). Inkjet-printing of indium tin oxide (ITO) films for transparent conducting electrodes. *Mater. Sci. Eng. B*, 176(14): 1128–1131.
50. Derby, B. (2011). Inkjet printing ceramics: from drops to solid. *J. Eur. Ceram. Soc.*, 31(14): 2543–2550.
51. Shi, C.W.P., Shan, X., Tarapata, G., Jachowicz, R., Weremczuk, J., and Hui, H.T. (2011). Fabrication of wireless sensors on flexible film using screen printing and via filling. *Microsyst. Technol.*, 17(4): 661–667.
52. Fjelstad, J. (2011). *Flexible Circuit Technology*, 4th ed. BR Publishing, Seaside, OR.
53. Aernouts, T., Cros, S., and Krebs, F.C. (2008). Processing and production of large modules. In: *Polymer Photovoltaics: A Practical Approach* (Krebs, F.C., Ed.), pp. 229–230. SPIE, Bellingham, WA.
54. Halonen, E., Kaija, K., Mäntysalo, M., Kemppainen, A., Österbacka, R., and Björklund, N. (2009). Evaluation of printed electronics manufacturing line with sensor platform application. In: *Proceedings of the European Microelectronics and Packaging Conference (EMPC 2009)*, pp. 1–8.

55. Schmidt, G., Kempa, H., Fuegmann, U., Fischer, T., Bartzsch, M., Zillger, T., Preissler, K., Hahn, U., and Huebler, A. (2006). Challenges and perspectives of printed electronics. In: *Proceedings of SPIE 6336: Organic Field-Effect Transistors V*, pp. 1–9.

56. Gili, E., Caironi, M., and Sirringhaus, H. (2010). Picoliter printing. In: *Handbook of Nanofabrication* (Wiederrecht, G., Ed.), pp. 183–208. Elsevier, Amsterdam.

57. Wallace, D. and Hayes, D. (2005). Ink-jet printing as a tool in manufacturing and instrumentation. In: *Nanolithography and Patterning Techniques in Microelectronics* (Bucknall, D., Ed.), pp. 267–298. Woodhead Publishing, Cambridge.

58. Romano, F.J. (1999). *Professional Prepress, Printing and Publishing*. Prentice Hall, Upper Saddle River, NJ, p. 89.

59. Freire, E.M. (2006). Ink jet printing technology (CIJ/DOD). In: *Digital Printing of Textiles* (Ujie, H., Ed.), pp. 29–52. CRC Press, Boca Raton, FL.

60. Mei, J., Lovell, M.R., and Mickle, M.H. (2005). Formulation and processing of novel conductive solution inks in continuous inkjet printing of 3-D electric circuits. *IEEE Trans. Electron. Packag. Manuf.*, 28(3): 265–273.

61. Kamyshny, A. and Magdassi, S. (2013). Inkjet printing. In: *Kirk–Othmer Encyclopedia of Chemical Technology*. Wiley Blackwell, Hoboken, NJ, pp. 1–21.

62. Schuetz, K., Hoerber, J., and Franke, J. (2014). Selective light sintering of aerosol-jet printed silver nanoparticle inks on polymer substrates. *AIP Conf. Proc.*, 1593(1): 732–735.

63. Zhao, D., Liu, T., Park, J.G., Zhang, M., Chen, J.M., and Wang, B. (2012). Conductivity enhancement of aerosol-jet printed electronics by using silver nanoparticles ink with carbon nanotubes. *Microelectron. Eng.*, 96: 71–75.

64. Hoey, J.M., Lutfurakhmanov, A., Schulz, D.L., and Akhatov, I.S. (2012). A review on aerosol-based direct-write and its applications for microelectronics. *J. Nanotechnol.*, 2012: 1–22.

65. Thakur, R.P.S. and Singh, R. (1988). Role of *in situ* rapid isothermal processing in advanced III–V technology. *J. Appl. Phys.*, 70(7): 3857–3861.

66. Nakos, J. and Shepard, J. (2008). The expanding role of rapid thermal processing in CMOS manufacturing. *Mater. Sci. Forum*, 573: 3–19.

67. Fiory, A.T. (2002). Recent developments in rapid thermal processing. *J. Electron. Mater.*, 31(10): 981–987.

68. Roozeboom, F. (1993). Manufacturing equipment issues in rapid thermal processing. In: *Rapid Thermal Processing, Science and Technology* (Fair, R.B., Ed.), pp. 349–423. Academic Press, San Diego, CA.

69. Ott, R.D., Blue, C.A., Dudney, N.J., and Harper, D.C. (2007). Pulse Thermal Processing of Functional Materials Using Directed Plasma Arc, U.S. Patent 7,220,936.

70. West, J., Carter, M., Smith, S., and Sears, J. (2012). Photonic sintering of silver nanoparticles: comparison of experiment and theory. In *Sintering: Methods and Products* (Shatokha, V., Ed.), pp. 173–188. InTech, Shanghai.

71. Schroder, K.A. (2011). Mechanisms of photonic curing: processing high temperature films on low temperatures substrates. *Nanotech*, 2: 220–223.

72. Courbat, J., Briand, D., and de Rooij, N.F. (2010). Ink-jet printed colorimetric gas sensors on plastic foil. In: *Proceedings of SPIE 7779: Organic Semiconductors in Sensors and Bioelectronics II*, pp. 1–6.

73. Maiwald, M., Günther, B., Ruttkowski, V., Werner, C., Müller, M., Godlinski, D., Wirth, I., Zöllmer, V., and Busse, M. (2008). Inktelligent printing of metal and metal alloys for sensor structures. In: *Proceedings of 2nd European Conference & Exhibition on Integration Issues of Miniaturized Systems— MOMS, MOEMS, ICS, and Electronic Components (SSI)*, pp. 1–6.

74. Dong, H., Barr, R., and Hinkley, P. (2010). Inkjet plating resist for improved cell efficiency. In: *Proceedings of the 35th IEEE Photovoltaic Specialists Conference (PVSC)*, pp. 2142–2146.

75. Singh, M., Haverinen, H.M., Dhagat, P., and Jabbour, G.E. (2010). Inkjet printing—process and its applications. *Adv. Mater.*, 22(6): 673–685.

76. Li, Y., Torah, R., Beeby, S., and Tudor, J. (2012). An all-inkjet printed flexible capacitor for wearable applications. In: *Proceedings of 2012 Symposium on Design, Test, Integration, and Packaging of MEMS/ MOEMS (DTIP)*, pp. 192–195.

77. Futera, K., Jakubowska, M., and Koziol, G. (2010). Printed electronic on flexible and glass substrates. In: *SPIE 7745: Photonics Applications in Astronomy, Communications, Industry, and High-Energy Physics Experiments*, pp. 1–6.

78. Choi, M.C., Kim, Y., and Ha, C.S. (2008). Polymers for flexible displays: from material selection to device applications. *Prog. Polym. Sci.*, 33(6): 581–630.

79. Cheng, I.C. and Wagner, S. (2009). Overview of flexible electronics technology. In: *Flexible Electronics: Materials and Applications* (Wong, W.S. and Salleo, A., Eds.), pp. 1–28. Springer, New York.

80. Yang, L., Rida, A., Vyas, R., and Tentzeris, M.M. (2007). RFID tag and RF structures on a paper substrate using inkjet-printing technology. *IEEE Trans. Microwave Theory Tech.*, 55(12): 2894–2901.

81. Miettinen, J., Kaija, K., Mäntysalo, M., Mansikkamäki, P., Kuchiki, M., Tsubouchi, M., Rönkkä, R., Hashizume, K., and Kamigori, A. (2009). Molded substrates for inkjet printed modules. *IEEE Trans. Comp. Packag. Technol.*, 32(2): 293–301.

82. Kim, D., Jeong, S., Park, B.K., and Moon, J. (2006). Direct writing of silver conductive patterns: improvement of film morphology and conductance by controlling solvent compositions. *Appl. Phys. Lett.*, 89(26): 264101.

83. Zhang, T., Wang, X., Li, T., Guo, Q., and Yang, J. (2014). Fabrication of flexible copper-based electronics with high-resolution and high-conductivity on paper via inkjet printing. *J. Mater. Chem. C*, 2: 286–294.

84. Määttänen, A., Ihalainen, P., Pulkkinen, P., Wang, S., Tenhu, H., and Peltonen, J. (2012). Inkjet-printed gold electrodes on paper: characterization and functionalization. *ACS Appl. Mater. Interfaces*, 4(2): 955–964.

85. Li, D., Sutton, D., Burgess, A., Graham, D., and Calvert, P.D. (2009). Conductive copper and nickel lines via reactive inkjet printing. *J. Mater. Chem.*, 19(22): 3719–3724.

86. Koskinen, S., Pykäri, L., and Mäntysalo, M. (2013). Electrical performance characterization of an inkjet-printed flexible circuit in a mobile application. *IEEE Trans. Comp. Packag. Manuf. Technol.*, 3(9): 1604–1610.

87. Mengel, M. and Nikitin, I. (2010). Inkjet printed dielectrics for electronic packaging of chip embedding modules. *Microelectron. Eng.*, 87(4): 593–596.

88. Baeg, K.J., Khim, D., Kim, J.H., Kang, M., You, I.K., Kim, D.Y., and Noh, Y.Y. (2011). Improved performance uniformity of inkjet printed *N*-channel organic field-effect transistors and complementary inverters. *Org. Electron.*, 12(4): 634–640.

89. Ho, C.C., Evans, J.W., and Wright, P.K. (2010). Direct-write dispenser-printed energy storage devices. In: *Proceedings of SPIE 7679: Micro- and Nanotechnology Sensors, Systems, and Applications II*, p. 76792A-1-9.

90. Kopola, P., Aernouts, T., Guillerez, S., Jin, H., Tuomikoski, M., Maaninen, A., and Hast, J. (2010). High efficient plastic solar cells fabricated with a high-throughput gravure printing method. *Sol. Energy Mater. Sol. Cells*, 94(10): 1673–1680.

91. Oprea, A., Courbat, J., Bârsan, N., Briand, D., de Rooij, N.F., and Weimar, U. (2009). Temperature, humidity and gas sensors integrated on plastic foil for low power applications. *Sens. Actuators B Chem.*, 140(1): 227–232.

92. Pelegrí-Sebastiá, J., García-Breijo, E., Ibáñez, J., Sogorb, T., Laguarda-Miro, N., and Garrigues. J. (2012). Low-cost capacitive humidity sensor for application within flexible RFID labels based on microcontroller systems. *IEEE Trans. Instrum. Meas.*, 61(2): 545–553.

93. Koo, M., Park, K.I., Lee, S.H., Suh, M., Jeon, D.Y., Choi, J.W., Kang, K., and Lee, K.J. (2012). Bendable inorganic thin-film battery for fully flexible electronic systems. *Nano Lett.*, 12(9): 4810–4816.

94. Zampetti, E., Maiolo, L., Pecora, A., Maita, F., Pantalei, S. et al. (2011). Flexible sensorial system based on capacitive chemical sensors integrated with readout circuits fully fabricated on ultra thin substrate. *Sens. Actuators B Chem.*, 155(2): 768–774.

95. Yin, Z., Huang, Y., Bu, N., Wang, X., and Xiong, Y. (2010). Inkjet printing for flexible electronics: materials, processes and equipments. *Chinese Sci. Bull.*, 55(30): 3383–3407.

96. Subramanian, V., Chang, P.C., Lee, J.B., Molesa, S.E., and Volkman, S.K. (2005). Printed organic transistors for ultra-low-cost RFID applications. *IEEE Trans. Comp. Packag. Technol.*, 28(4): 742–747.

97. Virtanen, J., Ukkonen, L., Björninen, T., Elsherbeni, A.Z., and Sydänheimo, L. (2011). Inkjet-printed humidity sensor for passive UHF RFID systems. *IEEE Trans. Instrum. Meas.*, 60(8): 2768–2777.

98. Azucena, O., Kubby, J., Scarbrough, D., and Goldsmith, C. (2008). Inkjet printing of passive microwave circuitry. In: *Proceedings of 2008 IEEE MTT-S International Microwave Symposium Digest*, pp. 1075–1078.

99. Virtanen, J., Bjorninen, T., Ukkonen, L., and Sydanheimo, L. (2010). Passive UHF inkjet-printed narrow-line RFID tags. *IEEE Trans. Antennas Wireless Propagat. Lett.*, 9: 440–443.

100. Koski, E., Koski, K., Ukkkonen, L., Sydanheimo, L., Virtanen, J., Bjorninen, T., and Elsherbeni, A.Z. (2011). Performance of inkjet-printed narrow-line passive UHF RFID tags on different objects. In: *Proceedings of 2011 IEEE International Symposium on Antennas and Propagation (APSURSI)*, pp. 537–540.

101. Shaker, G., Safavi-Naeini, S., Sangary, N., and Tentzeris, M.M. (2011). Inkjet printing of ultrawideband (UWB) antennas on paper-based substrates. *IEEE Trans. Antennas Wireless Propagat. Lett.*, 10: 111–114.

102. Shin, K.Y., Hong, J.Y., and Jang, J. (2011). Micropatterning of graphene sheets by inkjet printing and its wideband dipole-antenna application. *Adv. Mater.*, 23: 2113–2118.

103. Wang, W., Su, Y.W., and Chang, C.H. (2011). Inkjet printed chalcopyrite $CuIn_xGa_{1-x}Se_2$ thin film solar cells. *Sol. Energy Mater. Sol. Cells*, 95(9): 2616–2620.

104. Hoth, C.N., Schilinsky, P., Choulis, S.A., and Brabec, C.J. (2008). Printing highly efficient organic solar cells. *Nano Lett.*, 8(9): 2806–2813.

105. Umezu, S., Kawata, S., Ishii, A., Kunugi, Y., and Ohmori, H. (2012). Development of the dye-sensitized solar cell by micro digital fabrication. *J. Adv. Sci.*, 24(1–2): 16–20.

106. Rohr, S. (2013). Heliatek consolidates its technology leadership by establishing a new world record for organic solar technology with a cell efficiency of 12% [press release]. Heliatek, Dresden, Germany, January 16.

107. Kim, J., Na, S.I., and Kim, H.K. (2012). Inkjet printing of transparent InZnSnO conducting electrodes from nano-particle ink for printable organic photovoltaics. *Sol. Energy Mater. Sol. Cells*, 98: 424–432.

108. Eom, S.H., Senthilarasu, S., Uthirakumar, P., Yoon, S.C., Lim, J., Lee, C., Lim, H.S., Lee, J., and Lee, S.H. (2009). Polymer solar cells based on inkjet-printed PEDOT: PSS layer. *Org. Electron.*, 10(3): 536–542.

109. Okimoto, H., Takenobu, T., Yanagi, K., Miyata, Y., Shimotani, H., Kataura, H., and Iwasa, Y. (2010). Tunable carbon nanotube thin-film transistors produced exclusively via inkjet printing. *Adv. Mater.*, 22(36): 3981–3986.

110. Arias, A.C., Daniel, J., Sambandan, S., Ng, T.N., Russo, B., Krusor, B., and Street, R.A. (2008). All printed thin film transistors for flexible electronics. In: *Proceedings of SPIE 7054: Organic Field-Effect Transistors VII and Organic Semiconductors in Sensors and Bioelectronics*, pp. 1–7.

111. Han, S.Y., Lee, D.H., Herman, G.S., and Chang, C.H. (2009). Inkjet-printed high mobility transparent-oxide semiconductors. *J. Display Technol.*, 5(12): 520–524.

112. Sirringhaus, H., Kawase, T., Friend, R.H., Shimoda, T., Inbasekaran, M., Wu, W., and Woo, E.P. (2000). High-resolution inkjet printing of all-polymer transistor circuits. *Science*, 290(5499): 2123–2126.

113. Torrisi, F., Hasan, T., Wu, W., Sun, Z., Lombardo, A., Kulmala, T.S., Hshieh, G.W., Jung, S., Bonaccorso, F., Paul, P.J., Chu, D., and Ferrari, A.C. (2012). Inkjet-printed graphene electronics. *ACS Nano*, 6(4): 2992–3006.

114. Everaerts, K., Zeng, L., Hennek, J.W., Camacho, D.I., Jariwala, D., Bedzyk, M.J., Hersam, M.C., and Marks, T.J. (2013). Printed indium gallium zinc oxide transistors: self-assembled nanodielectric effects on low-temperature combustion growth and carrier mobility. *ACS Appl. Mater. Interfaces*, 5(22): 11884–11893.

115. Vyas, R., Lakafosis, V., Lee, H., Shaker, G., Yang, L., Orecchini, G., Traille, A., Tentzeris, M.M., and Roselli, L. (2011). Inkjet printed, self powered, wireless sensors for environmental, gas, and authentication-based sensing. *Sensors J.*, 11(12): 3139–3152.

116. Molina-Lopez, F., Briand, D., and de Rooij, N.F. (2012). All additive inkjet printed humidity sensors on plastic substrate. *Sens. Actuators B Chem.*, 166: 212–222.

117. Slotwinski, J.A. (2014). Additive manufacturing: overview and new challenges. *AIP Conf. Proc.*, 1581(1): 1173–1177.

118. Wohlers, T. (2012). *Wohlers Report 2012: Additive Manufacturing and 3D Printing State of the Industry*. Wohlers Associates, Fort Collins, CO.

119. Brücknera, F., Nowotnya, S., and Leyens, C. (2012). Innovations in laser cladding and direct metal deposition. In: *Proceedings of SPIE 8239: High Power Laser Materials Processing: Lasers, Beam Delivery, Diagnostics, and Applications*, pp. 1–6.

120. Gibson, I., Rosen, D.W., and Stucker, B. (2010). *Additive Manufacturing Technologies: Rapid Prototyping to Direct Digital Manufacturing*. Springer, New York, pp. 17–39.

121. Williams, C.B., Cochran, J.K., and Rosen, D.W. (2011). Additive manufacturing of metallic cellular materials via three-dimensional printing. *Int. J. Adv. Manuf. Technol.*, 53(1–4): 231–239.

122. Kirleis, M.A., Simonson, D. Charipar, N.A. Kim, H. Charipar, K.M. et al. (2014). Laser embedding electronics on 3D printed objects. In: *Proceedings of SPIE 8970: Laser 3D Manufacturing*, pp. 1–7.

123. O'Donnella, J., Ahmadkhanloub, F., Yoon, H.-S., and Washington, G. (2014). All-printed smart structures: a viable option? In: *Proceedings of SPIE 9057: Active and Passive Smart Structures and Integrated Systems 2014*, pp. 1–8.

124. Blumenthal, T., Fratello, V., Nino, G., and Ritala, K. (2013). Conformal printing of sensors on 3D and flexible surfaces using aerosol jet deposition. In: *Proceedings of SPIE 8691: Nanosensors, Biosensors, and Info-Tech Sensors and Systems 2013*, pp. 1–9.

125. Allen, J.W. and Wu, B.I. (2013). Design and fabrication of an RF GRIN lens using 3D printing technology. In: *Proceedings of SPIE 8624: Terahertz, RF, Millimeter, and Submillimeter-Wave Technology and Applications VI*, pp. 1–7.

126. Muth, J.T., Vogt, D.M., Truby, R.L., Mengüç, Y., Kolesky, D.B., Wood, R.J., and Lewis, J.A. (2014). Embedded 3D printing of strain sensors within highly stretchable elastomers. *Adv. Mater.*, 26(36): 6307–8312.

127. Gutierrez, C., Salas, R., Hernandez, G., Muse, D., Olivas, R., MacDonald, E., Irwin, M.D., and Wicker, R. (2011). Cubesat Fabrication Through Additive Manufacturing and Micro-Dispensing, paper presented at 7th International Conference and Exhibition on Device Packaging, Scottsdale, AZ, March 8–10.

128. Ladd, C., So, J.-H., Muth, J., and Dickey, M.D. (2013). 3D printing of free standing liquid metal microstructures. *Adv. Mater.*, 25(36): 5081–5085.

129. Mannoor, M.S., Jiang, Z., James, T., Kong, Y.L., Malatesta, K.A., Soboyejo, W.O., Verma, N., Gracias, D.H., and McAlpine, M.C. (2013). 3D printed bionic ears. *Nano Lett.*, 13(6): 2634–2639.

130. Van den Brand, J., Kusters, R., Barink, M., and Dietzel, A. (2010). Flexible embedded circuitry: a novel process for high density, cost effective electronics. *Microelectron. Eng.*, 87(10): 1861–1867.

131. Niittynen, J., Pekkanen, V., and Mantysalo, M. (2010). Characterization of ICA attachment of SMD on inkjet-printed substrates. In: *Proceedings of the 60th Electronic Components and Technology Conference (ECTC)*, pp. 990–997.

132. Ko, S.H., Chung, J., Hotz, N., Nam, K.H., and Grigoropoulos, C.P. (2010). Metal nanoparticle direct inkjet printing for low-temperature 3D micro metal structure fabrication. *J. Micromech. Microeng.*, 20(12): 125010.

133. Wang, X., Guo, Q., Cai, X., Zhou, S., Kobe, B., and Yang, J. (2014). Initiator-integrated 3D printing enables the formation of complex metallic architectures. *ACS Appl. Mater. Interfaces*, 6(4): 2583–2587.

134. Adams, J.J., Duoss, E.B., Malkowski, T.F., Motala, M.J., Ahn, B.Y., Nuzzo, R.G., Bernhard, J.T., and Lewis, J.A. (2011). Conformal printing of electrically small antennas on three-dimensional surfaces. *Adv. Mater.*, 23: 1335–1340.

135. Ehrenberg, I.M., Sarma, S.E., and Wu, B.I. (2012). A three-dimensional self-supporting low loss microwave lens with a negative refractive index. *J. Appl. Phys.*, 112(7): 073114.

6 Application of Radiometry in Laser Powder Deposition Additive Manufacturing

Joshua J. Hammell, Michael A. Langerman,
James L. Tomich, and Brett L. Trotter

CONTENTS

ABSTRACT

Laser powder deposition (LPD) is a directed energy deposition additive manufacturing (AM) process used for solid freeform fabrication, surface modification, and repair. LPD offers increased manufacturing flexibility to meet the demands of diverse markets. However, the connection between deposition parameters, thermal history, and final part quality is not sufficiently understood. Radiometric temperature measurements provide a real-time non-contact method for advancement of AM process understanding. These measurements can be used to augment process development, or process control, for the prodigious adoption of metal AM. Radiometric thermal data, collected during the deposition of AISI/SAE 1045, 4130, and 4140 steel thin-wall samples, were used to illustrate the connection between process input and final part quality. Several thermal zones were identified and compared to post-process metallographic analysis.

6.1   INTRODUCTION

The future of additive manufacturing (AM) is propitious; however, in order for AM to be accepted as a viable direct part manufacturing solution, several challenges must be addressed. Improvements in processing speed, initial investment, system availability, product traceability, and product reproducibility are essential.[1] It has been widely recognized that AM processes will transform the future of manufacturing, particularly for parts with complex geometries, low production volumes, and customized functionality. Within the AM community, an important long-range goal is the capability to manufacture components with tailored material properties, at the highest quality, with the most cost-effective use of time and materials, while ensuring traceability and reproducibility. As a result, industrial efforts are focusing on minimizing part manufacturing time, improving part quality, reducing costs, and producing "first-time-right" parts with the desired final dimensions and material properties. To address these challenges, advancements in process monitoring and control, strengthened with fundamental process understanding, are required. Furthermore, with increased AM awareness and the creation of institutions such as America Makes (National Additive Manufacturing Innovation Institute, NAMII), the AM industry is poised to experience large economic growth in the coming years. Therefore, the need for solutions to these challenges has never been greater.

6.1.1 Laser Powder Deposition Additive Manufacturing and Control

The fundamental AM processes are material extrusion, material jetting, binder jetting, vat photopolymerization, sheet lamination, powder bed fusion, and directed energy deposition. More information about each process can be found in the ASTM Standard F2792-12a.[2] Powder bed fusion and directed energy deposition (DED) will play a large role in the future of AM because of their ability to directly fabricate complex metal parts with unique material properties. DED is a process that fuses material using a focused energy source, such as a laser or an electron beam. The focused energy selectively melts a filler material layer by layer to form a desired shape that is typically defined by a computer-aided design (CAD) model.[2] Laser powder deposition (LPD) is a type of DED that uses a focused laser beam to melt blown powdered material into fully dense 3D structures, as well as coat the surface or build features on preexisting parts. Typically, an inert gas stream is used to blow the powdered material through a processing head assembly into the melt pool that is created by the laser beam. Computer-controlled robotics are used to manipulate both the processing head and part. A LPD process schematic can be found in Figure 6.1. LPD offers flexibility, accuracy, and precision of photon energy and material delivery. The concentrated laser beam heat source minimizes thermal effects when compared to more traditional processes, such as arc welding. Moreover, freedom in design for additive manufacturing (DFAM) allows LPD to be used for the production of complex structures that cannot be conventionally manufactured. As well, LPD can be used to create functionally graded components with tailored variations in material composition and properties.

Laser powder deposition also lends itself to the possibility of automation. Progression toward automation, for most manufacturing processes, is driven by the potential for cost reduction, increased productivity, increased quality, longer sustained production hours, and/or improved working conditions.[3] For LPD, automation can be achieved to varying degrees through open-loop control, closed-loop feedback control, feedforward control, a combination thereof, or through other advanced control methods. Regardless of the LPD application or control method employed, part quality can be defined as the difference between the actual part and the desired part. For a given application, part quality could be defined by placement, size, density, and finish of the as-deposited structure. It could also be defined by mechanical properties, such as hardness, yield strength, and ductility.

In the open-loop LPD process, parameters are typically selected based on prior operator experience. Path planning is achieved through generalized slicing algorithms that do not account for deposition material, part geometry, process parameters, thermal response, or the effect of base material and the surrounding environment. The selected parameters are input to the open-loop process, which is subject to the process dynamics and any disturbances that may be present. Lack of control of the process dynamics and/or disturbances can lead to error in the actual output of the process.[4] In some cases, this error can be accounted for by pseudo-feedback based on operator experience or post-process interpretation of sensor/quality measurements. Over time, this method could also be used to develop an experiential database that could be used for process planning; however, this "guess and check" method can be costly and time consuming.

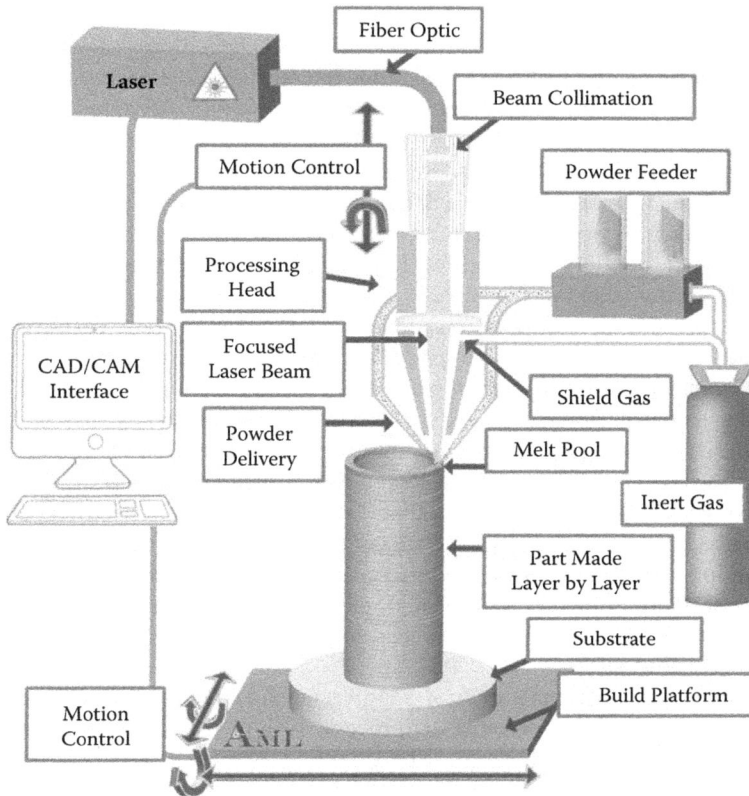

FIGURE 6.1 General schematic of a LPD system.

In closed-loop feedback control, a desired result is compared with sensor feedback to obtain real-time process error. The error is used to adjust process parameters for real-time error compensation.[4] If properly configured, this will lead to an actual output that more closely matches the desired output and is corrected in real time with minimal user input; therefore, this method readily applies to the goals of process automation. However, in some situations, feedback control cannot sufficiently account for process dynamics or disturbances and may not respond in sufficient time to correct certain errors. For example, a step decrease in laser power could have a longer response time than a step increase in laser power for a given material. When approaching a 90° corner, a reduction in laser power may be needed to account for the change in bead thickness and width caused by deceleration and acceleration through the corner. That is, if an unavoidable change in a process parameter, such as deceleration at a sharp corner, occurs faster than a material's response to a change in the control variables, then a feedback control system will lag in its ability to respond to the corresponding error. To account for this type of short-term error, the control system would need to be aware of its occurrence beforehand, so that the necessary changes in control variables could be implemented in sufficient time. Furthermore, in some applications feedback control may not be possible due to lack of sensor availability, lack of sensor capability, or the high cost and complexity of implementation.

Alternatively, feedforward control is a method that uses process models to predict the dynamic response, along with the effect of relevant disturbances, and uses a predefined method for error compensation. Feedforward control is commonly used when the process response and disturbances cannot be sufficiently controlled by feedback. When feedforward control is implemented without feedback control, error compensation occurs without real-time monitoring of the system's response to changes in the control variables. Over time, this could result in deviation of the control scheme from that which is required for actual error compensation. When feedforward control is coupled with feedback control, sensor feedback allows for real-time monitoring of the system's response to changes in the control variables and allows for additional error compensation. Because it is nearly impossible to obtain a perfect process model, feedforward control is usually accompanied by feedback control. This coupled feedforward/feedback control provides a more robust solution for process automation;[5] however, it is important to note that process flexibility may be sacrificed for more rigorous control.

There are many parameters that affect a given LPD part. Most of these parameters are fixed preprocess based on the configuration and specific type of processing equipment used. Only a select group of input parameters can be controlled in real time, predominantly laser power, beam diameter, velocity, acceleration/deceleration, position, standoff, and, in some cases, material type or material feed rate. The process plan is comprised of a series of changes in these parameters, which correspond to discrete segments of a given path plan. The desired output dictates the process plan, type of sensors, and control systems needed for error compensation during specific LPD builds; the more complex the desired output, the more complex the required control. For example, if the desired output is a single straight bead, then there is less to plan and control than if the desired output is to stack twenty straight beads on top of each other. Depending on the desired output, control could also be implemented locally or globally. A localized control system normally consists of one-to-one control; for example, temperature is sensed and temperature is controlled.[3] A globalized control system is based on more unified control of multiple subsystems, with the goal of more complete process control. Fuzzy logic and neural networks could be used for globalized control of LPD but have not sufficiently done so yet.[3]

For broader industrial implementation of additive manufacturing, control algorithms must be developed that more effectively account for process dynamics and disturbances. Furthermore, development of these control methods must be driven by part quality. It is well known that the quality of parts produced by LPD is directly related to material response. It is also well known that material response is directly related to thermal history.[6] Therefore, control of thermal history is vital to control of part quality. With respect to LPD additive manufacturing, radiometry is a tool that can be used for the development of process understanding as well as control methods that are based on thermal history.

6.1.2 Considerations for Application of Radiometry to LPD Additive Manufacturing

Radiometry, or radiation thermometry, is the optical measurement of radiation, primarily in the ultraviolet, visible, and/or infrared regions of the electromagnetic spectrum.[7] With knowledge of the target's emissive properties, the measured radiance can be used along with Planck's law to calculate surface temperatures. In general, temperature is a relative measure of "hotness" or "coldness" that is defined by predictable and repeatable fixed points that make up the International Temperature Scale of 1990 (ITS-90).[8] For example, temperatures between $-259.3467°C$ and $961.78°C$ are defined by platinum resistance thermometers calibrated to the triple point of equilibrium hydrogen, the triple point of water, and the freezing points of tin, zinc, aluminum, and silver. Above $961.78°C$, temperature is defined by monochromatic radiation thermometers and a ratio of measured blackbody radiance, defined by Planck's law, with respect to the freezing point of silver, gold, or copper.[8] A more in-depth discussion of the radiometric sensors, optics, measurement error, and radiometric methods used to realize ITS-90 above the freezing point of silver can be found in the ITS-90 supplementary information.[9]

For radiometric temperature measurements of real surfaces, knowledge of emittance is required. Emittance is defined as the ratio of radiance emitted from a real surface to that emitted from a blackbody at the same wavelength and temperature.[10] It is a function of temperature, wavelength, and direction and is influenced by material composition, microstructure, surface roughness, and the presence of films or oxides. In addition, the presence of reflections, extraneous radiation, variations in surface conditions, and participating media, along with the transmittance of surrounding gases, the effect of optics and filters, and specific detector response characteristics, must be taken into consideration when collecting radiometric data. An in-depth discussion of these topics can be found in Dewitt and Nutter.[11]

Several methods are commonly used for experimentally determining the emittance of opaque targets. Madding outlined four general methods.[12] In the first method, a material with a known emittance is added to the surface of the target. If both materials are at the same temperature, then the unknown emittance of the target can be calculated. For most LPD applications, this method does not readily apply to real-time temperature measurement due to the presence of large temperature ranges and thermal gradients. In the second method, the target is heated to two known temperatures and the unknown emittance is calculated. Due to the transient nature of the LPD process and the lack of known temperatures, this method is also not suitable for real-time application to the LPD process. In the third method, the reflectivity of the target is measured and a radiation energy balance is applied to calculate the unknown emittance. With the correct configuration, this method could be used for real-time emittance measurement during the LPD process; however, due to the presence of diffuse and spectral laser reflections, the presence of extraneous radiation, or the presence of participating media, this method may not provide dependable results. In the fourth method, a reference temperature, measured using an independent thermal sensor such as a thermocouple or pyrometer, is used along with Planck's law to calculate the unknown emittance of the target. Of the four methods described above, this method most readily applies to real-time temperature measurement during the LPD process. There are several other methods for obtaining emittance corrected thermal data which were not discussed by Madding. If the radiance emitted from the target surface can be assumed to be independent of wavelength (gray emission), then radiance can be measured at two or more wavelengths and temperature can be calculated using a ratio of radiance values defined by Planck's law. This method is commonly used for dual- or multi-wavelength spot or imaging pyrometers, which typically measure over relatively small spectral ranges.[13] Additionally, a blackbody cavity could be built into a test part. The unknown emittance could then be calculated using Planck's law to compare the radiance measured in the blackbody cavity to a surface temperature measured nearby.[14]

Along with emittance calibration, radiometry can be used to detect and record the thermal history of each part manufactured by LPD. The subsequent radiometric thermal response measurements provide insight into the connection between the process dynamics and the resulting part quality, without the necessity of direct contact with the part. Coupled with parametric studies and post-process analysis, radiometry can be used to establish the effect of influential process parameters on part quality for a range of LPD manufacturing scenarios. Traceability can then be achieved by using radiometry to detect, record, and establish the thermal history for each manufactured part. Repeatability can then be achieved by defining part quality requirements and using radiometry to monitor and guarantee quality of each part. It is important to note, however, that radiometry is not the only tool that could be used to achieve these goals for a given LPD application. A general discussion of laser additive manufacturing process monitoring and control can be found in Wiesemann and Steen and Mazumder.[3]

6.1.3 History of Thermal Measurements and Control in LPD Additive Manufacturing

In the past, thermal history has been controlled using open-loop and closed-loop feedback control methods. Melt pool controllers are commonly used to control melt pool size and temperature and are available on many modern LPD systems. For example, new Laser Engineered Net Shaping (LENS™) systems can be purchased with a multi-wavelength imaging pyrometer for melt pool control. Additionally, iterative open-loop development has been successfully used since the inception of LPD for specific application development; there are multitudes of companies and research facilities that apply this methodology successfully.

There has been a significant amount of work done in numerical process modeling, including simulation of melt pool dynamics, bulk-part thermal modeling, thermomechanical modeling, microstructural evolution modeling, powder flow interaction modeling, and computational fluid dynamics (CFD) modeling. To develop an accurate model of the LPD process, all existing models along with additional developments, must be considered. Attempts to model the process can become, seemingly without bounds, as complex and intricate as the modeler is determined to make it. Realistic models tend to be complicated and computationally intensive, often requiring super-computing capabilities to produce solutions in a reasonable time frame. Models of such computational magnitude can only be practically used for modeling short build times, on the order of a few minutes. With build times that could potentially reach hours or even weeks, it could take on the order of weeks or even years to compute;[16] therefore, for many industrial applications the computational cost is too high. Time- and temperature-dependent material phenomena as well as environmental interactions must be simplified to decrease the runtime, enough for numerical models to be considered as a viable process-planning option. Furthermore, models of high complexity still may not sufficiently account for the instabilities in the process dynamics or the presence of external disturbances. A significant amount of work is still necessary to combine the overall response into an efficient unified model that fully describes the LPD process dynamics. Nevertheless, these models could be used for feedforward and feedback control of the LPD process.

Historically, contact and non-contact thermal measurement techniques have been used to monitor, study, and control metal AM processes. Thermal measurements have been used for comparison with material properties, for model validation, and for closed-loop feedback control. It is important to note that portions of the following literature review were originally published by Hammell et al.[6]

In 1990, Li et al.[17] were among the first to use non-contact methods to measure melt pool radiance during the laser cladding AM process. They compared sensor response of a single photodiode to coating quality parameters such as adhesion, roughness, surface defects, thickness variation, and system failure. In 1996, Meriaudeau and Truchetet[18] used charge-coupled device (CCD) cameras to measure melt pool temperatures and geometric characteristics, as well as powder flow characteristics. They suggested that CCD cameras could be used for control optimization of the laser cladding process; however, the thermal results were reported in gray level intensity and were measured without emittance calibration. Two years later, Griffith et al.[19] compared thermocouple measurements to residual stress and hardness of laser-deposited H13 tool steel. They also used infrared imaging for bulk-part temperature measurements, and high-speed visible imaging for melt pool temperature and size measurements during LPD of 316 stainless steel. They determined that non-contact thermal measurements are important for reliability and repeatability of the process. They also determined that the microstructural evolution of laser-deposited H13 is highly dependent on the thermal history.

In 1999, Hofmeister et al.[20] used high-speed visible imaging pyrometry to study solidification during the LPD process. They also used high-speed visible imaging for closed-loop control of the LPD process based on melt pool geometry.[21] Griffith et al.[22] used high-speed visible imaging pyrometry to control deposition overhang, increasing it from 20° to 40°. They also compared the mechanical properties of various laser-deposited materials and discussed the formation of three distinct microstructural regions. Region I, the "solidification/supercritical region," was defined as the last pass that did not experience reheat thermal cycles. Region II, the "intercritical heating region," was defined as the section of the build that experienced reheat thermal cycles within the two-phase and three-phase regions on the equilibrium phase diagram. Region III, the "subcritical heating region," was defined as the portion of the build that already experienced Region I and II thermal cycles followed by reheat cycles below the Region II temperature range.

In 2001, Boddu et al.[23] used a dual-wavelength temperature sensor, a laser displacement sensor, and a standard CCD camera along with empirical modeling for adaptive control of the LPD process based on powder flow rate, laser power, and velocity. That same year, Hofmeister et al.[24] used high-speed visible imaging pyrometry to measure melt pool thermal gradients for various laser power and travel speed combinations, while depositing H13 tool steel and 316 stainless steel. They also used the mean intercept length between cells as a measure of microstructure scale for comparison to the thermal gradients. The next year, Wei et al.[25] used a dual-wavelength imaging pyrometer to compare stationary and moving melt pool temperatures to simple finite element method (FEM) simulations of thin walls. In 2003, Hu and Kovacevic[26] used high-speed infrared imaging, along with three-dimensional FEM, for closed-loop control of melt pool size based on laser power.

Three years later, Pinkerton et al.[27] reported using a thermal imager and thermocouples to characterize temperatures during the deposition of Waspaloy® for several combinations of laser power and powder mass flow rate. The resulting microstructure and grain morphology were also considered. In 2007, Wang et al.[28] compared dual-wavelength imaging results to a three-dimensional (3D) FEM model. The numerical and experimental temperatures were of the same approximate magnitude but did not share the same spatial profile. The next year, Zheng et al.[29,30] used a radiometry-based melt pool sensor to control melt pool temperature based on laser power and a height control sensor to control bead height based on velocity. They compared the experimental cooling rates to a finite difference model and studied the influence of real-time feedback control on the material properties of laser-deposited 316L stainless steel. They showed that radiometry-based feedback control can be used to produce fully dense parts with a more uniform microstructure and superior material properties, when compared to wrought 316L stainless steel.

In 2009, Xiong et al.[31] studied the thermal behavior of tungsten carbide–cobalt metal–matrix composite (MMC) during LPD of thick-wall samples. A high-speed CCD imager was used in combination with a 3D FEM model to control the melt pool based on laser power. Their results suggest that the melt pool size and temperature reach a steady state as the build progresses away from the base material. That same year, Hutter et al.[32] described the construction and testing of a radiometric sensor that addresses the emittance dependence of thermal imaging measurements. This sensor employs a method similar to that used in dual-wavelength measurements. They applied a spectral filter to compare the filtered and unfiltered radiance measurements from photodiodes arranged in pairs within the focal plane array. The measurements were then compared to a dual-wavelength pyrometer, which showed emittance compensation at various temperatures to be reasonable, with no apparent deviations over the temperature measurement range. Images of a tungsten filament lamp were reported to have an error of ~1.5% across the span of the filament, characterizing the resolution both spatially and in temperature.

In 2011, Craig et al.[33] reported the use of a dual-wavelength imaging pyrometer to study melt pool thermal profiles during LPD of small thin-wall samples. The dual-wavelength imaging pyrometer, known as ThermaViz™, has since been announced as an available option with new LENS systems.[34] In 2012, Song et al.[35] used multiple-input, single-output control to improve dimensional accuracy and layer height of LPD parts. Bead height was measured using three CCD cameras and triangulation, while melt pool temperature was monitored via dual-wavelength pyrometry. They used these sensors to control both the build height and the melt pool temperature simultaneously. They used a "rule-based" height controller to demonstrate build height stability and a model-based melt pool temperature controller to improve the thermal response, with a hierarchical preference to build height. This technology is currently deployed on modern DM3D Technology systems.

That same year, Muller et al.[36] used a two-dimensional (2D) single-band pyrometer and a spectral pyrometer to study laser melting of several materials in highly oxidizing environments, and Rodriguez et al.[14] integrated a thermal imaging feedback control system into an Arcam electron beam powder bed fusion system. They mounted a FLIR SC645 to an Arcam A2 EBM system and used it to measure the temperature distribution on the surface of the bed. They deposited blackbody cavities, which they used for emittance calibration. The next year, Köhler et al.[37] used temperature data measured with the thermal imaging system described by Hutter et al.[32] to validate FEM-based thermal simulations of a closed-loop laser cladding process for deposition of Stellite® 21 onto X5CrNi18-10 and 42CrMo4 steel-base materials. The peak melt pool temperature was also measured using a dual-wavelength pyrometer and controlled by adjusting laser power; however, the thermal imaging measurements were affected by reflections and participating media. Extraneous background radiation also affected the measurement of temperatures less than 800°C, and the peak temperatures measured by the pyrometer were at least 200°C lower than the imager. Between 800°C and 1900°C the melt pool and trailing profile could be seen clearly, and imager data corresponded well with the FEM simulations.

Also in 2013, Bi et al.[38] reported using a single-wavelength pyrometer to identify important factors that influence monitoring and control of the LPD process. They studied the effect of geometry, power density, and oxidation on temperature measurements. They determined that build geometry influences the melt pool temperature, particularly in regions where heat dissipation is limited. They also determined that radiometry-based temperature measurements can be used to control heat build-up and reduce oxidation. Zalameda et al.[39] deployed a near-infrared imager for closed-loop control based on melt pool area and changes in electron beam power in the NASA EBF3 system.[39] Data were collected during freeform deposition with 316 stainless steel and Ti6Al4V wire, which was used to define a set of image processing metrics to quantify melt pool area and "tail" area. Amine et al.[40] used thermocouples to measure the effect of varying laser power, velocity, and powder feed rate on the characteristics of 316 stainless steel deposition layers. They developed a laser power and velocity process map for prediction of hardness and secondary dendrite arm spacing. Gockel and Beuth[41] used process mapping to show how power/velocity relationships correspond to beta grain size and grain morphology in single-bead depositions of Ti6Al4V.

The literature outlined above provides one perspective of developments in the use of thermal response to understand the connection between process input and final part quality. It is important to note that there may be more sophisticated radiometric measurement and control systems currently in use, but there may not be any published literature on these systems due to their proprietary nature. There is also much more research, particularly in process modeling, that was not discussed. These have been topics of research for nearly a decade at the South Dakota School of Mines and Technology (SDSM&T), and many other institutions worldwide since the inception of LPD additive manufacturing. Some of the additional SDSM&T publications relating to these topics can be found in References 42 to 45. Nevertheless, the current understanding of LPD process phenomena is limited. In the following sections, radiometric analysis, coupled with post-process analysis of microstructure, is discussed with respect to characterization of thermal and microstructural zone formation during LPD of simple thin-wall structures.

6.2 EXPERIMENTAL APPLICATION OF RADIOMETRY TO LPD ADDITIVE MANUFACTURING

The experiment discussed in the following sections was designed to illustrate the application of radiometry to LPD additive manufacturing. Portions of the following text and experimental results were originally published in the 2013 proceedings of the ASME International Mechanical Engineering Congress and Exposition (IMECE).[6]

6.2.1 EXPERIMENT SETUP

The experiment was performed with a 3-kW Nd:YAG laser (1064-nm wavelength with a Gaussian distribution) and a custom LENS system in the Additive Manufacturing Laboratory at SDSM&T. Three thin-wall samples were deposited with gas-atomized 1045, 4130, and 4140 steel powders. All three powders were purchased from Carpenter Powder Products (Reading, PA) with the same particle size specification, −140+325 mesh (0.044- to 0.105-mm). Steel was selected because of the diverse microstructures that can form based on thermal history and chemistry. The microstructures of these alloys are also well documented with respect to traditional processing methods, which helps to establish a baseline for comparison.[46] This group of alloys was selected to demonstrate the connection between process input, thermal history, and output quality through the comparison of radiometric temperature measurements to post-process metallurgical analysis. Additionally, these three alloys exhibit different hardenability, which results in different hardness profiles for the same process plan. Each thin-wall sample was deposited using the same process plan and sensor configuration, which is described in more detail below. All process parameters were held constant throughout the deposition of each thin-wall sample.

For each thin wall, 100 layers were deposited in vertical succession to build a 51-mm tall by 102-mm long by ~2-mm thick wall. Each thin wall was deposited onto 1018 steel flat bar base material, which was 102 mm wide by 146 mm long by 6.4 mm thick. Each base plate was clamped directly to the steel build platform at two locations corresponding to the lengthwise endpoints of each thin wall. A laser power of 700 W and beam diameter of ~1 mm were used to deposit the thin walls at a laser-on travel speed of 8.47 mm/s, a laser-off travel speed of 33.87 mm/s, and with an acceleration/deceleration of 423.33 mm/s^2. Each layer was deposited unidirectionally, with a single powder nozzle leading the deposition direction at 57°, with respect to the vertical axes of the deposition, and a ~8-mm standoff distance. A powder mass flow rate of 16.75 g/min was directed into the melt pool through a 1.25-mm-diameter straight-bore powder nozzle with 2.2 L/min of argon carrier gas. All three thin walls were deposited in immediate succession in an argon environment with 7.8 ppm O_2 and 50 ppm H_2O, which was done to reduce the effect of oxidation on the deposition and radiometric measurement quality. The parameters described above were developed through trial and error to achieve the desired build geometry and uniformity.

An array of five radiometric sensors was used to monitor and record the deposition of each thin wall. A Mikron M9200 high-temperature imager (HTI) was used to measure melt pool temperatures, while a Mikron M7500 low-temperature imager (LTI) was used to measure bulk-part temperatures. The FAR Associates SpectroPyrometer, a high-temperature pyrometer (HTP), was used to measure average melt pool temperature and emittance. A Williamson Pro 92-29 low-temperature pyrometer (LTP) and 92-40 mid-temperature pyrometer (MTP) were used to measure bulk-part temperatures near the center of each thin-wall build. An image of the experimental setup can be found in Figure 6.2.

The five instruments listed above measure temperature based on radiometry and were calibrated to a blackbody over each sensor-specific wavelength and temperature ranges. The LTI is an uncooled focal plane array (UFPA) of microbolometer sensors, which was configured to measure temperatures between 200°C and 800°C. It measures radiance over a narrow spectral band of 3.7 to 3.9 μm.

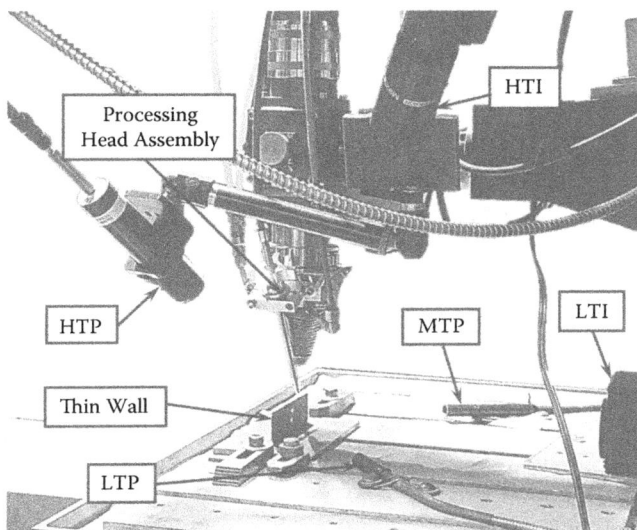

FIGURE 6.2 Image showing experimental setup.

The LTP is a dual-wavelength spot pyrometer that was configured to measure temperatures between 260°C and 700°C at an effective wavelength of 2 μm. Similarly, the MTP is a dual-wavelength spot pyrometer that was configured to measure temperatures between 482°C and 1482°C at an effective wavelength of 2 μm. The HTI is a UFPA comprised of quantum detectors which was configured to measure temperature between 800°C and 3000°C. It measures radiance over a spectral range of 0.65 to ~1 μm and has a notch filter at the laser wavelength (1.064 μm). The HTP is a multi-wavelength spectrophotometer that can measure temperature between 800°C and 4000°C over a spectral range of 0.5 to ~1.0 μm with a notch filter at 1.064 μm.

When collecting radiometric data with these instruments, it was assumed that the presence of extraneous radiation was negligible in comparison to the radiation emitted from the thin-wall structures. As well, optical filters were applied to each instrument to minimize the effect of laser radiation on the measurements. In addition, gray emission was assumed over the spectral range of each dual- and multi-wavelength pyrometer. Non-uniformity correction (NUC) was also used to account for internal changes in temperature of the LTI focal plane array (FPA). NUC is a function available on the LTI that corrects for non-uniform saturation of pixels by calibration to an internal blackbody.

6.2.2 Radiometric Analysis

Figure 6.3 shows the three thin-wall samples after deposition. From visual inspection, each build appeared to be fully dense and relatively uniform; however, an end effect was observed at the trailing end of each deposition. This was caused by deceleration and the orientation of the powder delivery nozzle. As well, slight undulation, on the order of ~600 μm from trough to peak, was observed at the top of the 4130 build. This was caused by instability in powder flow, which may be an indication of slightly too much carrier gas flow. A coarser finish was also observed at the bottom of the 4140 build, which was on the order of ~100 μm rougher than the rest of the build. In addition, small delamination fractures were observed at both ends of each thin wall.

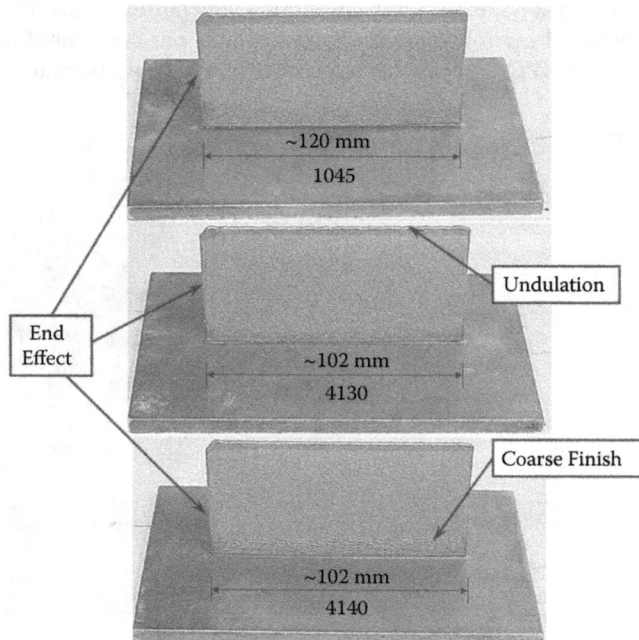

FIGURE 6.3   Images of as-deposited 1045, 4130, and 4140 thin walls showing macroscopic build deviations.

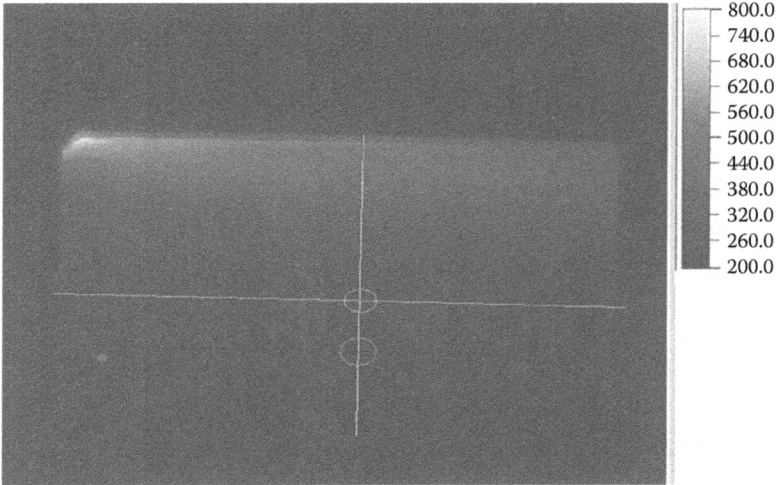

FIGURE 6.4 As-captured LTI image at the end of the 1045 deposition. The various ROIs used for real-time analysis are also shown. Temperatures are in °C.

Area averaged temperatures from the LTP, MTP, and LTI were recorded simultaneously using a programmable automation controller (PAC). Data were collected with the PAC for the entire duration of each thin-wall build. Average LTI temperature measured at an emittance of 1.0 within a small area, corresponding to that recorded by the LTP and MTP, was compared for each thin wall. Figure 6.4 is a representative thermal image showing the two elliptical regions of interest (ROIs) that correspond to the averaged thermal data. As each build progressed through these ROIs, only a portion of the recorded data was useful. These data described approximately 15% of the total build time for each alloy. As each build progressed through the measurement area of the LTP and MTP, there was a finite amount of time where the temperatures were within the measurement range of the instruments.

The LTP and LTI data, similar to those shown in Figure 6.5, were used to find an average calibration emittance for each alloy. Thirteen passes, centered at the midpoint of the build time (750 s), were used to calculate the average emittance for each alloy. The average emittances of 1045, 4130,

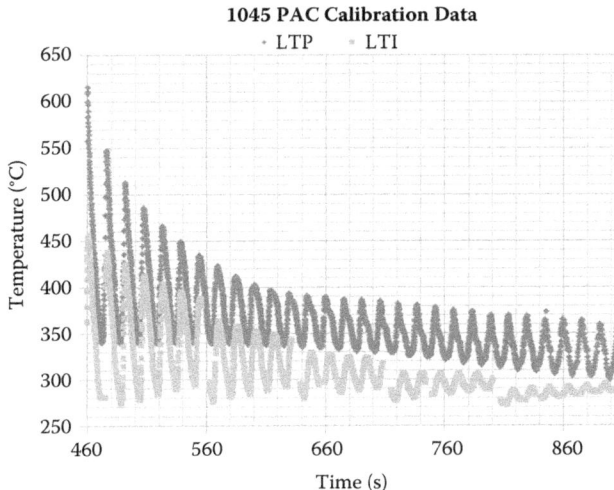

FIGURE 6.5 Plot of area averaged temperature with respect to time as measured by the LTP and LTI.

Emittance w.r.t Time for 13 Passes

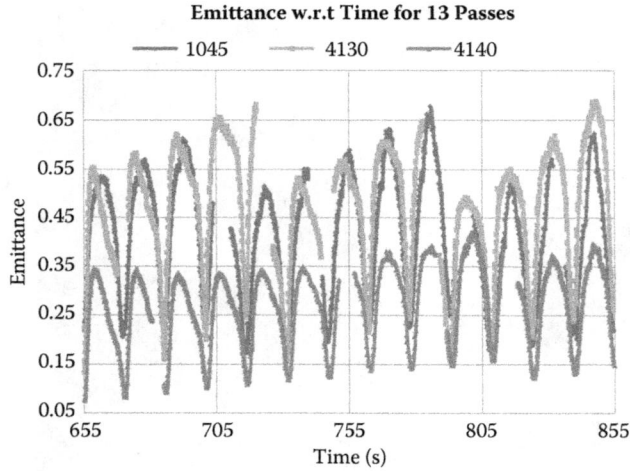

FIGURE 6.6 Emittance with respect to time calculated for each alloy.

and 4140 were determined to be 0.41, 0.47, and 0.27, respectively. The data used to find the average emittance values can be found in Figure 6.6, which shows primary oscillation due the laser-on heating and laser-off cooling cycles. It also shows a secondary oscillation of larger groups of data. This corresponds to the gradual increase in FPA internal temperature followed by an abrupt decrease due to NUC. The gaps in this data also correspond to NUC. Figure 6.7 shows MTP and LTI data recorded by the PAC showing discrepancies in the response of the two instruments. That is, the MTP data show a non-linear thermal response, which lagged the more linear LTI response. This was caused by improper configuration of the MTP with respect to the given application. For this reason, MTP/LTI emittance was not used for post-process thermal image calibration. However, this figure illustrates the importance of comparing the output from multiple sensors to help determine the validity of any given sensor's output.

1045 PAC Calibration Data

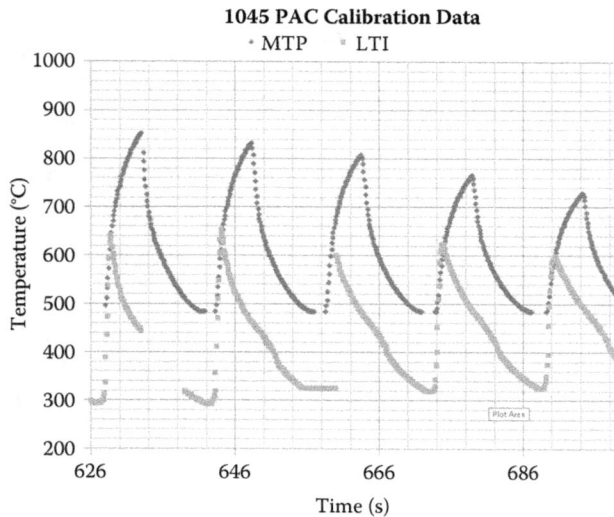

FIGURE 6.7 Plot of area averaged MTP and LTI temperature with respect to time.

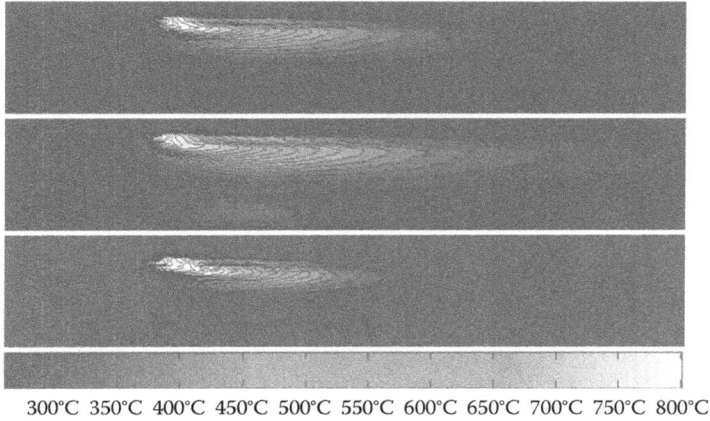

300°C 350°C 400°C 450°C 500°C 550°C 600°C 650°C 700°C 750°C 800°C

FIGURE 6.8   Representative LTI images near the bottom of the build for 1045 (top), 4130 (middle), and 4140 (bottom) after emittance calibration.

Equation 6.1, which was derived from Planck's law, and the emittance values listed above were used to calibrate representative thermal images at the bottom, middle, and top of the deposition for each thin wall. Figures 6.8, 6.9, and 6.10 show the calibrated thermal images near the bottom, middle, and top of each build, respectively. Three thermal zones of interest (ZOIs) were identified during deposition of the thin walls. ZOI-1 was identified near the bottom of the build where conduction is thought to be the dominant mechanism for heat transfer away from the deposition site. Figure 6.8 shows the thermal response for each alloy as the build progressed through ZOI-1. It also shows that 4140 had the highest spatial cooling rates, while 4130 had the lowest. ZOI-2 was identified as the quasi-steady-state region above ZOI-1 and below ZOI-3. This zone is thought to occur as a result of reduced conduction heat transfer through the thin-wall cross-section as the build progressed away from the base material. In this zone, heat loss from the deposition site is thought to be dominated by

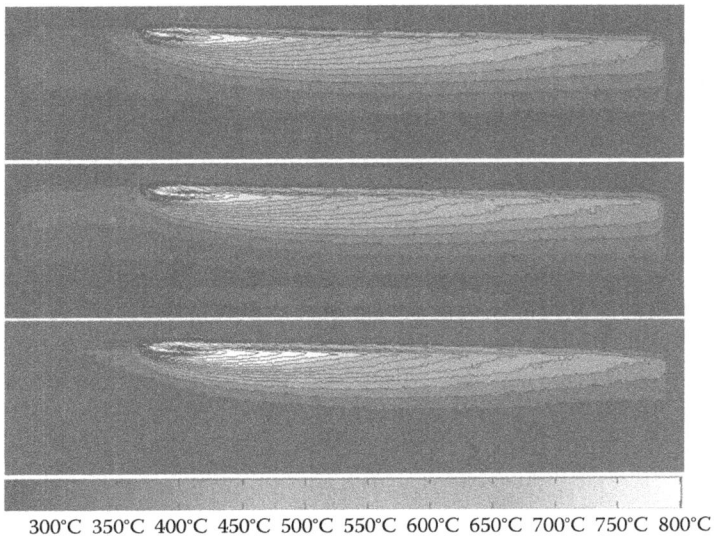

300°C 350°C 400°C 450°C 500°C 550°C 600°C 650°C 700°C 750°C 800°C

FIGURE 6.9   Representative LTI images near the middle of the build for 1045 (top), 4130 (middle), and 4140 (bottom) after emittance calibration.

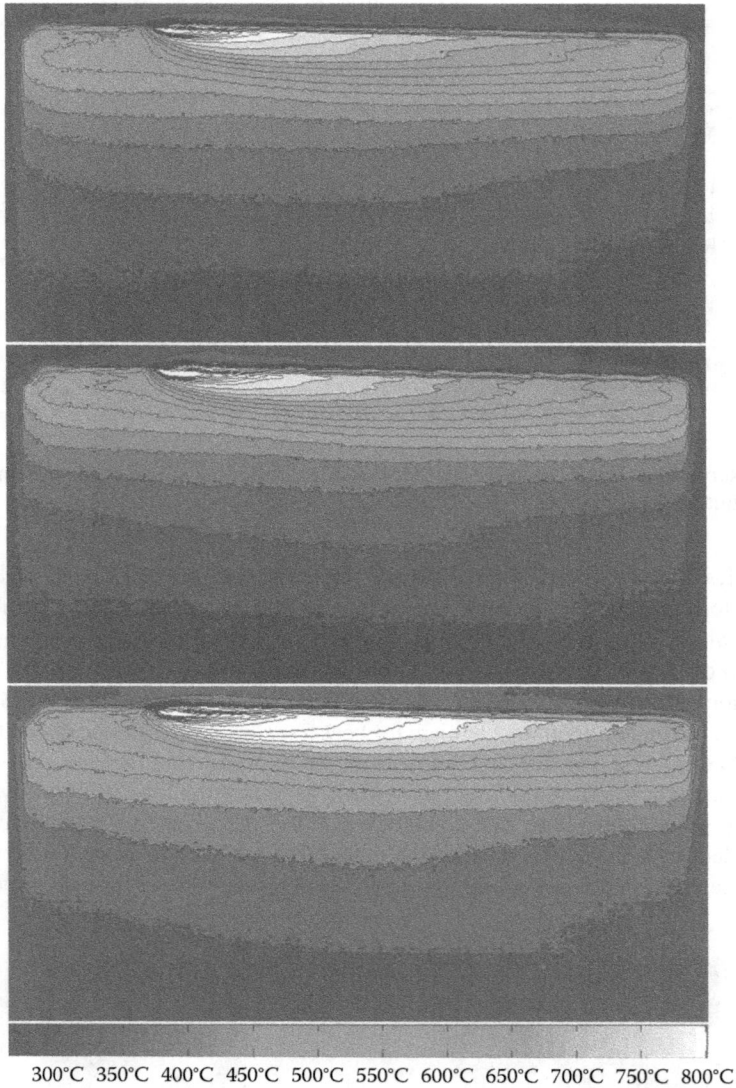

300°C 350°C 400°C 450°C 500°C 550°C 600°C 650°C 700°C 750°C 800°C

FIGURE 6.10 Representative LTI images near the top of the build for 1045 (top), 4130 (middle), and 4140 (bottom) after emittance calibration.

convection and radiation, which acts to balance the rate of heat transfer away from the deposition site with the rate of heat addition. ZOI-3 was identified as the top layers where an insufficient number of reheat cycles had occurred for transition into ZOI-2. ZOI-3 is shown at the top of each build in Figures 6.8 through 6.10. It is important to note that, as the builds progressed away from ZOI-1, the apparent spatial bulk-part cooling rates for the 4140 build were lower than those observed in the 1045 and 4130 builds, respectively.

$$T_c = \frac{c_2}{\lambda_e \cdot \ln\left[\varepsilon_c \cdot \left(\frac{c_2}{e^{\lambda_e \cdot T_m} - 1} \right) + 1 \right]} \tag{6.1}$$

HTP Temperature w.r.t. Time

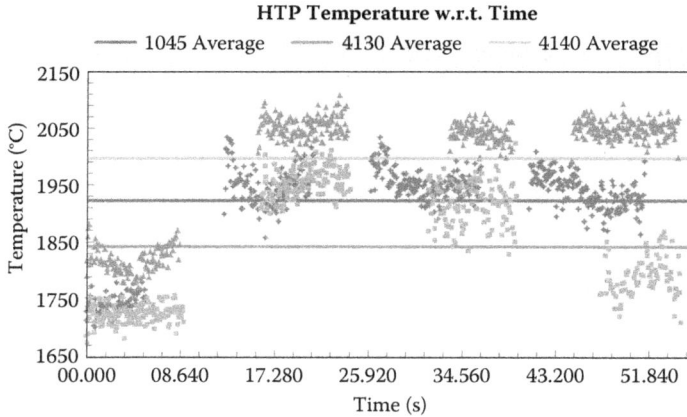

FIGURE 6.11 Plot of temperature with respect to time for the first four passes of each thin-wall build; lines of average temperature are also shown for each alloy.

where
   T_c = Corrected temperature.
   c_2 = Second radiation constant.
   λ_e = Effective wavelength.
   ε_c = Correction emittance.
   T_m = Measured temperature.

The HTP was used for melt pool image calibration of the HTI. A discussion of the use of spectral pyrometers for melt pool temperature measurements during LPD can be found in Muller et al.[36] Manual real-time comparison between the HTP and HTI temperatures was used to estimate the melt pool emittance prior to capturing images. This was accomplished at the beginning of each thin-wall build. Subsequent variations in emittance during the rest of the build were assumed to be negligible. This was done to determine the approximate magnitude and distribution of melt pool temperatures with respect to each alloy for a fixed emittance value.

Melt pool temperature was measured with the HTP for the first few minutes of each build. As each build progressed away from the substrate, increases in layer thickness led to a slight misalignment of the HTP. Because the HTP measurement area was approximately the same size as the melt pool, slight deviation in alignment resulted in large temperature fluctuations. Therefore, only data collected in the first four layers of each thin wall were used to determine average temperature and emittance. Figure 6.11 shows the corresponding HTP output along with the calculated average temperatures. The average melt pool temperatures of 1045, 4130, and 4140 were determined to be 1923°C, 1842°C, and 1997°C, respectively. The average measured HTP emittance values for 1045, 4130, and 4140 were 0.22, 0.19, and 0.24, respectively.

Due to current HTI system limitations, accurate real-time emittance calibration was not accomplished, nor was post-process emittance calibration applied. Because the analysis described herein was focused on bulk-part temperature measurements, only a few representative thermal images were captured with the HTI. Corresponding images captured at an assumed emittance of 0.25 can be found in Figure 6.12. Only slight differences in melt pool size, shape, and temperature distribution were observed. However, because the measured melt pool emittance was lower than the assumed value for each alloy, the resulting thermal images display slightly lower temperatures than actuality.

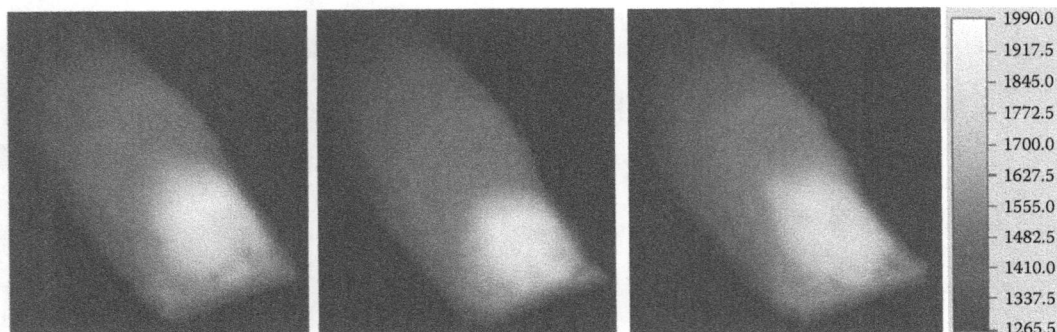

FIGURE 6.12　Representative HTI images captured during LPD of 1045 (left), 4130 (middle), and 4140 (right) at an emittance of 0.25. Temperatures are shown in °C.

6.2.3　Metallurgical Analysis

The resulting microstructure from most metal forming processes is a function of thermal history. Some metal alloys are less susceptible to microstructural variations, while others could have significant variations from one location to the next. This is a result of the reaction kinetics, microstructural phase transformations, diffusion, grain orientation/morphology, and nucleation and grain growth during the microstructural evolution. During LPD, the formation of the component may take place over long or short periods of time; this results in a long thermal history for some locations and a short thermal history for others. Some major considerations that play a part in the solid-state microstructure formation are initial grain size, grain growth, precipitate nucleation, precipitate growth, constituent diffusion, tempering, and stress state.[3,22,29]

For many LPD applications, final part quality is a function of the microstructures that form during the process; therefore, the part quality of the thin-wall structures described above was characterized through optical microscopy and hardness measurements. Hardness is a measure of a material's resistance to localized plastic deformation and is a mechanical property that is dependent on many factors, including yield strength, ultimate tensile strength, modulus of elasticity, and tendency to work harden. More importantly, hardness is a quality control measurement that is widely utilized by industry.[47] Additionally, the hardness of steel has been studied extensively, and in many cases can be used to help classify microstructure and to some degree the thermal history. Hardness is a functional indicator of bulk changes in material properties and microstructure.

The evolution of the microstructures during the LPD process is a result of rapid solidification, cyclic thermal behavior, solid-state transformation kinetics, and large thermal transients. It is well known by the metal AM community that the quality and uniformity of the powder material also influence the quality of the consolidated part. In turn, this has led to the recent development of the ASTM Standard F3049-14a, Standard Guide for Characterizing Properties of Metal Powders Used for Additive Manufacturing Processes.[48] To ensure that the particle size distribution, powder morphology, and composition met the specified conditions, particle size analysis, scanning electron microscopy (SEM), and electron diffraction spectroscopy (EDS) were used.

A Microtrac S3000 was used to characterize the size distribution of the 1045, 4130, and 4140 powders used in the experiment described above. The specified distribution of particle size was 0.044 to 0.105 mm. The mean particle size and standard deviation (sd) for the 1045, 4130, and 4140 powders were 0.05804 mm (sd, 0.0224 mm), 0.06709 mm (sd, 0.01744 mm), and 0.07651 mm (sd, 0.02045 mm), respectively. It is important to note that some of the 1045 powder was slightly smaller than the specified powder particle size. The chemical compositions for each alloy were evaluated through EDS and x-ray fluorescence (XRF) to determine if they complied with AISI/SAE standards. A list of the AISI/SAE compositions is shown in Table 6.1.

TABLE 6.1

Chemical Compositions of Selected Steel Alloys

SAE No.	C	Mn	Cr	Mo	Si	B	S	P	Iron
1045	0.43/0.50	0.60/0.90	—	—	0.10/0.35	0.0005/0.002	0.050 max	0.040 max	Balance
4130	0.28/0.33	0.40/0.60	0.80/10.10	0.15/0.25	0.15/0.35	0.0005/0.002	0.040 max	0.030 max	Balance
4140	0.38/0.43	0.75/10.00	0.80/10.10	0.15/0.25	0.15/0.35	0.0005/0.002	0.040 max	0.030 max	Balance

A Bruker Tracer IV-SD x-ray fluorescence spectrometer was used to compare the powder compositions to Bruker's standard alloys fundamental parameters (FP) database. Correlations of 9.5, 9.3, and 9.3 out of 10 were measured for 1045, 4130, and 4140, respectively. A Zeiss Supra® 40VP SEM with an Oxford Instruments X-Max 51-XMX1004 EDS was used to characterize the powder morphology and further evaluate their composition. The spectral bands from the EDS showed that the powder was composed of the expected constituents of the three alloys, yet the concentration of these constituents could not be resolved. Figure 6.13 shows that the 4140, 4130, and 1045 powder morphology was approximately spherical, with a small fraction of satellite particles attached to the gas atomized agglomerates.

Upon solidification of the laser-deposited steel alloys, described herein, there was a moment where delta-ferrite phase formed, which then transformed into austenite. Upon further cooling, the formation of bainite or martensite prevailed.[49] The subsequently deposited layers resulted in further solid-state reactions. Each layer exhibited microstructural variations that resulted from its unique thermal history, and microstructural zones were observed within individual beads.

Historically, steels have been characterized by their hardenability as a function of their chemical composition. ASTM Standard A255-89 describes a method for calculating hardenability from the composition of an alloy. This standard provides multiplication factors per constituent content for low alloy steels, which have been refined through thousands of steel samples encompassing a range of compositions.[50] Through the multiplication of these factors, an ideal critical diameter (DI) can be obtained. The DI represents the diameter of a steel round that will harden at the center to a microstructure of 50% martensite, when subject to an ideal quench. For the average AISI/SAE compositions of the 4140, 4130, and 1045, the DI values are 101.10 mm, 53.86 mm, and 24.13 mm, respectively; thus, 4140 is the most hardenable, followed by 4130 and 1045.

Optical microscopy is used extensively to determine the quality of metals and was used as a method to illustrate the connection between thermal history and part quality for the laser-deposited thin walls. A non-standard approach was used for mounting the samples. Instead of mounting standard 25.4-mm by 25.4-mm samples in Bakelite, each complete thin wall was mounted in acrylic resin. This was done to minimize material loss from cutting and to minimize any non-process heat treatment that could affect the microstructure. The samples were lapped to a 0.5-μm finish, then etched with a 3% nital

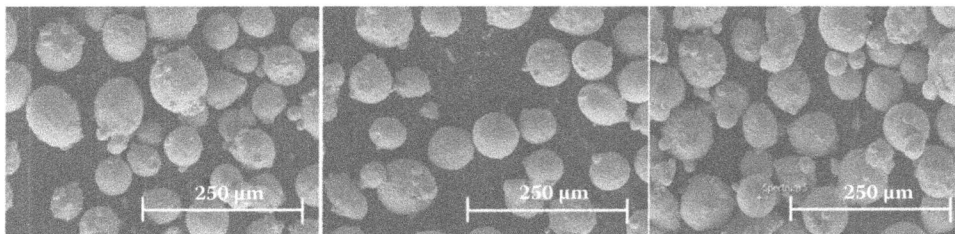

FIGURE 6.13 SEM image of AISI 4140, 4130, and 1045 (left to right) powder morphology showing spherical morphology with some satellite particles.

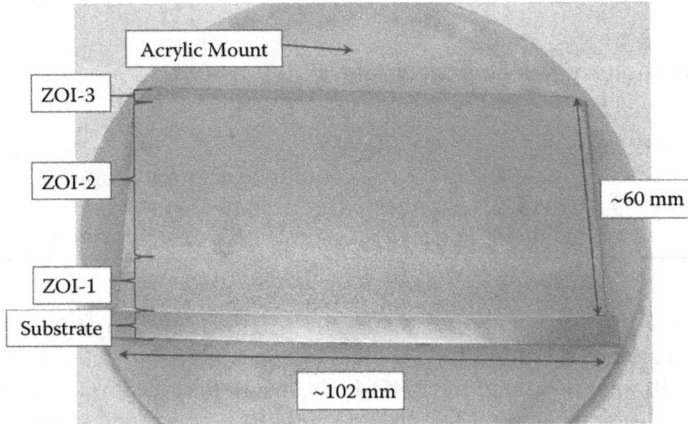

FIGURE 6.14 Image of the 4140 thin-wall structure after mounting, polishing, and etching. Note the visual difference of each macroscopic zone. Also, approximate dimensions are included to show relative size.

solution to reveal the microstructure. The distinct heterogeneity in the microstructure was obvious macroscopically. This is shown in Figure 6.14, which illustrates three major macroscopic zones. These are labeled ZOI-x, where x is the zone number with respect to distance away from the substrate.

Micrographs were captured at select magnifications for each of these macroscopic zones. A hardness profile for each sample was measured through Vickers microhardness indentation testing using a Buehler Micromet® 4 microhardness tester, 500-g load, and 10-s dwell time. These measurements were collected along the vertical axes of the samples near the center of the build to eliminate end effects. Figure 6.15 shows the measured hardness values as a function of distance from the top of the thin-wall build. It should be noted that the grain size of the structures was not measured, but the relative size was noted throughout the samples. The measured hardness draws an inference to the mechanical properties of the metal as a result of the microstructural evolution. Measured hardness data from the various steel alloys was in line with the values calculated for their ideal critical diameters (DIs). Note that 4140 has the largest DI and had the highest measured hardness, whereas 1045 has the smallest DI and had the lowest measured hardness.

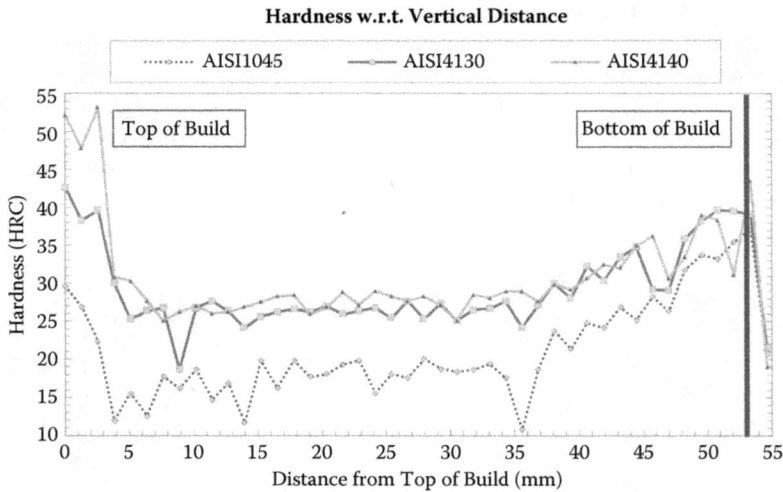

FIGURE 6.15 Plot of hardness with respect to vertical distance from the top of the 1045, 4130, and 4140 steel thin-wall LPD samples.

FIGURE 6.16   Micrographs showing substrate to build transition (ZOI-1): (A) 1045, (B) 4130, and (C) 4140.

Sample micrographs of the zones present in each alloy are shown in Figures 6.16 to 6.21. Figure 6.16 shows the micrographs depicting the interface between the substrate and the bottom of the build (ZOI-1). This figure also shows the difference in porosity found at the bottom of each build after polishing. The 1045 and 4140 samples were found to have a large amount of porosity at the bottom, and the 4130 sample had considerably less. This is likely the reason for the coarse finish observed in 4140 ZOI-1. According to the literature, the reasons for porosity could be inconsistent powder flow, powder with too low of a particle-to-void ratio, high oxidation kinetics, or unmelted powder.[51,52] In the case of the thin-wall structures, the likely cause was partial melting of the powder.

At the bottom of each sample (ZOI-1), the layers closest to the substrate experienced higher cooling rates and less time at higher temperatures. As a result, the structure was harder than the quasi-steady-state zone (ZOI-2). Figure 6.17 shows representative micrographs of ZOI-1 for each alloy. The 1045 micrograph shows a fine ferrite–pearlite structure with some indications of the onset of grain coarsening. The 4130 micrograph shows a mostly lower bainitic structure with the beginning of carbide precipitation and ferritic growth. Finally, the 4140 micrograph shows a mostly bainitic structure in ZOI-1. The measured hardness values agreed with these conclusions.

Figure 6.18 shows the transition between ZOI-1 and ZOI-2. A drastic change in appearance of microstructure in the 1045 and 4130 transition was not observed, but there was a noticeable change in the 4140. The middle of each build experienced periodic reheating and cooling cycles, which resulted in greater equilibrium or tempered microstructure.

Figure 6.19 shows micrographs taken in the middle zone (ZOI-2). This quasi-steady-state zone was similar in all three alloys. The 1045 micrograph shows small grains of ferrite and pearlite with an equiaxed grain structure. The 4130 and 4140 micrographs show a finer grain structure than the 1045 in this region. These micrographs also show fine carbide precipitates in the prior martensitic/bainitic structure. As well, the approximate prior delta-ferrite grain sizes were equal in these cases. This observation is further validated by the similar hardness values measured throughout this zone on the 4140 and 4130 samples, which was higher than the hardness of the 1045. Also, in ZOI-2 of 1045, small amounts of segregation were observed. This is depicted by the light lines passing

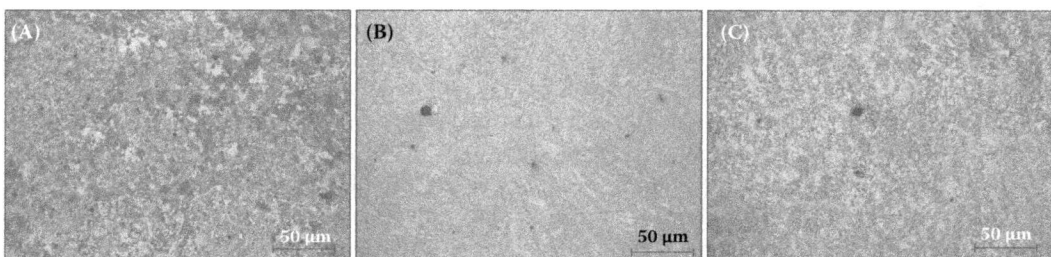

FIGURE 6.17   Micrographs showing the ZOI-1 of each alloy: (A) 1045, (b) 4130, and (C) 4140.

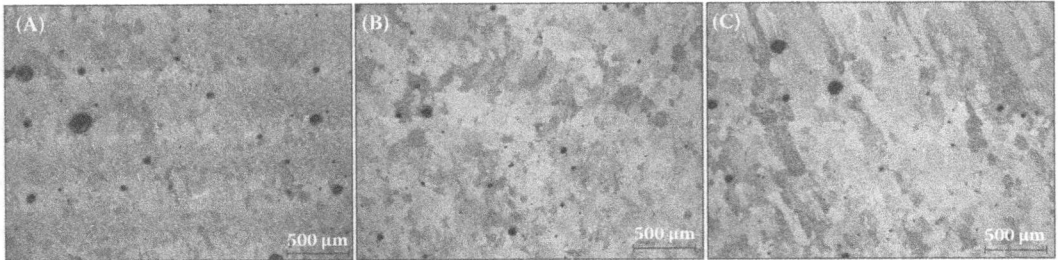

FIGURE 6.18 Micrographs showing the transition of ZOI-1 to ZOI-2: (A) 1045, (B) 4130, and (C) 4140.

FIGURE 6.19 Micrographs showing ZOI-2 of each alloy: (A) 1045, (B) 4130, and (C) 4140.

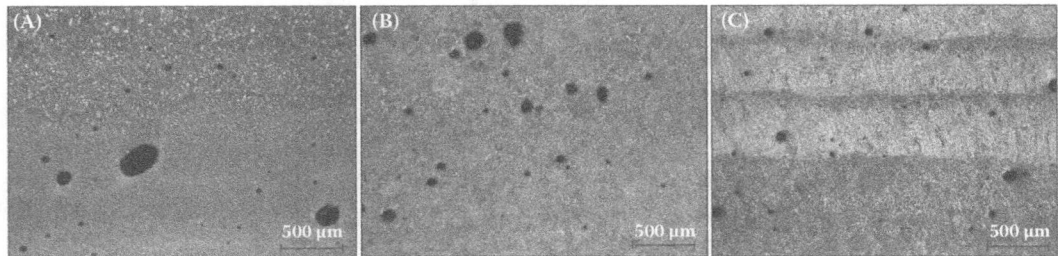

FIGURE 6.20 Micrographs showing transition of ZOI-2 to ZOI-3: (A) 1045, (B) 4130, and (C) 4140.

horizontally through the micrograph shown in Figure 6.18A. These lines of micro-segregation were likely caused by the reheating/remelting of prior deposited layers. Subsequent laser passes resulted in mixing, diffusion, and additional phase transformation. In the 4140 sample, ZOI-2 shows a columnar grain structure, whereas a more equiaxed structure was observed in ZOI-1. Also, the size of the grains increased in ZOI-2 with respect ZOI-1. This may explain the decrease in hardness between ZOI-1 and ZOI-2 for each alloy.

Figure 6.20 shows micrographs of the transition between the top zone (ZOI-3) and quasi-steady-state zone (ZOI-2) for each alloy. Major differences in the appearance of the top zones are evident in Figure 6.20. Also, micro-segregation was observed in ZOI-3 for 4140, which is shown as dark lines in the structure. Figure 6.21 shows micrographs of the top zone (ZOI-3). Again, each micrograph shows a greatly different microstructure. The 1045 micrograph shows a fine ferrite–pearlite structure with the formation of Widmanstätten ferrite at the prior delta-ferrite grain boundaries. The 4130 micrograph shows a fine grain structure that is likely a combination of upper and lower bainite. Finally, the 4140 micrograph also shows a fine grain structure, but the formation of needle-like stringers and the lighter color suggest the presence of martensite. According to Smith,[53] with the measured hardness values in this region and assuming a grain size number of 7, the structure was approximately 80% martensite.[53]

FIGURE 6.21 Micrographs showing ZOI-3 of each alloy: (A) 1045, (B) 4130, and (C) 4140.

6.2.4 EXPERIMENT CONCLUSIONS

It was found that the final part quality of laser-deposited steel can be related to *in situ* radiometric measurements. Examining the final metallurgical form of each part provides a measure of final part quality, while *in situ* radiometric monitoring provides a measure of process quality. In regard to the radiometric data collected during formation of ZOI-1, 4140 showed the highest spatial cooling rate followed by 1045 and 4130, respectively. 4140 had the most porosity in ZOI-1, followed by 1045 and 4130, respectively. This suggests that a higher cooling rate in ZOI-1 results in greater porosity, which is detrimental to build quality. For the transition out of ZOI-1 into ZOI-2, 4140 showed the lowest spatial cooling rate, while 1045 and 4130 were nearly identical. 4140, having a lower cooling rate, formed columnar grains, whereas 1045 and 4130 formed slightly smaller equiaxed grains. This suggests that the cooling rate during the transition between ZOI-1 and ZOI-2 affects grain size and morphology. As the build progressed away from ZOI-1, 4140 had the lowest spatial cooling rate followed by 1045 and 4130, respectively. These relative differences in cooling rates were observed throughout the formation of ZOI-2. This resulted in a quasi-steady-state thermal response, which is evident in the microstructures observed in ZOI-2 and corroborated with the measured hardness data.

The 4140 steel had the highest relative melt pool temperatures followed by 1045 and 4130, respectively. This may explain why 4140 had lower cooling rates than 1045 and 4130 for the bulk of each build; however, the differences in cooling rates between alloys could be caused by differences in absorptivity, differences in thermal properties, or random or systematic errors. Further analysis is necessary to determine the actual cause of differences in the observed cooling rates. It is important to note that 4140 had the highest measured melt pool emittance followed by 1045 and 4130, respectively; whereas, 4140 had the lowest bulk-part emittance followed by 1045 and 4130. This suggests that the amount of radiation heat transfer is significantly influenced by alloy composition, which may have contributed to the differences in observed cooling rates.

From the measured hardness values, 4140 was observed to have the highest hardness followed by 4130 and 1045, respectively. This is consistent with fundamental principles of steel hardenability. With respect to hardness, the relative size of ZOI-1 was the same for each alloy; however, the relative size of the 1045 ZOI-3 was smaller than the other two alloys. This suggests that the size of ZOI-1 is significantly influenced by heat transfer through the base plate, while the magnitude of the hardness in ZOI-1 is related to alloy composition. As well, the size of ZOI-3 and the magnitude of hardness in ZOI-3 are related to alloy composition. Further experimentation is necessary to fully characterize this behavior; however, the results presented herein suggest that the size of ZOI-3 increases with increasing hardenability.

6.3 CONCLUSIONS

The use of radiometry has been shown to be an effective method for controlling and monitoring metal AM processes. Many advancements in sensor design and their application have been made, spanning nearly 25 years of AM history, from the use of a single photodiode in 1990 to the current state of the art featuring high-speed, high-resolution, multi-wavelength thermal imaging and pyrometry. Sensor technology and measurement methods continue to advance, making the use of radiometry more reliable and less expensive while helping to refine processes such as LPD additive manufacturing.

Accelerating the use of new materials and processing technologies in metal AM will effectuate *in situ* radiometric analysis and control. It is important to note that the general methodologies, outlined herein, could be applied to a wide range of metals, ceramics, composites, and nano-phase materials in many AM applications. However, the specific use of radiometry as a tool for connecting process input to output quality may be application and material dependent. For example, radiometry could be used to obtain part quality measurements in laser brazing of copper joints with silver filler material by connecting wettability to measured radiance.[45] In addition, radiometry could be used to control bulk-part temperature for crack mitigation during internal bore laser cladding of hard surfacing alloys, such as the WC/Ni-Cr-Si-B metal–matrix composite, onto the internal surface of cylindrical components.[45] Radiometry could also be used to obtain part quality measurements during LPD of Ti6Al4V or Inconel structures for aerospace applications.

Regardless of the application specifics, connecting radiometric analysis to material response is one step toward building an auspicious future for metal AM. The example of radiometric and metallurgical analysis of the thin-wall builds, described herein, shows a complicated thermal and microstructural evolution. Further development of radiometric analysis and control for metal AM is essential to achieve efficient production of high-quality parts with tailored material properties that are traceable and reproducible. When these challenges have been bested, broader industrial implementation of metal additive manufacturing will follow.

ACKNOWLEDGMENTS

The authors acknowledge Nathan P. Saunders and Dr. Michael West for their invaluable contributions. This work was supported, in part, by Army Research Laboratories (Contract No. W911NF-08-2-0022).

REFERENCES

1. Kelbassa, I., Wohlers, T., and Caffrey, T. (2012). *Quo vadis*, laser additive manufacturing? *J. Laser Appl.*, 24(5): 050101.
2. ASTM (2012). *ASTM Standard F2792-12a: Standard Terminology for Additive Manufacturing Technologies*. ASTM International, West Conshohocken, PA.
3. Steen, W. and Mazumder, J. (2010). *Laser Material Processing*, 4th ed. Springer-Verlag, London.
4. Nise, N.S. (2008). *Control Systems Engineering*. John Wiley & Sons, New York.
5. Smith, C.A. and Corripio, A.B. (1997). *Principles and Practice of Automatic Process Control*. John Wiley & Sons, New York, pp. 487–544.
6. Hammell, J.J., Langerman, M.A., and Saunders, N.P. (2013). Radiometric and metallurgical analysis of zone formation in elementary laser deposited thin-wall structures. In: *Proceedings of ASME International Mechanical Engineering Congress and Exposition*, pp. 1-10.
7. Palmer J.M. and Grant, B.G. (2010). *The Art of Radiometry*. Society of Photo-Optical Instrumentation Engineers, Bellingham, WA.
8. -Thomas, H. (1990). The International Temperature Scale of 1990 (ITS-90). *Metrologia*, 27: 3–10.
9. Quinn, T.J. (1997). Supplementary information for the International Temperature Scale of 1990 (ITS-90). *Metrologia*, 34: 143–160.
10. Siegel, R. and Howell, J. (2002). *Thermal Radiation Heat Transfer*. Taylor & Francis, Boca Raton, FL.
11. Dewitt, D. and Nutter, G. (1988). *Theory and Practice of Radiation Thermometry*. Wiley, New York.
12. Madding, R. (1999). Emissivity measurement and temperature correction accuracy considerations. In: *Proceedings of SPIE 3700: Thermosense XXI*, pp. 1–9.
13. Mollmann, K.P., Pinno, F., and Vollmer, M. (2011). *Application Note for Research & Science: Two-Color and Ratio Thermal Imaging—Potentials and Limits*. FLIR, Boston, MA.

14. Rodriguez, E., Medina, F., Espalin, D., Terrazas, C., Muse, D. et al. (2012). Integration of a thermal imaging feedback control system in electron beam melting. In: *Proceedings of the 23rd Annual International Solid Freeform Fabrication Symposium—An Additive Manufacturing Conference*, pp. 1–17.
15. Wiesemann, W. (2004). Process monitoring and closed-loop control. In: *Landolt-Börnstein Numerical Data and Functional Relationships in Science and Technology*. Group VIII. *Advanced Materials and Technologies*. Vol. 1. *Laser Physics and Applications* (Poprawe, R. et al., Eds.), pp. 243–275. Springer, Berlin.
16. Stucker, B. (2014). Simulation of additive manufacturing technologies: enabling a 3D print preview. *Additive Manufacturing News*, April 10, http://additive244.rssing.com/chan-8421084/all_p12.html.
17. Li, L., Steen, W., Hibberd, R., and Brookfield, D. (1990). In-process clad quality monitoring using optical method. In: *Proceedings of SPIE 1279: Laser-Assisted Processing II*, pp. 1–12.
18. Meriaudeau, F. and Truchetet, F. (1996). Control and optimization of the laser cladding process using matrix cameras and image processing. *J. Laser Appl.*, 8: 317–324.
19. Griffith, M., Schlienger, M., Harwell, L., et al. (1998). Thermal behavior in the LENS process. In: *Solid Freeform Fabrication Symposium Proceedings*, pp. 89–96.
20. Hofmeister, W., Wert, M., Smugeresky J. et al. (1999). Investigation of solidification in the Laser Engineered Net Shaping (LENS) process. *JOM*, 51(7).
21. Hofmeister, W., MacCallum, D., and Knorovsky, G. (1999). Video monitoring and control of the LENS process. In: *Proceedings of the AWS 9th International Conference on Computer Technology in Welding*, p. 187.
22. Griffith, M., Ensz, M., Puskar, J., Robino, C., Brooks, J. et al. (2000). Understanding the microstructure and properties of components fabricated by Laser Engineered Net Shaping (LENS). *MRS Proc.*, 625: 9.
23. Boddu, M., Musti, S., Landers, R., Agarwal, S., and Liou, F. (2001). Empirical modeling and vision based control for laser aided metal deposition process. In: *Solid Freeform Fabrication Symposium Proceedings*, pp. 452–459.
24. Hofmeister, W., Griffith, M., Ensz, M., and Smugeresky, J. (2001). Solidification in direct metal deposition by LENS processing. *JOM*, 53(9): 30–34.
25. Wei, W., Zhou, Y., Ye, R., Lee, D., Craig, J., Smugeresky, J., and Lavernia, E. (2002). Investigation of the thermal behavior during the LENS process. In: *Proceedings of the International Conference on Metal Powder Deposition for Rapid Manufacturing*, pp. 128–135.
26. Hu, D. and Kovacevic, R. (2003). Sensing, modeling and control for laser-based additive manufacturing. *Int. J. Mach. Tools Manuf.*, 43: 51–60.
27. Pinkerton, A., Karadge, M., Syed, W., and Li, L. (2006). Thermal and microstructural aspects of the laser direct metal deposition of Waspaloy. *J. Laser Appl.*, 18(3): 216–226.
28. Wang, L., Felicelli, S., and Craig, J. (2007). Thermal modeling and experimental validation in the LENS™ process. In: *Solid Freeform Fabrication Symposium Proceedings*, pp. 100–111.
29. Zheng, B. Zhou, Y., Smugeresky, J., Schoenug J., and Lavernia, E. (2008). Thermal behavior and microstructural evolution during laser deposition with laser-engineered net shaping. Part I. Numerical calculations. *Metall. Mater. Trans. A*, 39A: 2228–2236.
30. Zheng, B., Zhou, Y., Smugeresky, J., Schoenug, J., and Lavernia, E. (2008). Thermal behavior and microstructural evolution during laser deposition with laser-engineered net shaping. Part II. Experimental investigation and discussion. *Metall. Mater. Trans. A*, 39A: 2237–2245.
31. Xiong, Y., Hofmeister, W.H., Cheng, Z., Smugeresky, J.E., Lavernia, E.J., and Schoenung, J.M. (2009). *In situ* thermal imaging and three-dimensional finite element modeling of tungsten carbide–cobalt during laser deposition. *Acta Mater.*, 57: 5419–5429.
32. Hutter, F.X., Brosch, D., Graf, H.-G., Klingler, W., Strobel, M., and Burghartz, J.N. (2009). A 0.25μm logarithmic CMOS imager for emissivity-compensated thermography. In: *Proceedings of the IEEE International Solid-State Circuits Conference (ISSCC 2009)*, p. 354.
33. Craig, J., Wakeman, T., Grylls, R., and Bullen, J. (2011). On-line imaging pyrometer for laser deposition processing. In: *Proceedings of Sensors, Sampling, and Simulation for Process Control, TMS 2011 Annual Meeting and Exhibition*.
34. Watson, S. (2012). *Enhanced LENS Thermal Imaging Capabilities Introduced by Optomec*. Optomec, Inc., Albuquerque, NM (http://www.optomec.com/enhanced-lens-thermal-imaging-capabilities-introduced-by-optomec/).
35. Song, L., Bagavath-Singh, V., Dutta, B., and Mazumder, J. (2012). Control of melt pool temperature and deposition height during direct metal deposition process. *Int. J. Adv. Manuf. Technol.*, 58: 247–256.
36. Muller, M., Fabbro, R., El-Rabii, H., and Hirano, K. (2012). Temperature measurement of laser heated metals in highly oxidizing environment using 2D single-band and spectral pyrometry. *J. Laser Appl.*, 24(2): 022006.

37. Köhler, H., Jayaraman, V., Brosch, D., Hutter, F., and Seefeld, T. (2013). A novel thermal sensor applied for laser materials processing. *Phys. Proc.*, 41: 502–508.

38. Bi, G., Sun, C., and Gasser, A. (2013). Study of the influential factors for process monitoring and control in laser aided additive manufacturing. *J. Mater. Proc. Technol.*, 213(3): 463–468.

39. Zalameda, J.N., Burke, E.R., Hafley, R.A. et al. (2013). Thermal imaging for assessment of electron-beam freeform fabrication (EBF3) additive manufacturing deposits. In: *Proceedings of SPIE 8705: Thermosense: Thermal Infrared Applications XXXV*, pp. 1–8.

40. Amine, T., Newkirk, J., and Liou, F. (2014). An inverstigation of the effect of direct metal deposition parameters on the characteristics of the deposited layers. *Case Stud. Therm. Eng.*, 3: 21–34.

41. Gockel, J. and Beuth, J. (2013). Understanding Ti-6Al-4V microstructure control in additive manufacturing via process maps. In: *Solid Freeform Fabrication Symposium Proceedings*, pp. 12–14.

42. Langerman, M.A., Buck, G.A., Korde, U.A., and Kalanovic, V.D. (2004). Thermal control of laser powder deposition—heat transfer considerations. In: *Proceedings of the 2004 ASME International Mechanical Engineering Congress and Exposition*, pp. 185–190.

43. Koester, J.J., Langerman, M.A., Korde, U.A., Sears, J.W., and Buck, G.A. (2005). Preliminary design of a calorimeter for experimental determination of effective absorptivity of metal substrates during laser powder deposition. In: *Proceedings of the 2005 ASME International Mechanical Engineering Congress and Exposition*, pp. 857–862.

44. Muci-Küchler, K., Tirukovelluri, P.K., and Langerman, M.A. (2009). Numerical simulation of laser glazing and laser deposition processes using coupled temperature-displacement FEM models. In: *Proceedings of the 2009 ASME International Mechanical Engineering Congress and Exposition*, pp. 51–60.

45. Hammell, J., Ludvigson, C., Langerman, M., and Sears, J. (2011). Thermal imaging of laser powder deposition (LPD) for process diagnostics. In: *Proceedings of the 2011 ASME Internation Mechanical Engineering Congress and Exposition*, pp. 41–48.

46. Honeycombe, R. and Bhadeshia, H. (2006). *Steels: Microstructure and Properties*, 3rd ed. Elsevier, Oxford.

47. NDT Resource Center. (2014). *NDT Course Material: Hardness*, https://www.nde-ed.org/Education Resources/CommunityCollege/Materials/Mechanical/Hardness.htm.

48. ASTM. (2014). *ASTM Standard F3049-14: Standard Guide for Characterizing Properties of Metal Powders Used for Additive Manufacturing Processes*. ASTM International, West Conshohocken, PA.

49. El Kadiri, H., Wang, L., Horstemeyer, M., Yassar, R., Berry, J. et al. (2008). Phase transformations in low-alloy steel laser deposits. *Mater. Sci. Eng. A*, 494: 10–20.

50. ASTM Standard A255-07e1, Historical Standard: ASTM A255-02 Standard Test Method for Determining Hardenability of Steel, ASTM International, West Conshohocken, Pa, 2007.

51. Wang, L., Pratt, P., Felicelli, S., El Kadiri, H., Berry, J. et al. (2009). Experimental analysis of porosity formation in laser-assisted power deposition process. In: *Supplemental Proceedings*. Vol. 1. *Fabrication, Materials, Processing, and Properties, TMS 2009*, pp. 389–396.

52. Susana, D., Puskwa, J., Brooks, J., and Robinoa, C. (2000). Porosity in stainless steel LENS powders and deposits. *Mater. Char.*, 57(1): 36–43.

53. Smith, W.F. (1993). *Structure and Properties of Engineering Alloys*, 2nd ed. McGraw-Hill, New York.

7 Powder and Part Characterizations in Electron Beam Melting Additive Manufacturing

Xibing Gong, James Lydon, Kenneth Cooper, and Kevin Chou

CONTENTS

ABSTRACT

Ti-6Al-4V powder and parts from the electron beam melting (EBM) additive manufacturing (AM) process have been characterized. In the EBM AM process, preheating is applied and serves to aggregate the precursor powder and may affect the subsequent melting stage. Specimens with preheated Ti-6Al-4V powder enclosed and solid parts were fabricated and prepared for microstructural and morphological examinations. In addition, micro-CT scan analysis was conducted to study the powder porosity and powder size distributions. Moreover, the process parameters in EBM AM were investigated in build part microstructures. The results can be summarized as follows. Preheating results in metallurgical bonds or even partial melting of the powder and neck formations are clearly evident. Micro-CT scans show a porosity of 50% for the preheated powder in EBM AM. The microstructure of the Ti-6Al-4V build parts is characterized by a columnar structure of prior β phase along the build direction, and fine Widmanstätten structures and martensites are presented inside of the prior β. Uneven microstructures are noted along the build height; the top layers show finer microstructure while the bottom layers display a high percentage of α'. Both the prior β grain size and α-lath thickness decrease with an increase of the scanning speed.

7.1 INTRODUCTION

Additive manufacturing (AM), a layer-based fabrication technology, has emerged as an enabling technology for a variety of novel applications adopted by different industries. Electron beam melting (EBM) AM, using a high-energy electron beam to melt and fuse powder, is a relatively new AM process that can produce full-density metallic parts directly from the digital data of a design.[1] Despite a short history of EBM AM technology, it has attracted the growing interest of aerospace, military, and biomedical industries because of the several advantages it offers, such as unique geometries and structures, rapid scan speeds, and moderate operation costs, among others. Research subjects in EBM cover a wide spectrum including material and microstructural characterization,[1,2] process modeling and simulation,[3] process metrology,[4] and part accuracy and surface finish.[5]

The EBM process involves powder spreading, preheating, and contour and hatch melting. The preheating is intended to lightly sinter the precursor powder layer by using an electron beam at a low power but rather high scanning speed. Preheating-induced sintering can hold metal powder in place during subsequent melting and reduce the thermal gradient between the melted layer and the rest of the build part.[6] Syam et al.[7] reported that preheating helps the subsequent melting process to maintain a high power density and a very rapid scanning speed. Sintered powder surrounding the part also may support down-facing surfaces during the building process. Kahnert et al.[8] observed that preheating the powder in EBM AM is an important step to avoid the spreading effect, and it affects the effectiveness of the subsequent melting and solidification steps. Yang[9] pointed out that preheating reduces the amount of energy required in melting and therefore enhances heat distribution. Although preheating in EBM AM is critical to subsequent melting processes, there has been little systematic research on powder characteristics from preheating. In addition, the mechanism of the layer-building process affected by preheating may not be clearly understood.

Ti-6Al-4V is one of the most commonly used alloys in EBM AM. Microstructures of Ti-6Al-4V samples from EBM contain a mixture of phases such as α (hexagonal closed packed, or HCP), β (body-centered cubic, or BCC), and α' martensite (HCP). A columnar prior β structure formed during initial solidifications has been reported which is a result of very high temperature gradients along the build direction.[10,11] Safdar et al.[11] observed typical Widmanstätten (α+β) inside of the prior β grains. Facchini et al.[12] investigated EBM Ti-6Al-4V microstructures and showed that the main constituent is α with only a small fraction of β. Christensen et al.[13] compared the EBM Ti-6Al-4V microstructures to counterparts from cast specimens. For the cast specimen, a coarse acicular (α+β) and thick prior β grain boundaries were observed, while the microstructure of the EBM specimen consisted of fine acicular (β) and thin prior β grain boundaries. The thickness of the α-lath is around 1.4 to 2.1 μm for different EBM samples.[1] Due to high cooling rates during the solidification, α' martensitic platelets exist in EBM parts, which may contribute to increased strength and hardness but lower ductility.[14,15] In summary, Ti-6Al-4V samples from EBM show a fine Widmanstätten (α+β) microstructure combined with α', which can be expected from the thermal characteristics of the EBM process: small melt pool and rapid cooling.[16] Because the EBM parts are built layer by layer, it is of great interest to investigate the possible build heights and orientation effects on built part microstructures.

It has been reported that EBM process parameters may have significant effects on the part quality.[17,18] Murr et al.[17] indicated that variations in the melt scan, beam current, and scan speed affect the EBM built defects such as porosity, which may be related to the microstructure/property variations in the final product. The electron beam scanning speed is a critical parameter of the EBM process that affects process conditions.[3,19] Jamshidinia et al.[20] developed a finite element (FE) model to study the heat distribution in EBM using Ti-6Al-4V powder. They reported that the scanning speed of 1000 mm/s had a much greater cooling rate than did the speeds of 100 mm/s and 500 mm/s. In research conducted by Bontha et al.,[21] thermal process maps were developed to predict solidification microstructures in the wire-feed electron beam freeform fabrication process. It was suggested that increasing the scanning speed may result in a predominant decrease in grain size of Ti-6Al-4V build parts. Despite several research efforts aimed at modeling and studying EBM, the relationship between the scanning speed and the microstructure of EBM parts is still not fully understood.

In spite of many advantages and potential benefits of EBM AM, there exist several challenges for effective usage and widespread applications. For example, a common process deficiency of EBM AM is the delamination of build layers. When the residual stresses exceed the binding abilities between the adjacent layers, layer delaminations occur and significantly degrade the part quality.[3,22] In order to avoid EBM part defects, understanding the powder characteristics and parts microstructure is of great importance.

The objective of this study was to gain an understanding of Ti-6Al-4V powder characterizations and part microstructures in EBM AM. A commercial EBM system was used to fabricate solid specimens and specimens with preheated powder enclosed. The morphology, porosity, and size distribution of Ti-6Al-4V powder in EBM AM were studied experimentally. The microstructures influenced by the build height and orientation in EBM-built Ti-6Al-4V parts were also studied. In addition, the microstructural variations, in terms of phases and characteristic length, affected by the scanning speed were investigated as well. The intent was to correlate the metal powder characteristics and part microstructures with the thermal process in EBM.

7.2 EXPERIMENTAL DETAILS

7.2.1 POWDER ANALYSIS

7.2.1.1 Raw Powder

Pre-alloyed Ti-6Al-4V raw powder was used as the feedstock material. The chemical composition is listed in Table 7.1.

TABLE 7.1
Chemical Analysis for Ti-6Al-4V Powder

	Element							
	Al	V	C	Fe	O	N	H	Ti
EBM (wt%)	6	4	0.03	0.1	0.15	0.01	0.003	Balance
Standard (wt%)	5.5–6.75	3.5–4.5	<0.1	<0.3	<0.2	<0.05	<0.01	Balance

Source: Data from Gong X. and Chou, K., in *Proceedings of ASME 2013 International Manufacturing Science and Engineering Conference/41st North American Manufacturing Research Conference*, pp. 1–8.

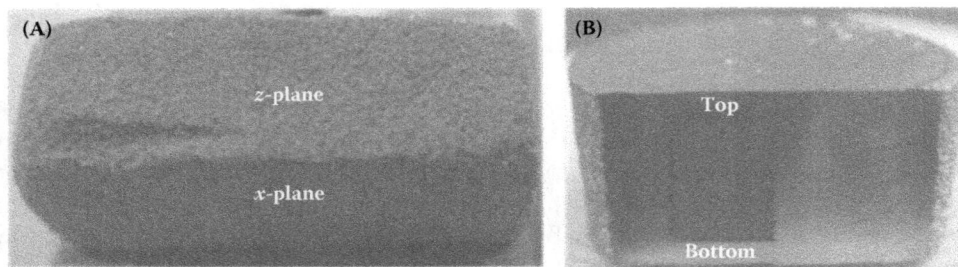

FIGURE 7.1 Samples for preheated powder characterization study: (A) powder-bed sample, and (B) powder-enclosed sample.

7.2.1.2 Machine, Material, and Fabrication Parameters and Conditions

The experimental specimens were fabricated based on designed part models and using an EBM AM system (Arcam S12) at NASA's Marshall Space Flight Center (Huntsville, AL) with Ti-6Al-4V powder and default process parameters (70-μm layer thickness). The scan speed in preheating was on the order of 10 m/s with a beam current of about 30 mA. The target preheat temperature for Ti-6Al-4V powder is 730°C, and the preheating stage lasts for about 5 s.

7.2.1.3 Powder-Bed and Powder-Enclosed Samples

Two kinds of samples with sintered powder particles were prepared for microstructural characterization: (1) powder-bed samples, which were obtained from the powder bed after the build, and (2) powder-enclosed samples, as shown in Figure 7.1.[23] Both the z-plane (scanning surface) and x-plane (side surface) were investigated (Figure 7.1A).

7.2.1.4 Metallographic Method

Samples were prepared for microstructural observations with standard metallographic procedures including sectioning, mounting, grinding with SiC papers up to the grit size of 1000, and then polishing using diamond suspensions down to 0.5 μm. Kroll's reagent (92 mL distilled water, 6 mL nitric acid, and 2 mL hydrofluoric acid) was applied for etching.[10] The samples were immersed in the solution for about 30 s, immediately rinsed with water, and then air dried. The metallographic samples were examined using a Leitz optical microscope (OM) and a Philips XL-30 scanning electron microscope (SEM).

7.2.1.5 Micro-Computed Tomography Analysis

To examine particle sizes and distributions as well as porosity from preheating, specimens from powder-enclosed parts (about 4.8 mm in length and width and 6.6 mm in height) were further fabricated using the same EBM AM system. The specimens, with sintered powder inside, were scanned using a micro-computed tomography (μCT) system (Micro Photonics SkyScan 1172) by conducting multiple continuous scans with the following parameters: image pixel size of 2 μm and source voltage of 100 kV. The image analysis using μCT data was conducted to determine the porosity of the sintered powder, measured by ImageJ software. Moreover, the particle size distributions were quantified from image analyses using Image-Pro Plus software.

7.2.2 ANALYSIS OF PART MICROSTRUCTURES

7.2.2.1 Fabrication Parameters and Conditions

Several simple blocks were modeled from CAD software and then fabricated using the EBM system, with Ti-6Al-4V powder and the same process parameters described in the previous section.

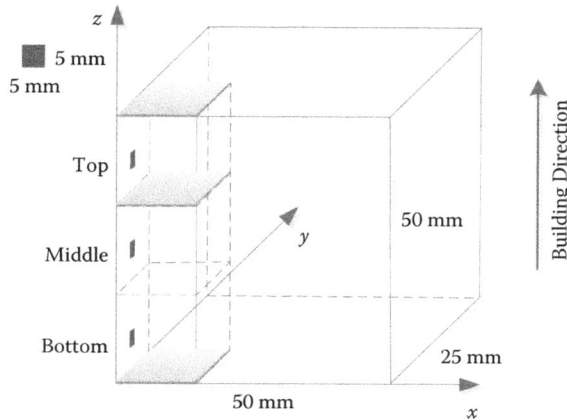

FIGURE 7.2   Illustration of specimen locations in reference to the overall part.

7.2.2.2   Build Height and Orientation Effects

A solid part with the size of 50 mm × 25 mm × 50 mm was fabricated. Then, several specimens from the EBM part were prepared (Figure 7.2) for microstructural analysis. Both the z-plane and x-plane specimens were sectioned from the part to investigate the effect of build height. For the samples with the z-plane surface, the observed planes were about 0 mm, 34 mm, and 50 mm from the bottom of the part. For the samples exposing the x-plane surface, the focus was in a localized area (5 mm × 5 mm) near the center parallel to the build direction, as can be noted from Figure 7.2.

7.2.2.3   Beam Speed Effect

For the Arcam EBM AM systems, the speed function (SF) is a process parameter setting related to the actual beam scanning speed.[24] In this study, four different levels of SF were tested (20, 36, 50, and 65) to investigate the beam speed effect on the build part microstructures. Figure 7.3 shows the EBM samples (60 mm long, 5.5 mm wide and 25 mm high) fabricated with the corresponding speed function.

7.2.2.4   Metallographic Method

Fabricated Ti-6Al-4V samples were prepared for microstructural observations with standard metallographic procedures. Specimens close to the top surface of the build parts were used for analysis. Specimens of different cross-sections (z-plane and x-plane) were prepared to examine the anisotropic conditions in microstructures. To reveal microstructures, polished specimens were then etched

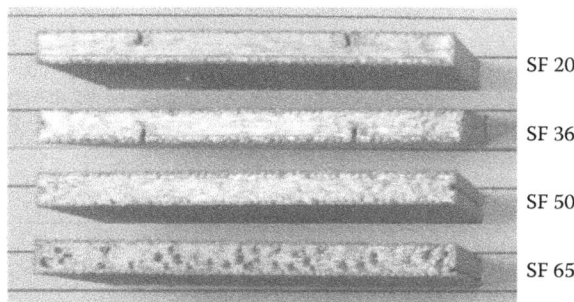

FIGURE 7.3   EBM parts built with different speed functions.

using a hydrofluoric-acid based solution. The etched metallographic samples were examined by OM and SEM. In order to quantify the size of columnar β, equiaxed β, and α-lath, a measurement methodology from Wang et al.[51] was applied.

7.3 RESULTS AND DISCUSSIONS

7.3.1 POWDER

7.3.1.1 Raw Powder

In general, fine raw powder used in EBM AM has a diameter in the range of 10 to 100 μm. Figure 7.4 presents the SEM images of Ti-6Al-4V loose powder.[23] Figure 7.4A shows different particle sizes from a backscatter electron (BSE) image, and Figure 7.4B shows a few particles from a secondary electron (SE) image at a higher magnification. Very small-sized particles, also called satellite powder, were found attached to some larger particles. During the EBM AM process, a layer of metal powder is formed using a rake to uniformly distribute powder supplied from hoppers, followed by preheating for powder sintering, prior to melting and fusing of each layer build. The typical layer thickness in EBM AM is in the range of 0.05 to 0.15 mm.[23]

7.3.1.2 Sintered Powder

After preheating, a certain degree of interparticle cohesion or aggregation occurs. Figures 7.5 and 7.6 show SEM images of the aggregated powder from both the z- and x-planes.[23] The neck formations are clearly evident in both planes. The sample shows elongated powder caused by connected

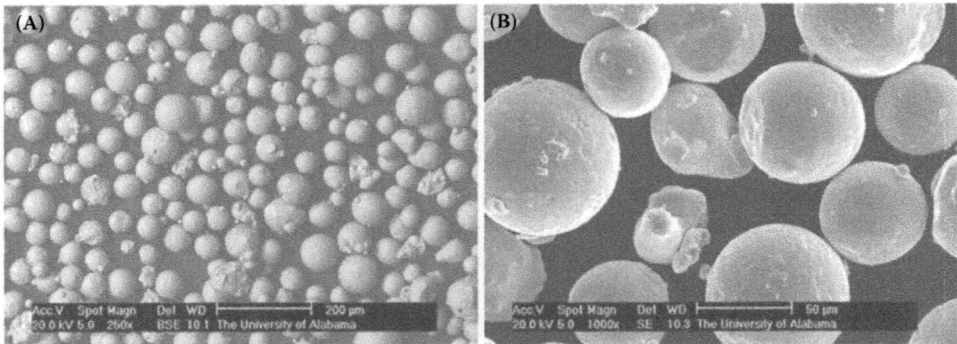

FIGURE 7.4   SEM images of Ti-6Al-4V powder particles: (A) BSE image of powder population, and (B) SE image of a few particles.

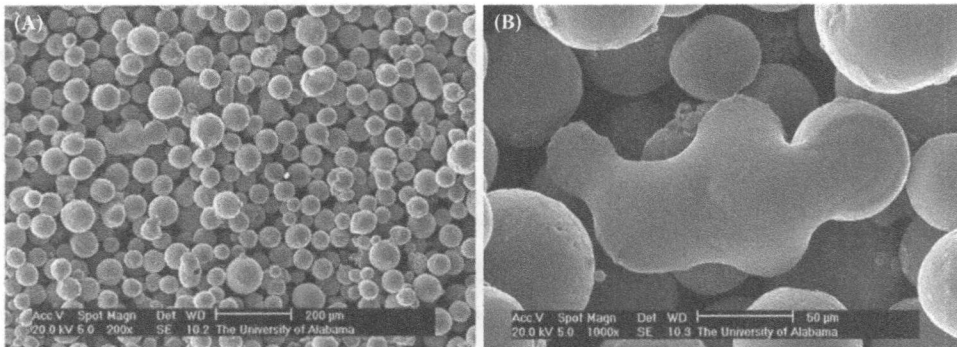

FIGURE 7.5   SEM images of preheated powder on z-plane: (A) low magnification, and (B) high magnification.

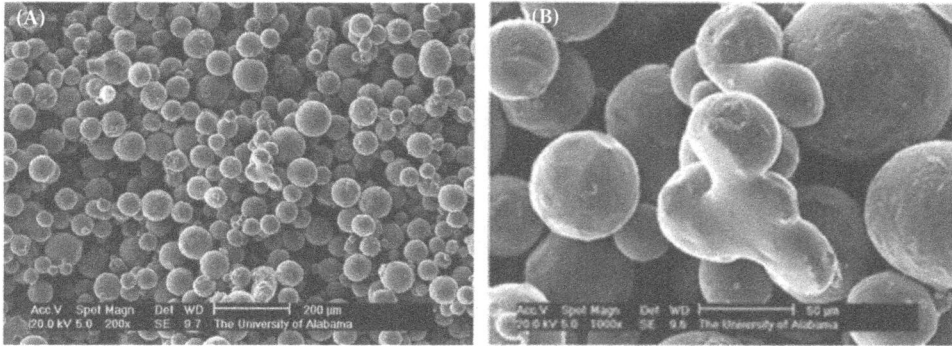

FIGURE 7.6   SEM images of preheated powder on *x*-plane: (A) low magnification, and (B) high magnification.

powder with necks or partial melting. He et al.[26] investigated the microstructure of the Ti-6Al-4V powder in EBM AM and found that the satellite particles in the interspaces are connected to large spherical particles, forming a skeleton that bonds the large particles together. Also, the authors observed that the connection between particles is frangible but tough enough to hold powder and resist the impact from the electron beam during the EBM AM process. The diameter of the necks is on the order of 1 to 10 μm. Tolochko et al.[27] found that the sintering kinetics affect heat transfer in the powder bed through the dependence of thermal conductivity on the neck size. It can be expected that the size of the necks may also affect the subsequent melting.

Figure 7.7 presents morphologies of the preheated powder of the powder-bed sample after fine polishing. The necks (circled area) can be noted on samples from both the *x*- and *z*-planes. Some powder particles are connected with surrounding powder, and the chain-like structure is formed due to the preheating. Generally, it has been observed that more neck formations are present on the *z*-plane than on samples from the *x*-plane. This is consistent with the fact that the electron beam power density will quickly diminish along the depth from the build surface. He et al.[26] also found that preheating only heats the upper thin layers of powder and causes them to bond with the formed aggregation. Figure 7.8 shows the morphologies of preheated powder from the powder-enclosed sample. The interface of the powder-solid sample (top section) is also presented. The neck formations can be observed from both the central and top sections. It is also interesting to note that the powder particles are connected to the undulated boundary of the down-facing surface of the solid part.

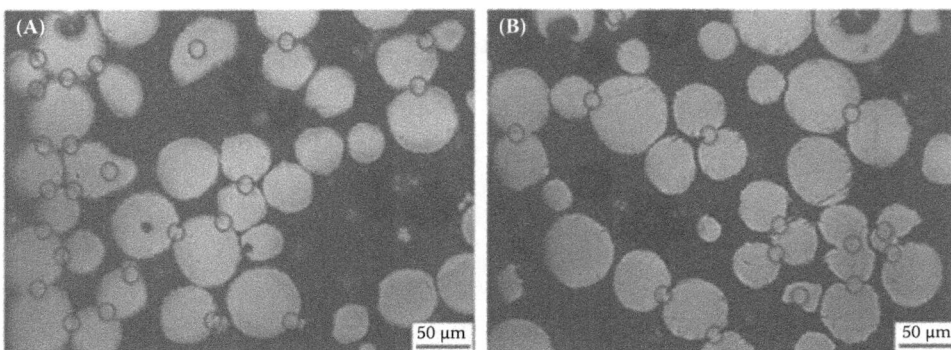

FIGURE 7.7   OM images of preheated powder from powder-bed sample after polishing: (A) *z*-plane, and (B) *x*-plane.

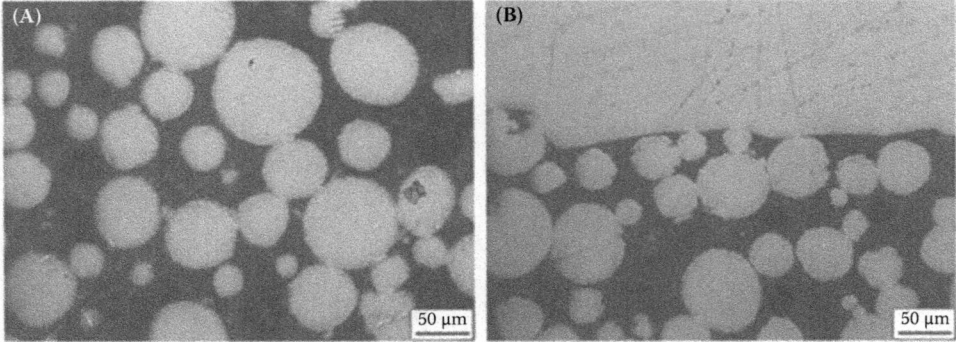

FIGURE 7.8   OM images of particles from powder-enclosed sample after polishing: (A) central section, and (B) top (solid-powder interface).

The prepared metallographic powder samples were also etched to reveal their microstructures. In general, the microstructure of Ti-6Al-4V powder (raw or sintered) is the common Widmanstätten (α+β) morphology, as can be observed in Figure 7.9.[23] Ti-6Al-4V is an α–β alloy, and its microstructure is strongly dependent on the process thermal history.[28,29] The Widmanstätten morphology is a typical feature of a Ti-6Al-4V alloy.[30] During solidification, the primary phase is BCC β. When the temperature is below β transus temperature, the solid phase transformation, into the HCP α phase, begins.[31] In the metallographic images, the light phase is the α phase and the dark phase is the β phase under an OM, while the contrast in an SEM is the opposite. For the Ti-6Al-4V powder, the

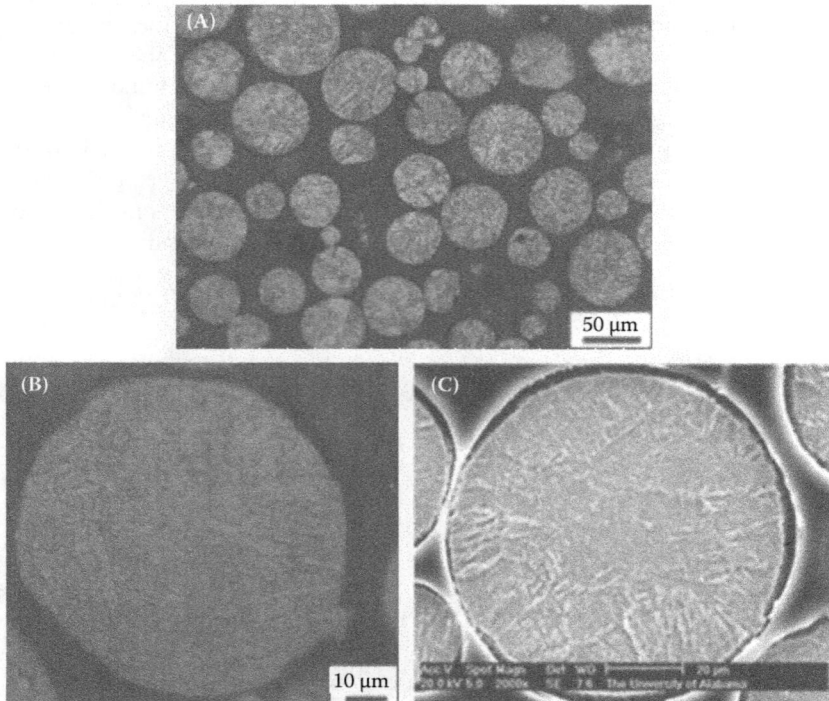

FIGURE 7.9   Microstructures of powder-bed sample after etching: (A) powder with different sizes (OM), (B) single particle (OM), and (C) single particle (SEM).

primary phase is the bulk α phase with a small amount of β in the α-lath boundaries. Figures 7.9B and 7.9C show that the intergranular structure is also present, and the microstructures observed by OM (Figures 7.9B) and SEM (Figures 7.9C) are consistent.

Qi and Yang[32] studied preheating in EBM AM and reported that preheating may result in the increase of critical scanning speed required based on the sintering neck theory. In a single particle, there will always exist some surfaces with higher surface energy that will be more easily melted than other portions by electron beam preheating. If the preheating scheme can ensure partial melting of the powder, then sintering necks between neighboring particles will be formed.[32] Although the effects of preheating parameters have not yet been investigated, a longer preheating time is expected to produce more neck formations in the sample. Qi and Yang[32] investigated the effects of preheating time on 316L stainless steel from 60 s to 120 s. The authors observed that sintering necks become visible and then particle necks start to form continuous networks, and eventually many powder particles connect and form a mass clump.

On the other hand, some researchers have reported different results for preheating. With regard to the spreading effect of the EBM process, Eschey et al.[31] found that a powder layer is able to be exposed to the electron beam at room temperature without spreading when a beam spot diameter of 0.4 mm or less is applied. The preheating can be neglected, with an associated saving of fabrication time. Furthermore, efforts in terms of the removal of unmelted powder can be reduced, as the formation of sintering bonds is eliminated. Although these results are different from other reports described earlier, they are only specific for some designed process parameters.

In order to observe and quantify the sintered powder resulting from preheating, μCT scanning was applied to fabricated powder-containing blocks. μCT scanning is a nondestructive inspection technique that is able to characterize internal features. An example of the CT scan image from a horizontal section (z-plane) is shown in Figure 7.10A. ImageJ was used to determine the porosity of the sintered powder in EBM AM. The powder portion image was then cropped from Figure 7.10A and applied with a threshold value to distinguish between porosity and powder material (Figure 7.10B). The porosity of powder was calculated based on the area ratio between the void portion and the entire area in Figure 7.10B. The result shows that the powder porosity was about 52.0% for the z-plane section. In addition, different locations (top, middle, and bottom) of the z-plane sections were also analyzed, showing very similar results (less than a 3% relative difference). The CT scan images from the z-plane and x-plane sections were comparable, as shown in Figure 7.11. The particle morphologies of both sections look similar. The analyzed porosity of the x-plane section is 50.6%, which is close to

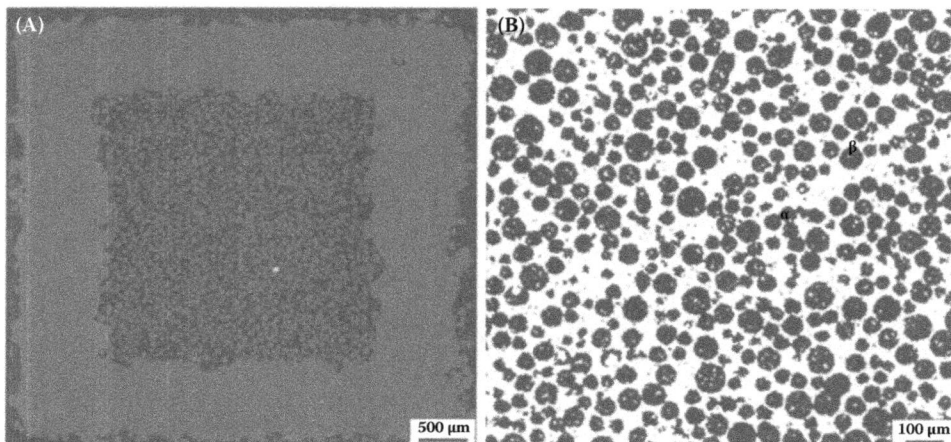

FIGURE 7.10 Images of sintered powder from z-plane section: (A) CT-scan image, and (B) threshold binary image of enclosed powder.

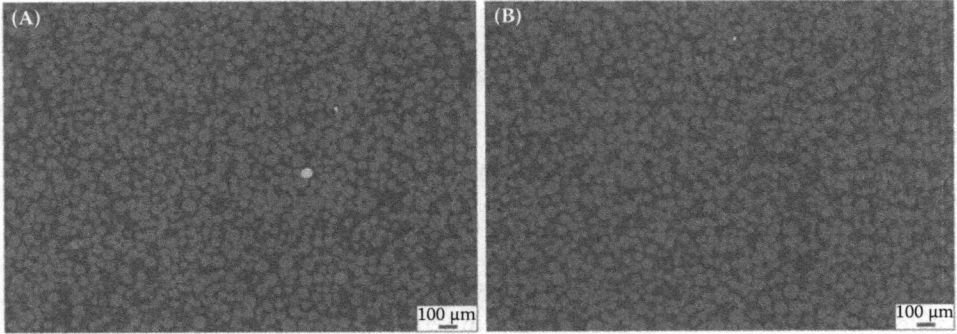

FIGURE 7.11 CT scan images: (A) z-plane section, and (B) x-plane section.

the porosity of the z-plane sections. It can be inferred that the enclosed powder particles are distributed homogeneously. It has been argued earlier that the thermal properties of Ti-6Al-4V are porosity dependent. The result from μCT image analysis of preheated powder specimens implies that the powder from preheating may significantly affect the thermal properties of the powder bed in EBM AM.[33]

To achieve quantitative analysis of the powder particle size and size distribution, Image-Pro Plus was applied to study the powder distribution in both z-plane and x-plane sections. The minimum diameter was set as 10 μm in order to exclude the attached tiny particles[17] and noise. The distribution of the particles was obtained as shown in Figure 7.12. For powder diameters in the range of 20 to 90 μm, the particles show a normal distribution. It can also be noted that the major range of powder diameters is about 30 to 50 μm. The mean overall diameters of the powder particles are 36.6 μm from the z-plane section and 34.1 μm from the x-plane section. The similar mean particle

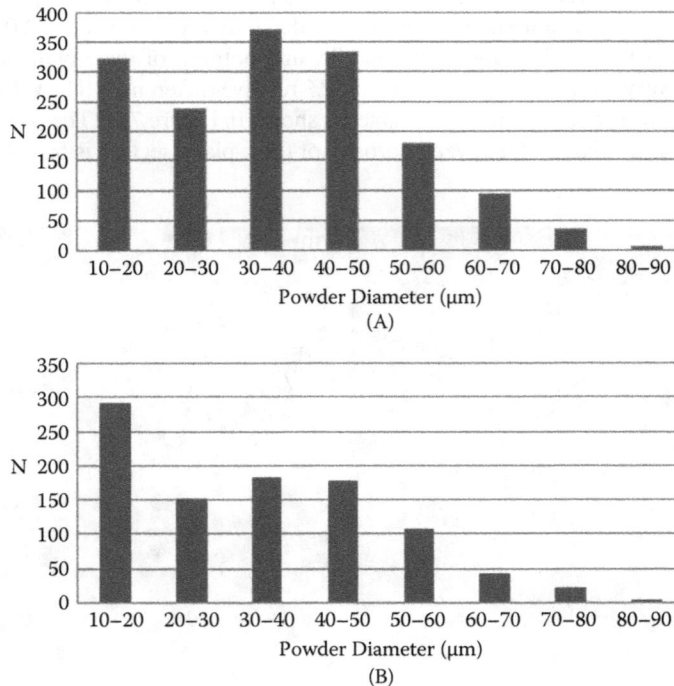

FIGURE 7.12 Histograms of powder particles: (A) z-plane section, and (B) x-plane section.

diameters between both planes indicate a homogeneous distribution of powder particles. Murr et al.[17] also studied the distribution of the powder diameters (Ti-6Al-4V and EBM AM) and reported that they ranged from 1 to 100 μm. The powder diameters show a bimodal distribution; the mean overall powder diameter is about 30 μm, while the mean diameter of large particles is about 60 μm.

7.3.2 PART MICROSTRUCTURES

7.3.2.1 Typical Microstructures of EBM AM Parts

Figure 7.13 shows typical microstructures from the x-plane of an EBM sample, SF 36. It is obvious that the prior β grains grew along the build direction and across multiple layers in the sample. The solidification of Ti-6Al-4V alloy involves two steps: liquid to primary solid phase of β and solid phase transformation (β to α or α'), depending on the cooling rate. The nucleation and growth of columnar grains of prior β take place during the initial rapid solidification when the temperature is above the β transus temperature (about 980°C). The columnar prior β grains are typical in high-energy materials processing, and upon rapid cooling from the melt the growing grains align themselves with the steepest temperature gradients[34,35] and result in a columnar-shaped morphology.[36,37] According to Antonysamy et al.,[38] in EBM the nucleation of β grains occurred heterogeneously from the boundary layers at the build plate or the part surfaces. The authors also studied the prior β grain texture of EBM parts, and the columnar structure showed a strong fiber texture of <001> β along the build direction, which could be attributed to the elongated shape of the moving melt pool.

Another feature in typical EBM Ti-6Al-4V microstructures is the martensitic phase, α', which appears as plates, as can be observed in Figure 7.13.[23] The α' is transformed from the β phase due to a very high cooling rate. Ahmed and Rack[39] reported that a cooling rate for Ti-6Al-4V of more than 410°C/s, from the single β to the ($\alpha+\beta$) region, will induce α' formation. In addition, to trigger the formation of martensites, the temperature must be lower than the martensite starting temperature (M_S): 575°C for Ti-6Al-4V.[34] Other researchers, however, reported the M_S as 650°C and the martensite finishing temperature as 400°C.[40] The α' phase is commonly observed in Ti-6Al-4V alloy subject to rapid solidifications such as selective laser melting[41,42] and electron beam welding.[43,44]

The width of the columnar prior β grains was measured in this study. Figure 7.13A illustrates an example of the width measurements. The image is overlaid with straight lines that are normal to the boundaries of the columnar structures, and the intersections of the lines with grain boundaries were examined and used for width estimates. Different images from different specimen areas were measured to obtain statistical data. The average width of the columnar structure is about 41.6 μm for this case (SF 36). This result is smaller than the columnar width of 75 to 150 μm reported by Al-Bermani et al.[10]

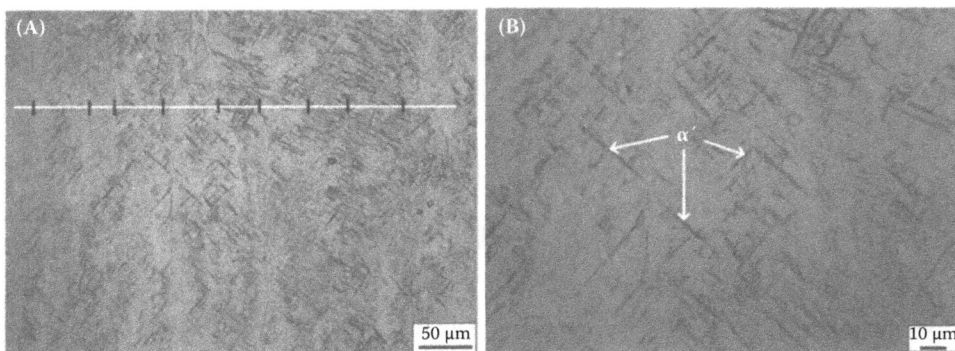

FIGURE 7.13 Microstructure of x-plane specimen from an EBM sample (SF 36): (A) low magnification showing columnar width analysis, and (B) high magnification image.

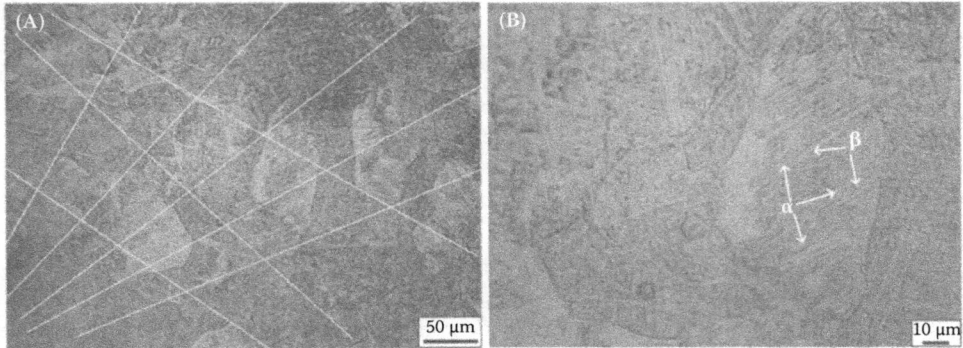

FIGURE 7.14 Microstructure of z-plane specimen from an EBM sample (SF 36): (A) low magnification showing grain size analysis, and (B) high magnification image.

Figures 7.14 and 7.15 illustrate typical microstructures of the z-plane specimen, also from the SF 36 sample.[23] In contrast to the x-plane microstructures (Figure 7.13), equiaxed grains are noted on the z-plane; thus, it can be concluded that the prior β grains have a rod shape. Similar equiaxed grains at the z-plane have also been reported by other researchers for Ti-6Al-4V alloy[10] and other alloys.[17,45] Figures 7.14B and 7.15 show higher magnification images of the z-plane specimen from OM and SEM, respectively; both α and β phases can be identified. Upon cooling from the β transus temperature, the initial α that nucleates is grain boundary α (α_{GB}) because of its location on a β boundary. Eventually, the β boundaries will be replaced by α_{GB} in a continuous fashion,[34] as can be noted from Figure 7.15A. In addition, fine Widmanstätten ($\alpha+\beta$) structures are shown inside of equiaxed grains, indicating a rapid cooling rate in EBM AM. Widmanstätten ($\alpha+\beta$) is a typical microstructure of Ti-6Al-4V alloys produced by EBM, as can be clearly identified from high magnification images (Figures 7.14B and 7.15B). In solid phase transformation, the prior β columnar grains are transformed into fine α-laths. The ($\alpha+\beta$) structure is formed by diffusion-controlled solid phase transformation, in which V diffuses into β while Al diffuses into α.[11,47] Further, the classical α-lath structure is surrounded by a very small amount of β in α boundaries. The result is similar to studies from Safdar et al.[11] and Murr et al.[17] Compared with the wrought or cast Ti-6Al-4V, which shows coarse α-laths or equiaxed α/β,[1,49] EBM AM Ti-6Al-4V parts show a finer α. Al-Bermani et al.[10] reported that fine α-laths have no preferred texture, a finding that differs from the strong texturing of prior β grains.

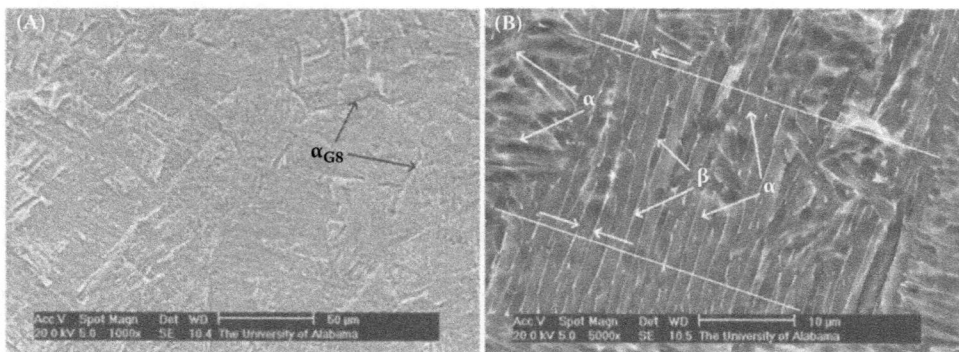

FIGURE 7.15 Scanning electron micrographs from the z-plane of an EBM sample: (A) 1000× image, and (B) 5000× image showing α-lath analysis.

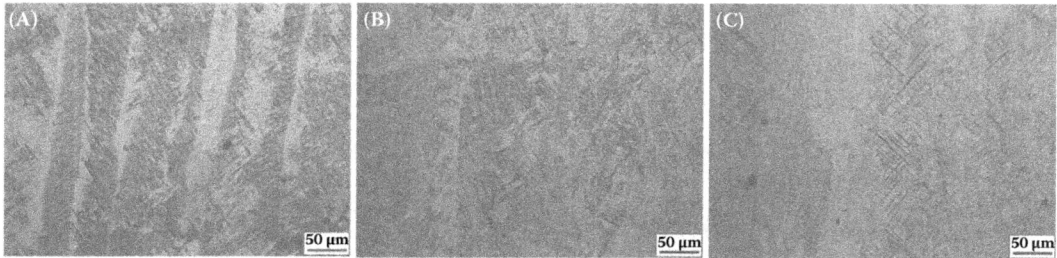

FIGURE 7.16 OM images of *x*-plane specimens at different build heights: (A) top, (B) middle, and (C) bottom.

Similar to the measurement method for the columnar width, intersections of the lines with prior β grain boundaries were marked and used for grain size estimates. The microstructural image was overlaid with several random lines, as shown in Figure 7.14A. From the measurements, the estimated size of the equiaxed grains was 50.1 μm in this case. The width of α-laths was also quantitatively analyzed using SEM images; an example is shown in Figure 7.15B. The average thickness of the α-laths was 1.1 μm for this particular sample, SF 36.

7.3.2.2 Height and Orientation Effects on Microstructures

Generally, the top layers present a smaller columnar structure, while the middle and bottom layers show larger columnar grains, as can be noted in Figure 7.16. The difference of the microstructures could be attributed to different thermal cycling that occurs during the layer building process. For the bottom layers, because the build layer has direct contact with the substrate plate, it contributes to a higher cooling rate in the bottom layers, and the higher cooling rate subsequently results in a larger percentage of α′ phase.[23] On the other hand, the middle layers of the part are significantly influenced by the thermal cycle of repeated layer building; it is obvious that the microstructure is coarser, and very little α′ is found. The top layers occur near the end of the EBM AM process, thus they are less affected by the repeated heating. The less repeated heating results in a lesser amount of growth of the columnar structure, and then subsequently smaller columnar grains on the top layers are obtained.

As can be seen from the microstructure of the *z*-plane section (Figure 7.17), the grain size of equiaxed structures is comparable to the width of the columnar grains. The top layers show a smaller size of equiaxed grains, and the bottom layers display the combined microstructure of equiaxed grains and a large percentage of α′ because of a higher cooling rate.

7.3.2.3 Scanning Speed Effect on Microstructures

The influence of the scanning speed on the microstructures from the *x*-plane specimen is shown in Figure 7.18. Generally, the width of columnar structure decreases with an increase in the scanning speed, 109.7 μm at SF 20 (214 mm/s) vs. 37.1 μm at SF 50 (529 mm/s), as shown in Table 7.2.

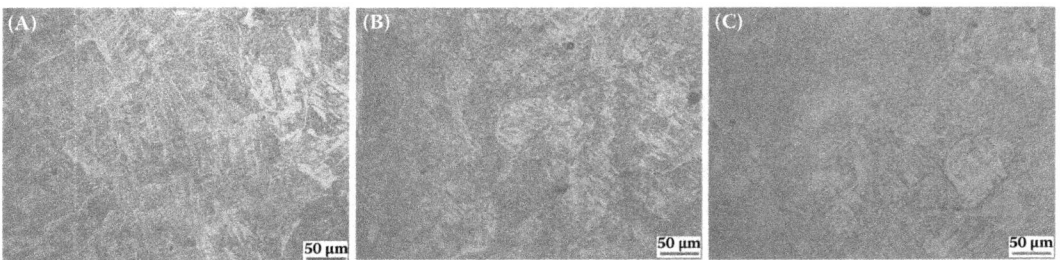

FIGURE 7.17 OM images of *z*-plane specimens at different build heights: (A) top, (B) middle, and (C) bottom.

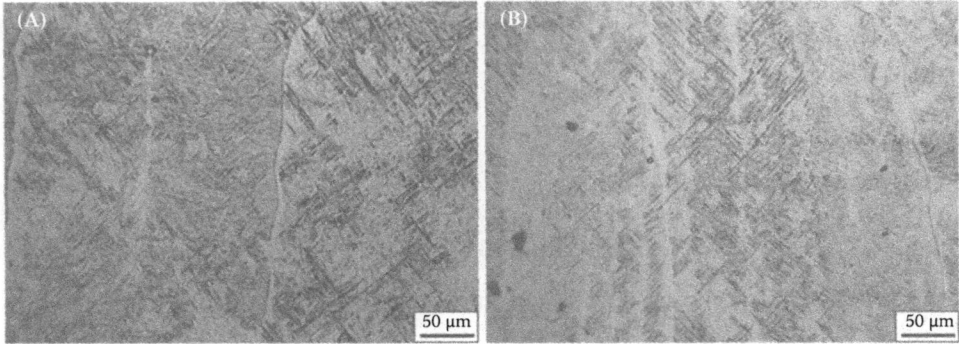

FIGURE 7.18   OM microstructures from *x*-plane specimens: (A) SF 20, and (B) SF 50.

However, the exception is the SF 65 sample, in which the width of columnar structure is slightly larger than that of SF 50. For a given beam power, increasing the scanning speed would increase the cooling rate and the thermal gradient, which will form smaller columnar β grains.[21] Gockel and Beuth[50] also reported that increasing the scanning speed will result in a higher tendency to form columnar grain structures. Wu et al.[36] investigated the effect of scanning speed on the grain morphology in laser AM of Ti-6Al-4V. At a higher scan speed, more nuclei can be generated due to more rapid cooling with finer grains in a long and narrow columnar morphology.

Figure 7.19 shows microstructures from OM for the *z*-plane specimen, which is characterized by equiaxed grains for all SF values tested. The estimated size of equiaxed grains is also listed in Table 7.2. The SF 20 (lowest) sample has the largest grain size of 85.2 μm, while the grain sizes for SF 36, SF 50, and SF 65 are similar, all less than 50 μm. Gil et al.[51] studied the effects of the cooling rate on the grain size in Ti-6Al-4V. The authors observed that the grain size is generally smaller at a faster cooling rate; however, further increases over a certain cooling rate would not further reduce the grain size. In this study, it is also noted that the SF 65 sample shows a larger percentage of α′, as is evident in Figure 7.20 (SEM micrographs from the *z*-plane specimen). Upon cooling, the transformation of equiaxed β grains that exist at high temperatures produces the Widmanstätten (α+β) structure. The α_{GB} is also found to form along the grain boundaries of prior β grains.

As illustrated in Figure 7.20, SEM micrographs exhibit α colony morphologies.[23] The size of the α-lath in Ti-6Al-4V ranges from 0.5 μm to 5 μm depending on different process conditions.[10] When Ti-6Al-4V alloy is cooled down from the β transus, the key factor to controlling the dimensions of α-laths is the cooling rate.[21] The average thicknesses of the α-lath from various SFs are listed in Table 7.2. The SF 20 sample had an α-lath thickness of 1.5 μm, and the α-lath thickness

TABLE 7.2

Measured Characteristic Sizes from Samples with Various Speed Functions

Sample	Columnar β Size (μm)		Equiaxed β size (μm)		α-Lath Size (μm)	
	Average	Standard Deviation	Average	Standard Deviation	Average	Standard Deviation
SF 20	109.7	30.2	85.2	34.6	1.5	0.3
SF 36	41.6	6.1	50.1	16.1	1.1	0.3
SF 50	37.1	6.3	48.7	16.7	1.1	0.1
SF 65	48.6	11.6	50.8	14.5	1.0	0.3

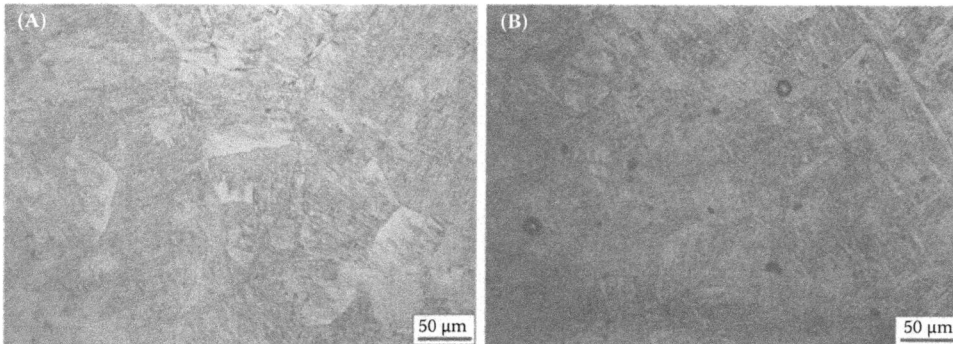

FIGURE 7.19 OM images of microstructures from z-plane specimens: (A) SF 20, and (B) SF 50.

was around 1.0 μm for other SF samples. Murr et al.[17] reported that the thickness of the α-lath is around 1.4 to 2.1 μm for different EBM parts. It is evident that higher cooling processes result in a smaller lath size. At a higher cooling rate, a larger undercooling may lead to the formation of many α nuclei and result in smaller α-laths. Moreover, the size of the α-lath is also affected by the size of the prior β.[52] The larger the prior β columnar structure, the larger the α-lath during the subsequent solid state transformation. Further, the α phase grows in a lamellar fashion inside of β grains; thus, the final length is limited by the prior β grain sizes. On the other hand, the thickness of the α-lath is controlled by diffusion, and additional coarsening is possible for a slower cooling rate.[53] Because a higher cooling rate will result in an increased amount of α′, as expected, more α′ is observed in the samples with a higher scanning speed (SF 65), as shown in Figure 7.20.

The scanning speed also affects the build defects such as the part porosity. In this study, the SF 65 sample showed a much higher percentage of pores, which were found in both x-plane and z-plane samples, with an example shown in Figure 7.21. Pores are also noticeable on the final build surface (as in Figure 7.3). The void regions were also accompanied with unmelted particles and unconsolidated materials. The porous condition was also observed by Gaytan et al.[54] and Puebla et al.[55] Puebla et al.[55] observed that the percentage of porosity is related to the cooling rate and the melt pool size, all affected by the beam speed. A higher scanning speed results in a lower build temperature, thus producing a smaller melt pool and a higher cooling rate.[56] Gong and Chou[23] statistically evaluated the effects of the process parameters on the porosity of EBM Ti-6Al-4V parts. The authors reported that the scanning speed has the most significant effect on the build part porosity, which increases greatly with increasing scanning speed.

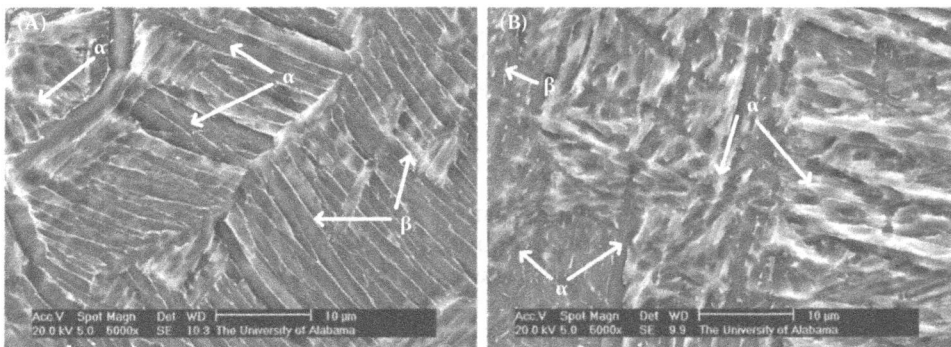

FIGURE 7.20 SEM micrograph from z-plane specimens: (A) SF 20, and (B) SF 65.

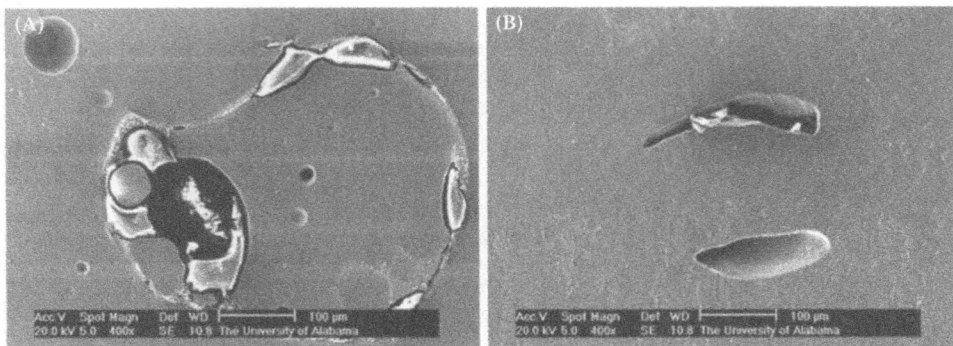

FIGURE 7.21 SEM micrograph showing porosity and unmelted powder in SF 65 sample: (A) unmelted powder, and (B) pores.

7.4 CONCLUSIONS

Powder-bed EBM AM is capable of making full-density metallic components of complex geometries and structures, greatly enabling design freedom and manufacturing flexibility. In this study, preheated Ti-6Al-4V powder and parts fabricated from the EBM AM process have been characterized. Metallographic techniques were applied to reveal and characterize sintered powder and EBM AM build parts by both OM and SEM. The porosity of the preheated powder was analyzed using images from μCT scanning of the specimens containing preheated powder inside. The powder particle sizes and size distributions were also analyzed. For solid parts, the build height and orientation effect on the microstructures were also investigated. In addition, the effect of the beam scanning speed on build part microstructures from EBM was quantified. The major findings are summarized as follows:

1. Preheating results in metallurgical bonds or even partial melting of the powder during the EBM AM process. The phenomenon of neck formations is evident in both the z-plane and x-plane sections. The diameter of the necks is on the order of 1 to 10 μm. In addition, the x-plane seems to have fewer particles being sintered because of the energy penetration limit.
2. The μCT scan images of preheated powder show a similar porosity level between the z-plane and x-plane surfaces. The calculated porosity of the preheated powder is about 50%. Moreover, the major diameter range of the Ti-6Al-4V powder used ranges from 30 to 50 μm on both planes.
3. The x-plane surface of fabricated specimens shows columnar prior β grains with α′ martensitic structures, while the z-plane surface shows equiaxed grains. The microstructure inside of equiaxed grains is Widmanstätten (α+β).
4. For the build height effect, the top layers have a finer columnar structure while the middle and bottom show wider and curved columnar structures. In addition, the middle layers generally show a Widmanstätten structure, while the bottom layers have a high percentage of α′ due to more rapid cooling.
5. The EBM AM process is very sensitive to the beam scanning speed. With an increase in scanning speed, the width of the columnar grains and size of equiaxed grains tend to be smaller because of a higher cooling rate during solidifications. In addition, increasing the scanning speed will also result in finer α-lath and a higher percentage of α′. With an increase in scanning speed, the porosity defects in the build part are also more noticeable. From this study, both SF 36 and SF 50 may be workable beam speeds that may result in fine microstructures without severe porosity.

ACKNOWLEDGMENTS

The materials presented in this paper are primarily supported by NASA, under award no. NNX11AM11A; additional support from NSF (CMMI 1335481) is also acknowledged. The research is in collaboration with the Marshall Space Flight Center, Advanced Manufacturing Team. In addition, Brandon Walters, Rajaram Manoharan, and Benjamin Ache at Micro Photonics, Inc., assisted with micro-CT scans of Ti-6Al-4V EBM AM specimens. XG also acknowledges the AL EPSCoR GRSP for financial support.

REFERENCES

1. Murr, L.E., Esquivel, E.V., Quinones, S.A., Gaytan, S.M., Lopez, M.I. et al. (2009). Microstructures and mechanical properties of electron beam-rapid manufactured Ti-6Al-4V biomedical prototypes compared to wrought Ti-6Al-4V. *Mater. Charact.*, 60: 96–105.
2. Karlsson, J., Snis, A., Engqvist, H., and Lausmaa, J. (2013). Characterization and comparison of materials produced by electron beam melting (EBM) of two different Ti-6Al-4V powder fractions. *J. Mater. Process. Technol.*, 213: 2109–2018.
3. Zäh, M.F. and Lutzmann, S. (2010). Modelling and simulation of electron beam melting. *Prod. Eng.*, 4: 15–23.
4. Price, S., Cooper, K., and Chou, K. (2014). Evaluations of temperature measurements in powder-based electron beam additive manufacturing by near-infrared thermography. *Int. J. Rapid Manuf.*, 4: 1–13.
5. Koike, M., Martinez, K., Guo, L., Chahine, G., Kovacevic, R., and Okabe, T. (2011). Evaluation of titanium alloy fabricated using electron beam melting system for dental applications. *J. Mater. Process. Technol.*, 211: 1400–1408.
6. Cormier, D., Harrysson, O., and West, H. (2004). Characterization of H13 steel produced via electron beam melting. *Rapid Prototyping J.*, 2004: 35–41.
7. Syam, W.P., Al-Ahmari, A.M., Mannan, M.A., Al-Shehri, H.A., and Al-Wazzan, K.A. (2012). Metallurgical, accuracy and cost analysis of Ti6Al4V dental coping fabricated by electron beam melting process. In: *Proceedings of the 5th International Conference on Advanced Research in Virtual and Rapid Prototyping (VRAP 2012)*, pp. 375–383.
8. Kahnert, M., Lutzmann, S., and Zaeh, M.F. (2007). Layer formations in electron beam sintering. In: *Solid Freeform Fabrication Symposium Proceedings*, pp. 88–99.
9. Yang, L. (2011). Structural Design, Optimization and Application of 3D Re-entrant Auxetic Structures, PhD thesis, North Carolina State University, Raleigh.
10. Al-Bermani, S.S., Blackmore, M.L., Zhang, W., and Todd, I. (2010). The origin of microstructural diversity, texture, and mechanical properties in electron beam melted Ti-6Al-4V. *Metall. Mater. Trans. A*, 41A: 3422–3434.
11. Safdar, A., Wei, L.Y., Snis, A., and Lai, Z. (2012). Evaluation of microstructural development in electron beam melted Ti-6Al-4V. *Mater. Charact.*, 65: 8–15.
12. Facchini, L., Magalini, E., Robotti, P., and Molinari, A. (2009). Microstructure and mechanical properties of Ti-6Al-4V produced by electron beam melting of pre-alloyed powders. *Rapid Prototyping J.*, 15: 171–178.
13. Christensen, A., Kircher, R., and Lippincott, A. (2007). Qualification of electron beam melted (EBM) Ti6AI4V-ELI for orthopaedic applications. In: *Proceedings of the Materials and Processes for Medical Devices Conference*, pp. 48–53.
14. Gong, X., Anderson, T., and Chou, K. (2014). Review on powder-based electron beam additive manufacturing technology. *Manuf. Rev.*, 1: 1–12.
15. Li, S.J., Murr, L.E., Cheng, X.Y., Zhang, Z.B., Hao, Y.L., Yang, R., Medina, F., and Wicker, R.B. (2012). Compression fatigue behavior of Ti-6Al-4V mesh arrays fabricated by electron beam melting. *Acta Mater.*, 60: 793–802.
16. Price, S., Lydon, J., Cooper, K., and Chou, K. (2013). Experimental temperature analysis of powder-based electron beam additive manufacturing. In: *Solid Freeform Fabrication Symposium Proceedings*, pp. 162–173.
17. Murr, L.E., Quinones, S.A., Gaytan, S.M., Lopez, M.I., Rodela, A. et al. (2009). Microstructure and mechanical behavior of Ti-6Al-4V produced by rapid-layer manufacturing for biomedical applications. *J. Mech. Behav. Biomed. Mater.*, 2: 20–32.
18. Ramirez, D.A., Murr, L.E., Martinez, E., Hernandez, D.H., Martinez, J.L. et al. (2011). Novel precipitate-microstructural architecture developed in the fabrication of solid copper components by additive manufacturing using electron beam melting. *Acta Mater.*, 59, 4088–4099.

19. Gong, X., Lydon, J., Cooper, K., and Chou, K. (2014). Beam speed effects on Ti-6Al-4V microstructures in electron beam additive manufacturing. *J. Mater. Res.*, 29: 1951–1959.
20. Jamshidinia, M., Kong, F., and Kovacevic, R. (2013). The coupled CFD-FEM model of electron beam melting (EBM). *Early Career Tech. Conf.*, 12: 163–171.
21. Bontha, S., Klingbeil, N.W., Kobryn, P.A., and Fraser, H.L. (2009). Effects of process variables and size-scale on solidification microstructure in beam-based fabrication of bulky 3D structures. *Mater. Sci. Eng. A*, 513–514: 311–318.
22. Qian, Z., Chumbley, S., and Johnson, E. (2011). The effect of specimen dimension on residual stress relaxation of carburized and quenched steels. *Mater. Sci. Eng. A*, 529: 246–252.
23. Gong X. and Chou, K. (2013). Characterization of sintered Ti-6Al-4V powders in electron beam additive manufacturing. In: *Proceedings of ASME 2013 International Manufacturing Science and Engineering Conference/41st North American Manufacturing Research Conference*, pp. 1–8.
24. Mahale, T.R. (2009). Electron Beam Melting of Advanced Materials and Structures, PhD thesis, North Carolina State University, Raleigh.
25. Wang, K., Zeng, W., Shao, Y., Zhao, Y., and Zhou, Y. (2009). Quantification of microstructural features in titanium alloys based on stereology. *Rare Metal Mater. Eng.*, 38: 398–403.
26. He, W., Jia, W., Liu, H., Tang, H., Kang, X., and Huang Y. (2011). Research on preheating of titanium alloy powder in electron beam. *Rare Metal Mater. Eng.*, 40: 2072–2075.
27. Tolochko, N.K., Arshinov, M.K., Gusarov, A.V., Titov, V.I., Laoui, T., and Froyen, L. (2003). Mechanisms of selective laser sintering and heat transfer in Ti powder. *Rapid Prototyping J.*, 9: 314–326.
28. Seshacharyulu, T., Medeiros, S.C., Frazier, W.G., and Prasad, Y.V.R.K. (2000). Hot working of commercial Ti-6Al-4V with an equiaxed α/β microstructure: materials modeling considerations. *Mater. Sci. Eng. A*, 284: 184–194.
29. Ding, R., Guo, Z.X., and Wilson, A. (2002). Microstructural evolution of a Ti-6Al-4V alloy during thermomechanical processing. *Mater. Sci. Eng. A*, 327: 233–245.
30. Fan, Y., Shipway, P.H., Tansley, G.D., and Xu, J. (2011). The effect of heat treatment on mechanical properties of pulsed Nd:YAG welded thin Ti6Al4V. *Adv. Mater. Res.*, 189: 3672–3677.
31. Eschey, C., Lutzmann, S., and Zaeh, M.F. (2009). Examination of the powder spreading effect in electron beam melting (EBM). In: *Solid Freeform Fabrication Symposium Proceedings*, pp. 308–319.
32. Qi, H. and Yang, L. (2010). Powder blowing and preheating experiments in electron beam selective melting. In: *Proceedings of the 2nd International Conference on Computer Engineering and Technology (ICCET)*, pp. 540–543.
33. Neira Arce, A. (2012). Thermal Modeling and Simulation of Electron Beam Melting for Rapid Prototyping on Ti6Al4V Alloys, PhD thesis, North Carolina State University, Raleigh.
34. Kelly, S.M. (2004). Characterization and Thermal Modeling of Laser Formed Ti-6Al-4V, PhD thesis, Virginia Polytechnic Institute and State University, Blacksburg.
35. Baufeld, B., Biest, O.V.d., and Gault, R. (2010). Additive manufacturing of Ti-6Al-4V components by shaped metal deposition: microstructure and mechanical properties. *Mater. Des.*, 31: S106–S111.
36. Wu, X., Liang, J., Mei, J., Mitchell, C., Goodwin, P.S., and Voice, W. (2004). Microstructures of laser-deposited Ti-6Al-4V. *Mater. Des.*, 25: 137–144.
37. Hrabe, N. and Quinn, T. (2013). Effects of processing on microstructure and mechanical properties of a titanium alloy (Ti-6Al-4V) fabricated using electron beam melting (EBM). Part 1. Distance from build plate and part size. *Mater. Sci. Eng. A*, 573: 264–270.
38. Antonysamy, A.A., Meyer, J., and Prangnell, P.B. (2013). Effect of build geometry on the β-grain structure and texture in additive manufacture of Ti6Al4V by selective electron beam melting. *Mater. Charact.*, 84: 153–168.
39. Ahmed, T. and Rack, H.J. (1998). Phase transformations during cooling in $\alpha + \beta$ titanium alloys. *Mater. Sci. Eng. A*, 243: 206–211.
40. Elmer, J.W., Palmer, T.A., Babu, S.S., Zhang, W., and Debroy, T. (2004). Phase transformation dynamics during welding of Ti-6Al-4V. *J. Appl. Phys.*, 95: 8327–8339.
41. Baufeld, B., Brandl, E., and Biest, O.V.d. (2011). Wire based additive layer manufacturing: comparison of microstructure and mechanical properties of Ti-6Al-4V components fabricated by laser-beam deposition and shaped metal deposition. *J. Mater. Process. Technol.*, 211: 1146–1158.
42. Thijs, L., Verhaeghe, F., Craeghs, T., Humbeeck, J.V., and Kruth, J.-P. (2010). A study of the microstructural evolution during selective laser melting of Ti-6Al-4V. *Acta Mater.*, 58: 3303–3312.
43. Lu, W., Shi, Y., Lei, Y., and Li, X. (2012). Effect of electron beam welding on the microstructures and mechanical properties of thick TC4-DT alloy. *Mater. Des.*, 34: 509–515.

44. Wang, S. and Wu, X. (2012). Investigation on the microstructure and mechanical properties of Ti-6Al-4V alloy joints with electron beam welding. *Mater. Des.*, 36: 663–670.
45. Amato, K., Hernandez, J., Murr, L.E., Martinez, E., Gaytan, S.M., Shindo, P.W., and Collins, S. (2012). Comparison of microstructures and properties for a Ni-base superalloy (alloy 625) fabricated by electron beam melting. *J. Mater. Sci. Res.*, 1: 3–41.
46. Murr, L.E., Gaytan, S.M., Ceylan, A., Martinez, E., Martinez, J.L. et al. (2010). Characterization of titanium aluminide alloy components fabricated by additive manufacturing using electron beam melting. *Acta Mater.*, 58: 1887–1894.
47. Murgau, C.C., Pederson, R., and Lindgren, L.E. (2012). A model for Ti-6Al-4V microstructure evolution for arbitrary temperature changes. *Modell. Simul. Mater. Sci. Eng.*, 20: 055006.
48. Murr, L.E., Gaytan, S.M., Medina, F., Martinez, E., Martinez, J.L. et al. (2010). Characterization of Ti-6Al-4V open cellular foams fabricated by additive manufacturing using electron beam melting. *Mater. Sci. Eng. A*, 527: 1861–1868.
49. Kobryn, P.A. and Semiatin, S.L. (2003). Microstructure and texture evolution during solidification processing of Ti-6Al-4V. *J. Mater. Process. Technol.*, 135: 330–339.
50. Gockel, J. and Beuth, J. (2013). Understanding Ti-6Al-4V microstructure control in additive manufacturing via process maps. In: *Solid Freeform Fabrication Symposium Proceedings*, pp. 666–674.
51. Gil, F.J., Ginebra, M.P., Manero, J.M., and Planell, J.A. (2001). Formation of α-Widmanstätten structure: effects of grain size and cooling rate on the Widmanstätten morphologies and on the mechanical properties in Ti6Al4V alloy. *J. Alloy. Compd.*, 329: 142–152.
52. Safdar, A. (2012). A Study on Electron Beam Melted Ti-6Al-4V, MS thesis, Lund University, Sweden.
53. Tiley, J.S. (2002). Modeling of Microstructure Property Relationships in Ti-6Al-4V, PhD thesis, The Ohio State University, Columbus.
54. Gaytan, S.M., Murr, L.E., Medina, F., Martinez, E., Lopez, M.I., and Wicker, R.B. (2009). Advanced metal powder based manufacturing of complex components by electron beam melting. *Mater. Technol.*, 24: 180–190.
55. Puebla, K., Murr, L.E., Gaytan, S.M., Martinez, E., Medina, F., and Wicker, R.B. (2012). Effect of melt scan rate on microstructure and macrostructure for electron beam melting of Ti-6Al-4V. *Mater. Sci. Appl.*, 3: 259–264.
56. Gong, X., Lydon, J., Cooper, K., and Chou K. (2015). Characterization of Ti-6Al-4V powder in electron beam melting additive manufacturing. *Int. J. Powder Metall.*, 51: 1–10.
57. Gong, H., Rafi, K., Starr, T., and Stucker, B. (2013). The effects of processing parameters on defect regularity in Ti-6Al-4V parts fabricated by electron beam melting. In: *Solid Freeform Fabrication Symposium Proceedings*, pp. 424–439.

8 Simulation of Powder-Based Additive Manufacturing Processes

Deepankar Pal, Chong Teng, and Brent Stucker

CONTENTS

ABSTRACT

With the advent of new manufacturing technologies such as additive manufacturing (AM) processes, it has become increasingly possible to test design ideas and decrease the time between design, concept improvement iterations, and market-ready finished products, along with reducing costs, to produce and deliver these products to the desired customers. In particular, metals-based AM products are of significant interest because they find applications in the aircraft, automotive, medical implant, power tool, and traditional manufacturing tooling fields. Current metal AM processing suffers from significant inefficiencies such as designing and pre-distorting the parts based on the high cooling and strain rates experienced during these processes, in addition to challenges related to process monitoring, repeatability, and quality of the finished parts which can result in costly post-processing solutions, decreased reliability, and increased time between the raw material and market-ready stage for these parts. In order to mitigate these challenges and better understand nonlinear thermomechanical material responses, nonlinear, three-dimensional, discretized numerical approaches such as finite element and finite volume approaches are required. The major obstacle to simulating these processes is the complicated phase and grain morphology evolution that occurs at multiple length and time scales. Such evolutions are triggered by energy sources that are orders of magnitude smaller and faster than the dimensions and time scales for building the fabricated parts. This chapter presents a novel three-dimensional, multi-spatial, multi-temporal, and multi-physics continuum-based thermomechanical finite element process modeling approach and a dislocation density-based crystal plasticity finite element method. The process modeling framework predicts *in situ* thermomechanical fields much more quickly than commercially available state-of-the-art modeling software platforms. Some of the salient features of this process modeling framework include predicting grain morphology, crystal structure, thermal fields, phase and dislocation density evolutions, residual stress, and distortion warpage. Based on these inputs, the dislocation density-based crystal plasticity finite element framework predicts the uni-/multi-axial static and dynamic mechanical properties and in-service performance of fabricated parts. These frameworks and their various pre-processing, solution, and post-processing modules are discussed throughout the chapter.

8.1   HISTORY AND MOTIVATION

Manufacturing has been an integral part of human history from prehistoric times. The industrial revolution[1] resulted in major advancements due to the increasing use of steam power, smelting of iron ore, and most importantly the introduction of new machine tools that supplanted skilled artisanship. Manufacturing was no longer practiced as a local business catering to local end users but as mass production for the population at large.

Since the inception of computers in the 1930s[2] and the availability of affordable personal computers in the mid-1980s, inventors have sought to bring the flavor of customization and local production back into manufacturing. The advent of additive manufacturing (AM), beginning with stereolithography in 1986,[3] has opened up amazing opportunities for customized and local production. Although manufacturing methods in general have advanced significantly in the past 10 to 15 years, the simulation architectures used to predict the performance of those new machines are based on the architectures used to predict older manufacturing technologies. The speed of these simulation architectures is negligible compared to the speed of production for traditional manufacturing but become quite considerable when simulating additive manufacturing processes which are multi-scale in nature and constrained to low-volume, mass-customized part production.

All modern AM machines utilize at least three integral components: an energy source, materials, and geometry representation. The inter- and intralayer strengths of fabricated parts are a function of the energy source and the materials. Therefore, evaluation of these bonds as a function of process parameters and materials answers questions about the reliability of these parts.

Layer-by-layer addition of materials enables inclusion of internal features such as channels and embedment of objects such as fibers and sensors. The inclusion of these features, physical state changes as phase transitions evolve, and the highly dynamic nature of AM processes lead to residual stresses and dimensional warping which may lead to build failure or premature failure of these parts in service. This calls for evaluation of the residual stresses and dimensional warping of the parts as a function of process parameters, materials, and build geometry. In addition, the surface finish and geometrical accuracy of AM-produced parts are not necessarily good enough for precision applications, thus post-processing operations are often required. Therefore, evaluation of surface finish and geometrical accuracy as a function of two-dimensional (2D) slice information, part orientation, materials, and process characteristics is also important.

To fully capitalize on the future potential for AM for precision engineered components, a theoretical understanding of the physics involved in AM processing and validated software tools that can predict process effects are needed. These software tools will enable future machines to be more efficient, scientists and researchers to develop new material alloys specifically tweaked for cooling rates experienced during additive manufacturing, and designers to more fully explore the geometric freedom and functionality desired by the end user while still maintaining essential strength, fatigue, corrosion, or other attributes.

8.2 INTRODUCTION

The mechanical properties, microstructure, dimensional accuracy, and surface finish of the parts created using metal melting-based additive manufacturing (AM) technologies are a function of the resolution of the energy source, geometry of the desired part, orientation of the part during the build, material, process parameters, and their coupled interactions. The resolution of the energy source is generally on the order of $\sim 10^{-2}$ to 10^{-3} times less than part feature dimensions in metal laser powder bed AM. If the accuracy required is greater than the beam diameter ($\sim 10^{-4}$ m), then achieving that accuracy does not pose a significant challenge. However, if higher accuracies are required, then excellent control of laser position and a scan strategy are required.

Unlike for powder bed fusion-based AM, the resolution of the energy sources in most directed energy deposition processes, such as a Sciaky wire-fed electron beam system, pose accuracy problems because the beam diameters are much larger ($\sim 10^{-2}$ m), and the accuracy required may be a small fraction of the beam diameter. Yet another source of inaccuracies in part dimensions occurs due to the stair-stepping effect induced as a result of transformation of a three-dimensional (3D) solid model into 2D slices for fabrication. This effect can be minimized by performing careful calculations and optimization of the build orientation, leading to a smoother texture on desired surfaces along with better control and tailored use of thinner layers, if possible. However, some of the remedies such as a smaller beam size and thinner layers are counterproductive in terms of build times.

Better control of AM processes requires fast and accurate simulations. This is particularly difficult for laser-based AM because the cause-and-effect relationships are not linear and therefore require evaluation of the full factorial process-parameter space and material variable space (a time-consuming simulation that, due to its difficulty, has not been frequently used for furthering machine development to date).

Over the past 4 years, a comprehensive set of modeling tools has been developed at the University of Louisville for quickly and accurately predicting the effects of changes in process parameters on mechanical properties, residual stress/strain, crystal structure, and other micro and macro features of components made using metal-based additive manufacturing techniques.[4–17] This research has resulted in the development of multi-scale, multi-physics finite element solvers that have been shown to be many orders of magnitude faster than commercially available tools while achieving comparable or better solution accuracy. In the next few sections, we will be discussing the various algorithms developed in-house to fully understand their capabilities and some results and validations to show their usefulness in the area of metal melting-based AM.

8.3 BROADER OBJECTIVES

The objective of this modeling effort is to build new software tools that can efficiently simulate highly dynamic, coupled manufacturing and metallurgical processes, including the high strain and cooling rates exhibited by additive manufacturing. Some of the key objectives of this effort are to

1. Formulate process map response surfaces that correlate parameters and outcomes such as beam power and scan speed to grain size and shapes or melt pool geometry. This enables optimization of materials and geometry *a priori*. Other predictive outputs can include residual stress, dynamic warping, and mechanical properties.
2. Enable closed-loop control where on-the-fly compensation for out-of-spec thermal or microstructural defects can be achieved. A possible path for doing so includes using optical sensors that record layer-by-layer information such as temperatures, geometry, distortion, porosity, etc. and their spatiotemporal locations so this information can be fed for an instantaneous next-layer, feedforward simulation using the real, current build geometry so that a local tweak of the process parameters can be achieved in the *x-y* vicinity of the defect in the previous layer to counteract or fix the defect. With the increasing speed of simulations achievable by porting codes to a general-purpose graphical processing unit (GPGPU), these defects may even be fixable within a layer by changing, for instance, beam power and scan speed in the next scan vector.
3. Accurately predict mechanical properties of the parts as a function of process parameters, materials, and geometry.
4. Formulate better scan strategies and optimize support structure placement for metal melting-based AM, leading to lower residual stresses and less post-processing of the fabricated part.

8.4 SOFTWARE CAPABILITIES

Finite element analysis (FEA) software suites are generally comprised of three integral components—preprocessing, solver, and post-processing capabilities. In this paper, we focus on the preprocessing and solver capabilities of our software. It should be noted that all of these capabilities were written as Fortran and/or MATLAB® routines and can be run independently from commercial FEA software tools.

8.5 LITERATURE REVIEW OF THREE-DIMENSIONAL MULTI-SCALE DYNAMIC MESHING CAPABILITIES (PREPROCESSING)

The adaptive meshing strategies are commonly applied in computational fluid dynamics (CFD) and offer promising solutions of dynamic fluid/wind flow simulations. The unique adaptive feature and calculation accuracy generated from the corresponding mesh density alterations with respect to the flow movement offer the ability to solve history-dependent, unsteady dynamic problems induced by fast movement of fluid/wind flow. One mesh generation strategy is the unstructured grid-based meshing method, which has been developed following the tendency of classical triangular or tetrahedral grids, quadrilateral or hexahedral grids, or mixed or hybrid grids. In the early stages of adaptive meshing development, the triangular elements were widely applied to easily capture the boundary geometry.[18-27] A common issue with the mesh created using triangular or tetrahedral grids is that the quality of the mesh near highly dense regions is generally not guaranteed and the mesh Jacobian diverges in those regions. To overcome this, the Delaunay triangulation mapping method has been used to refine the triangular mesh by maximizing the minimum inscribed angle of all the elements,[20,28] and the elastic solid method (ESM) and layered elastic solid method (LESM) were used to calculate the new node positions from elasticity constraint equations.[21]

Later, quadrilateral and hexahedral elements were studied and developed further with patch testing methodologies[29] to investigate better solutions, as these elements do not have element locking issues when compared to their triangular element counterparts.[30] The radial basis function (RBF) interpolation method was applied to describe the adaptive mesh deformation with consideration of the connections between global and local scales.[31-35] However, the forward sweep mesh deformation used in RBF results in the selection of complicated nodal points followed by a reduction (cluster selection) procedure that makes the calculations very expensive computationally.[32] Several other approaches reported in the literature have been designed to alleviate the mesh quality problem of the transition area between the cluster region and the coarse mesh region. The boundary motion approach basically changes the entire mesh boundary with respect to the flow boundary.[36] The hierarchical refinement approach typically improves the mesh quality by decreasing the mesh density hierarchically away from the cluster region.[37,38] The arbitrary Lagrangian–Eulerian–adaptive mesh refinement (ALE–AMR) method implements an arbitrary Lagrangian–Eulerian hydrodynamics into an adaptive mesh refinement framework which offers additional resolution in regions with rapidly changing dynamics.[39] A hybrid dynamic mesh has been developed for the moving boundary problem where the quadrilateral grids are first generated near the solid bodies, the adaptive Cartesian mesh is then generated to encompass the entire computational domain, and the triangular grids are used to fill the gap between the quadrilateral grids and Cartesian mesh.[40] This mesh is generally better than the other methods mentioned above; however, it still has a mesh overlapping problem, and it is very difficult to deal with the frequent movement of energy source in additive manufacturing.

8.6 RESULTS

Novel dynamically adaptive meshing strategies have been formulated with mesh movement tied to the energy sources and sinks employed in AM. The adaptive meshes are generally refined near the energy source or energy dispersive mechanisms, as high thermal or stress/strain gradients at these locations require fine-scale accuracy to minimize the errors that would occur if the mesh remained at a coarser scale. The required refinement regimes are multi-dimensional in nature. Some examples of energy sources include Gaussian point energy sources (Figure 8.1) for selective laser melting (SLM) and electron beam melting (EBM) and line energy ultrasonic vibration compressive forces (Figure 8.2A) for ultrasonic consolidation (UC). An example of superficial or area-based refinement occurs at the interface of mating surfaces of foils (Figure 8.2B) undergoing ultrasonic consolidation or friction stir welding. An example of volume-based refinements is the modeling of Inconel alloys, as they contain secondary and tertiary precipitates (Figure 8.3), which respond differently than the bulk to the thermal and deformation fields experienced in AM. Figure 8.4 demonstrates a microstructurally informed hexagonal mesh for predicting the mechanical properties of Ti6Al4V parts produced by EBM.

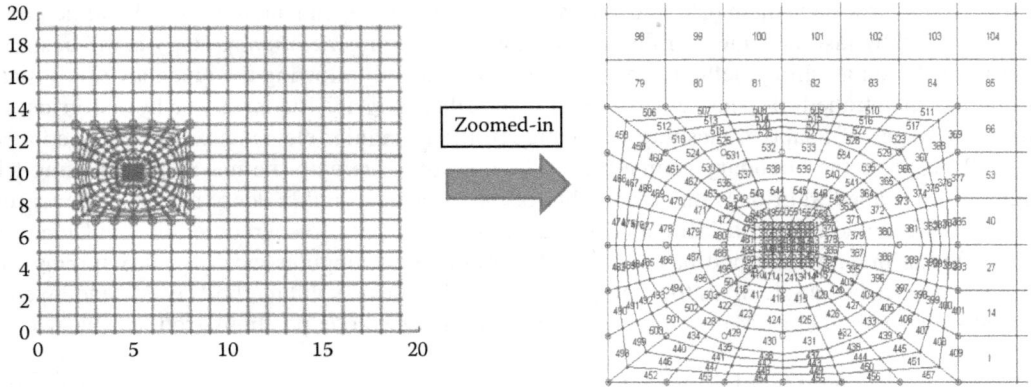

FIGURE 8.1 Meshing configuration for a point energy source located within the fine mesh region. The point source could be a Gaussian heat source such as a concentrated laser or electron beam, and the fine mesh region moves as the point energy source moves.

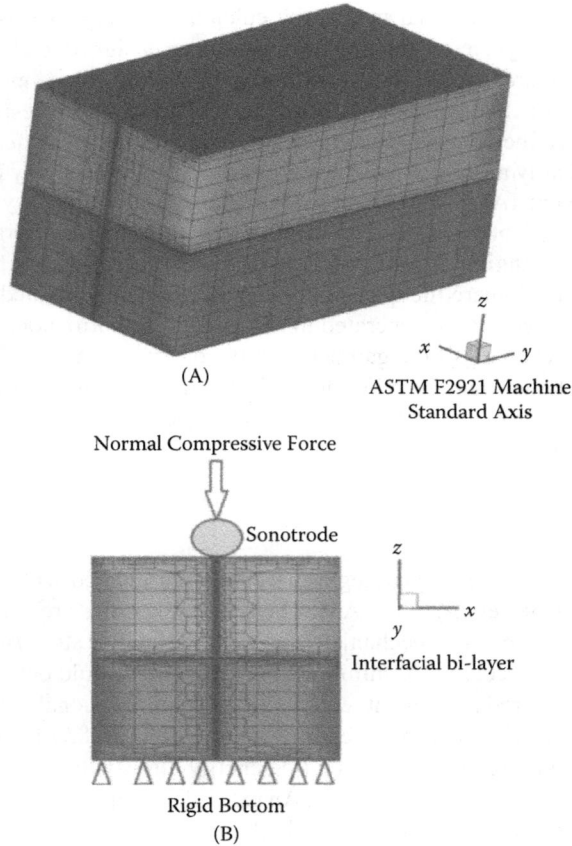

FIGURE 8.2 Multi-scale mesh for ultrasonic consolidation showing: (A) the full 3D mesh with selective adaptive refinement near the sonotrode and the interfacial mating surfaces and (B) a close-up showing how the hexahedral elements are preserved throughout refinement. Light gray and dark gray denote the top and bottom foils, respectively. The vertical refined region moves with sonotrode motion while the horizontal fine mesh region is indexed up layer by layer as new foils are added.[4,15]

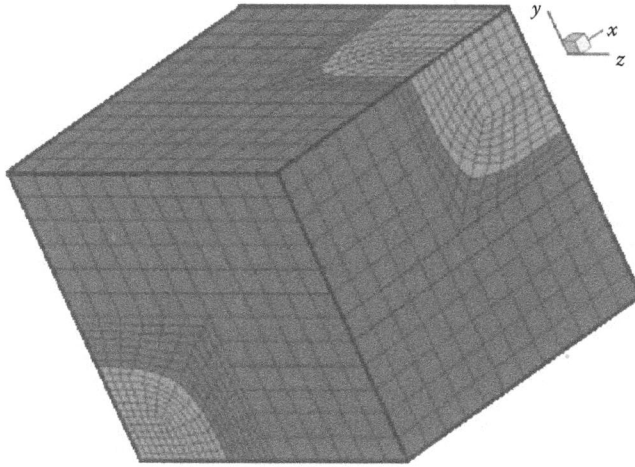

FIGURE 8.3 One-eighth view of a volumetric refined mesh of two secondary precipitates (light gray) diagonally opposite to each other placed in an Inconel 718 matrix (dark gray) with different curvatures.[12]

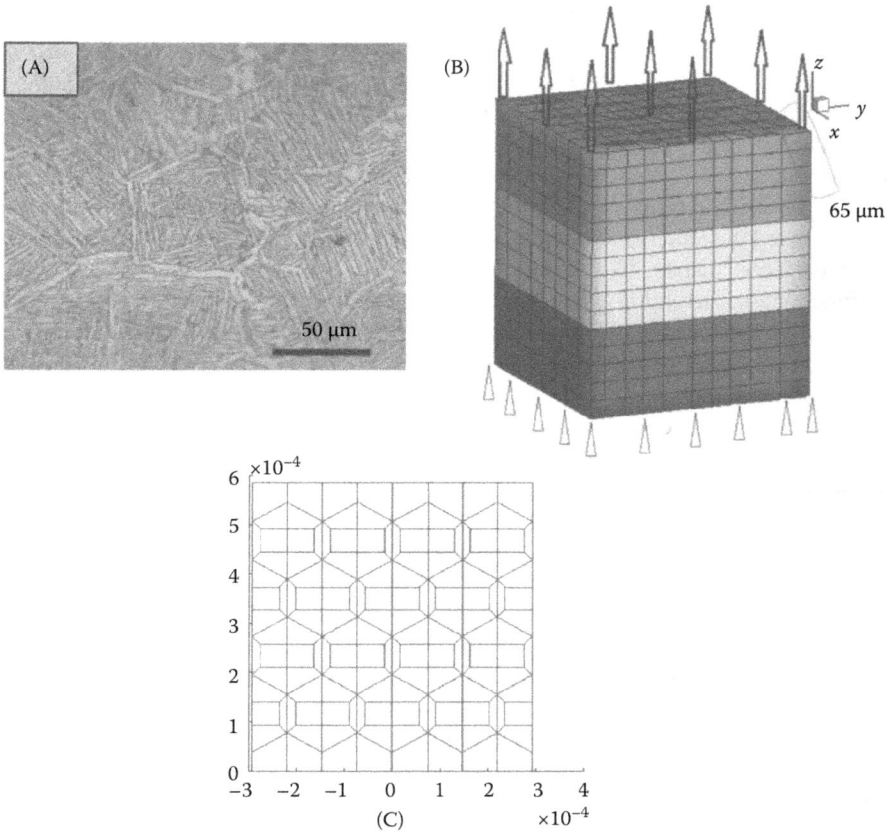

FIGURE 8.4 (A) Optical micrograph of an EBM-made Ti6Al4V microstructure showing hexagonal prior β grain motifs; (B) microstructurally informed mesh; and (C) in-plane mesh complexity.[13]

8.7 SOLUTION CAPABILITIES

8.7.1 Literature Review of Thermal Finite Element Method for Selective Laser Melting of Metals and Their Alloys

Selective laser melting (SLM), also known as metal laser sintering, is a widely used additive manufacturing process derived from selective laser sintering (SLS).[41,42] SLM produces high-density functional metallic part and tools. To achieve high-density functional part fabrication, optimal process parameters are required such that the powder bed can melt and solidify while achieving the desired surface roughness, design geometry, and mechanical properties. Poor parameter selection leads to distortion and defects due to the residual stresses that result from high thermal gradients. In order to avoid these defects, an experimental trial-and-error method is typically used across a range of process parameters. This is time consuming and may not include all the process parameters of interest because these experiments are expensive. If a set of verified computational tools could be built to quickly test process parameter combination, this would greatly aid in the optimization of SLM.

Various predictive models have been developed and applied to AM processes. Most of these studies used numerical methods such as the finite element method. Analytical solutions are typically not used, as their derivation requires ignoring constraints/boundary mismatch and simplifying the model to be too simplistic for real problem sets. Numerical methods that take advantage of high-performance computers can enable the inclusion of more realistic physics and boundary condition constraints at ever-increasing efficiencies, but current simulation tools still enable only very small problems to be solved.

In prior studies, both 2D and 3D numerical models have been developed. The SLS thermal problem for polymers and metals has been described by Childs.[54] These models have the ability to solve the transient thermal problem with the material being treated as a state variable. The mesh in the 3D model counterpart was coarsened as a function of increasing distance from the heat source and is represented by 10×10 elements in a layer in the matrix zone. The local fine mesh was created by increasing the mesh density in the high heat flux zone. This localized fine discretization also inevitably increased the number of elements in the surrounding area, causing the assembled thermal conductivity (stiffness matrices) to be huge.

A new functionality involving element death and birth was developed for modeling powder deposition processes by Ibraheem et al.[44] The model uses ANSYS software to simulate the thermal problem for a 20 mm × 20 mm × 9 mm part made from H13 tool steel. The size of each element depends on the melt depth and track width of the fused powder. In this study, the latent heat was ignored due to the small melt pool dimensions. Another coarse- and fine-mesh-based FE study focused on SLM processing of titanium powder.[45] The computational domain of the model was 5 mm × 5 mm × 2 mm, and the fine mesh domain was 2 mm × 2 mm × 0.5 mm; the associated mesh discretization sizes for the coarse and fine mesh were 0.1 mm and 0.01 mm, respectively. However, the mesh used in this strategy was nonconforming along the interface between the coarse and fine mesh regions, which could produce unphysical stiffness matrices compared to their conformal mesh counterparts.[45] An ANSYS model for simulation of the melting of W–Ni–Fe powder was studied by Zhang et al.[46] The dimensions of the substrate and powder bed were 2 mm × 3 mm × 1.5 mm and 1 mm × 2 mm × 0.5 mm, respectively. The substrate was meshed using tetrahedral elements, whereas the powder was meshed using hexahedral elements. The Gaussian heat flux zone was represented using 2 × 2 mesh matrices. A more comprehensive simulation of the temperature field for SLM processing was developed by Roberts et al.[47] The computational domain for this simulation study was comprised of five layers of 1 mm × 1 mm × 0.03 mm titanium powder deposited over a 3 mm × 3 mm × 3 mm mild steel substrate. The powder is uniformly divided using a mesh discretization of 0.025 mm × 0.025 mm × 0.03 mm. Compared to the model in Zhang et al.,[46] the elements used in this model were of a hexahedral nature for both the substrate and the powder bed. This model was also built in ANSYS, and the functionality of element death and birth was applied for a new layer of deposition on top of the substrate. A recent FE model developed for simulation of SLM processes offers the functionality to simulate only single scan tracks of titanium powder and was developed by Song et al.[48] Other modeling efforts on SLM processing have been reported;[49–52] however, these models do not focus on the efficiency and comprehensiveness of simulating thermal phenomena in the SLM/SLS process.

Most SLM machines, such as the EOSINT M 270, focus the laser energy onto a relatively small spot size of 100 μm with energy intensity values reaching 25 kW/mm^2 for a 200-W source.[53] The material thermal response to the heat source tracks the energy source as a function of both space and time. The powder exposed by the laser quickly heats up and then cools down due to the dynamics of the process. Heat dissipation is dependent on the thermal conductivity (which increases with temperature) of the previously solidified volume, the previously solidified scan area lying next to the current scan track, the recently solidified portion of the current scan track, support structures, and the base plate in the −z direction from the melt pool. Regarding heat dissipation, evolutionary solidification regions may not be present simultaneously at various locations of the build, and generally the ~25.4-mm base plate is much greater than the ~10- to 50-μm layers, which can significantly affect heat dissipation and counteract the solidification-related residual stresses. Thus, high energy intensity, short irradiation times, and small melt pools, along with metastable (usually diffusionless transformations) microstructures that add sharp discontinuities to the nonlinear temperature-dependent thermal property curves (such as thermal conductivity, specific heat, and density), make this a difficult and computationally expensive process to simulate. For the purpose of this study, discontinuities were smoothed to allow convergence of the ANSYS framework, and computational time was reduced using a multi-scale mesh.

The concept of simultaneous coarse and fine meshing for the simulation of SLM processes using FEM has been shown in the literature.[45,54] In these studies, the fine mesh region remained stationary as a function of laser movement, and the entire region of laser movement per layer had been fine meshed. However, the ability to move meshes as a function of laser movement in an efficient manner (with no renumbering of the fine mesh inscribed in its boundaries and intelligent matrix assembly) has not been illustrated until recently.[4,15,16] In this study, the fine and coarse mesh motifs used by Patel et al.[15,16,55] were borrowed for analysis, although the intelligent matrix assembly part has been discarded because ANSYS does not enable such functionality. The main objective of this study was to obtain accurate and efficient thermal solutions using the dynamic mesh strategy, to compare these attributes against a uniform fine mesh model, and to validate the simulation results using experiments. Finally, the time for solving this ANSYS model was compared against the time for solving a similar model in MATLAB using intelligent matrix assembly.

Compared to the model in Zhang et al.,[46] the elements used in this model were of a hexahedral nature for both the substrate and the powder bed. This model was also built in ANSYS, and the functionality of element death and birth was applied for a new layer of deposition on top of the substrate. A recent FE model developed to simulate SLM processes has the functionality to simulate only single scan tracks of titanium powder and was developed by Song et al.[48] Other modeling efforts on SLM processing have been reported;[49–52] however, these models do not focus on the efficiency and comprehensiveness of simulating the thermal phenomena in SLM/SLS processes.

8.7.2 FORMULATION

8.7.2.1 Thermal Phenomena in SLM

When the laser beam begins its movement in SLM, the energy is transferred in four major modes: reflection, conduction, convection, and radiation. Because the metallic materials are lustrous in nature, a large part of the incident energy is reflected. Absorptivity of the incident energy depends on the material and powder morphology. Heat energy absorbed in the powder is subsequently conducted through surrounding previously solidified regions, the powder bed, or the substrate. The ambient environment in SLM includes an inert gas such as argon or nitrogen, which results in heat loss due to convection. Heat energy losses due to radiation also occur during SLM processing.[50] Heat reflection, conduction, and convection are considered in this model, but radiation has been ignored.

8.7.2.2   Governing Equations and Boundary Conditions

In SLM, the heat transfer process can be described using the Fourier thermal equation:[55]

$$\lambda(T) \quad {}^2T + q = \rho(T)c(T)\frac{\partial T}{\partial t} \tag{8.1}$$

The substrate is preheated to a certain temperature prior to fabrication (typically 353 K in an EOS M 270 machine). The prescribed temperature is described in Equation 8.2 as an initial condition of the problem:

$$T(x,y,z,0) = T_0 \tag{8.2}$$

Heat convection between the powder bed surface and the surrounding environment occurs at the top surface and is described using Equation 8.3:

$$-\lambda(T)\frac{\partial T}{\partial z}\bigg|_{z=H} = h\left(T - T_e\right) \tag{8.3}$$

Because the bottom of the base plate is often controlled to a constant temperature, a fixed thermal boundary condition has been assumed at the bottom surface of the simulation domain:

$$T(x,y,0,t) = T_0 \tag{8.4}$$

In the prior equations, T is temperature, λ is scalar thermal conductivity, ρ is density, c is heat capacity, q is internal heat, T_0 is the powder bed initial temperature, T_e is the ambient temperature, and h is the convective heat transfer coefficient. The x, y, and z directions in the above-mentioned equations are oriented according to the ASTM F2921 standard for denoting directions inside an additive manufacturing machine.

8.7.2.3   Material Properties

In this paper, commonly used Ti6Al4V powder has been studied. Material properties such as λ, ρ, and c are temperature dependent. The latent heat of fusion and vaporization are considered during phase changes. The values for thermophysical properties such as density, thermal conductivity, enthalpy, and heat capacity of Ti6Al4V (Ti64) have been adopted from existing literature values.[56] For the powder bed, a ratio is used to describe the difference between bulk and powder bed thermal conductivity. This ratio for a powder bed of pure titanium with 30-µm-diameter particles was reported to be 0.646 in one study[57] and 0.602 in another study.[43] In this model, the ratio of 0.48[14] has been used for simulation purposes.

8.7.2.4   Laser Beam Approximation

Laser beam profiles are generally approximated as Gaussian. The Gaussian mathematical form most widely used in the literature[6,43,46,48,58] is described in Equations 8.5 and 8.6:[47]

$$q(r) = \frac{2P}{\pi r_0^2}e^{\frac{2r^2}{r_0^2}} \tag{8.5}$$

where P is the laser power, r_0 is the laser beam radius, and r is the radial distance of a point from the center of laser exposure.

Equation 8.6 is the average heat flux:

$$q_m = \frac{1}{\pi r_0^2} \int_0^{r_0} q(2\pi r)dr = \frac{0.865P}{\pi r_0^2} \tag{8.6}$$

The laser power P and the laser beam radius r are two of the most important parameters to characterize the Gaussian model. SLM machines are able to focus the energy into a circular region with a diameter of ~100 μm. In order to account for the fraction of energy that is absorbed, an absorptivity constant (α) is usually incorporated. For this study a value of 0.7 has been assumed for the top surface comprised of the powder and a value of ~0.36[59] for the solid placed beneath the top layer of powder.

8.7.3 Results

8.7.3.1 Case Study with One-Dimensional Line as a Simulation Domain

Two subcases were attempted with constant and temperature-dependent thermal parameters. They have been described as follows:

Subcase 1: Thermal material parameters are assumed to be constant.
Subcase 2: Nonlinear temperature-dependent thermal material parameters[4] are considered.

8.7.3.1.1 Temperature Distribution

Figure 8.5 shows the temperature distribution of a one-dimensional bar considered here for both the subcases described above and at three equidistant time instances at 0.0025 s, 0.005 s, and 0.0075 s. All three time instances are plotted when melt-pool distribution has achieved its steady state over time. Temperatures in all cases are normalized by dividing by the maximum temperature in the linear case over time. The rationale for doing this normalization is to make the results more generalized and meaningful to compare the response between the linear and nonlinear simulations. It can be observed from Figure 8.5 that the temperatures while using nonlinear parameters are significantly lower than while using linear parameters. The reason for this behavior can be attributed to the increased thermal conductivity and volumetric heat capacity near the laser beam spot as shown in Figure 8.6 and Figure 8.7, respectively, which leads to faster heat dissipation near the melt pool area. Further, the melt pool diameter is comparatively smaller in subcase 2.

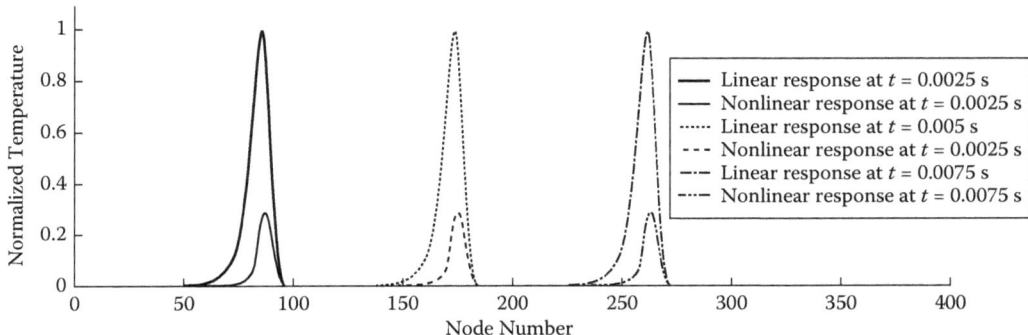

FIGURE 8.5   Temperature distribution for one-dimensional simulation domain.

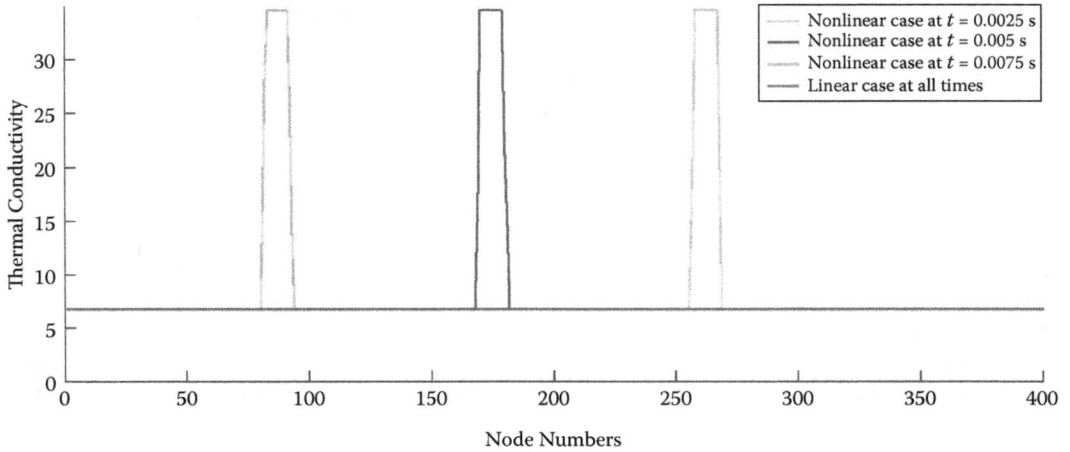

FIGURE 8.6 Thermal conductivity distribution for nonlinear and linear cases of the one-dimensional problem at three different times plotted against nodes.

8.7.3.1.2 Thermal Conductivity

The thermal conductivity remains constant as assumed over time and space in subcase 1, whereas in subcase 2 it shows a top-hat distribution due to constant thermal conductivity at temperatures higher than 1923 K (the melting temperature), as shown in Figure 8.6.

8.7.3.1.3 Volumetric Heat Capacity

The nonlinear phenomenon[15] has led to a reverse top-hat distribution near the melt pool, as shown in Figure 8.7. The reason behind this distribution is that the volumetric heat capacity increases with temperature and then suddenly converges to a lower value with temperature post-melting.

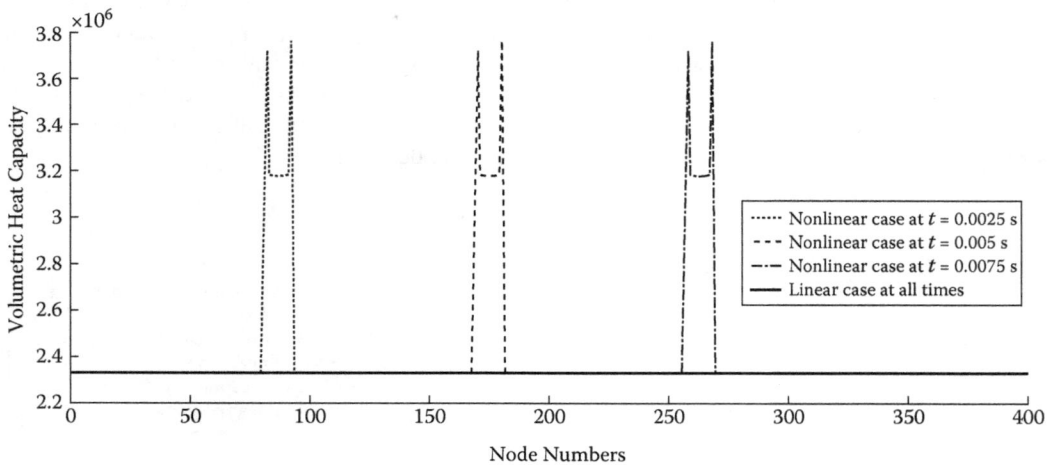

FIGURE 8.7 Volumetric heat capacity (ρc) for nonlinear and linear cases of the one-dimensional problem at three different times plotted against nodes.

(A)

Thermal Contour

(B)

$x = 0.0010735$
$y = 0.0011127$
Level = 362.48

$x = 0.001098$
$y = 0.0010588$
Level = 2063.739

125 μm

353 μm

$x = 0.001098$
$y = 0.00094118$
Level = 2071.2604

$x = 0.0010735$
$y = 0.0007598$
Level = 367.9272

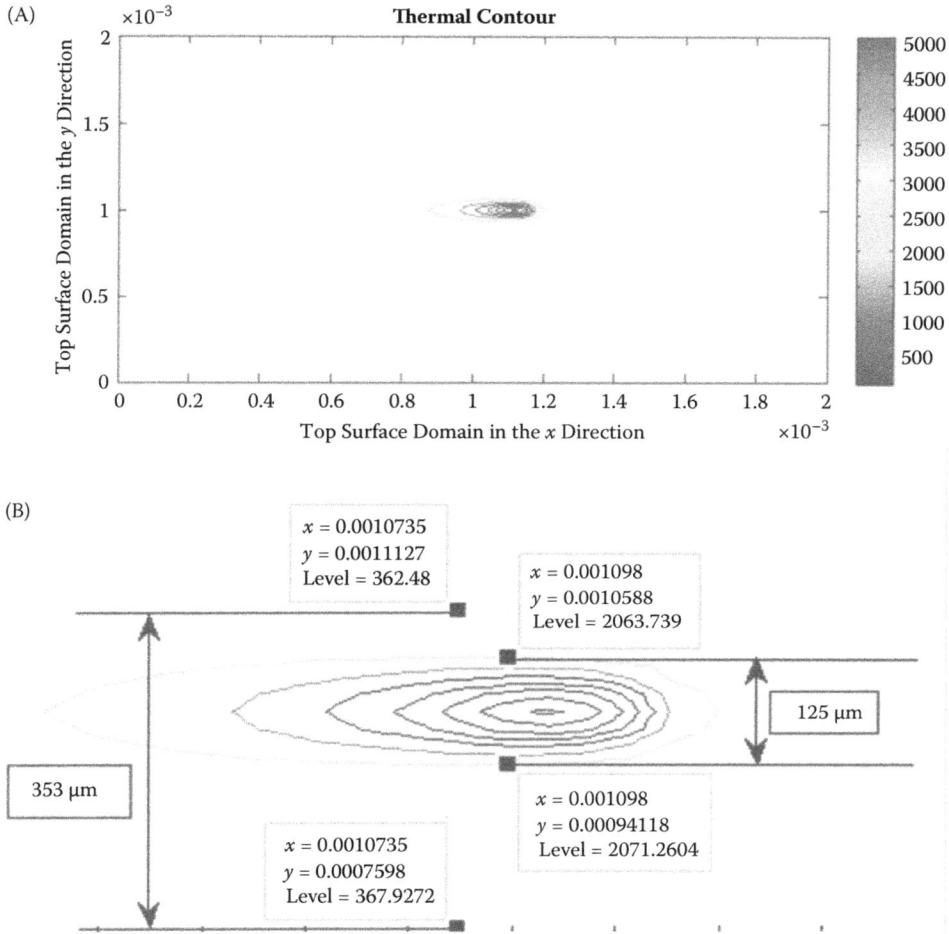

FIGURE 8.8 Comparison of thermal contours. The melt pool diameter in (B) is 125 μm.

8.7.3.2 Case Study on FEM Simulations of Metal Laser Sintering of Ti6Al4V with Constant Thermal Properties

A three-dimensional simulation using the FFD–AMRD algorithm was performed for this case study. Figure 8.8 shows the surface temperature contours for the stabilized melt pool. It can be seen that the melt pool diameter is 125 μm.

8.7.3.2.1 Linear FEM Validations with Ti6Al4V Microstructures Fabricated Using Metal Laser Sintering

Microstructural samples were created using a commercial EOS M 270 metal laser sintering machine. These samples were fabricated using the same set of parameters used in the simulation and then cut to illuminate the transverse section. A set of cylindrical samples fabricated horizontally in the x-y plane of the machine (normal direction $+x$) were made. It can be seen in Figure 8.9 that the melt pool diameter, illustrated by the grain boundary size, is around 100 μm. The difference between the simulation and analysis can be attributed to the simplifying linearity assumption made in the analysis.

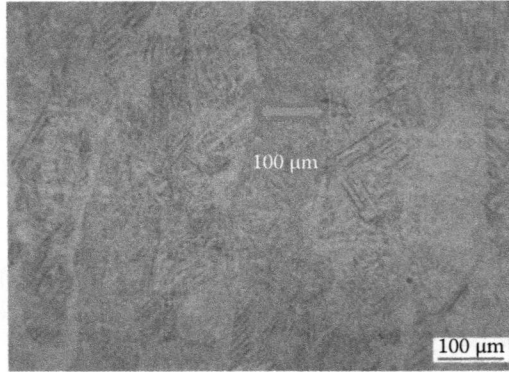

FIGURE 8.9 Optical microscopy image of microstructure showing grain boundaries along the melt pool boundary.

8.7.3.3 Case Study on FEM Simulations of Metal Laser Sintering of Ti6Al4V with Nonlinear Thermal Properties

The thermal contours for linear and nonlinear thermal parameters during metal laser sintering of Ti6Al4V are compared in Figure 8.10 during a straight pass. It can clearly be seen that the melt pool diameter in the linear scenario overestimates the melt pool diameter compared to the more accurate nonlinear solution. In the nonlinear scenario, the melt pool diameter closely matches the beam diameter (~100 μm). In Figure 8.11, the stable and unstable thermal contours have been plotted as a function of scan location. It was observed during plotting the thermal contours that the thermal contours were not able to stabilize because the turn length (hatch spacing) during the serpentine was less than the length of the melt pool.

The martensitic α' and solidified volume fractions have also been captured as shown in Figure 8.12. In Figure 8.12A, the thermal contours have been plotted as a function of location in the powder bed. Figures 8.12B and 8.12D illustrate the region transformed from the initial α phase of the

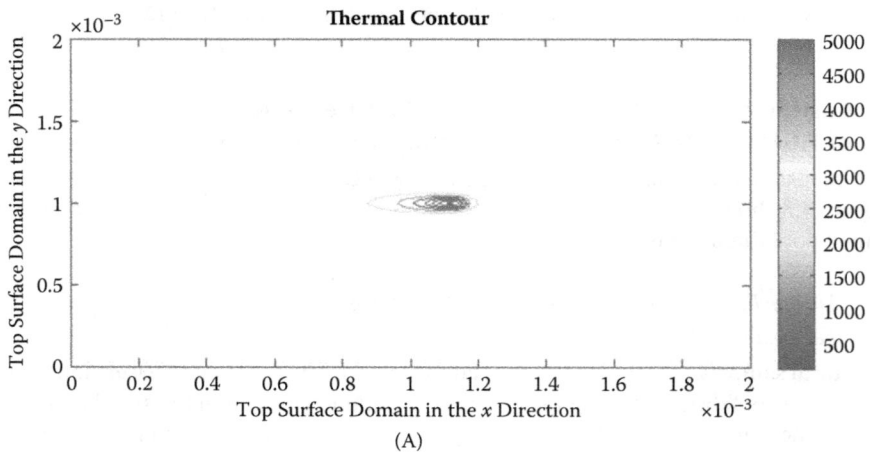

FIGURE 8.10 Comparison of thermal contours between linear (A,B) and nonlinear (C,D) scenarios. The melt pool diameter in (B) is 125 μm, whereas in (D) it is 107 μm. The spatial gradient due to the laser spot decays steeply in the y direction for the nonlinear scenario (~230 μm) compared to 353 μm observed in the linear case. *(Continued)*

x = 0.0010735
y = 0.0011127
Level = 362.48

x = 0.001098
y = 0.0010588
Level = 2063.739

125 μm

353 μm

x = 0.0010735
y = 0.0007598
Level = 367.9272

x = 0.001098
y = 0.00094118
Level = 2071.2604

(B)

Thermal Contour

×10⁻³

Top Surface Domain in the y Direction

Top Surface Domain in the x Direction

×10⁻³

(C)

x = 0.0010784
y = 0.0011127
Level = 361.1758

x = 0.0011029
y = 0.0010539
Level = 2013.5634

~230 μm

107 μm

x = 0.0010784
y = 0.00088235
Level = 367.5866

x = 0.0011029
y = 0.00094608
Level = 2127.9942

(D)

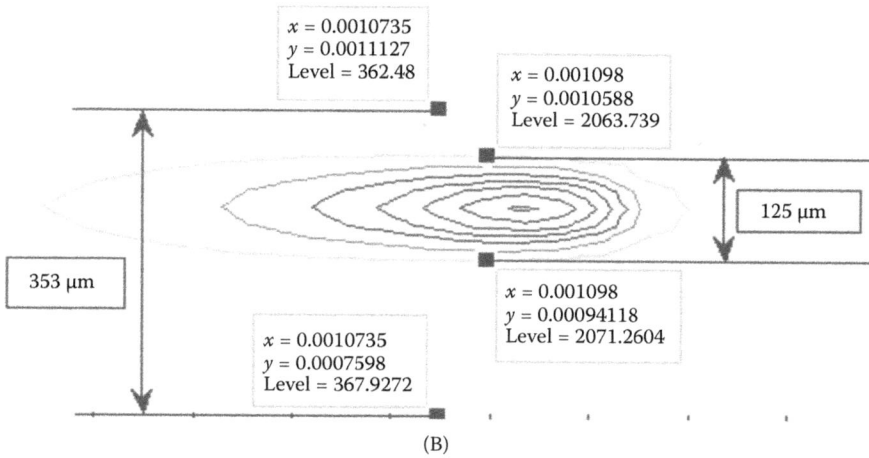

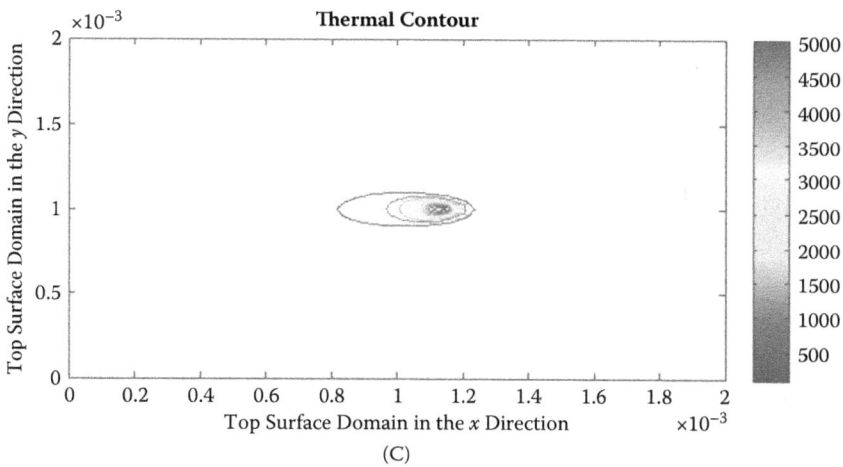

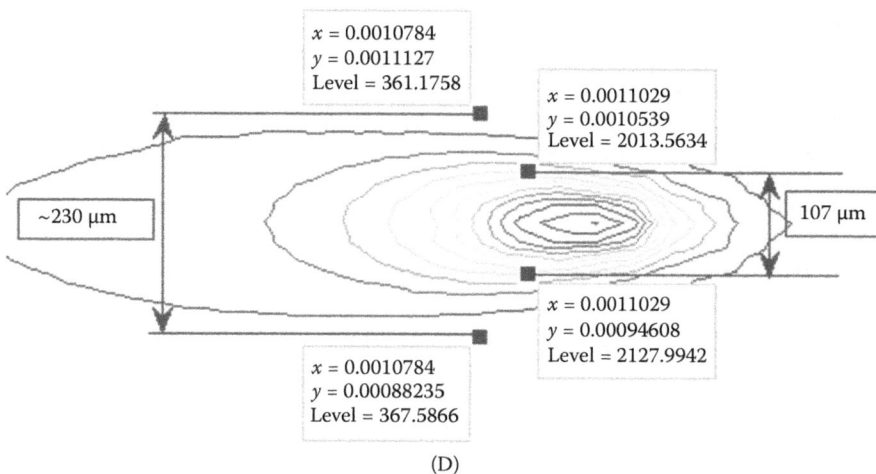

FIGURE 8.10 (Continued) Comparison of thermal contours between linear (A,B) and nonlinear (C,D) scenarios. The melt pool diameter in (B) is 125 μm, whereas in (D) it is 107 μm. The spatial gradient due to the laser spot decays steeply in the y direction for the nonlinear scenario (~230 μm) compared to 353 μm observed in the linear case.

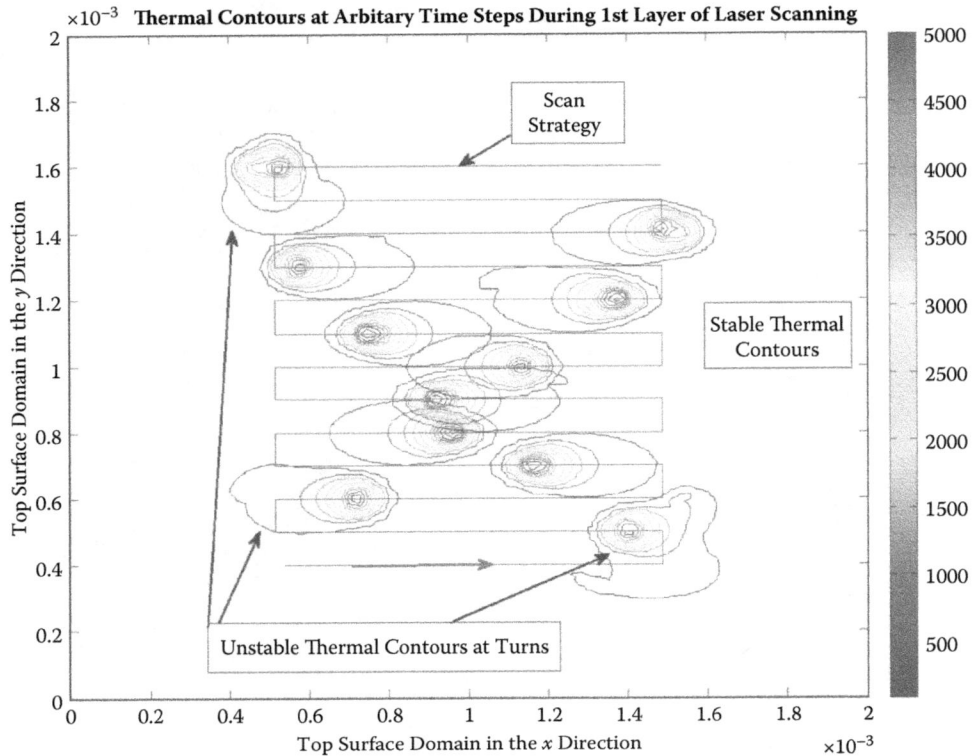

FIGURE 8.11 Stable and unstable thermal contours as a function of scan location. The lowest temperature contour in this image is ~1000 K, hence thermal asymmetry toward the solid with respect to the powder on the symmetrically opposite side of the melt pool is not visible.

powder to the α' phase and total bulk solidified area, respectively. Figure 8.12C and 8.12E show the evolution of the α' phase and bulk solidified area as a fraction of total domain area. It can be clearly observed from Figure 8.12C and 8.12E that the area fraction of martensitic α' is slightly greater than the bulk solidified area fraction, as the powders at the periphery of the part will be transformed to α' without fully melting, leading to thermally modified but still unmelted powder particles at the periphery of the build slice.

In order to illustrate the effect of unmelted particles at the periphery of a build slice, internal designed defects have been introduced in a continuously tapered cylindrical build as shown in Figure 8.13A. Figure 8.13B clearly shows the unmelted particles at the interior periphery of the designed defect. The external periphery of the build was accessible for surface preparation and it was sand blasted to remove loosely bound unmelted particles. The sand-blasted external surface is shown in Figure 8.13C.

The thermal history at the center of the domain held over the base plate has been captured and plotted as a function of time and number of build layers in Figure 8.14. It can be clearly observed that the heat does not significantly affect more than three prior layers, as temperatures seen by the fourth and consequent layers will be much less than the β transus temperature. The thermal contours on subsequent layers at the center of the domain held at the topmost layer of powder have been plotted in Figure 8.15. It can be clearly observed that the thermal contours become more and more diffused as the +z distance of the powder layer (distance from the large thermal mass of the base plate) increases.

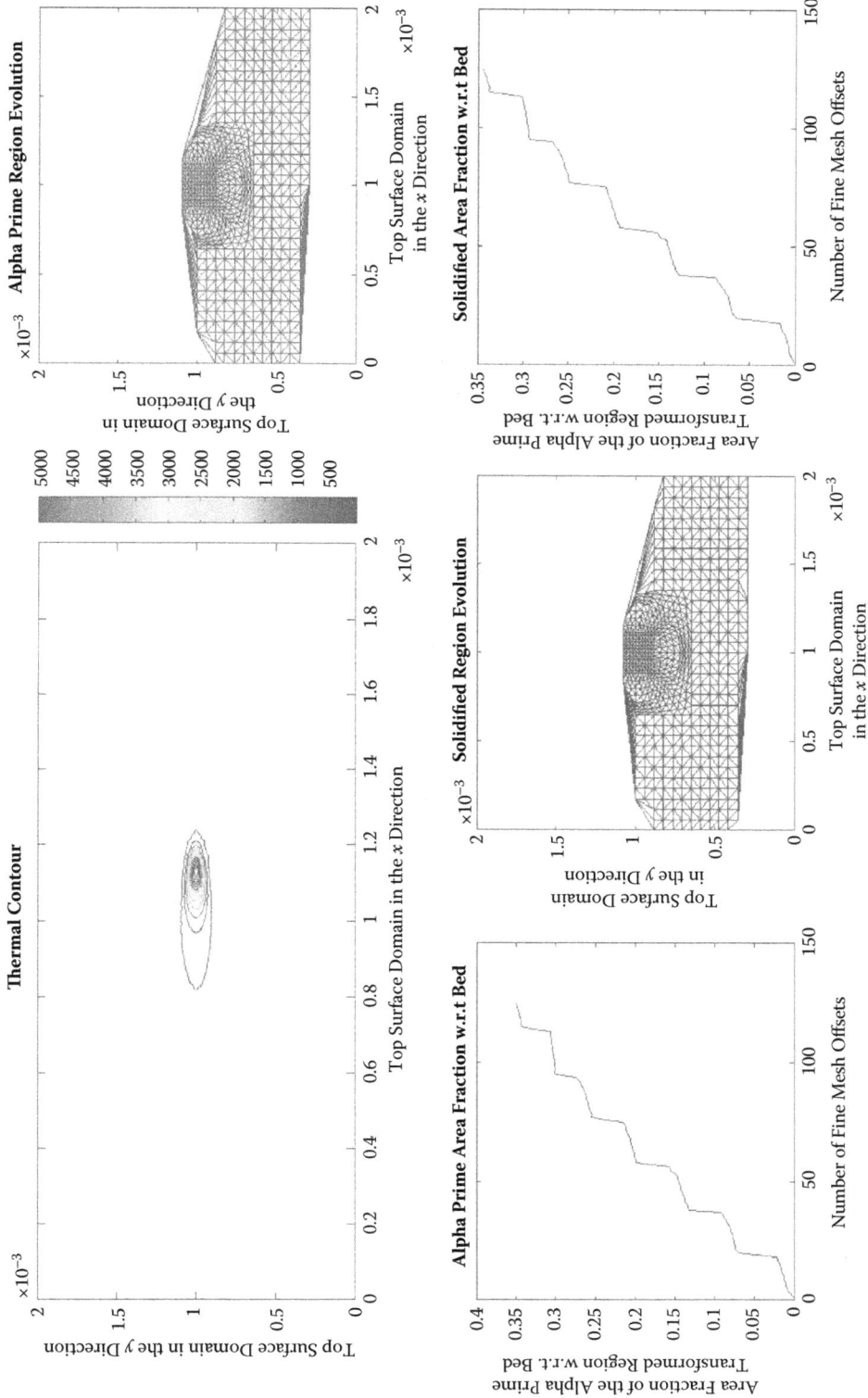

FIGURE 8.12 Plots showing the nonlinear thermal distribution, martensitic α′ region evolution, solidified region evolution, and solidified area phase fraction with respect to the total area of the inset, martensitic α′ area phase fraction with respect to the total area of the inset. It can be observed that the solidified area fraction (34.8%) is slightly smaller than that of the martensitic α′ area fraction (35.2%).

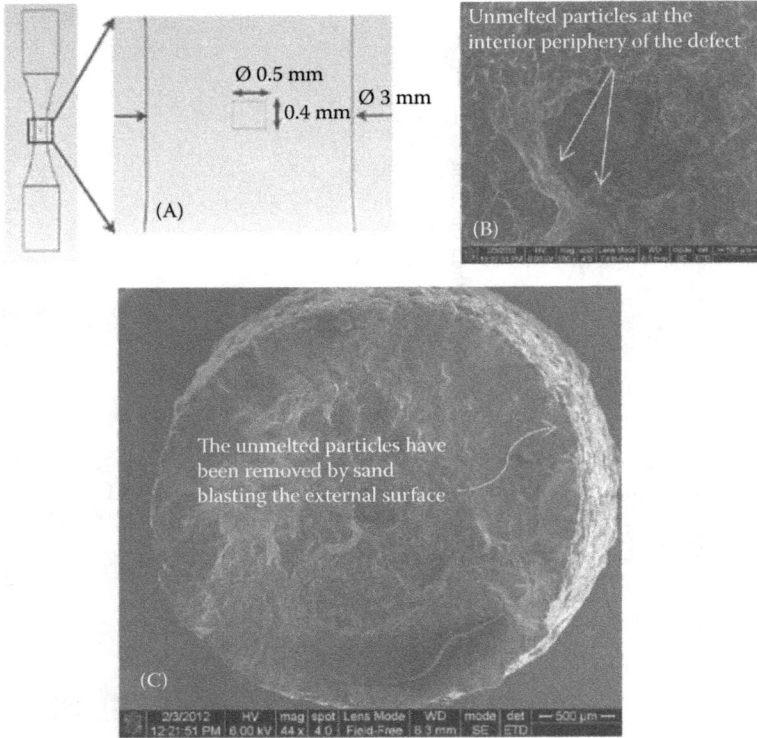

Ø 0.5 mm

0.4 mm

Ø 3 mm

(A)

Unmelted particles at the interior periphery of the defect

(B)

The unmelted particles have been removed by sand blasting the external surface

(C)

FIGURE 8.13 (A) Internal defect design, (B) unmelted particles at an internal surface pertaining to the central cylindrical defect of the build, and (C) unmelted particle removal using sanding operation at the external surface of the build.[61]

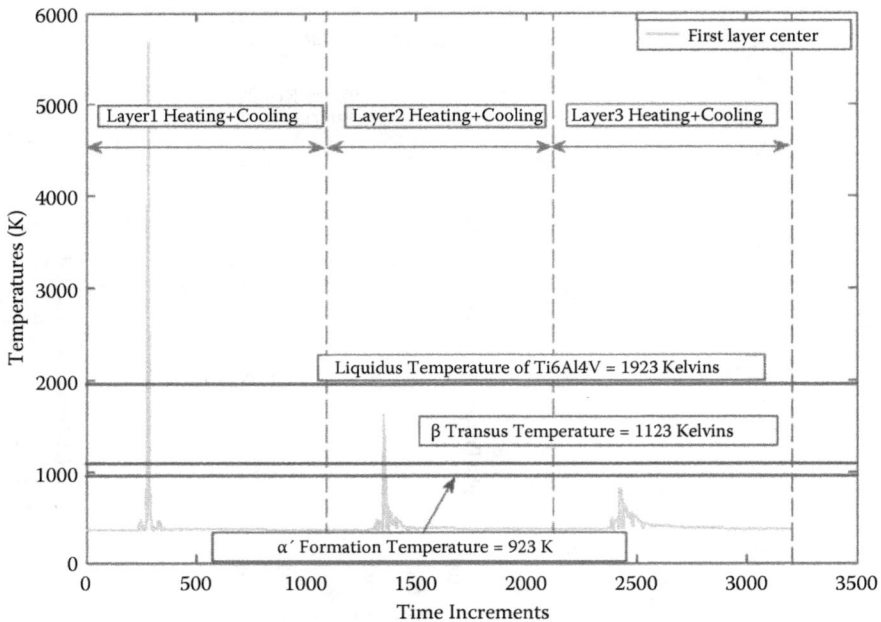

First layer center

Layer1 Heating+Cooling

Layer2 Heating+Cooling

Layer3 Heating+Cooling

Liquidus Temperature of Ti6Al4V = 1923 Kelvins

β Transus Temperature = 1123 Kelvins

α′ Formation Temperature = 923 K

Temperatures (K)

Time Increments

FIGURE 8.14 Thermal history at the center-top of the base plate as additional layers are built over it.

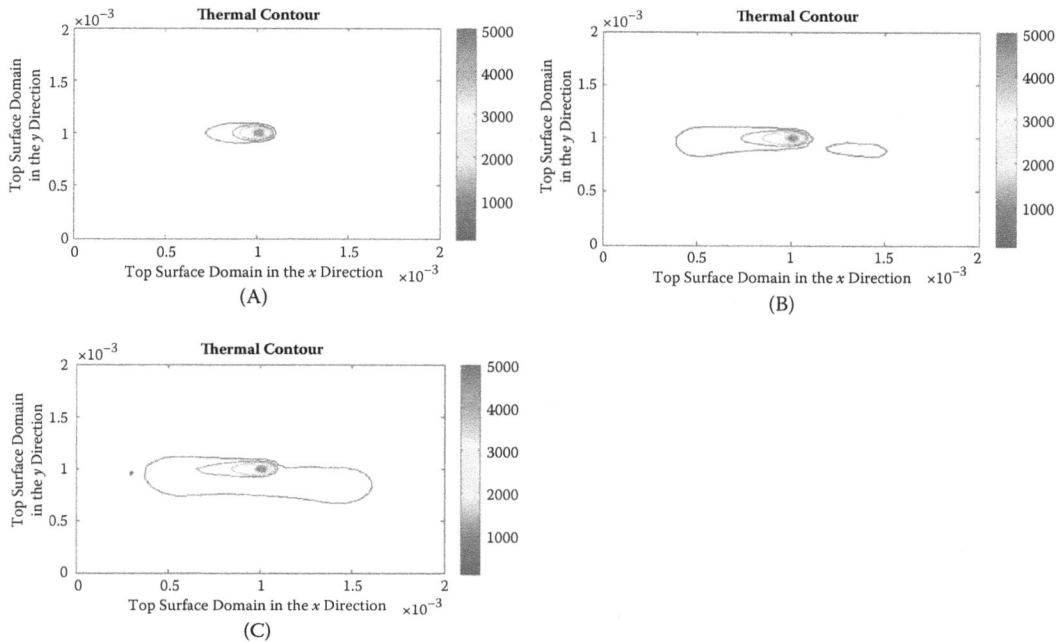

FIGURE 8.15 Layer-by-layer thermal evolution in metal laser sintering: (A) first layer, (B) second layer, and (C) third layer.

8.7.3.3.1 Nonlinear FEM Validations with Experimentally Obtained Thermal Contours and Ti6Al4V Microstructures Fabricated Using Metal Laser Sintering

The thermal contours were found to be in good agreement with experimentally observed thermal contours for Ti6Al4V when subjected to a laser beam diameter of 200 μm and at a scan speed of 5 mm/s. The experiment was conducted on a laser process development cell at a research partner facility using a forward looking infrared (FLIR) camera. The simulated and experimentally obtained thermal contours match for the exact same input processing parameters and are shown in Figure 8.16 and Figure 8.17, respectively. A large beam diameter and slow scan speed were chosen to ease the ability of the camera to capture the frames smoothly during the experiment. The slow scan speed also provides an opportunity for the simulation to show its versatility in both high and low scan speed regimes. The results for a laser beam diameter of 100 μm and the exact same input processing parameters as prescribed earlier[4] were indirectly verified using a suitable micrograph as shown in Figure 8.18. It could be clearly observed that the melt pool diameter from the simulations matches well the transverse prior β grain diameter.

8.7.3.4 Case Study on FEM Simulations of Metal Laser Sintering of 17-4 PH Stainless Steel with Nonlinear Thermal Properties

Figure 8.19 is an optical micrograph along the longitudinal cross-section of an SLM as-built 17-4 PH stainless steel sample. Melt pool boundaries are distinctly visible, which is in contrast with the microstructure of SLM-processed Ti6Al4V, where the melt pool boundaries are not visible.[61] The average depth of a melt pool is about 60 μm, even though the powder layer thickness is 40 μm. This can be attributed to the higher energy density of the laser beam and the low powder bed density, which is typically about 68%. Due to the higher energy input, the top surface of the previously solidified layer is remelted and forms part of the new melt pool. The vertical separation of two overlapping melt pools can be estimated as the actual layer thickness.

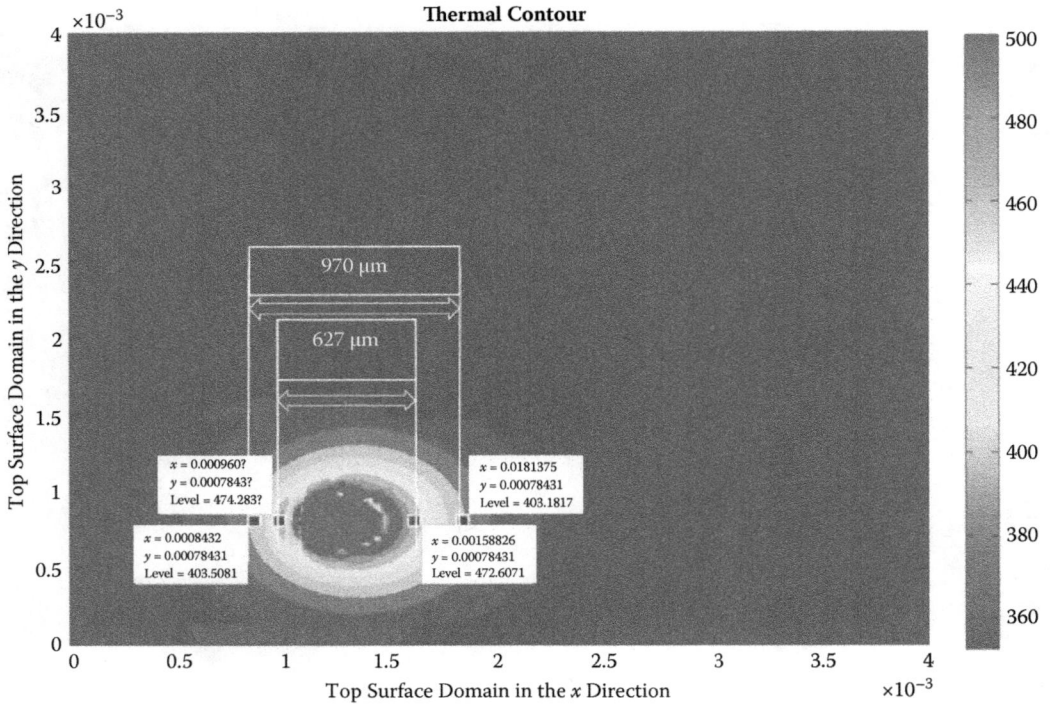

FIGURE 8.16 Simulated thermal contours at the same location and process parameters as those used in the experiment (Figure 8.13). The temperature color bar is shown in Kelvin.

Melt pool formation during selective laser melting is modeled using a three-dimensional thermomechanical in-house finite element method (FEM). This model uses a novel meshing scheme that has multi-resolution capability. The input material data for the models were taken from an EOS material data sheet[62] and from the open literature.[63] Figure 8.20 shows the shape of the melt pool,

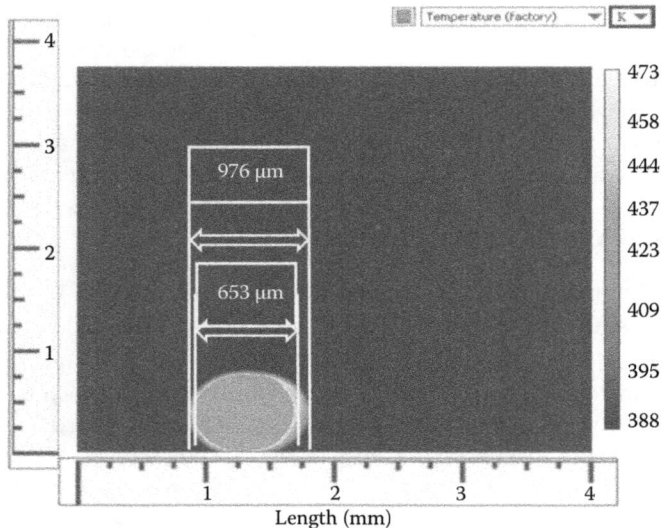

FIGURE 8.17 Experimentally obtained thermal contours.

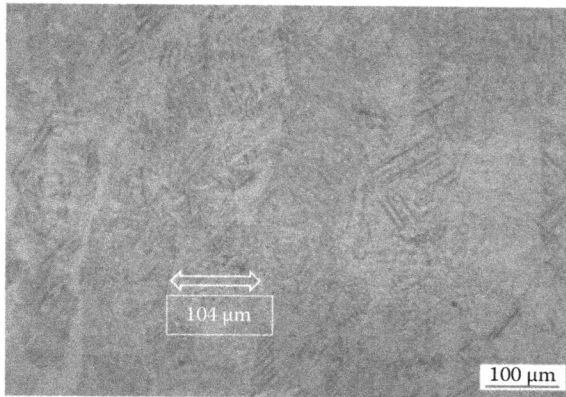

FIGURE 8.18 Transverse prior β grain diameter (~100 μm) matches the predicted melt pool diameter (~100 μm).

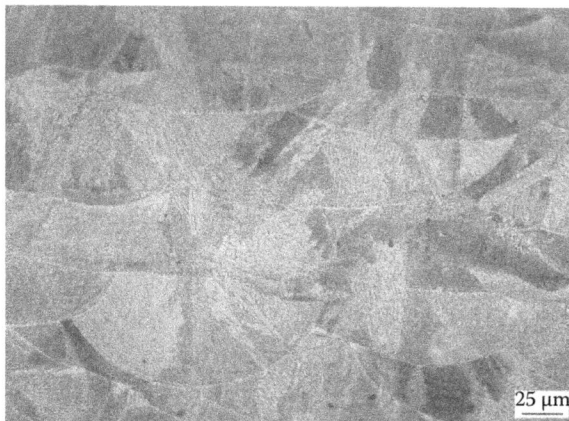

FIGURE 8.19 Transverse cross-section of SLM PH steel sample showing distinct melt pools.

the top view (*x-y* cross-section), and the side view (*x-z* cross-section) of the thermal field generated during the SLM process. The model predicts the melt pool dimensions to be 150 μm wide by 60 μm deep, which is similar to what is observed in the microstructure. Figure 8.20A shows the larger thermal contours on the solidified metal side as compared to the unmelted powder side due to the difference in thermal diffusivity between the solidified metal and the powder.

8.7.4 LITERATURE REVIEW ON ULTRASONIC CONSOLIDATION AND DISLOCATION DENSITY-BASED FINITE ELEMENT METHOD OF ULTRASONIC CONSOLIDATION

Ultrasonic consolidation (UC)[8,10,11,64] is a solid-state fabrication process that combines ultrasonic metal welding, additive manufacturing techniques, and milling to produce three-dimensional free-form objects. The process uses the power of high-frequency ultrasonic vibration at low amplitude to bond thin foils of materials to form solid objects. It combines the compressive normal and oscillating shear forces on mating foils, and the resulting friction forces between the materials fracture and displace surface oxides from the materials. These surfaces bond by direct contact under modest pressure and temperatures that are less than half of the melting point of the materials. The materials are thus metallurgically bonded.[10, 65–67] Fractured oxides and surface impurities in the materials are

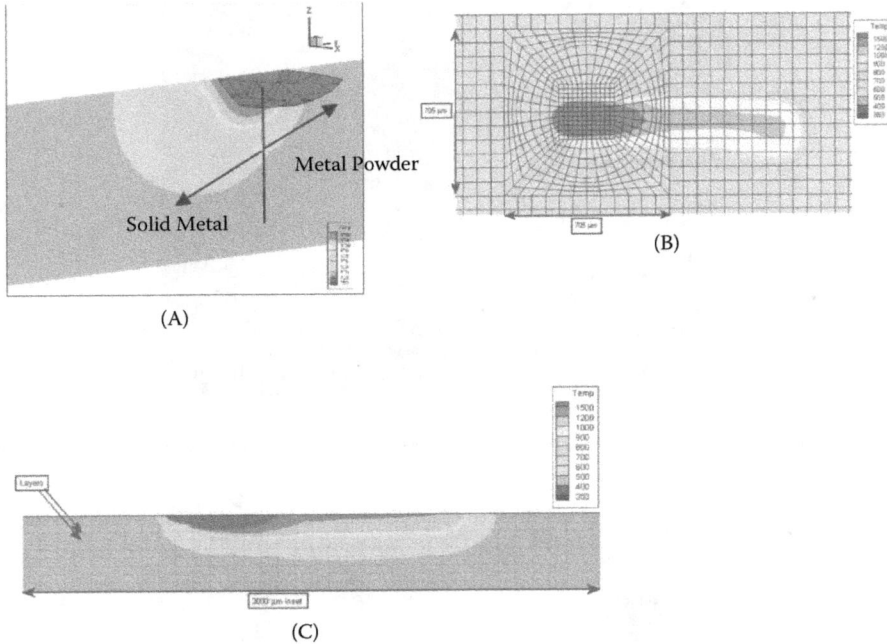

(A)

(B)

(C)

FIGURE 8.20   Three-dimensional FEM of the melt pool: (A) 3D view of the melt pool, (B) x-y cross-section, (C) x-z cross-section.

distributed in the bond zone. The process combines the layer-by-layer addition of foils with contour milling using the integrated three-axis computer numerical control (CNC) machining facilities to produce the desired component geometry. It is therefore both an additive and subtractive process. Thus, fabrication using UC involves the generic AM process in which a solid computer-aided design (CAD) model is numerically sliced into thin horizontal layers that are sequentially sent to the UC machine to build the part from the bottom up.

The foil material thickness is used as the layer thickness for the component in the machine code. Apart from removing the substrate upon which the deposition is made after fabrication is completed, no further machining of the part is required, making it a net shape fabrication process. Some notable advantages of the solid-state UC process are as follows:[65]

- No process is associated with high temperatures or airborne powder safety hazards. In selective laser melting, flares come out from the melt pool and land on various locations of the powder bed.
- No atmospheric control is required.
- As a low temperature is required for the small volume of material affected, less energy is needed.
- Embrittlement, residual stress, distortion, and dimensional changes are greatly reduced with the low processing temperatures (much lower compared to the melting points of the mating materials).

The UC machine consists of a welding horn, also known as a sonotrode, which exerts normal force and oscillatory high-frequency vibration on the materials to be welded. Welding takes place on a substrate fixed on a heated plate. The UC machine is designed for automatic foil material feed, but materials can also be fed manually, as well. Figure 8.21 provides a schematic view of the ultrasonic consolidation process.

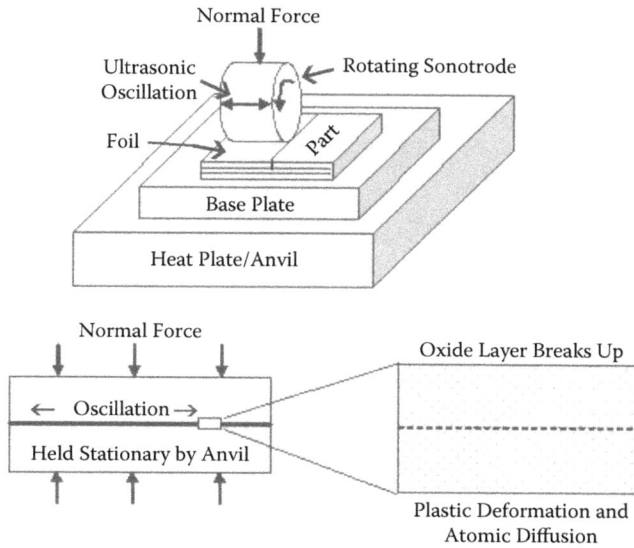

FIGURE 8.21 Schematic view of UC process.[69]

Ultrasonic consolidation is applicable for rapid tooling in injection molding, extrusion, vacuum forming tools, and other uses. It is also used for fabricating tools with conformal cooling channels.[65] Previous work has demonstrated other potential applications of UC. These include honeycomb structure fabrications,[69,70] embedding shape memory alloy (SMA) fibers and silicon carbide fibers in an aluminum matrix,[65,69,71-75] and embedded electronic structures.[76] The multi-material capabilities of UC were demonstrated by Janaki Ram et al.[77] in their work in which copper, brass, nickel, Inconel 600, AISI 347 stainless steel, AISI 304 stainless steel wire mesh, MetPreg, and aluminum alloy 2024 were individually welded to aluminum 3003 H18 materials. Domack and Baughman[78] fabricated graded nickel and titanium materials using UC. Obielodan and Stucker[68,79] also demonstrated UC multi-material capabilities by welding different combinations of molybdenum, tantalum, titanium, copper, silver, nickel, boron powder, and the aluminum alloys 1100, 3003, and 6061.

The primary process parameters in UC fabrications are vibration amplitude, temperature, welding speed, and normal force.[71,80-82] Other parameters that can affect weld quality include welding sonotrode roughness, materials surface finish,[83] and sonotrode displacement relative to machine-specified material width in an automated material feed system.[84] The optimum process parameters for different materials such as the aluminum alloys 3003 and 6061, 316L stainless steel, and the Al/SiC metal–matrix composite have been experimentally determined in earlier work.[66,69,71,73-75,79,83-85]

The bonding mechanisms of ultrasonically consolidated foils as explained by Janaki Ram et al.[83] highlighted the dominant factors influencing good bonding between two foils of materials. The ability to plastically deform the foils under the action of the normal and oscillating shear forces acting at the interface of the mating foils is of paramount importance as it helps in breaking the hard surface oxides and repeatedly deforming the surface asperities, thereby exposing atomically clean surfaces for metallurgical bonding between the mating foils during the weld cycle. Successful welding between two mating foils can be a measure of how well the surface oxides of the foil materials can be removed as well as the ease of surface deformation. According to Obielodan and Stucker,[68] at least one of the two materials being bonded at any time must be plastically deformable under the action of the normal and oscillating shear forces of the sonotrode. Figure 8.22 illustrates the primary UC bonding mechanism.

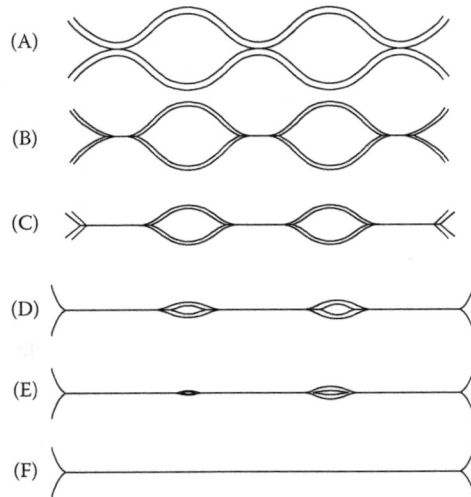

FIGURE 8.22 Bonding mechanism in UC: (A) surface asperities makes contact, (B) breaking of surface oxides begins, (C) exposed clean surfaces in intimate contact are bonded, (D) and (E) growth occurs in the bonding area, and (F) 100% linear weld density is achieved.

Macroscopic strength parameters can be used to quantify the quality of the bond obtained between material foils during UC processing, such as measurement of tensile and shear strengths using tensile and three-point bend tests. Additionally, there are two microstructural parameters that quantify the amount and quality of UC bonding: the amount of grain fragmentation (Figure 8.23)[86] and `. The LWD can be measured as the ratio of bonded length to the total interface length (Figure 8.24).[83] The higher the amount of subgrain formation, the more embrittlement will take place at the interface. Similarly, a high amount of LWD will ensure that the materials undergoing UC are adhered metallurgically. Thus, a higher amount of LWD and smaller amount of subgrain formation, which should be mostly near the interface, support better bonding at the UC interface. It can be inferred from Figure 8.24 that a lower LWD results in the interlayer region, as the previously deposited foils are damaged by the sonotrode horn while traversing which causes difficulty in interfacial void closure. In contrast, the base plate is surface machined and its smooth top surface provides ideal conditions for the virgin foil to be deposited over it during UC.

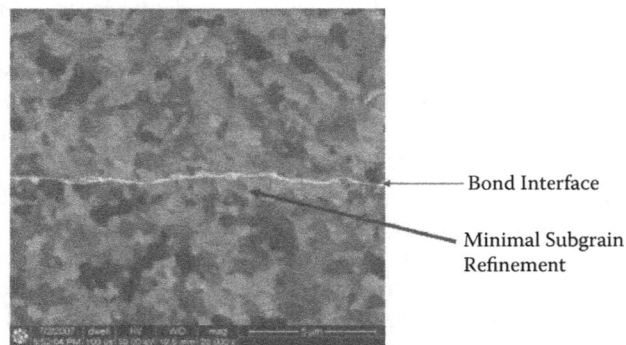

FIGURE 8.23 Ion beam-induced secondary electron micrograph showing smooth-to-smooth interface (polished top of the bottom foil) following ultrasonic welding showing minimal subgrain refinement in and around the interface. The bond is primarily due to the plastic deformation involved.[87]

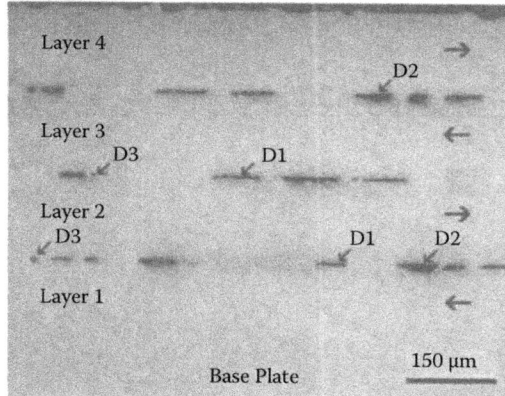

FIGURE 8.24 Microstructures of UC deposits (longitudinal section). Arrows indicate the welding direction for each layer.[84] The black artifacts show lack of joining during UC. The figure clearly indicates a higher amount of surface damage (plastic work caused by the sonotrode horn) in the previously deposited foils causing less LWD in the interlayer region, whereas the base plate has been already machined in the beginning and shows almost 100% bonding and no signs of interfacial defects in between layer 1 and the baseplate.

8.7.5 Results

The UC problem as a function of normal force, weld speed, and surface roughness has been modeled using a dislocation density-based crystal plasticity finite element method (DDCPFEM). The model has been shown to accurately predict average grain size at the interface. Figure 8.25 illustrates the ability of the model to predict trends for surface-finish-based differences in subgrain refinement at the interface. The effect of the microstructure on mechanical properties has also been predicted using DDCPFEM. The analysis provides both global scale and local scales results. Results in Figures 8.26 and 8.27 illustrate the detailed stress/strain analysis using DDCPFEM in regimes where commercial software packages fail to accurately predict the grain specific response for plastic deformation.

8.7.6 Literature Review of Thermal EigenSolver Using Finite Element Self and Mutual Stiffness Matrices

All SLM process models are computationally expensive to solve using traditional FEM methods, as they require fine-scale resolution across a part volume that is many orders of magnitude larger than the fine-scale lengths. This has led to efforts to solve the problem using asymptotic behaviors, modal space methodologies, or beam theories. The intent behind this new EigenSolver strategy is to solve for dominant modes in the macroscopic domain and local high-frequency modes in the microscopic domain. These can be accurately applied at certain locations (e.g., five layers) below the melt pool.

Beam and plate theories are meant to benefit from asymptotic theories involved in dimensionally reducible structures. Beam theories for complicated geometries such as aircraft wings and ship hulls and for material variations across the cross-section and length of a component have been areas of intense research. An analytical approach known as the variational asymptotic method (VAM)[87] has been used extensively to derive these beam theories. One of the limitations of this methodology lies in the difficulty of deriving beam theories for very complicated beams. To overcome this problem, we have developed a new EigenSolver beam theory derivation for use with FEA (Figure 8.28). This novel method can consider any complicated shape and cross-sectional variation and derives beam or plate theories involved in it. Further it has applications in problems that do not fall in the categories of beams or plates (e.g., prismatic bodies or any general structural configuration).

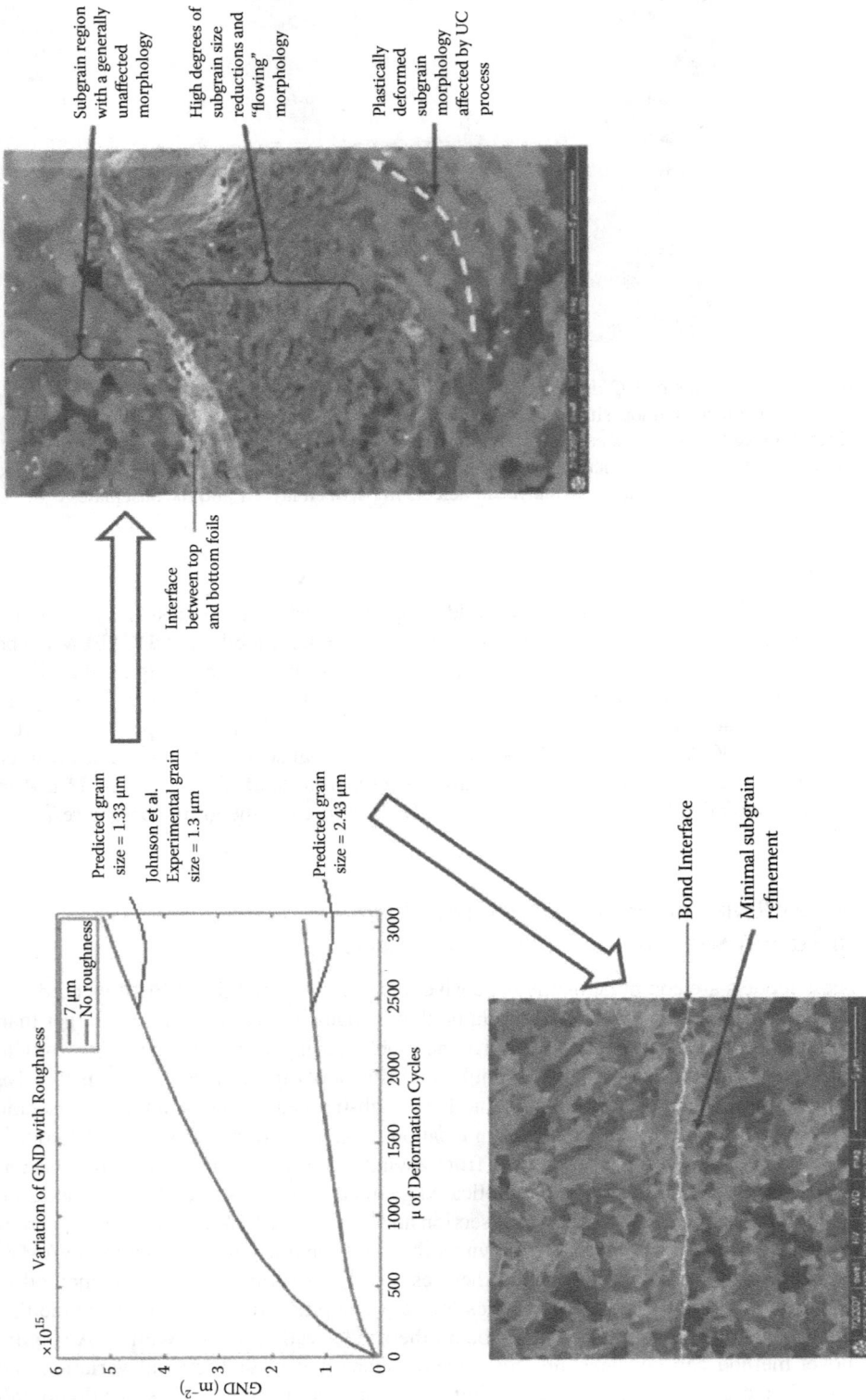

FIGURE 8.25 Variation of GND with surface roughness. The simulated average grain size at the mating interface for the rough sample was ~1.3 μm, whereas the experimentally obtained and weight averaged value was found to be ~1.33 μm. The simulated average grain size at the mating interface for the smooth sample was ~2.43 μm, which is in good agreement with Dr. K.E. Johnson's doctoral work.[87]

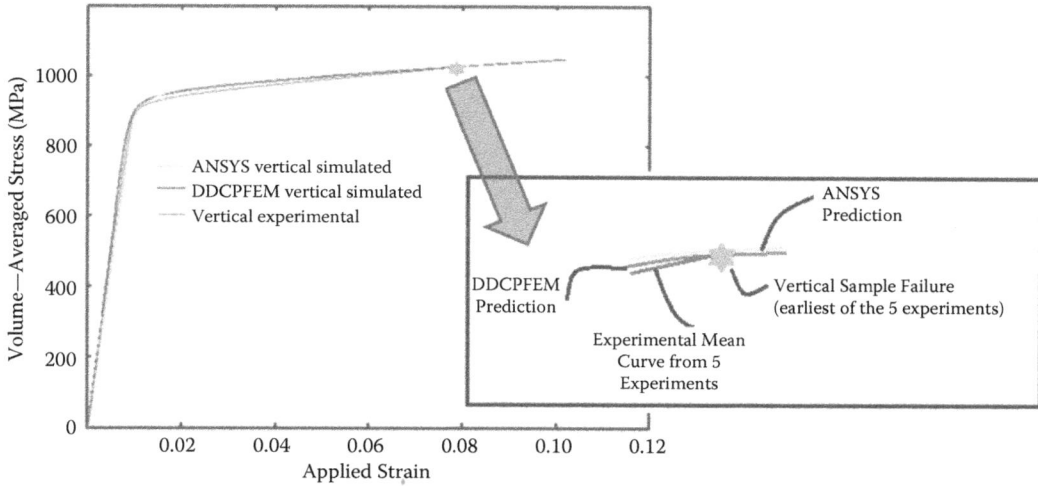

FIGURE 8.26 Volume averaged true stress-strain plot for a vertically built EBM Ti6Al4V sample computed by ANSYS (pale gray) and DDCPFEM (dark gray). These average stress/strain evolutions have been compared against experiments (medium gray).[13]

The importance of this methodology lies in calculating eigenvectors of the cross-section or group of nodes in FEM matrices (such as the stiffness matrix). It is well known that eigenvalues and eigenvectors are computationally expensive to calculate. To overcome this difficulty, any set of orthogonal vectors can be employed for this purpose, and their coupling is calculated to derive simplified simultaneous equations. Various material and geometrical nonlinearities can be incorporated in this beam theory with the help of our other existing solver tools.

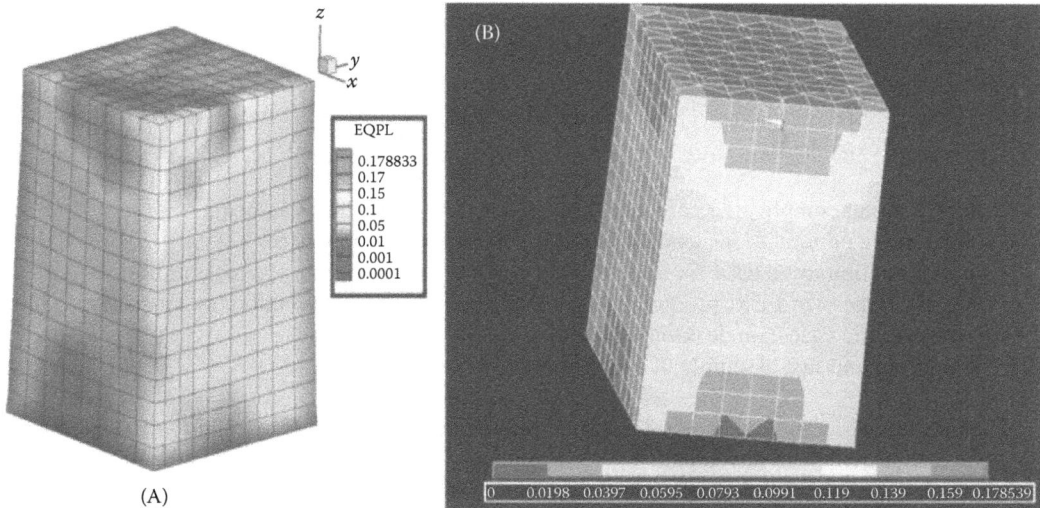

FIGURE 8.27 Plastic strain distribution at 10% total average strain for the stress/strain curves from Figure 8.26: (A) DDCPFEM simulations, and (B) ANSYS anisotropic multilinear continuum plasticity model. The plastic strain evolution in DDCPFEM simulations is grain-orientation specific in (A), whereas the plastic strain is symmetric about the top and bottom centers of the simulated model in (B), which is physically incorrect. The value of the maximum plastic strain at the top center is ~17.88% in both the ANSYS and DDCPFEM simulations.[13]

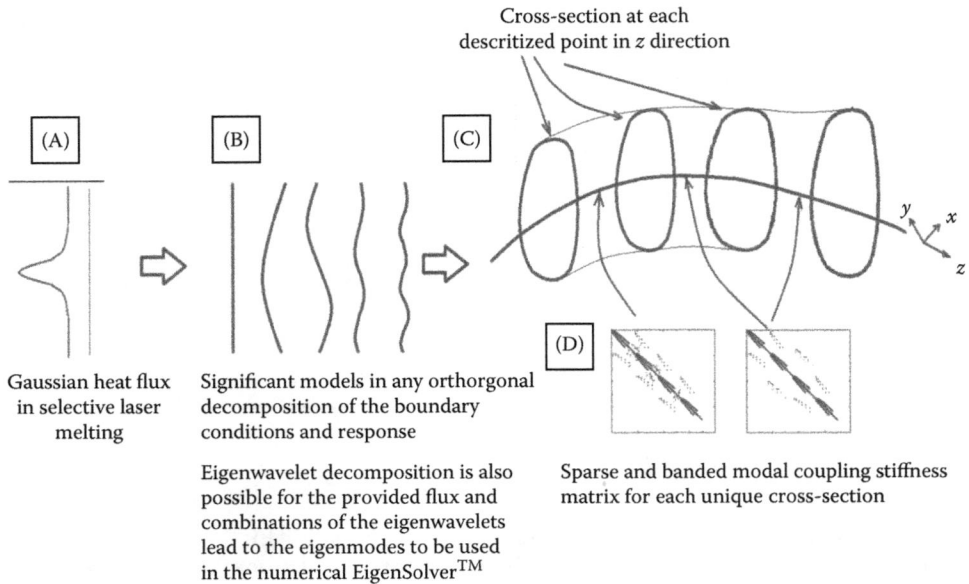

FIGURE 8.28 Schematic diagram illustrating the concept of an EigenSolver. In (A), first a Gaussian heat flux is applied to the material followed by the split of the flux in orthogonal modes in (B). These orthogonal modes are a function of geometry and the material (inhomogeneity and nonlinearity) that traverses forward through the cross-sections shown in (C) followed by their temperature counterparts, which traverse backward toward the source as the flux modes hit the fixed temperature boundary. In (D), the stiffness modal coupling of flux and temperature modes between changing cross-sections is shown. These matrices will assume an identity matrix or integral multiples of the identity matrix along the matrix diagonal in case the consecutive cross-sectional material properties or geometry are assembled in a fashion such that they are integral multiples with respect to one another, whereas if the material properties or geometry are continuously changing this will lead to off-diagonal terms comprising an interaction of one cross-section with respect to the other.

This methodology can be further generalized to work with groups of nodes in banded FEM meshes for large-size problems. These can then be solved on desktop computers. In summary this methodology is capable of

- Representing complicated geometry and material inhomogeneity in beams
- Inexpensive computations leading to global–local coupled responses
- Generalization to be used for any problem with some hidden unknown asymptotics
- Solving large-size FEM problems (e.g., 10 million degrees of freedom or much larger in some specific cases) on desktop computers
- Simplified and fast assembly of FEM matrices

In addition, an analytical method using a multi-dimensional, multi-resolution wavelet-based eigentheory has also been developed. This methodology can solve both linear and nonlinear problems provided a space transformation between the cross-sections is available. The theory has been extended to account for inhomogeneous material distributions, which are very common among powder-based additive manufacturing technologies where solid and powder continuously interact in a nonlinear mode with the applied thermal and resulting deformation-based boundary conditions.

The above-mentioned strategy works with polynomial basis functions such as orthonormal polynomial shape functions formulated for finite element methods resulting in creation of the stiffness matrix first, followed by finding the eigenvalue and eigenvector of the thermomechanical propagator matrix extracted from the stiffness matrix. The eigenmodes transform the problem in its frequency domain. Some other methods of interest in this case and which may require further investigation are the spectral finite element methods (SFEM). Although the method and its variants have not been attempted in this description, the research in this direction has been reviewed.

For general wave propagation-related problems, the traditional finite element method is the most used and trustworthy numerical technique due to its ability to analyze complex structures and capture the dynamic response with high accuracy. However, for reliable predictions of dynamic response of wave propagation, the number of elements required is typically large, which induces enormous computational costs. The SFEM uses exponential functions instead of polynomial functions as basis functions, and the finite element method is implemented based on variational integration in a frequency-dependent domain. The SFEM has been studied in stationary vibrations of fluid-filled pipes[88] and railway car structures,[89] wave propagation of composite tubes,[90] composite plates[91,92] and composite beams,[93–95] and the diagnosis of cracking and damage of structures,[96–99] where its good reliability and efficiency have been proved and validated when compared against traditional FEM.

The fast Fourier transform (FFT)-based spectral finite element (FSFE) method is known for its ability to simulate inverse wave motion problems; however, due to the Fourier series approximation in the spatial domain and discrete assumption in the temporal domain, the lateral boundary responds with an inaccurate wrap around response for short length waves.[100] The wavelet-based spectral finite element (WSFE), which is similar to the FSFE method, alleviates this problem by using compactly orthonormal and supported Daubechies scaling function[101] approximation in both time and space. The WSFE method allows the imposition of initial values with no periodicity assumption and thus is free from wrap-around problems.[102] The wave equations are reduced to the ordinary differential equations (ODEs) in both temporal and spatial domains and are typically decoupled using eigenvalue analysis and solved to obtain the element shape functions.

8.7.7 Formulation

A novel solution strategy for prediction of thermomechanical variables away from the point of laser exposure has been formulated. The strategy has been borrowed from structural vibrations and image analysis algorithms where modal contributions (eigenmodes) are computed in terms of orthogonal functions. The typical approach for determination of modes using sine or cosine functions has been extended to fit a finite element framework where an eigenvalue problem of the thermomechanical propagator matrices (which compute the propagation of the thermomechanical field from one cross-sectional layer to another along the $-z$ direction) is computed in order to determine the orthogonal basis functions (eigenvectors) for thermomechanical fields. It has been observed that very near to the point of laser exposure the number of modes required to reconstruct the solutions is very high (the number of modes is limited by the number of nodal points in each cross-section) and as the distance from the point of laser exposure increases, the number of modes required to accurately predict the solution drops very quickly. In the case of Ti6Al4V, accurate thermomechanical solutions can be calculated almost instantaneously four or more layers away from the point of laser exposure using a limited number of modes. The error introduced using this approach for a point energy source is negligibly small (see Figure 8.29), while the speed improvement is substantial. For cases of a line energy source (see Figure 8.30), the match is good but not excellent. For cases of a constant area source, due to unrefined orthogonal modes the trend of solutions is replicated but the error is much worse when compared to a point energy source solution (see Figure 8.31). Arnoldi and other methods for refinement of modes[103] are used for this scenario, which is relevant for manufacturing processes in which point sources of energy are not used. For focused energy sources such as SLM and EBM, this approach completely eliminates the need to recompute thermomechanical fields using fine-scale feed forward dynamic adaptive mesh

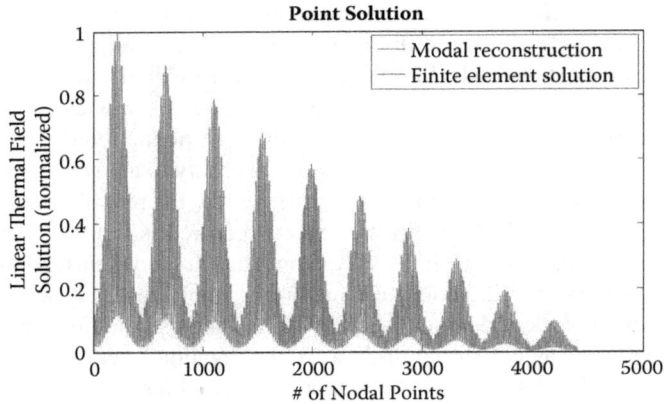

FIGURE 8.29 Nodal point-by-point match of a finite element solution for a thermal field with a modal reconstructed (EigenSolver) solution for a Gaussian point energy source. The error is negligibly small (<0.1%), and the results lie right on top of each other.

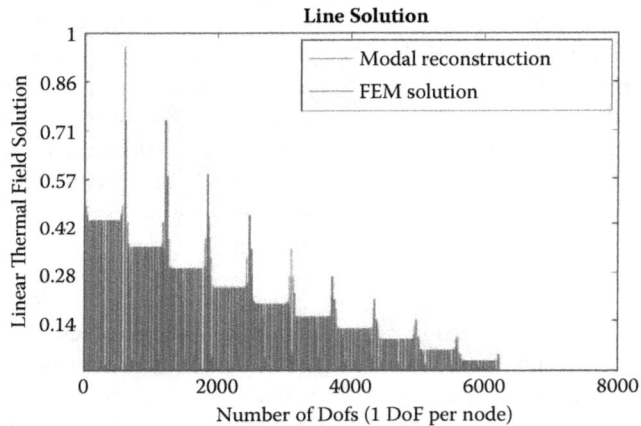

FIGURE 8.30 Nodal point-by-point comparison of a finite element solution with a modal reconstructed solution for a constant line energy source. The match is good but not excellent when compared to Figure 8.29.

refinement and derefinement (FFDAMRD) for the problem domains that are more than a few layers away from the laser, thus eliminating the vast majority of the computational time necessary to simulate real parts (which are made up of thousands or tens of thousands of layers).

It should be noted that so far eigenvalues have been identified and reconstructed for the thermal case only. A parallel architecture is required for solving the deformation problem using our dislocation density-based crystal plasticity (DDCP) finite element solver. For DDCP, the eigenvalue problem has orders of magnitude higher complexity than for the FFDAMRD thermal problem and will thus require a very fine refinement to compute the dispersion matrix and penetration matrices correctly in order to obtain the propagator matrix for its eigenvalue determination. The basic derivation has been presented here. The derivation on eigenmode computation for changing material properties with cross-sections will be presented in a future revision of this book chapter with the commission and release of a new patent describing the invention. The steps involved in the basic derivation are as follows:

Area Solution

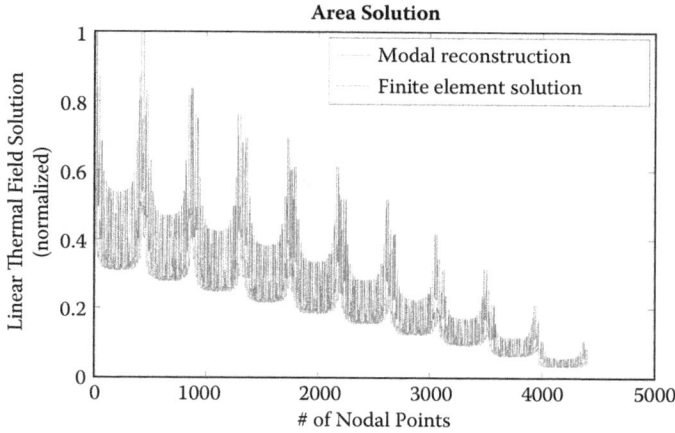

FIGURE 8.31 Nodal point-by-point comparison of a finite element solution with a modal reconstructed solution for a constant area energy source. The modally reconstructed solution matches the trend but not the magnitude of the solution due to unrefined orthogonal modes. This result is irrelevant for moving point energy problems associated with SLM or EBM but should be taken into account when seeking the correct solution for area energy sources.

1. *Construction of a stiffness matrix.* The thermal stiffness matrix is computed using the thermal conductivity of the material at each integration point. For this basic derivation, assume the geometry is 2.5 dimensions in nature with extrusion of the two-dimensional slice in the 3rd or $-z$ direction away from the laser source. The matrix format appears as shown below:

$$[K] = \begin{bmatrix} A/2 & B & 0 & 0 & 0 & 0 \\ B & A & B & 0 & 0 & 0 \\ 0 & B & A & B & 0 & 0 & \cdots & \cdots & \cdots & \cdots & \cdots \\ 0 & 0 & B & A & B & 0 \\ 0 & 0 & 0 & B & A & B \\ 0 & 0 & 0 & 0 & B & A \\ & & & & & & \cdot \\ & & & & & & \cdot \end{bmatrix} \tag{8.7}$$

Here, the submatrix B (also the dispersion matrix) denotes the stiffness matrix between any two consecutive layers and positions itself at an off-diagonal location in $[K]$, whereas the submatrix A (also the penetration matrix) denotes the self-stiffness of each layer. It should be noted that the submatrix $[K]_{11}$ is $A/2$, as the geometry in Figure 8.32 does not have any continuation above the topmost layer.

2. The second step is to solve for the equation

$$[K]\{T\} = \{flux\} \tag{8.8}$$

where $\{T\}$ denotes the layer-by-layer temperature and $\{flux\}$ denotes the layer-by-layer flux vectors.

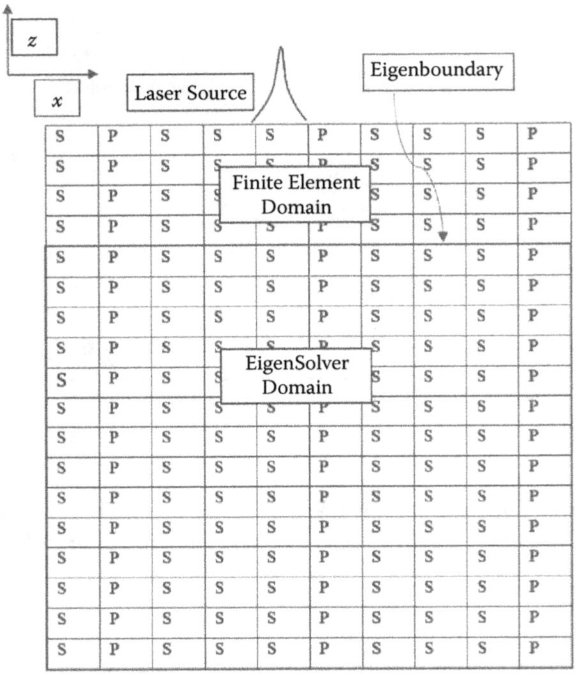

z

x

Laser Source

Eigenboundary

Finite Element Domain

EigenSolver Domain

S	P	S	S	S	P	S	S	S	P
S	P	S	S	S	P	S	S	S	P
S	P	S	S	S	P	S	S	S	P
S	P	S	S	S	P	S	S	S	P
S	P	S	S	S	P	S	S	S	P
S	P	S	S	S	P	S	S	S	P
S	P	S	S	S	P	S	S	S	P
S	P	S	S	S	P	S	S	S	P
S	P	S	S	S	P	S	S	S	P
S	P	S	S	S	P	S	S	S	P
S	P	S	S	S	P	S	S	S	P
S	P	S	S	S	P	S	S	S	P
S	P	S	S	S	P	S	S	S	P
S	P	S	S	S	P	S	S	S	P
S	P	S	S	S	P	S	S	S	P
S	P	S	S	S	P	S	S	S	P
S	P	S	S	S	P	S	S	S	P
S	P	S	S	S	P	S	S	S	P

FIGURE 8.32 The *x-z* plane cross-section elements viewed from the $-y$ direction. The solid powder pattern in the $+x$ direction is repeated and extruded in the $-z$ direction, providing the part with 2.5 dimension extrusion. **S** denotes the portion of the powder bed that has converted from powder to solid, and **P** denotes the portion of the powder bed that did not change its state and remained powder.

3. Setting the entire problem in the inverse B space, the following modification of Equation 8.8 is obtained:

$$
\begin{bmatrix}
B^{-1}A/2 & I & & & & \\
I & B^{-1}A & I & & & \\
 & I & B^{-1}A & I & & \\
 & & I & B^{-1}A & I & & \cdots & \cdots & \cdots & \cdots & \cdots \\
 & & & I & B^{-1}A & I & \\
 & & & & I & B^{-1}A & \\
 & & & & & & \cdot \\
 & & & & & & \cdot
\end{bmatrix}
\begin{Bmatrix}
T_{eigenbound-layer} \\
T_{layer\,2} \\
T_{layer\,3} \\
T_{layer\,4} \\
T_{layer\,5} \\
T_{layer\,6} \\
\cdot \\
\cdot
\end{Bmatrix} =
$$

$$
\begin{Bmatrix}
Flux_{eigenbound-layer} \\
0 \\
0 \\
0 \\
0 \\
0 \\
\cdot \\
\cdot
\end{Bmatrix}
\tag{8.9}
$$

It should be noted that the {*flux*} is only prescribed at the eigenboundary topmost layer using the finite element method followed by no −z directional flux in the succeeding layer below the topmost eigenboundary layer.

4. In order to make identity matrices on the left off-diagonal part zero, the following permutations (P_i) and new diagonal submatrices (D_i) have been computed:

$$P_1 = \left(\frac{A}{s}\right)^{-1}$$
$$P_2 = (D_2)^{-1}$$
$$P_3 = (D_3)^{-1}$$

(8.10)

$$D_2 = B^{-1}A - P_1$$
$$D_3 = B^{-1}A - P_2$$
$$D_4 = B^{-1}A - P_3$$

(8.11)

5. Generally, the permutation matrices converge very quickly and quadratically for thermal FE problems. From recursion, Equations 8.10 and 8.11 can be rewritten as

$$P_{i+1} = (D_{i+1})^{-1} = (B^{-1}A - P_i)^{-1}$$

(8.12)

6. Because the eigenvectors converged quickly and usually within two rows, the following assumptions can be made:

$$P_i = \phi D_i \phi^T$$
$$P_{i+1} = \phi D_{i+1} \phi^T$$

(8.13)

7. Inserting Equation 8.13 back into Equation 8.12, we get

$$\phi D_{i+1} \phi^T = (B^{-1}A - \phi D_i \phi^T)^{-1}$$
$$= (\phi K_s \phi^T - \phi D_i \phi^T)^{-1}$$
$$= (\phi^T)^{-1}(K_s - D_i)^{-1}\phi^{-1}$$

Using the identity that $\phi^T \phi = I$ (inner product of ϕ with itself is an identity matrix, meaning ϕ is a unit vector) or $\phi^T = \phi^{-1}$:

$$\phi D_{i+1} \phi^T = \phi(K_s - D_i)^{-1}\phi^T$$

(8.14)

8. From Equation 8.14, it is very straightforward that

$$D_{i+1} = (K_s - D_i)^{-1}$$

(8.15)

9. Equation 8.15 can be used to generalize for any κth eigenvalue of λ_κ:

$$\lambda_{\kappa,i+1} = \left(\kappa - \lambda_{\kappa,i}\right)^{-1} \qquad (8.16)$$

$$\lambda_{\kappa,i+1}\left(\kappa - \lambda_{\kappa,i}\right) = 1 \qquad (8.17)$$

$$\lambda_{\kappa,i+1}\lambda_{\kappa,i} - \kappa\lambda_{\kappa,i+1} + 1 = 0 \qquad (8.18)$$

$$\lambda_{\kappa,i+1} = \kappa - \frac{1}{\lambda_{\kappa,i}} \qquad (8.19)$$

where κ is the κth eigenvalue of $B^{-1}A$.

8.7.8   Results

8.7.8.1   Basic Eigenmodes for Prismatic Thermal Problems

The first few cross-sectional eigenmodes for reconstructing the nodal degrees of freedom for prismatic thermal problems are shown in Figures 8.33 and 8.34. These and other eigenmodes with increasingly higher eigenvalues are used for reconstructing the nodal degrees of freedom for a variety of energy sources, as shown in Figures 8.29, 8.30, and 8.31. Cross-sectional eigenvalues are shown in Figure 8.35 for up to 3600 degrees of freedom.

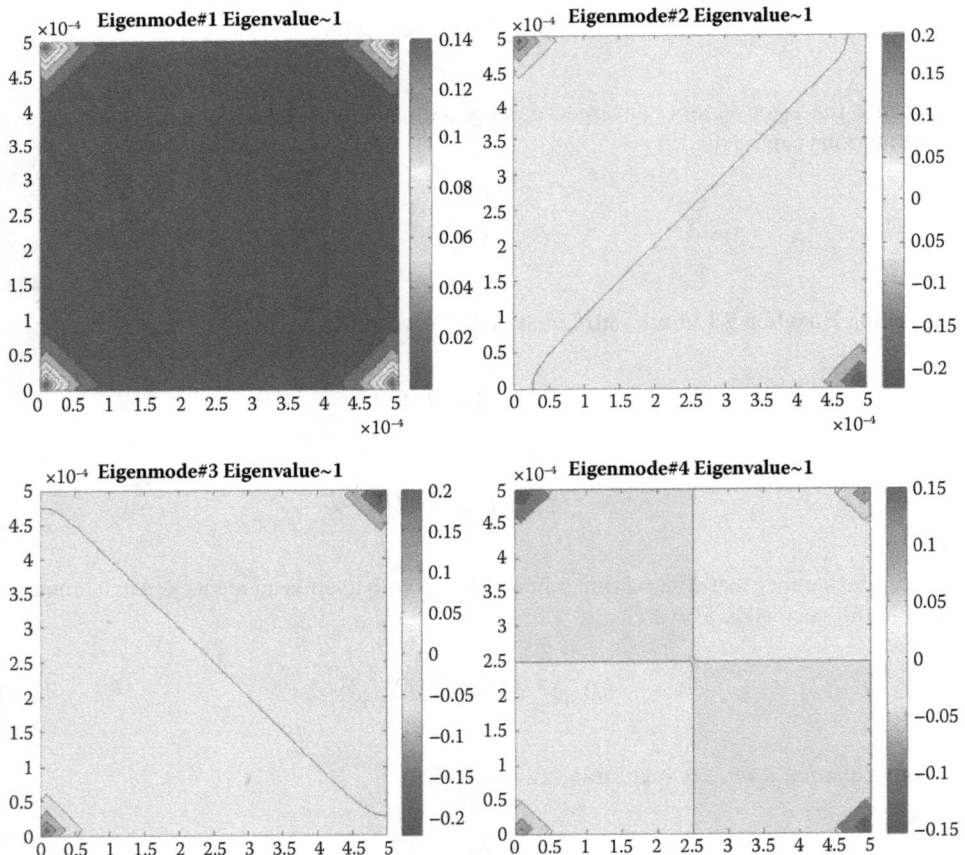

FIGURE 8.33   First four cross-sectional eigenmodes of the prismatic thermal problem.

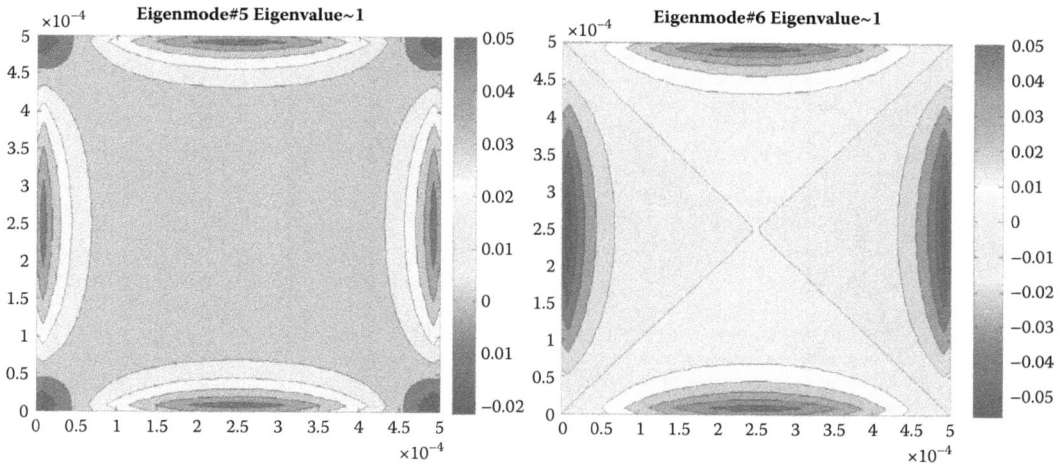

FIGURE 8.34 First fifth and sixth cross-sectional eigenmodes of the prismatic thermal problem.

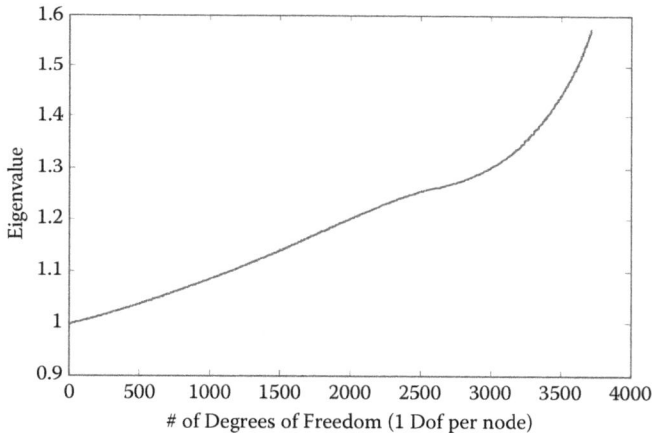

FIGURE 8.35 Cross-sectional eigenvalues of the thermal prismatic problem.

8.7.8.2 Basic Eigenmodes for Prismatic Structural Problems

The structural problem is orders of magnitudes more complex (3 degrees of freedom per node) compared to the thermal problem in terms of determination of eigenmodes and reconstruction of the nodal degrees of problem. Nevertheless, eigenvalues have been obtained for this problem. The reconstruction to obtain nodal degrees for this problem is still under development (Figure 8.36).

8.7.8.3 Large Matrix Eigenvalue Problems

With increasing number of degrees of freedom it becomes increasingly difficult to solve the eigenvalue problem due to the complexity involved in solving the corresponding characteristic equation as the order of the polynomial equation increases tremendously. The powder bed in metal laser sintering requires a huge amount of discretization to populate the entire powder bed. In order to solve the problem efficiently it is necessary to predict the evolution of eigenmodes with an increasing number of degrees of freedom in the cross-section (Figure 8.37). It was observed that with an increasing number of degrees of freedom a 10-parameter Fourier expansion could be used for fitting the evolution of eigenvalues. The reason why a 10-parameter Fourier expansion could fit this evolution is due to

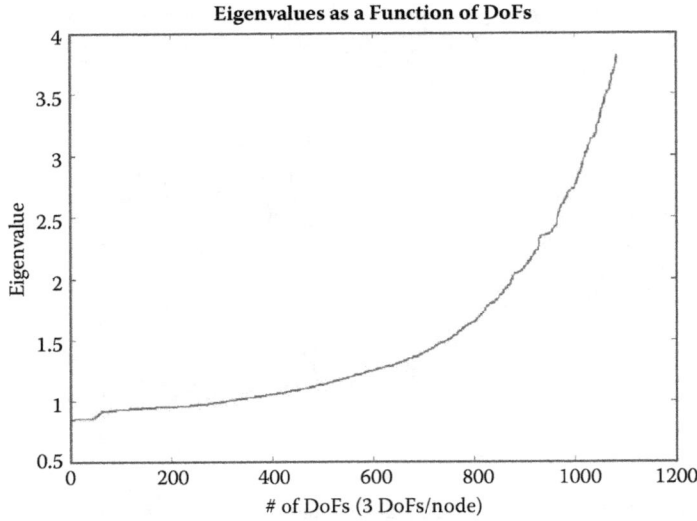

FIGURE 8.36   Cross-sectional eigenvalues of the structural prismatic problem.

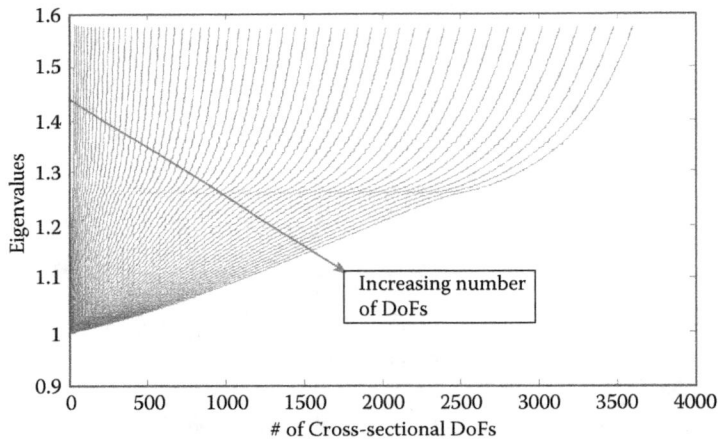

FIGURE 8.37   Eigenvalues as a function of cross-sectional degrees of freedom.

the fact that 10 parameters are required to extract the eigenvalues—namely, the thermal conductivity, discretizations, and physical dimensions of the problem in the x, y, and z directions along with a constraint that the first eigenvalue essentially needs to be 1 (meaning the flux travels straight through all the cross-sections in the corresponding mode). The fit using a 10-parameter Fourier expansion is shown with an increasing number of nodal density in the cross-section in Figure 8.38. For metal laser sintering, with combinations of unmelted powder and regions of solidified material, the number of parameters will be 13 (an additional 3 for powder thermal conductivities).

A number of simplified cases for cross-sections containing powder and solidified material are shown in Figure 8.39. Figure 8.40 shows regions where a 10-parameter Fourier expansion could be used to fit the eigenvalues to different cases shown in Figure 8.39. The nonlinear effects close to the melt pool when solved using the FFDAMRD algorithm[15] do not affect the linear solution

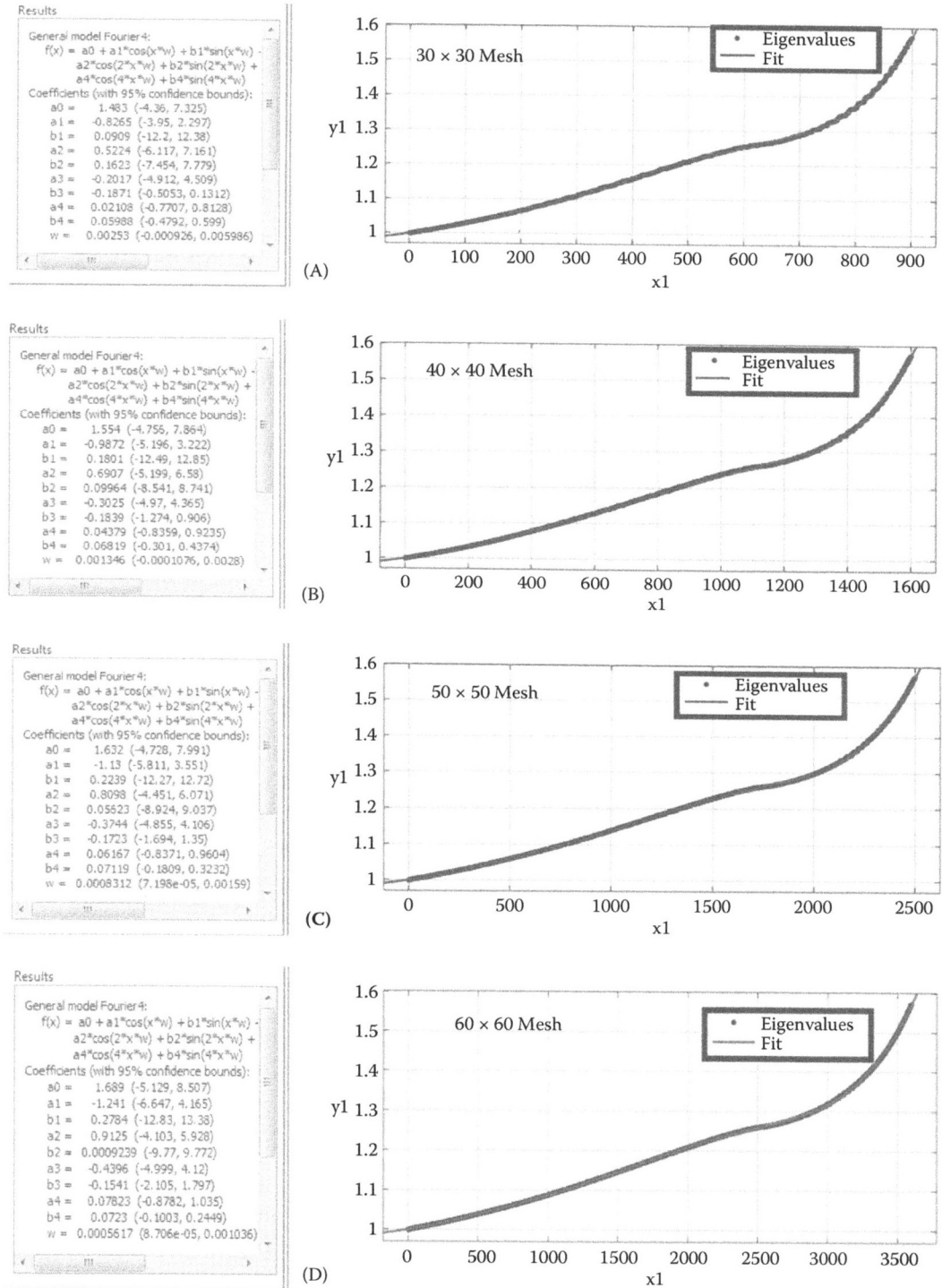

FIGURE 8.38 Eigenvalue fitting as a function of a seven-parameter Fourier expansion.

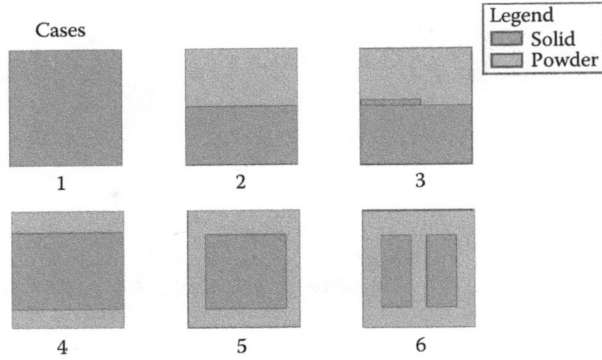

FIGURE 8.39 Six cases with the solidified portion (in dark gray) and powder portion (in light gray) in a given cross-section of the powder bed.

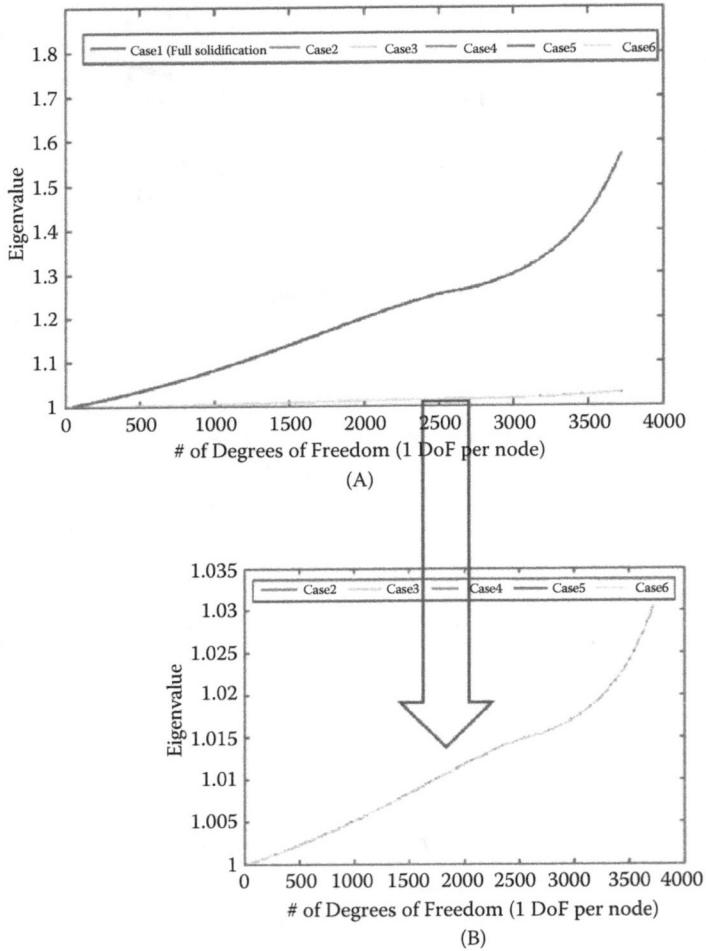

FIGURE 8.40 Eigenvalue evolution in cross-sections for the cases shown in Figure 8.39. The trend is exactly the same as the completely solidified bed and has been observed to be fit individually with a 10-parameter Fourier expansion. The evolution using a 10-parameter Fourier expansion can be used to solve large eigenvalue problems.

space outside the fine-scale domain. Due to the existence of only linear solutions far from the melt pool, the macroscopic coarse-scale powder bed eigenvalues should correctly be obtained using only 13 parameters. The change in eigenvalues in the coarser problem with transformation of a given volume from powder to solid could be accommodated by the addition of two more parameters correlated to the location and volume fraction change from powder to the solidified domain, thus increasing the total number of parameters to 15 to accurately predict the powder bed transitions at a coarser length scale.

8.7.9 Literature Review of Truncated Cholesky Algorithm

This solver[16] is based on the fact that any sparse stiffness matrix obtained using a finite element framework could be subdivided into lower triangular and upper triangular matrices. The lower triangular matrices (lower layers) act as wave propagators from one cross-section to another cross-section in the direction of energy propagation from the applied boundary conditions and involve in-plane wave scattering as well. This wave propagator generally solves for equilibrium, whereas the upper triangular solves for the wave coming back from a fixed boundary condition at the base of a matrix and redistributes the response variable (such as deformation or temperature). Thus, the lower triangular matrix is of primary importance for solving FEA problems, and its importance is multiplied for AM problems because the layer-by-layer build cross-sections are normal to the applied thermomechanical boundary conditions.

Some other methods have been reviewed here, as well, that work on similar concepts of formation of lower, diagonal, and upper triangular matrix decompositions of the full matrix in order to solve for the response variables such as temperature and displacement. A brief review of these methods has been provided in this section.

In finite element-based numerical analyses, the stiffness matrices that govern the problems are normally populated as sparse matrices that are positive definite in nature and comprised of lots of zero entries and few non-zero entries. The resulting linear system with unknown degrees of freedom is usually solved by using direct inverse methods known for most efficiently solving problems of this nature. For problems with matrix orders that are not so huge, the LU decomposition introduced by Alan Turing[104] is widely used to obtain the matrix inverse by expressing the matrix as a product of a lower triangular matrix L and an upper triangular matrix U. The LU decomposition can be viewed as the matrix form of Gaussian elimination.[105] The LDU decomposition is an extended LU decomposition where a diagonal matrix, D, is used to obtain unity along the diagonals of the lower and upper triangular matrices.[106] If the matrix is sparse, symmetric, and positive definite, Cholesky decomposition (or LL* decomposition) is usually applied, as the upper triangular matrix, U, can be represented as a conjugate transpose matrix of L and requires one-half of the calculations and one-half of the storage space involved in the LU decomposition. The implementation of this strategy proves to be much more efficient than the LU decomposition.[107]

Typical solution algorithms built on the basis of LU or Cholesky decomposition for solving the response variables in finite element analysis are available, such as the frontal[108] and the multi-frontal solvers.[109] The frontal solver assembles only on a subset (front) of the total number of elements and constructs one LU/Cholesky decomposition of the assembled front matrix at a time in order to save the disk cache and memory; the entire element matrix is never assembled explicitly. The multi-frontal solver is a multi-front analog of the frontal solver, and its implementation is favored in a distributed computing environment or on graphical processing units (GPUs).[110] For large problems, LU decomposition often meets a computational bottleneck such that the calculation of the L or a U triangular matrix requires a huge amount of flops because the bandwidth of the parent matrix gets completely or mostly filled in the L or U counterpart, causing memory issues and disk storage issues. In this case, incomplete LU decomposition is used to seek a sparse approximation of the LU decomposition with the sparsity pattern matching the square of the parent matrix.[111] There are also several iterative methods used in finite element solvers. One worth mentioning is the

Broyden–Fletcher–Goldfarb–Shanno (BFGS) algorithm,[112] which applies the rank 2 counterpart of the Sherman–Morrison method[113] to calculate the matrix inverse in the $(t + 1)$th time step based on the matrix inverse computed in the tth time step and the rank 2 approximation of the change in the parent stiffness matrix.

8.7.10 METHODOLOGY AND RESULTS

A new methodology has been formulated to compute the required number of essential lower triangular values which preserves the accuracy of the solution as well as the matrix values required to solve the problem while eliminating a large number of flops. The addresses of these values are stored in a banded vectorized fashion for quasistatic and dynamic problems to solve for deformation or temperatures using the solvers described above.

In Figure 8.41, the MATLAB CHOL (Cholesky) algorithm has been used to compute the lower triangular matrix of a thermal stiffness matrix of a 3D prismatic thermal problem with a Gaussian flux applied on its top surface. The base of the domain has been kept at a temperature of 353 K with opposing faces perpendicular to x and y directions comprising periodic boundary conditions. The periodic boundary conditions involve constraints that affect the bandwidth of the thermal stiffness matrix; therefore, the lower triangular matrix in Figure 8.40 is highly populated. When performing banded vectorization using L(row number1,:) * L(row number2,:) capping for calculation of updated L(row number1, :) with increasing thresholds, and checking the resulting changes in quality of the solution vector in terms of error compared to the original solution using first and second norms of the difference vector, it has been found that the solution vector has not changed beyond 0.026% while the number of the values required to compute the lower triangular matrix has decreased to only 6.7% of that required by the MATLAB CHOL algorithm. The optimized matrix at 10^{-4} is shown in Figure 8.42. In Figure 8.43, the number of flops calculated with respect to the logarithm of $L*L^T$ capping threshold is shown. It has been found that for the matrix represented in Figure 8.42 the number of flops required to calculate that matrix is around 1.57% of the flops required for calculating the lower triangular matrix by the Cholesky algorithm. Also, the efficiency and speed statistics of the modified Cholesky method have been completely tracked for solution and flops as a function of discrete values of the logarithm of the $L*L^T$ capping threshold.

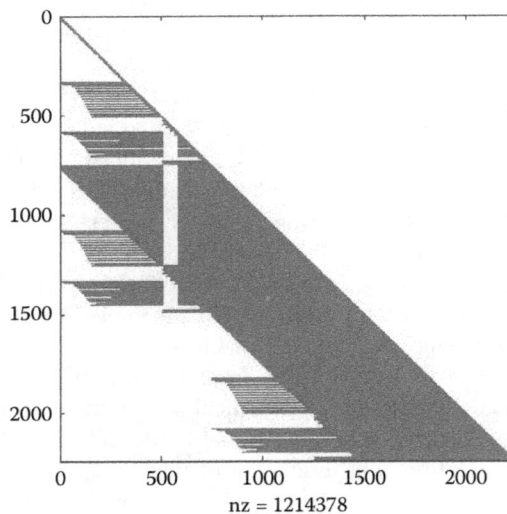

FIGURE 8.41 MATLAB®-generated lower triangular matrix for a sample thermal problem with periodic boundary conditions.

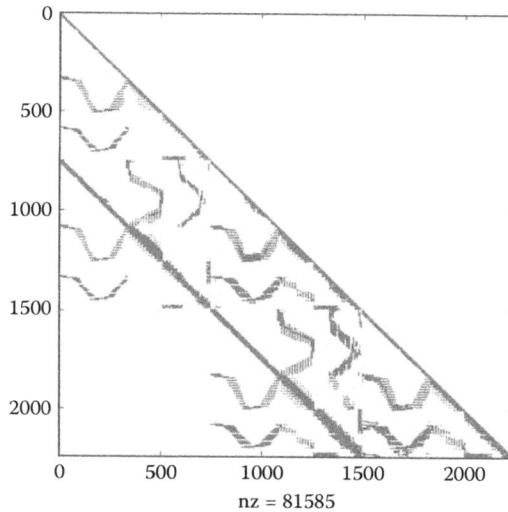

FIGURE 8.42   Preserved solution optimized lower triangular matrix by the modified CHOL (Cholesky factorization) method. Only 6.7% of the values were required to efficiently solve the problem with solution quality preservation.

8.7.11   LITERATURE REVIEW OF EFFECTIVE THERMAL CONDUCTIVITY USING COMPUTATIONAL HOMOGENIZATION

The homogenization technique[114] is recognized as an efficient way of finding effective material properties of composite materials. Although some of the effective properties can be determined and tested directly by experiments, in some cases, especially composites with complicated geometry in one or more directions, the effective properties are not easily obtained using direct testing or measurement, such as the thermal conductivity of the support structure in additive manufacturing process. Recognizing the necessity and importance of homogenization, different models of homogenization have been studied and proposed since the early 1960s, and typical reviews on

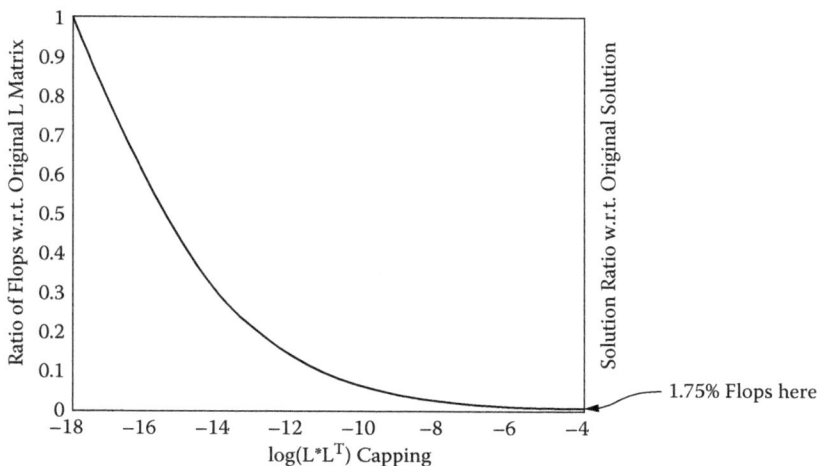

FIGURE 8.43   Efficiency and speed statistics for modified CHOL method.

thermomechanical problems can be found in the literature.[114–119] Here, we can classify them into six categories: (1) mechanics of materials method,[120–124] (2) fiber substructuring method,[125–127] (3) self-consistent method,[128–136] (4) Mori–Tanaka method,[137–144] (5) method of cells,[145–147] and (6) variational asymptotic method.[88,148,149]

8.7.12 Methodology

As shown in Figure 8.44, the support structure generation module is being incorporated into the current thermomechanical process solver. The goal is to create a force/strength map on the lower surfaces of a part to be built that guides the fitting of an optimized support structure to that force/strength map. One approach currently under investigation includes placing an initial support structure into the FFDAMRD model along with the part geometry. An example of this initially assumed support structure could be a block structure with powder in the intercellular spaces, as shown in Figure

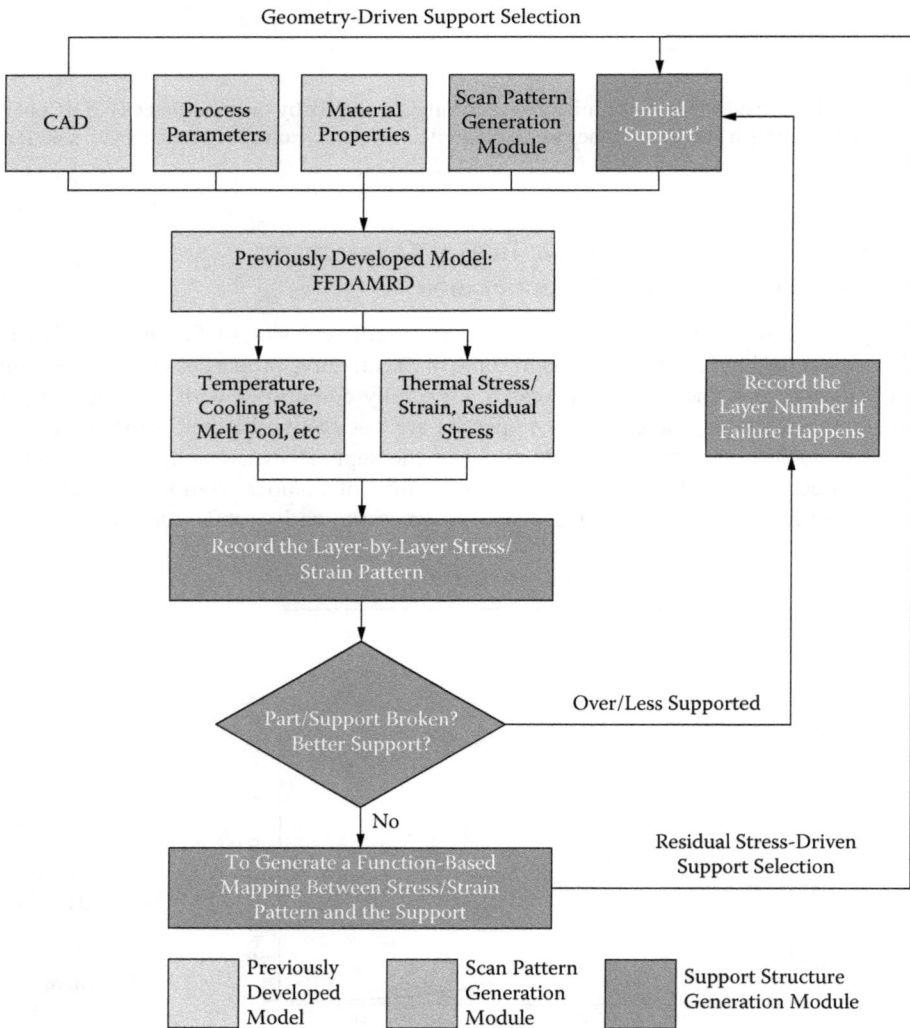

FIGURE 8.44 Schematic of incorporating support structure and scan pattern generation modules to the current FFDAMRD model.

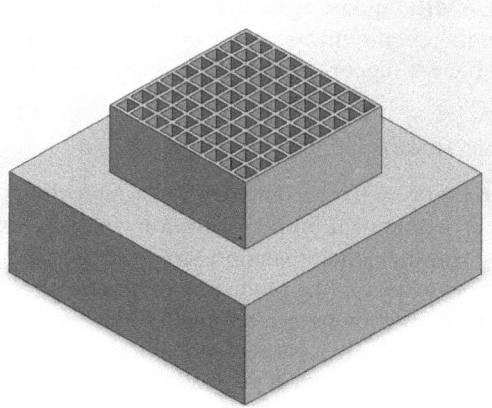

FIGURE 8.45 Three-dimensional model with support structure.

8.45. This block structure will be used to support the fabricated part on the substrate. The solidified support and the powder trapped inside the support structure will be modeled using appropriate thermophysical properties. A cross-section of the mesh with solidified struts and trapped powder inside the intercellular space is shown in Figure 8.46. This model will be solved using a thermomechanical eigenmodal strategy layer by layer with a contact model[150] to account for snapping of supports at their interface with the baseplate and/or the part. The dimensions of the solidified struts will be periodically updated as a function of the layer-by-layer buildup of the fabricated structure, resulting in updated element properties from powder to solid with an updated total stiffness matrix.

Accumulated residual stress/thermal strain will be recorded until the end of fabrication and a force/strength map will be created in three dimensions that represents the force required in a particular location to hold the part in place. That force/strength map will then be used to create a second efficient support structure. This efficient truss structure will be used as the real support structure to be made in the machine. Early on in development of this code we anticipated a possible need to

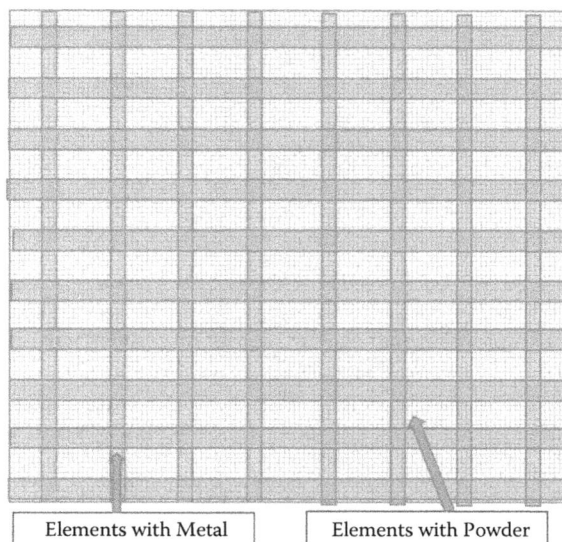

| Elements with Metal | Elements with Powder |

FIGURE 8.46 Meshing in the $-z$ direction.

iterate by running the FFDAMRD model again using the updated support structure to make sure it meets the anticipated loading requirements, as the change in thermal history between the initial support structure and the efficient support structure may be significant.

8.7.13 FORMULATION AND RESULTS

The addition of support structures to the current 3DSIM architecture makes the model more complicated and increases its computational burden. However, in order to include the effects of the residual powder left trapped in the cellular support structure, an equivalent model must be considered. The equivalent model results in the prediction of an effective thermal conductivity of the support structure in the z-axis direction by performing a finite element simulation, and the results are validated by comparing the results to a parallel model for springs.

Figure 8.47 shows the initially chosen model geometry to represent the support structure for evaluating the effective thermal conductivity. A constant temperature of 100 K and 0 K was prescribed at the center nodes of the top and bottom surfaces, respectively, as boundary conditions in a finite element model. Other nodes on the top and bottom surfaces were constrained to maintain a temperature difference of 100 K between the top and bottom surfaces. The intent of this simulation was to create a representative model and to perform a finite element simulation with periodic boundary conditions such that the opposite faces of the model domain perpendicular to the x- and y-axis have exactly the same temperatures. A node-by-node thermal solution has been obtained in ANSYS by subjecting the model geometry to the above-mentioned boundary conditions and surface constraints. An effective thermal conductivity was then calculated based on the heat equation shown below:

$$q = -k\frac{dT}{dz} \tag{8.20}$$

where q is the heat flux, k is the thermal conductivity, T is the temperature, and z is the distance in the z direction.

As seen in Figure 8.48, the temperature difference in the z direction is 100 K, which is predefined as the boundary condition. The thermal gradient in the z direction can be obtained through integration of Equation 8.20. As shown in Figure 8.49, the temperature distribution along the z direction is

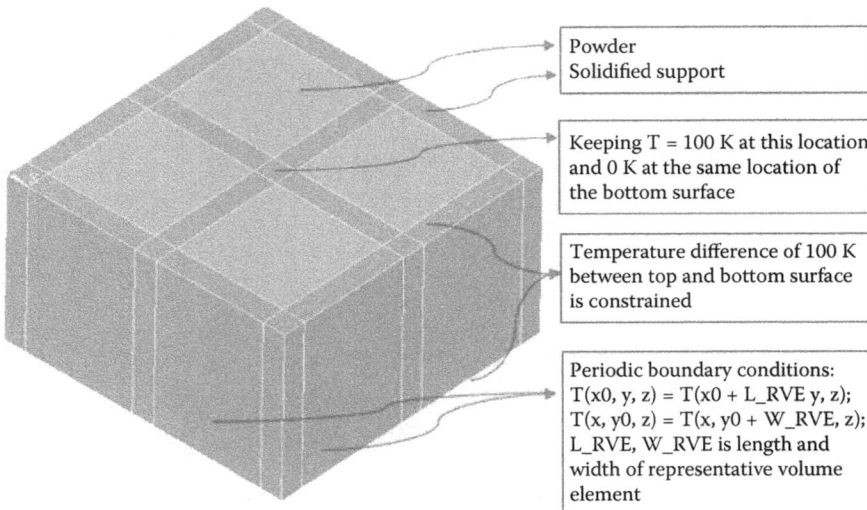

FIGURE 8.47 Representative volume element for the support structure with defined boundary conditions.

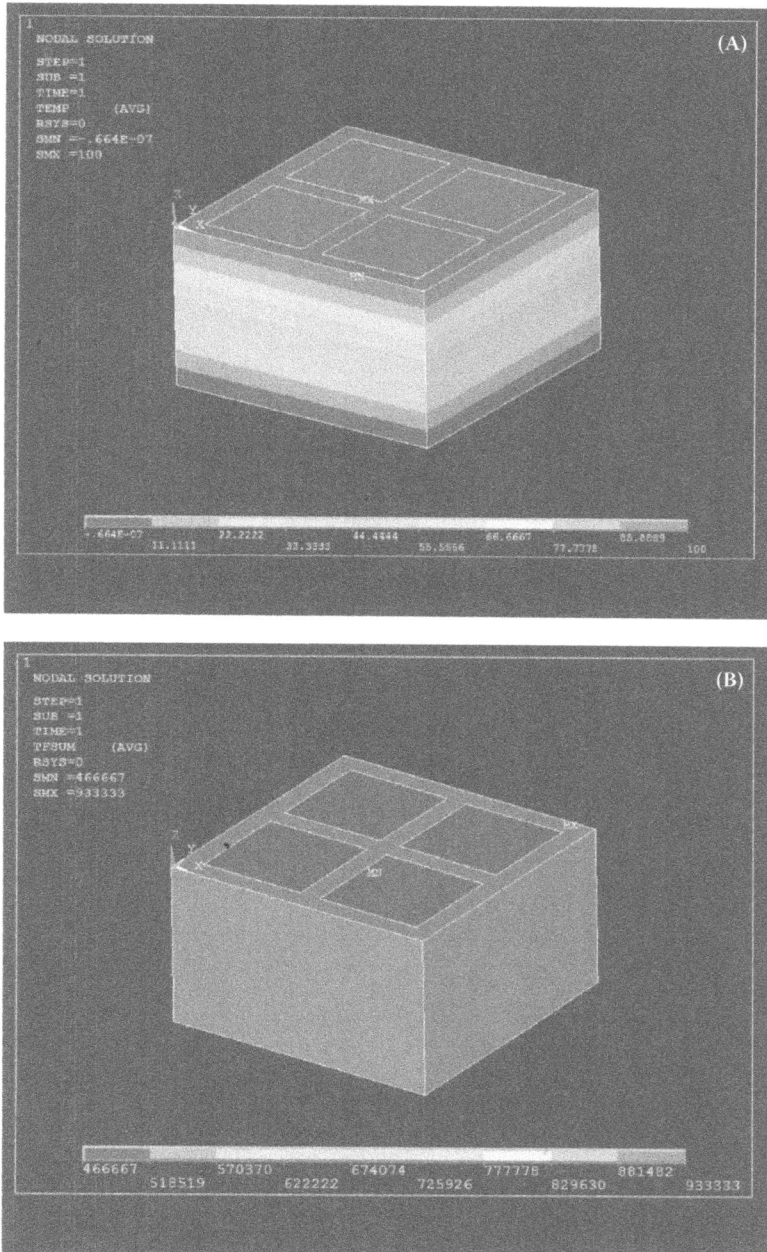

FIGURE 8.48 (A) Temperature distribution and (B) heat flux results.

linearly distributed, thus the thermal gradient is easily calculated in this case. Because the heat flux is also available for each group of elements (powder and solidified sections) through the z direction, the effective thermal conductivity is calculated for each group of elements through the z direction by using the average thermal gradient. The effective thermal conductivity of the representative volume element is then obtained by averaging the thermal conductivity for each group (powder and solid) of elements in the z direction.

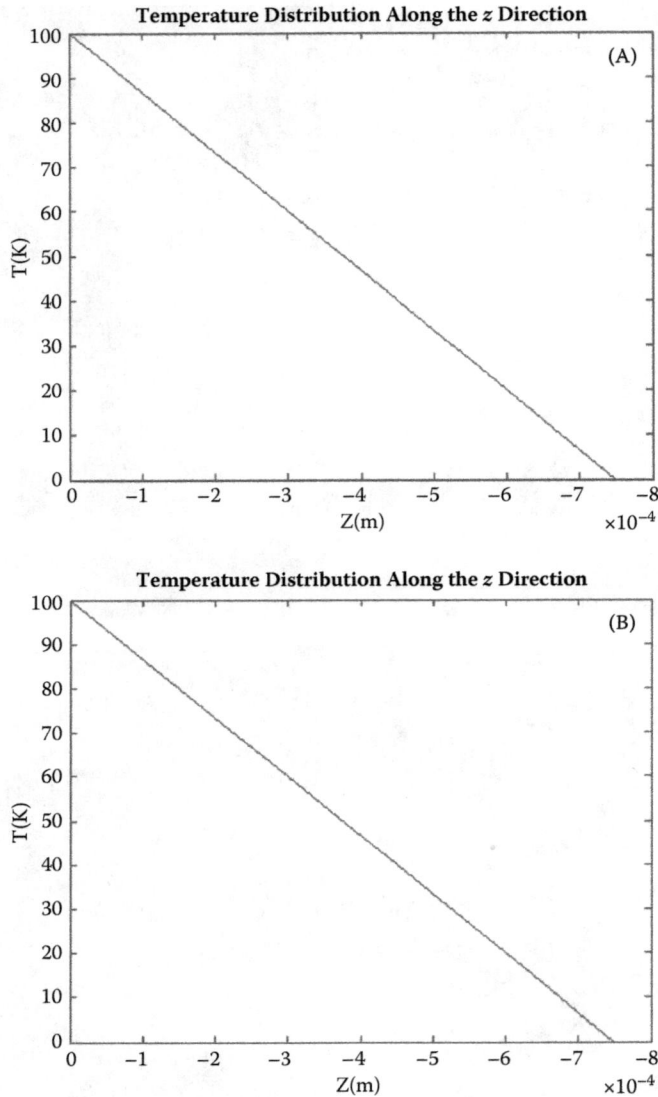

FIGURE 8.49 Temperature distribution along the z direction: (A) temperature distribution along the z direction for the center surface, and (B) temperature distribution along the z direction for the powder area.

The influence of the total number of cells in the support structure and the percentage of the solidified material in the representative volume element were studied. According to the literature, the series and parallel models for studying the effective properties of composite material have been used. The series model is chosen to verify the numerical model in this study. By comparison to an effective spring calculation, a parallel model has been implemented as shown in Equation 8.21:

$$K_{effective} = K_{solid} \times \theta + K_{powder} \times (1 - \theta) \tag{8.21}$$

where K_{solid} and K_{powder} are the thermal conductivity for the solidified support and powder, respectively, and θ is the percentage of solidified support.

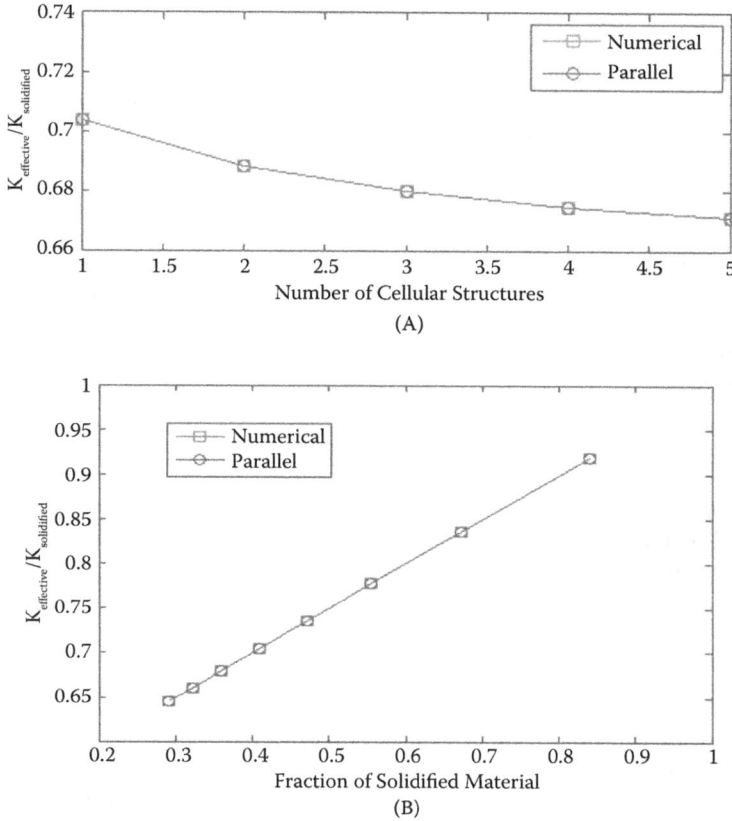

FIGURE 8.50 Effective thermal conductivity under different scenarios: (A) $K_{effective}/K_{solidified}$ vs. the number of cellular structures, (B) $K_{effective}/K_{solidified}$ vs. the fraction of solidified material.

Figure 8.50A shows the trend for effective thermal conductivity converging with an increasing number of support structure cells. When increasing the percentage of solidified support material, the effective thermal conductivity increases, as shown in Figure 8.50B. Both results show that the numerical effective thermal conductivity matches the series model results.

8.8 FUTURE WORK ON THERMAL CONDUCTIVITY

Future work on thermal conductivity will focus on

- Obtaining effective thermal conductivity as a function of temperature difference (for a varying set of set temperatures) between the top and bottom nodes, taking into account thermal conductivity as a nonlinear function of temperature and power density evolution as a function of temperature
- For a set temperature difference, predicting the effective thermal conductivity as a function of base-referencing the bottom node temperatures
- Determining effective thermal conductivity as a function of an initial ratio of powder to bulk density

- Repeating the previous three numerical experiments by setting the temperature difference between opposite faces perpendicular to the x and y directions to obtain geometry-based anisotropy for thermal conductivity
- Studying the convergence of the effective thermal conductivity as a function of number of unit cells for the previous numerical experiments

8.9   CONCLUSIONS

In response to an increasing need for optimization of parameters, materials, and geometries for AM-fabricated parts, an integrated approach to accurately simulate the physics and materials science aspects of AM has been undertaken. Because AM is a particularly time- and resource-consuming problem to solve due to the moving energy sources and material additions over time and space, new numerical algorithms and software tools must be developed. Homogenization strategies applied to some of these tools may lead to better machine architectures in the near future, such as for modeling the multi-beam capabilities of electron beam-based and solid-state additive manufacturing. The numerical algorithms and tools are generic for highly dynamic processes with high strain and thermal gradient environments and can be flexibly adapted for relatively quasistatic and macroscopic traditional manufacturing processes, as well. The process of integrating these software tools to achieve a fully functional software suite for additive manufacturing processes is underway.

REFERENCES

1. Ashton, T. (1948). *The Industrial Revolution*. Oxford University Press, London.
2. Zuse, K. (1993). *The Computer—My Life*. Springer, New York.
3. Hull, C.W. (1986). Apparatus for Production of Three-Dimensional Objects by Stereolithography, Google Patent US 4575330.
4. Pal, D., Patil, N., and Stucker, B.E. (2015). A generalized feed forward dynamic adaptive mesh refinement and de-refinement finite element framework for metal laser sintering. Part II. Non-linear thermal simulations and validations. *J. Manuf. Sci. Eng.*, submitted for review.
5. Zeng, K. et al. (2013). A new dynamic mesh method applied to the simulation of selective laser melting. In: *Solid Freeform Fabrication Symposium Proceedings*, pp. 549–559.
6. Zeng, K., Pal, D., and Stucker, B. (2012). A review of thermal analysis methods in laser sintering and selective laser melting. In: *Solid Freeform Fabrication Symposium Proceedings*, pp. 796–814.
7. Pal, D. and Stucker, B. (2013). A study of subgrain formation in Al 3003 H-18 foils undergoing ultrasonic additive manufacturing using a dislocation density based crystal plasticity finite element framework. *J. Appl. Phys.*, 113(20): 203517.
8. Pal, D. and Stucker. B. (2011). Dislocation density based finite element modeling of ultrasonic consolidation. In: *Solid Freeform Fabrication Symposium Proceedings*.
9. Pal, D. (2011). Dislocation Density-Based Finite Element Method Modeling of Ultrasonic Consolidation, PhD dissertation, Utah State University, Logan.
10. Pal, D. and Stucker. B.E. (2012). Time homogenization of Al3003 H-18 foils undergoing metallurgical bonding using ultrasonic consolidation. In: *Solid Freeform Fabrication Symposium Proceedings*, pp. 373–384.
11. Pal, D. and Stucker, B.E. (2012). Modelling of ultrasonic consolidation using a dislocation density based finite element framework: in this paper a dislocation density based constitutive model is developed and implemented into a crystal plasticity quasi-static finite element. *Virtual Phys. Prototyping*, 7(1): 65–79.
12. Pal, D., Behera, S., and Ghosh. S. (2009). Crystal plasticity modeling of creep and microtwinning in nickel based superalloys. In: *Proceedings of the 10th U.S. National Congress of Computational Mechanics*, pp. 15–19
13. Pal, D., Patil, N., and Stucker, B.E. (2012). Prediction of mechanical properties of electron beam melted Ti6Al4V parts using dislocation density based crystal plasticity framework. In: *Solid Freeform Fabrication Symposium Proceedings*, 19 pp.
14. Pal, D. et al. (2013). An integrated approach to cyber-enabled additive manufacturing using physics based, coupled multi-scale process modeling. In: *Solid Freeform Fabrication Symposium Proceedings*, 18 pp.

15. Patil, N. et al. (2015). A generalized feed forward dynamic adaptive mesh refinement and de-refinement finite element framework for metal laser sintering. Part I. Formulation and algorithm development. *J. Manuf. Sci. Eng.*, submitted for review.

16. Patil, N., Pal, D., and Stucker. B. (2013). A new finite element solver using numerical eigen modes for fast simulation of additive manufacturing processes. In: *Solid Freeform Fabrication Symposium Proceedings*, pp. 535–543.

17. Nikoukar, M. et al. (2013). Methods for enhancing the speed of numerical calculations for the prediction of the mechanical behavior of parts made using additive manufacturing. In: *Solid Freeform Fabrication Symposium Proceedings*, pp. 525–534.

18. Vatandas, E. and Özkol, I. (2008). Coupling dynamic mesh technique and heuristic algorithms in 3-D-tapered wing design. *Int. J. Numer. Meth. Eng.*, 74(12): 1771–1794.

19. Vatandas, E. and Özkol, I. (2006). Dynamic mesh and heuristic algorithms for the design of a transonic wing. *Aircr. Eng. Aerosp. Technol.*, 78(1): 39–44.

20. Liu, X., Qin, N., and Xia, H. (2006). Fast dynamic grid deformation based on Delaunay graph mapping. *J. Comput. Phys.*, 211(2): 405–423.

21. Huo, S. et al. (2010). Layered elastic solid method for the generation of unstructured dynamic mesh. *Finite Elem. Anal. Des.*, 46(10): 949–955.

22. Farhat, C. et al. (1998). Torsional springs for two-dimensional dynamic unstructured fluid meshes. *Comput. Methods Appl. Mech. Eng.*, 163(1): 231–245.

23. Lin, P.T. et al. (2006). Two-dimensional implicit time-dependent calculations on adaptive unstructured meshes with time evolving boundaries. *Int. J. Numer. Meth. Fluids*, 50(2): 199–218.

24. Jahangirian, A. and Hadidoolabi, M. (2005). Unstructured moving grids for implicit calculation of unsteady compressible viscous flows. *Int. J. Numer. Meth. Fluids*, 47(10–11): 1107–1113.

25. Vatandas, E., Özkol, I., and Kaya, M.O. (2004). Using the dynamic mesh method for genetically obtained wing structures. *Aircr. Eng. Aerosp. Technol.*, 76(3): 314–319.

26. Vatandas, E., Hacioglu, A., and Özkol, I. (2007). Vibrational genetic algorithm (VGA) and dynamic mesh in the optimization of 3D wing geometries. *Inverse Probl. Sci. Eng.*, 15(6): 643–657.

27. Montalvo-Urquizo, J., Akbay, Z., and Schmidt, A. (2009). Adaptive finite element models applied to the laser welding problem. *Comput. Mater. Sci.*, 46(1): 245–254.

28. Qin, N., Liu, X., and Xia, H. (2005). An efficient moving grid algorithm for large deformation. *Mod. Phys. Lett. B*, 19(28n29): 1499–1502.

29. Brezzi, F. (1974). On the existence, uniqueness and approximation of saddle-point problems arising from Lagrangian multipliers. *ESAIM: M2AN*, 8(R2): 129–151.

30. Wang, Z. (2000). A fast nested multi-grid viscous flow solver for adaptive Cartesian/quad grids. *Int. J. Numer. Meth. Fluids*, 33(5): 657–680.

31. Jakobsson, S. and Amoignon, O. (2007). Mesh deformation using radial basis functions for gradient-based aerodynamic shape optimization. *Comput. Fluids*, 36(6): 1119–1136.

32. Poirier, V. (2012). An Efficient Radial Basis Function Mesh Deformation Scheme within an Adjoint-Based Aerodynamic Optimization Framework, MEng thesis, McGill University, Montreal.

33. De Boer, A., Van der Schoot, M., and Bijl, H. (2007). Mesh deformation based on radial basis function interpolation. *Comput. Struct.*, 85(11): 784–795.

34. Rendall, T. and Allen, C. (2010). Parallel efficient mesh motion using radial basis functions with application to multi-bladed rotors. *Int. J. Numer. Meth. Eng.*, 81(1): 89–105.

35. Rendall, T. and Allen, C. (2008). Unified fluid–structure interpolation and mesh motion using radial basis functions. *Int. J. Numer. Meth. Eng*, 74(10): 1519–1559.

36. Wei, D. et al. (2014). A structured mesh boundary motion approach for simulating wind effects on bluff bodies with changing boundaries. *J. Wind Eng. Indust. Aerodyn.*, 126: 118–131.

37. Cai, J., Tsai, H.M., and Liu, F. (2006). A parallel viscous flow solver on multi-block overset grids. *Comput. Fluids*, 35(10): 1290–1301.

38. Xiao, T. and Ang, H. (2009). Combination of mesh deformation and overset grid for simulating unsteady flow of insect flight. *Mod. Phys. Lett. B*, 23(03): 525–528.

39. Masters, N. et al. (2010). Laser ray tracing in a parallel arbitrary Lagrangian–Eulerian adaptive mesh refinement hydrocode. *J. Phys. Conf. Ser.*, 244: 032022.

40. Zhang, L. and Wang, Z. (2004). A block LU-SGS implicit dual time-stepping algorithm for hybrid dynamic meshes. *Comput. Fluids*, 33(7): 891–916.

41. Kruth, J.-P. et al. (2004). Selective laser melting of iron-based powder. *J. Mater. Process. Technol.*, 149(1): 616–622.

42. Kruth, J.-P. et al. (2005). Binding mechanisms in selective laser sintering and selective laser melting. *Rapid Prototyping J.*, 11(1): 26–36.
43. Taylor, C.M. (2004). Direct Laser Sintering of Stainless Steel: Thermal Experiments and Numerical Modelling, PhD dissertation, University of Leeds.
44. Ibraheem, A.K., Derby, B., and Withers. P.J. (2003). Thermal and residual stress modelling of the selective laser sintering process. In: *Materials Research Society Symposium Proceedings*, pp. 47–52.
45. Kolossov, S. et al. (2004). 3D FE simulation for temperature evolution in the selective laser sintering process. *Int. J. Mach. Tools Manuf.*, 44(2): 117–123.
46. Zhang, D. et al. (2010). Select laser melting of W–Ni–Fe powders: simulation and experimental study. *Int. J. Adv. Manuf. Technol.*, 51(5–8): 649–658.
47. Roberts, I. et al. (2009). A three-dimensional finite element analysis of the temperature field during laser melting of metal powders in additive layer manufacturing. *Int. J. Mach. Tools Manuf.*, 49(12): 916–923.
48. Song, B. et al. (2012). Process parameter selection for selective laser melting of Ti6Al4V based on temperature distribution simulation and experimental sintering. *Int. J. Adv. Manuf. Technol.*, 61(9–12): 967–974.
49. Cervera, G.B.M. and Lombera, G. (1999). Numerical prediction of temperature and density distributions in selective laser sintering processes. *Rapid Prototyping J.*, 5(1): 21–26.
50. Dai, K. and Shaw, L. (2005). Finite element analysis of the effect of volume shrinkage during laser densification. *Acta Mater.*, 53(18): 4743–4754.
51. Dong, L. et al. (2009). Three-dimensional transient finite element analysis of the selective laser sintering process. *J. Mater. Process. Technol.*, 209(2): 700–706.
52. Gusarov, A. et al. (2007). Heat transfer modelling and stability analysis of selective laser melting. *Appl. Surf. Sci.*, 254(4): 975–979.
53. Shellabear, M. and Nyrhilä, O. (2004). DMLS: development history and state of the art. In: *Proceedings of the 4th International Conference on Laser Assisted Netshape Engineering*, pp. 21–24.
54. Childs, T., Hauser, C., and Badrossamay, M. (2004). Mapping and modelling single scan track formation in direct metal selective laser melting. *CIRP Ann. Manuf. Technol.*, 53(1): 191–194.
55. Dowden, J.M. (2001). *The Mathematics of Thermal Modeling: An Introduction to the Theory of Laser Material Processing*. CRC Press, Boca Raton, FL.
56. Mills, K.C. (2002). *Recommended Values of Thermophysical Properties for Selected Commercial Alloys*. Woodhead Publishing, Cambridge, U.K.
57. Fischer, P. et al. (2004). Temperature measurements during selective laser sintering of titanium powder. *Int. J. Mach. Tools Manuf.*, 44(12): 1293–1296.
58. Gibson, I., Rosen, D.W., and Stucker, B. (2010). *Additive Manufacturing Technologies*. Springer, New York.
59. Verhaeghe, F. et al. (2009). A pragmatic model for selective laser melting with evaporation. *Acta Mater.*, 57(20): 6006–6012.
60. Gong, H. (2013). Generation and Detection of Defects in Metallic Parts Fabricated by Selective Laser Melting and Electron Beam Melting and Their Effects on Mechanical Properties. PhD dissertation, University of Louisville.
61. Thijs, L. et al. (2010). A study of the microstructural evolution during selective laser melting of Ti-6Al-4V. *Acta Mater.*, 58(9): 3303–3312.
62. *Material Data Sheet: EOS Stainless Steel 17-4 for EOSINT M 270*. EOS GmbH, Munich.
63. Neira Arce, A. (2012). Thermal Modeling and Simulation of Electron Beam Melting for Rapid Prototyping on Ti6Al4V Alloys. PhD dissertation, North Carolina State University, Raleigh.
64. Pal, D. and Stucker. B. (2011). Some studies on dislocation density based finite element modeling of ultrasonic consolidation. In: *Proceedings of the 5th International Conference on Virtual and Rapid Prototyping*, pp. 667–676.
65. White, D.R. (2003). Ultrasonic consolidation of aluminum tooling. *Adv. Mater. Process.*, 161(1): 64–65.
66. Gonzalez, R. and Stucker, B. (2012). Experimental determination of optimum parameters for stainless steel 316L annealed ultrasonic consolidation. *Rapid Prototyping J.*, 18(2): 172–183.
67. Gu, Z. et al. (2013). Fabrication, characterization and applications of novel nanoheater structures. *Surf. Coat. Technol.*, 215: 493–502.
68. Obielodan, J. and Stucker, B. (2009). Further exploration of multi-material fabrication capabilities of ultrasonic consolidation technique. In: *Solid Freeform Fabrication Symposium Proceedings*, pp. 354–373.
69. Kong, C., Soar, R., and Dickens, P.M. (2004). Ultrasonic consolidation for embedding SMA fibres within aluminium matrices. *Compos. Struct.*, 66(1): 421–427.
70. Zhu, Z., Lee, K.Y., and Wang, X. (2012). Ultrasonic welding of dissimilar metals, AA6061 and Ti6Al4V. *Int. J. Adv. Manuf. Technol.*, 59(5–8): 569–574.

71. Kong, C., Soar, R., and Dickens, P.M. (2004). Optimum process parameters for ultrasonic consolidation of 3003 aluminium. *J. Mater. Process. Technol.*, 146(2): 181–187.
72. George, J.L. (2006). Utilization of Ultrasonic Consolidation in Fabricating Satellite Decking, PhD dissertation, Utah State University, Logan.
73. Kong, C. and Soar, R. (2005). Fabrication of metal–matrix composites and adaptive composites using ultrasonic consolidation process. *Mater. Sci. Eng. A*, 412(1): 12–18.
74. Yang, Y., Ram, G.J., and Stucker, B. (2007). An experimental determination of optimum processing parameters for Al/SiC metal matrix composites made using ultrasonic consolidation. *J. Eng. Mater. Technol.*, 129(4): 538-549.
75. Yang, Y., Janaki Ram, G., and Stucker, B. (2009). Bond formation and fiber embedment during ultrasonic consolidation. *J. Mater. Process. Technol.*, 209(10): 4915–4924.
76. Siggard, E.J. (2007). *Investigative Research into the Structural Embedding of Electrical and Mechanical Systems Using Ultrasonic Consolidation (UC)*. ProQuest, Ann Arbor, MI.
77. Janaki Ram, G. et al. (2007). Use of ultrasonic consolidation for fabrication of multi-material structures. *Rapid Prototyping J.*, 13(4): 226–235.
78. Domack, M. and Baughman, J. (2005). Development of nickel-titanium graded composition components. *Rapid Prototyping J.*, 11(1): 41–51.
79. Obielodan, J. and Stucker. B. (2009). Effects of post processing heat treatments on the bond quality and mechanical strength of Ti/Al3003 dual materials fabricated using ultrasonic consolidation. In: *Solid Freeform Fabrication Symposium Proceedings*, pp. 406–427.
80. Kelly, G.S. et al. (2013). A model to characterize acoustic softening during ultrasonic consolidation. *J. Mater. Process. Technol.*, 213(11): 1835–1845.
81. Kelly, G. (2012). A Thermo-Mechanical Finite Element Analysis of Acoustic Softening During Ultrasonic Consolidation of Aluminum Foils, PhD dissertation, University of Delaware, Newark.
82. Kelly, G.S. et al. (2014). Energy and bond strength development during ultrasonic consolidation. *J. Mater. Process. Technol.*, 214(8): 1665–1672.
83. Janaki Ram, G., Yang, Y., and Stucker, B. (2006). Effect of process parameters on bond formation during ultrasonic consolidation of aluminum alloy 3003. *J. Manuf. Syst.*, 25(3): 221–238.
84. Obielodan, J. et al. (2010). Minimizing defects between adjacent foils in ultrasonically consolidated parts. *J. Eng. Mater. Technol.*, 132(1): 011006.
85. Kong, C., Soar, R., and Dickens, P.M. (2003). Characterisation of aluminium alloy 6061 for the ultrasonic consolidation process. *Mater. Sci. Eng. A*, 363(1): 99–106.
86. Johnson, K.E. (2008). Interlaminar Subgrain Refinement in Ultrasonic Consolidation, PhD dissertation, Loughborough University.
87. Berdichevskii, V. (1979). Variational asymptotic method of constructing a theory of shells. *Prikl. Mat. Mekh.*, 43(4): 664–687.
88. Finnveden, S. (1997). Spectral finite element analysis of the vibration of straight fluid-filled pipes with flanges. *J. Sound Vibrat.*, 199(1): 125–154.
89. Finnveden, S. (1994). Exact spectral finite element analysis of stationary vibrations in a rail way car structure. *Acta Acustica*, 2: 461–482.
90. Mahapatra, D.R. and Gopalakrishnan, S. (2003). A spectral finite element for analysis of wave propagation in uniform composite tubes. *J. Sound Vibrat.*, 268(3): 429–463.
91. Birgersson, F., Ferguson, N.S., and Finnveden, S. (2003). Application of the spectral finite element method to turbulent boundary layer induced vibration of plates. *J. Sound Vibrat.*, 259(4): 873–891.
92. Chakraborty, A. and Gopalakrishnan, S. (2006). A spectral finite element model for wave propagation analysis in laminated composite plate. *J. Vibrat. Acoust.*, 128(4): 477–488.
93. Roy Mahapatra, D. and Gopalakrishnan, S. (2003). A spectral finite element model for analysis of axial–flexural–shear coupled wave propagation in laminated composite beams. *Compos. Struct.*, 59(1): 67–88.
94. Wang, G. and Wereley, N.M. (2002). Spectral finite element analysis of sandwich beams with passive constrained layer damping. *J. Vibrat. Acoust.*, 124(3): 376–386.
95. Palacz, M., Krawczuk, M., and Ostachowicz, W. (2005). The spectral finite element model for analysis of flexural-shear coupled wave propagation. Part 2. Delaminated multilayer composite beam. *Compos. Struct.*, 68(1): 45–51.
96. Kumar, D.S., Mahapatra, D.R., and Gopalakrishnan, S. (2004). A spectral finite element for wave propagation and structural diagnostic analysis of composite beam with transverse crack. *Finite Elem. Anal. Des.*, 40(13–14): 1729–1751.
97. Krawczuk, M. (2002). Application of spectral beam finite element with a crack and iterative search technique for damage detection. *Finite Elem. Anal. Des.*, 38(6): 537–548.

98. Ostachowicz, W.M. (2008). Damage detection of structures using spectral finite element method. *Comput. Struct.*, 86(3): 454–462.

99. Mahapatra, D.R. and Gopalakrishnan, S. (2004). Spectral finite element analysis of coupled wave propagation in composite beams with multiple delaminations and strip inclusions. *Int. J. Solids Struct.*, 41(5): 1173–1208.

100. Samaratunga, D., Jha, R., and Gopalakrishnan, S. (2014). Wavelet spectral finite element for wave propagation in shear deformable laminated composite plates. *Compos. Struct.*, 108: 341–353.

101. Daubechies, I. (1992). *Ten Lectures on Wavelets*. Society for Industrial and Applied Mathematics, Philadelphia, PA.

102. Mitra, M. and Gopalakrishnan, S. (2006). Wavelet based spectral finite element modelling and detection of de-lamination in composite beams. *Proc. R. Soc. Lond. Philos. Trans. Ser. A Math. Phys. Eng. Sci.*, 462(2070): 1721–1740.

103. Saad, Y. (1992). *Numerical Methods for Large Eigenvalue Problems*, Vol. 158. Manchester University Press, Manchester, U.K.

104. Poole, D. (2011). *Linear Algebra: A Modern Introduction: A Modern Introduction*. Cengage Learning, Boston, MA.

105. Grcar, J.F. (2011). Mathematicians of Gaussian elimination. *Notices Amer. Math. Soc.*, 58(6): 782–792.

106. Chu, M.T., Funderlic, R.E., and Golub, G.H. (1995). A rank-one reduction formula and its applications to matrix factorizations. *SIAM Rev.*, 37(4): 512–530.

107. Press, W.H. (2007). *Numerical Recipes: The Art of Scientific Computing*, 3rd ed. Cambridge University Press, Cambridge, U.K.

108. Irons, B.M. (1970). A frontal solution program for finite element analysis. *Int. J. Numer. Meth. Eng.*, 2(1): 5–32.

109. Duff, I.S. and Reid, J.K. (1983). The multifrontal solution of indefinite sparse symmetric linear. *ACM Trans. Math. Software*, 9(3): 302–325.

110. Lucas, R.F. et al. (2011). Multifrontal computations on GPUs and their multi-core hosts. In: *Proceedings of High Performance Computing for Computational Science—VECPAR 2010*, pp. 71–82.

111. Saad, Y. (2003). *Iterative Methods for Sparse Linear Systems*. Society for Industrial and Applied Mathematics, Philadelphia, PA.

112. Nocedal, J. and Wright, S.J. (2006). *Numerical Optimization*, 2nd ed. Springer-Verlag, Heidelberg.

113. Sherman, J. and Morrison, W.J. (1949). Adjustment of an inverse matrix corresponding to changes in the elements of a given column or a given row of the original matrix. *Ann. Math. Stat.*, 20: 621.

114. Teng, C. (2014). Variational Asymptotic Method for Unit Cell Homogenization of Thermomechanical Behavior of Composite Materials, PhD dissertation, Utah State University, Logan.

115. Chamis, C. and Sendeckyj, G. (1968). Critique on theories predicting thermoelastic properties of fibrous composites. *J. Comp. Mater.*, 2(3): 332–358.

116. Peterson, G. and Fletcher, L. (1987). A review of thermal conductivity in composite materials. *Rep. AIAA*, 87: 1586.

117. Dvorak, G.J. (1991). Plasticity theories for fibrous composite materials. In: *Metal Matrix Composites, Mechanisms, and Properties* (Everett, R.K. and Arsenault, R.J., Eds.), Vol. 2, pp. 1–77. Academic Press, Boston, MA.

118. Noor, A.K. and Shah, R.S. (1993). Effective thermoelastic and thermal properties of unidirectional fiber-reinforced composites and their sensitivity coefficients. *Compos. Struct.*, 26(1): 7–23.

119. Agarwal, B.D., Broutman, L.J., and Chandrashekhara, K. (2006). *Analysis and Performance of Fiber Composites*. John Wiley & Sons, New York.

120. Greszczuk, L. (1965). Thermoelastic properties of filamentary composites. In: *6th AIAA Structures and Materials Conference Proceedings*, pp. 285–290.

121. Abolin'sh, D. (1965). Compliance tensor for an elastic material reinforced in one direction. *Polym. Mech.*, 1(4): 28–32.

122. Chamis, C.C. (1983). Simplified composite micromechanics equations for hygral, thermal and mechanical properties. *SAMPE Q.*, 15: 14–23.

123. Caruso, J.J. and Chamis, C.C. (1986). Assessment of simplified composite micromechanics using three-dimensional finite-element analysis. *J. Compos. Technol. Res.*, 8: 77–83.

124. Hopkins, D.A. and Chamis, C.C. (1988). A unique set of micromechanics equations for high-temperature metal matrix composites. In: *Testing Technology of Metal Matrix Composites*, STP 694 (DiGiovanni, P.R. and Adsit, N.R., Eds.), pp. 159–176. American Society for Testing and Materials, Philadelphia, PA.

125. Jones, R.M. (1998). *Mechanics of Composite Materials*. CRC Press, Boca Raton, FL.

126. Christensen, R.M. (2012). *Mechanics of Composite Materials*. Dover, New York.

127. Mital, S., Murthy, P., and Chamis, C. (1995). Micromechanics for ceramic matrix composites via fiber substructuring. *J. Compos. Mater.*, 29(5): 614–633.
128. Eshelby, J.D. (1957). The determination of the elastic field of an ellipsoidal inclusion, and related problems. *Proc. R. Soc. Lond. Ser. A. Math. Phys. Sci.*, 241(1226): 376–396.
129. Hershey, A. (1954). The elasticity of an isotropic aggregate of anisotropic cubic crystals. *J. Appl. Mech. Trans. ASME*, 21(3): 236–240.
130. Kröner, E. (1958). Berechnung der elastischen Konstanten des Vielkristalls aus den Konstanten des Einkristalls. *Zeitschrift für Physik*, 151(4): 504–518.
131. Hill, R. (1965). A self-consistent mechanics of composite materials. *J. Mech. Phys. Solids*, 13(4): 213–222.
132. Hill, R. (1965). Theory of mechanical properties of fibre-strengthened materials. III. Self-consistent model. *J. Mech. Phys. Solids*, 13(4): 189–198.
133. Budiansky, B. (1965). On the elastic moduli of some heterogeneous materials. *J. Mech. Phys. Solids*, 13(4): 223–227.
134. Budiansky, B. (1970). Thermal and thermoelastic properties of isotropic composites. *J. Comp. Mater.*, 4(3): 286–295.
135. Laws, N. (1973). On the thermostatics of composite materials. *J. Mech. Phys. Solids*, 21(1): 9–17.
136. Banerjee, B. and Adams, D.O. (2003). Micromechanics-based prediction of thermoelastic properties of high energy materials. In: *Constitutive Modeling of Geomaterials* (Ling, H.I. et al., Eds.), pp. 158–164. CRC Press, Boca Raton, FL.
137. Mori, T. and Tanaka, K. (1973). Average stress in matrix and average elastic energy of materials with misfitting inclusions. *Acta Metall.*, 21(5): 571–574.
138. Benveniste, Y. (1987). A new approach to the application of Mori–Tanaka's theory in composite materials. *Mech. Mater.*, 6(2): 147–157.
139. Weng, G. (1990). The theoretical connection between Mori–Tanaka's theory and the Hashin–Shtrikman–Walpole bounds. *Int. J. Eng. Sci.*, 28(11): 1111–1120.
140. Berryman, J.G. and Berge, P.A. (1996). Critique of two explicit schemes for estimating elastic properties of multiphase composites. *Mech. Mater.*, 22(2): 149–164.
141. Kuster, G.T. and Toksöz, M.N. (1974). Velocity and attenuation of seismic waves in two-phase media. Part I. Theoretical formulations. *Geophysics*, 39(5): 587–606.
142. Norris, A. (1989). An examination of the Mori–Tanaka effective medium approximation for multiphase composites. *J. Appl. Mech.*, 56(1): 83–88.
143. Benveniste, Y., Chen, T., and Dvorak, G. (1990). The effective thermal conductivity of composites reinforced by coated cylindrically orthotropic fibers. *J. Appl. Phys.*, 67(6): 2878–2884.
144. Böhm, H.J. and Nogales, S. (2008). Mori–Tanaka models for the thermal conductivity of composites with interfacial resistance and particle size distributions. *Compos. Sci. Technol.*, 68(5): 1181–1187.
145. Aboudi, J. (1989). Micromechanical analysis of composites by the method of cells. *Appl. Mech. Rev.*, 42(7): 193–221.
146. Aboudi, J. (1982). A continuum theory for fiber-reinforced elastic-viscoplastic composites. *Int. J. Eng. Sci.*, 20(5): 605–621.
147. Paley, M. and Aboudi, J. (1992). Micromechanical analysis of composites by the generalized cells model. *Mech. Mater.*, 14(2): 127–139.
148. Yu, W. and Tang, T. (2007). Variational asymptotic method for unit cell homogenization of periodically heterogeneous materials. *Int. J. Solids Struct.*, 44(11): 3738–3755.
149. Teng, C., Yu, W., and Chen, M.Y. (2012). Variational asymptotic homogenization of temperature-dependent heterogeneous materials under finite temperature changes. *Int. J. Solids Struct.*, 49(18): 2439–2449.
150. Pal, D. et al. (2014). An integrated approach to additive manufacturing simulations using physics based, coupled multi-scale process modeling. *J. Manuf. Sci. Eng.*, 136(6): 061022.

9 Advances in Additive Manufacturing

Effect of Process Parameters on Microstructure and Properties of Laser-Deposited Materials

Mohsen Eshraghi and Sergio D. Felicelli

CONTENTS

ABSTRACT

The layer-by-layer method of additive manufacturing (AM) offers many advantages over other conventional manufacturing methods. However, various issues have yet to be addressed, especially related to materials and properties, before AM technologies are more widely adopted. In order to take advantage of AM technologies, engineers and scientists must have a good understanding of material microstructures, defect formation, and proper adjustment of process parameters to achieve the properties required for the intended applications. In this chapter, some of the recent numerical and experimental studies on the laser deposition of metallic materials are presented. The focus is placed on how process parameters can affect microstructural evolution, defect formation, and properties of the laser-deposited materials. The topics include thermal behavior during layer depositions, residual stresses, porosity formation, and solidification microstructure. These studies show that the combination of numerical simulations with experimental validation can help develop a better understanding of the AM processes and their effect on material structures and properties. They can also contribute to optimize the AM processes in order to successfully implement these technologies for commercial applications.

9.1   INTRODUCTION

Additive manufacturing (AM) has become an important technology for building complex shape components with high strength and ductility directly from a computer-aided design (CAD) solid model. This technology has the potential to be an energy-efficient alternative or complement to current manufacturing processes, including forging and investment casting to fabricate components, by reduction of material waste, process steps, and environmental impact. If the current challenges of the technologies (e.g., microstructure homogeneity, geometry accuracy, porosity defects, throughput) are overcome, the additive manufacturing technology can save time, material, energy, and money for manufacturers. Even when secondary machining is required, the near net shape produced by the processes reduces machining operations to minor surface cleanup. This cascades into savings on cutting fluid processing and disposal energy costs, chip collection and disposal, cutting tool insert manufacturing, machine tool operation, and facility energy utilization. The short-term result would be the capability to test form, fit, and function of castings earlier in the design cycle. Fatigue, thermally induced stresses, and performance can be validated before committing to conventional casting/forging tooling. This could potentially save several hundred megawatts and potentially gigawatts per application. Given the thousands of new castings produced annually in the United States alone, each one requiring multiple iterations, it is estimated that this would save one to ten trillion BTU/yr. Globally, the same impact could be expected.[1]

If the challenges (shrink and throughput) of additive fabrication can be overcome, enabling equivalent replacement of existing components, then one can envision leveraging the capability of the additive processes to fabricate the "unmanufacturable" equivalent. This would reduce the weight of the component by using truss-like structures that were

1. Too expensive to machine conventionally
2. Too complex to cast
3. Not part of the basic geometric design rules

The Laser Engineered Net Shaping (LENS™) process is a particular example of AM technologies that show considerable potential for rapid manufacturing and repair applications. The LENS process utilizes a laser, metallic powder, and a CAD solid model to fabricate three-dimensional (3D) fully dense and fully functional components.[2,3] It has been applied to fabricate components for a large class of metal alloys, such as low-alloy steels,[4] stainless steel,[5,6] nickel-based alloys,[7,8] and titanium alloys.[9,10] The advantage of the LENS process is that it allows the fabrication of complex three-dimensional components with high strength and ductility directly from a CAD solid model. The process can reduce the cost and time significantly through a one-step operation. Unlike other laser processing techniques, LENS uses low-power lasers that produce a very small heat-affected zone (HAZ). In typical applications to steel alloys, the laser power is 300 to 500 W, the platform traveling velocity is ~8 mm/s, and the volume of the liquid pool is ~0.5 mm^3.[11] These conditions can produce large temperature gradients (100 to 500 K/mm) and cooling rates (100 to 6000 K/s), with microstructures that can approach those of rapid solidification.[7,12]

In spite of their many attractive features, LENS and other AM methods share similar difficulties in mastering the technology. Process control, repeatability, surface roughness control, residual stress, porosity, lack of fusion, and fatigue performance are among the issues affecting these techniques. This is partly because the rapid emergence of these technologies has not been accompanied by enough study of the fundamentals behind them.

The layer-by-layer fabrication approach of additive manufacturing technologies such as direct metal laser sintering (DMLS) and laser metal deposition (LMD) (e.g., LENS) avoids many of the geometric constraints imposed by traditional manufacturing processes. To take advantage of AM technologies, engineers not only must know the manufacturing capabilities and constraints of AM but must also be able to synthesize their economic and environmental impacts on a manufacturing value chain.

Generally speaking, AM technologies provide new opportunities for customization, improvements in product performance, multifunctionality, and lower overall manufacturing costs due to their unique capabilities. The AM technologies greatly help manufacturers to tailor their designs to best utilize manufacturing process capabilities to achieve the desired performance and life-cycle objectives.

The combination of advanced computational and experimental methods provides a new route to materials discovery. The focus would be on the materials and processes—that is, understanding how process parameters produce material structures, how those structures give rise to material properties, and how to tune materials and processes for a given application. A National Academies report[13] describes the need for using multiscale materials modeling to capture the process, structures, properties, and performance of a material. Computational materials engineering reduces the costs of failure and redesign in the production of high-performance components for defense, aerospace, automotive, and other commercial applications. By combining bedrock computational physics and informatics with systematic experiments and advanced manufacturing, we can reduce the cost, risk, and cycle time for new product development. Such an approach drastically reduces the time-consuming cycle of experimentation and testing needed to move new materials to the marketplace.[14] Numerical models have been undertaken by several authors to simulate the thermal behavior and microstructure evolution in layer-additive deposition processes. Although these models can provide some insight into how processing parameters can be manipulated to obtain favorable metallurgical structures and mechanical properties, the fabrication of good quality parts remains largely an art, where many tricks and recipes are zealously kept undisclosed by independent manufacturers. Despite claims of success of some material deposition methods in obtaining "good quality" parts, the fact is that the capability of microstructure control is very limited, the deviation from learned practice produces unpredictable results,[15] and a closer examination of so-claimed good parts reveals porosity and other defects that may seriously affect performance in many applications.

In this chapter, some of the recent numerical and experimental studies on the laser deposition of metallic materials will be presented. The focus will be on how process parameters affect microstructural evolution, defect formation, and properties. Six main topics will be discussed: experimental study of thermal behavior in laser deposition process, process modeling of laser deposition, residual stresses in laser deposition process, porosity in laser-deposited materials, solidification microstructure in laser-deposited materials, and potential benefits of laser deposition processes for manufacturing industries.

9.2 EXPERIMENTAL STUDY OF THERMAL BEHAVIOR IN LASER DEPOSITION PROCESSES

Several issues are involved in the laser deposition process, including repeatability, geometry accuracy, surface roughness, and microstructural uniformity.[16–18] Several studies have been done to address these issues and investigate the effects of process parameters on the properties of the fabricated parts. Laser power, travel velocity, and powder flow rate are the main parameters involved in the process that influence the thermal history of each point in the deposited material. Accordingly, they control the molten pool size, microstructural evolution, solid-state phase transformations, pore formation, distortion, residual stresses, and the final physical and mechanical properties in the manufactured parts. Therefore, it is important to study the thermal behavior during the laser deposition process in order to investigate the relationship between the process parameters and the resultant properties, and consequently optimize the process.

The thermal phenomena during the laser deposition processes have been studied both numerically and experimentally by various researchers. On the experimental side, two different thermal measurement methods—radiation pyrometers[19–23] and thermocouples[19,20]—have been employed. Thermocouples are used to measure the temperature history as a function of time at locations far from the molten pool,[19] whereas radiation pyrometers are used to measure the temperature inside or nearby

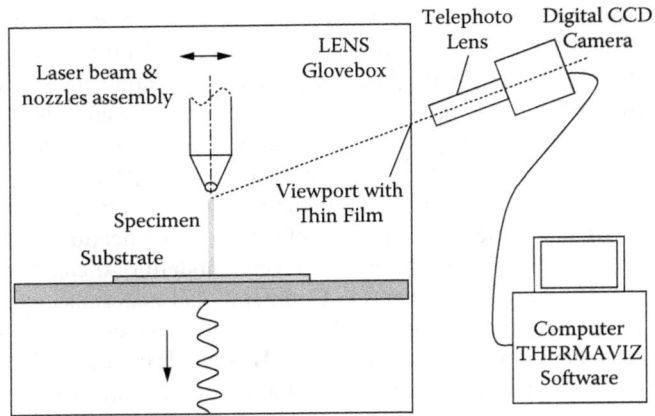

(A)

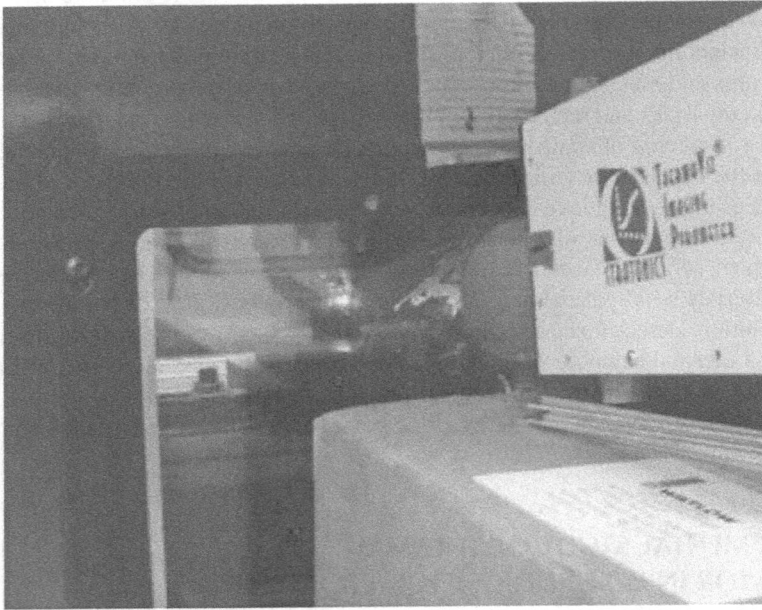

(B)

FIGURE 9.1 (A) Schematic and (B) side view of the experimental setup.[24]

the pool. Charge-coupled device (CCD) cameras have been used to capture thermal images from the molten pool.[11,17,18,21–23] The outcome can provide the temperature profile, maximum temperature, temperature gradient, cooling rate, and geometry of the molten pool. Because studies have found that laser power is the most influential factor for pool temperature, a closed-loop control based on the regulation of laser power was implemented. The results showed that the closed-loop control system can improve the geometry accuracy and microstructural uniformity of the deposited parts.[16–18,21]

In Wang et al.,[24] a THERMAVIZ™ two-wavelength imaging pyrometer system was used to capture images from the molten zone and surrounding area. The temperature was determined from the ratio of the relative radiation intensities at two different wavelengths and independent of emissivity, which makes the temperature measurements more accurate compared to the one-wavelength systems. Figure 9.1 shows the experimental setup. The powder nozzles and laser assembly move in the x-y plane. After

FIGURE 9.2 Typical process of depositing a single wall using LENS.[24]

each pass in the *x-y* plane, the substrate and the part move down and the next layer is deposited. To capture a side view of the molten zone, a digital Si-based CCD camera is mounted outside the glovebox at an angle of approximately 15° with the *x-y* plane. The pyrometer can record a dynamic range of temperature between 1450°C and 1860°C and captures one frame every two seconds. The two-wavelength intensity images and the temperature images are then transferred to a computer.

Thermal experiments were conducted using the LENS 850 M machine with a 3-kW IPG laser and SS410 stainless steel powder. The thermal measurements were performed on nine samples using different laser powers and travel velocities. Figure 9.2 illustrates a typical process of depositing a single wall using LENS. Figure 9.3 shows two specimens that were deposited using (1) laser power of 300 W with laser travel velocity of 2.5 mm/s, and (2) laser power of 600 W with laser travel velocity of 4.2 mm/s. It can be seen that a uniform width was obtained in both cases. However, higher power resulted in a thicker wall, because more powder was melted and deposited during the process.

The *in situ* short- and long-wavelength images and temperature distributions around the molten zone of the case deposited by the 600-W power and 2.5-mm/s travel velocity are presented in Figure 9.4. The shape of the molten pool corresponds to the 1450°C isotherm that is the melting point of SS410 and is depicted by dashed lines. The molten pool is approximately 1.0 mm long by 0.5 mm wide with a maximum temperature of about 1650°C.

The maximum temperature gradient in the *x* direction is observed at the solid–liquid interface and is about 400°C/mm. Due to the substrate heat sink, heat dissipates in the depth direction much faster than in the travel direction, leading to a much higher maximum temperature gradient of 1000°C/mm in the *y* direction. The cooling rate at the solid–liquid interface can be calculated using the temperature gradient and travel speed and ranges from 1000°C/s to 2500°C/s.

Increasing the laser power increases the size of the molten zone and the maximum temperature, but the temperature profiles and gradients outside the pool are similar for different laser powers. There is a significant superheating in the molten zone. The maximum temperature increases with increasing laser power and decreasing travel velocity. Although the effect of travel speed on the maximum temperature is not significant, a slower travel speed leads to a higher heat input and a higher maximum temperature. Because of the effect of the substrate as a heat sink, the size of the molten zone increases as the number of deposited layers increases at a constant laser power. This shows that having a closed-loop control for the molten pool is essential in order to ensure microstructural uniformity and dimensional accuracy. The depth of the molten pool is not much affected by the number of layers but increases with laser power and slightly with travel velocity. The maximum cooling rate increases with increasing travel velocity and decreasing laser power. The highest cooling rate was found to be around 6000°C/s which corresponds to the highest travel velocity and the lowest laser power.

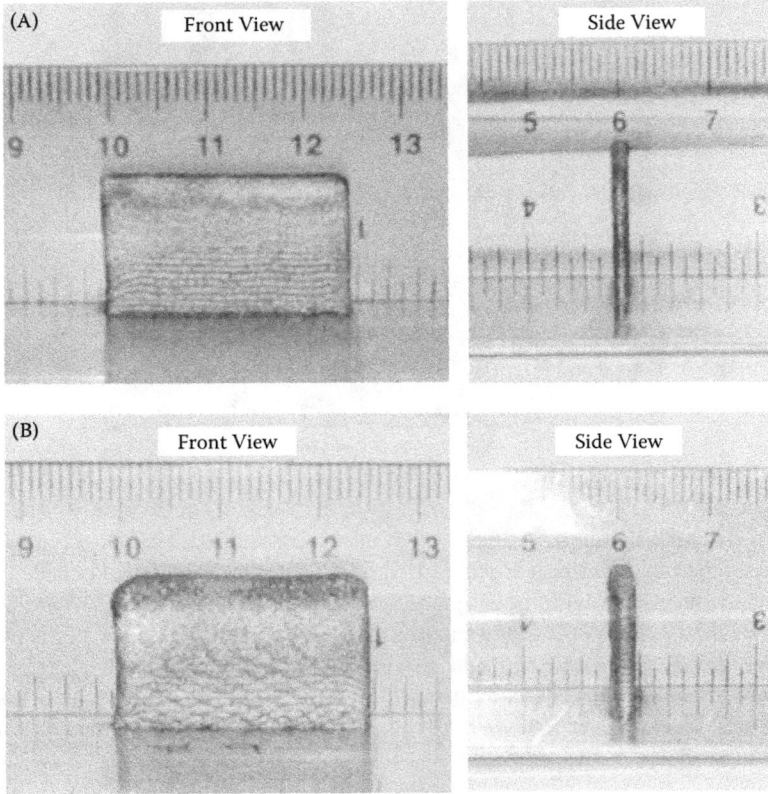

FIGURE 9.3 Single-wall specimens deposited using (A) laser power of 300 W with laser travel velocity of 2.5 mm/s, and (B) laser power of 600 W with laser travel velocity of 4.2 mm/s.[24]

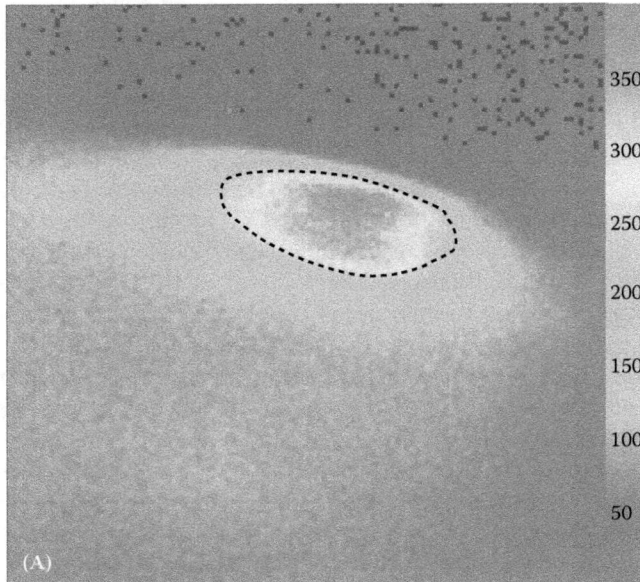

FIGURE 9.4 (A) Short-wavelength intensity image; (B) long-wavelength intensity image; and (C) temperature image and molten pool size for the case deposited by 600-W power and 2.5-mm/s velocity.[24] *(Continued)*

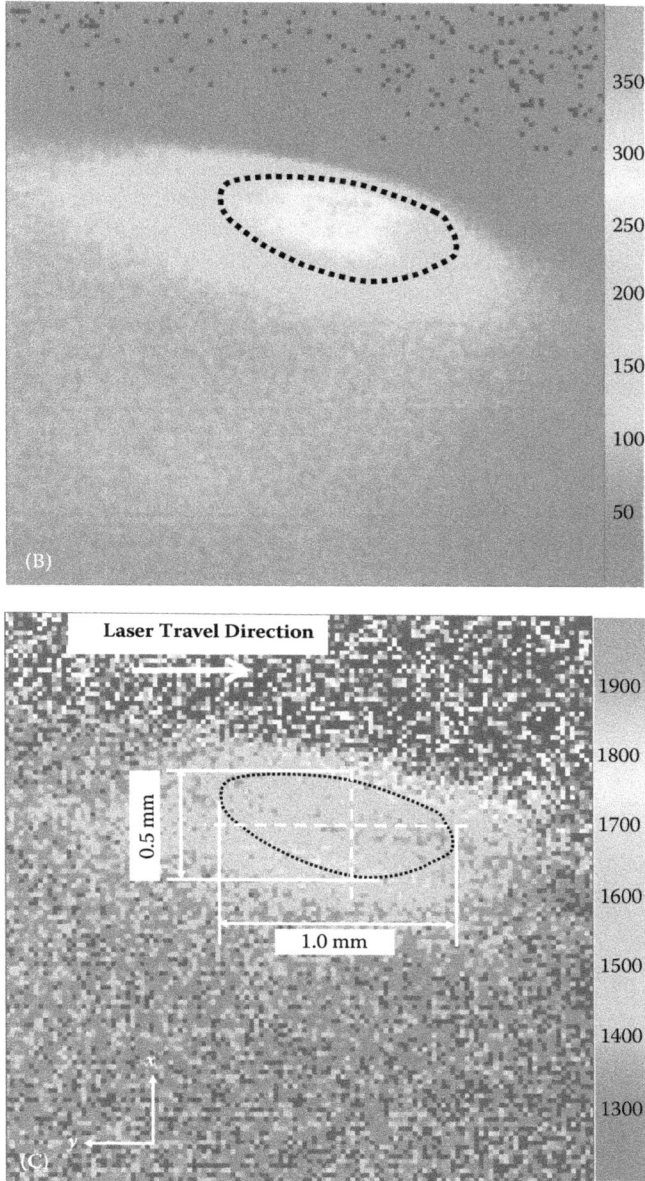

FIGURE 9.4 (Continued) (A) Short-wavelength intensity image; (B) long-wavelength intensity image; and (C) temperature image and molten pool size for the case deposited by 600-W power and 2.5-mm/s velocity.[24]

9.3　PROCESS MODELING OF LASER DEPOSITION

As every point in the deposited part experiences several thermal cycles, which is not common in traditional manufacturing processes, the microstructure of the laser-deposited materials is very complex. In order to analyze microstructural evolution, and predict mechanical properties and distortions in the parts, it is important to know the thermal history of each point of the deposited materials, something that is not experimentally possible due to the difficulty of temperature measurement during laser deposition processes. On the other hand, numerical simulations can provide a better understanding of the thermal behavior of laser-deposited materials. However, it is essential to calibrate and validate the numerical models in order to obtain reliable predictions. The validated numerical models can be coupled to metallurgical and mechanical models in order to predict microstructural evolution, phase transformations, residual stresses, and mechanical properties. The finite element method (FEM) has been used to simulate the thermal profiles in the molten pool[25,26] and predict the pool size.[27–29] Thermal profiles have been predicted in multilayer depositions of a single wall using LENS.[24,30–32]

One of the most important parameters for maintaining the optimal build conditions is the size of the molten pool.[33] The effect of process parameters on the size of the molten pool has been studied by both experiments[27] and simulations.[28] If the laser traveling speed is kept constant, the molten pool geometry will depend on the heat input distribution. To simulate the laser deposition of SS410 stainless steel, a finite element model was developed using SYSWELD software. Details of the numerical model are available in Wang et al.[31] The effect of solid-state phase transformations was taken into account by using continuous cooling transformation (CCT) diagrams and temperature-dependent material properties. Ten passes of laser deposition (each 10.0 mm long, 0.5 mm thick, and 1.0 mm wide) were performed to build a single-wall structure. The substrate was a plate with a length of 20.0 mm, thickness of 5.0 mm, and width of 10.0 mm. The laser beam moves the path (left to right) for each pass with a travel speed of 7.62 mm/s.

To simulate the heat input distribution for a conical moving heat source, a Gaussian heat flux profile was considered. Not all of the energy generated by the laser beam is absorbed by the deposited material. Studies[34] have shown that the laser energy transfer efficiency is about 30 to 50%. Various factors may contribute to the loss of laser energy, such as reflection from the deposited metal, absorption by in-flight powder, absorption by evaporating metal from the pool, and dependence of the absorptivity of the material on temperature and laser wavelength. It should be noted that there are other phenomena such as evaporation or melt convection that are not considered in this study.

The temperature contour for the time when the laser beam is in the middle of 10th layer is shown in Figure 9.5A. The simulation and experimental results for temperature and cooling rate profiles are shown in Figures 9.5B and 9.5C, respectively. Profiles are for the top layer and are plotted from the center of the molten pool in the travel direction of the fabricated part, which is opposite to the

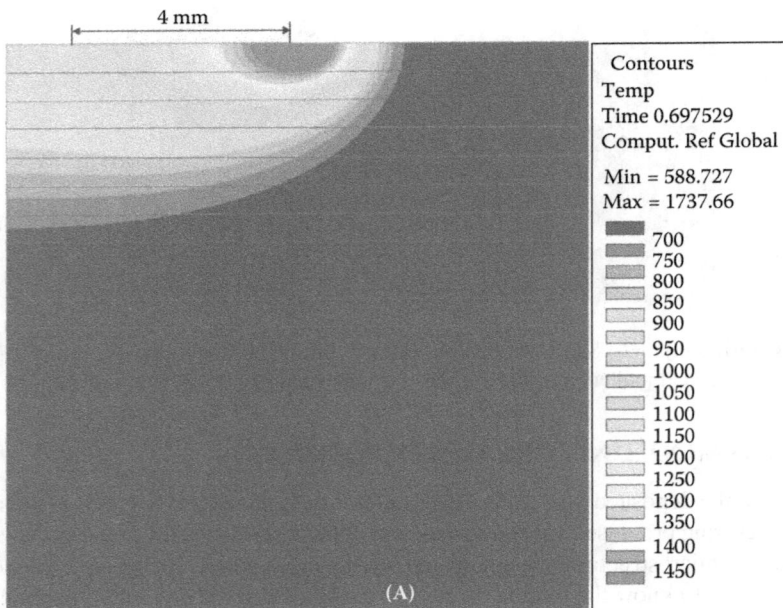

FIGURE 9.5 (A) Predicted temperature distribution during laser deposition of SS410, and comparison of experimental and modeling results for (B) temperature profile and (C) cooling rate.[31]   *(Continued)*

moving direction of the laser beam. The absorbed laser power and travel speed used in the simulations were 100 W and 7.62 mm/s, respectively. The initial temperature for the substrate and the deposited part was assumed to be 600°C before the 10th layer was deposited. The experimental results are for a nominal laser power of 275 W.[11] This means that a laser energy transfer efficiency of 36.4% has been considered for the simulations which is consistent with the studies[34] reporting 30 to 50% efficiency. The figures show a good agreement between experimental and simulation results with an error of less than 8%. However, as we get closer to the pool, the simulations show a higher cooling rate compared to the experiments. The highest predicted cooling rate is for the area next to the solidification interface. There is very little experimental data available for this area, which does not show a well-defined trend and makes comparisons difficult.

After calibration, the model was used to simulate ten passes of the laser deposition process. The laser power was adjusted in a way to achieve a steady molten pool size for each pass. Considering a 36.4% laser energy transfer efficiency, the nominal laser power applied to each pass was obtained. As more layers are deposited, the laser power needed to maintain a steady molten pool size decreases. To compensate for the heat dissipation by the cold substrate, a higher laser power is required to deposit

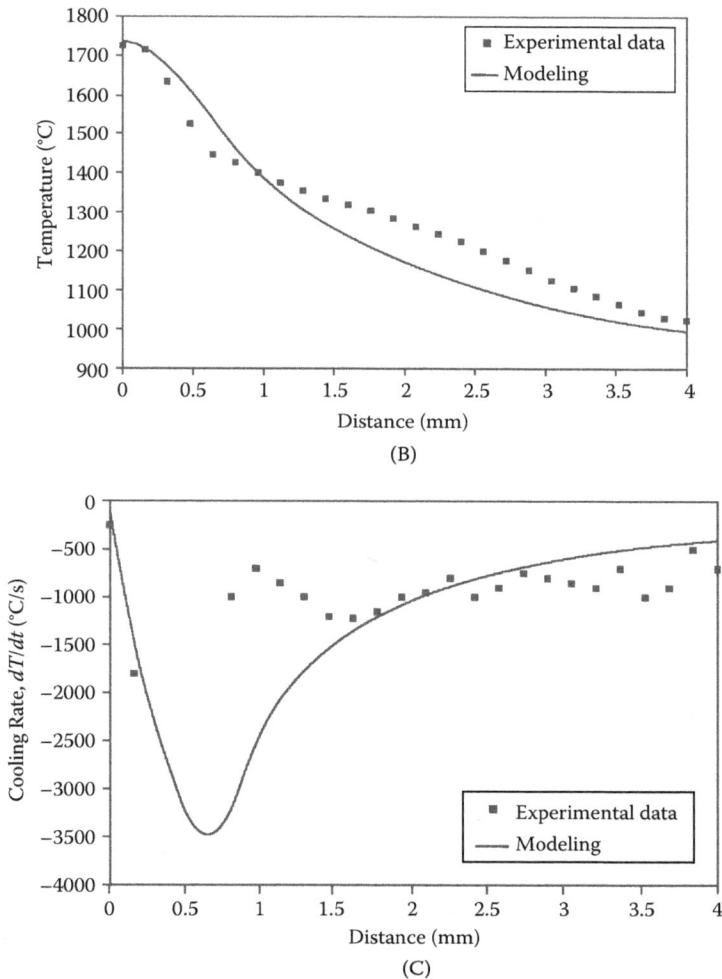

FIGURE 9.5 (Continued)   (A) Predicted temperature distribution during laser deposition of SS410, and comparison of experimental and modeling results for (B) temperature profile and (C) cooling rate.[31]

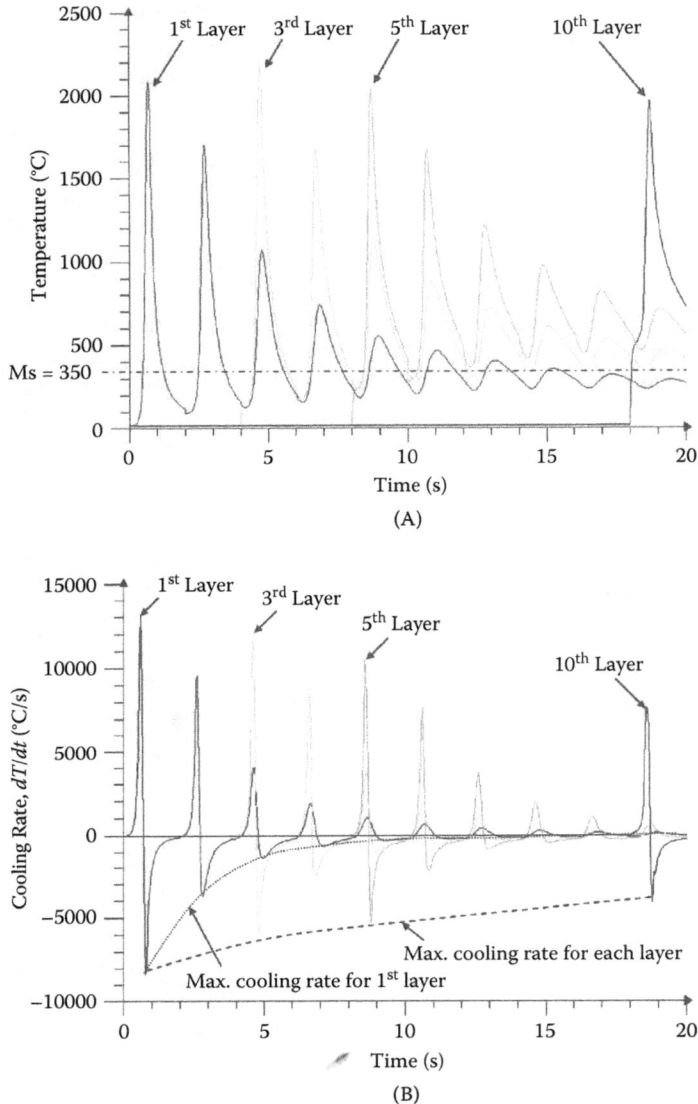

FIGURE 9.6   (A) Thermal cycles, and (B) cooling rates for the mid-points of the 1st, 3rd, 5th, and 10th layers.[31]

the first few layers. However, the required laser power decreases for the subsequent layers, as the part becomes hotter. The laser power required to maintain a steady molten pool decreases linearly after the fifth layer, showing a transient behavior that only affects the initial layers. During each pass, the laser melts approximately one layer and a half and the size of the molten pool is almost constant for all layers. This means that the temperature distributions, the resultant phase transformations, and the microstructure that forms during each pass are rather uniform in the final part.

The thermal histories for the points located at the center of the 1st, 3rd, 5th, and 10th layers are plotted in Figure 9.6A. During the first pass, the temperature reaches a maximum of 2100°C and then cools down to a temperature of around 100°C at $t = 2$ s. Because of the rapid cooling, the solidification and solid-state phase transformations that occur during the first pass should result in a hard martensitic microstructure. However, the following deposition passes reheat the previous layers to higher temperatures where martensite can be tempered. Each layer experiences cycles of heating and cooling which affect the microstructure and mechanical properties of the deposited parts. The subsequent layers go through similar thermal cycles with comparable maximum temperatures. However, after the 5th layer, the temperature never drops below the starting temperature for the martensitic transformation which is 350°C for SS410.[35] For the upper layers, martensitic transformation occurs at the end of the process, resulting in greater hardness at the top layers compared to the softer tempered martensite microstructure of the lower layers. This agrees with the finding of other researchers.[20,30]

Cooling rates for the 1st, 3rd, 5th, and 10th layers are plotted in Figure 9.6B. A positive value means that the material is heating up and a negative value means that it is cooling down. The maximum cooling rate for the first layer during the first pass is about −8000°C/s. The maximum cooling rate during the next passes decreases for the first layer. Still, even after three passes, the first layer experiences a maximum cooling rate about −1000°C/s. The subsequent layers undergo cooling curves similar to the first layer. However, because of the continuous heating of the substrate and previous layers, the maximum cooling rate decreases as more layers are deposited.

The influence of laser travel speed on the thermal behavior of the layer deposits was also studied. It is assumed that the powder feed rate is adjusted in such a way that the layer thickness remains constant for different travel speeds. The model is then used to find the laser power that keeps a steady molten pool size. To obtain the same molten pool size, a higher laser power should be applied as the travel speed increases. Higher temperatures are reached when higher travel speeds are used because higher laser powers are applied for those cases. Higher travel speeds lead to much higher cooling rates but somewhat lower peak temperatures. In addition to higher production rates, the higher travel speeds are expected to yield a finer microstructure due to the higher cooling rates. However, the influence of travel speed on other elements such as the formation of pores or incomplete melting of powder should be investigated as well. The molten pool is somewhat elongated at the highest speed and barely enters the previous layer. In order to obtain the penetration observed at lower travel speeds, a higher power should be applied at higher speeds.

9.4 RESIDUAL STRESSES IN LASER DEPOSITION PROCESS

Previous studies have shown that the amount of macroscopic residual stress in the laser-deposited parts is significant;[20,36] however, the effect of process parameters on the magnitude and distribution of residual stresses within the parts is not yet clear. In Pratt et al.,[37] the High Flux Isotope Reactor Neutron Residual Stress Mapping Facility (NRSF2) at Oak Ridge National Laboratory was used to study residual stresses in seven single-wall AISI 410 laser deposits. The samples were 22 to 38 mm long, 15 mm wide, and 1 to 3 mm thick. The process parameters used for deposition of the samples are listed in Table 9.1.

TABLE 9.1
LENS Sample Process Parameters

Sample No.	Laser Power (W)	No. of Layers	Laser Speed (mm/s)	Sample Length (mm)	Powder Flow Rate (cm³/s)
1	300	25	2.54	38.1	37.85
2	300	25	2.54	22.1	37.85
3	300	25	4.23	25.4	44.16
4	600	25	2.54	25.4	37.85
5	600	25	4.23	25.4	44.16
6	450	25	2.54	25.4	37.85
7	450	25	4.23	25.4	50.47

Source: Pratt, P. et al., *Metall. Mater. Trans. A*, 39(13), 3155–3163, 2008.

TABLE 9.2

Maximum and Average Compressive Stress in Samples

Sample No.	Max Stress (MPa)	Average Stress (MPa)	Laser Power (W)	Laser Speed (mm/s)
1	−379.75	−151.43	300	2.5
2	−70.65	−39.88	300	2.5
3	−73.08	−43.77	300	4.2
4	−503.16	−266.13	600	2.5
5	−199.14	−118.31	600	4.2
6	−248.69	−143.35	450	2.5
7	−67.00	−41.78	450	4.2

Source: Pratt, P. et al., *Metall. Mater. Trans. A*, 39(13), 3155–3163, 2008.

Past studies[36] showed that strain in laser-deposited samples was only significant in the z direction of the samples which corresponds to the growth direction in the manufacturing process. Residual stress measurements in welds also yielded similar results.[38] Measurements of strains for AISI 410 laser deposits showed that residual stresses are not considerable in x and y directions compared to the z direction. The maximum and average stress values (all compressive) measured in each of the samples are listed in Table 9.2. The results do not show any certain trends between process parameters and the measured residual stresses. At least for the range of process parameters used in the study, there is no evidence showing that residual stresses can be reduced by adjusting laser power and travel speed. However, when looking at average values, there seems to be a trend to overall lower values of the stress for lower laser powers. To investigate this observation, a numerical model was employed to simulate the process and predict residual stresses for parameters used in the experiments.

The model was based on the thermal model presented in the previous sections.[24,26,31,32] The mechanical model was based on the SYSWELD's material library and was coupled to the thermal model in order to calculate mechanical behavior during laser deposition and predict residual stresses in the final part. The mechanical model used an additive constitutive relation for strains due to elastic, thermal, and plastic deformation, including transformation-induced plasticity. Details about the mechanical model can be found in Wang et al.[39]

The laser power was optimized as described above to compensate for the heat-sink effect of the substrate and get a steady molten pool size. For three values of travel speed (2.5, 4.2, and 8.5 mm/s), the corresponding laser powers were 254, 285, and 344 W, respectively. Figure 9.7 shows numerical predictions for residual stress distribution, σ_z, at room temperature for different samples. Figure 9.7A shows the stress profile along the laser travel direction for the third deposited layer, and Figure 9.7B shows the profile at the vertical center line along the plate measured from the free end toward the substrate. The average value of the overall stress level is also shown in Figure 9.7A. A trend similar to the experimental results is observed in the simulation results.

In the horizontal direction, a compressive stress in the center region of the plate is observed, shifting to tensile or less compressive values near the side edges. In the vertical direction, the magnitude of the compressive stress increases as the distance from the top free end increases and becomes less compressive at locations close to the substrate. These results are similar to what is observed in experiments. A small increase can be seen at the center of the stress profile along the horizontal direction (Figure 9.7A), which is comparable to some of the measured samples. Like experiments, the simulation results presented in Figure 9.7 do not show any specific trends for how residual stress may change with changing laser power and travel speed.

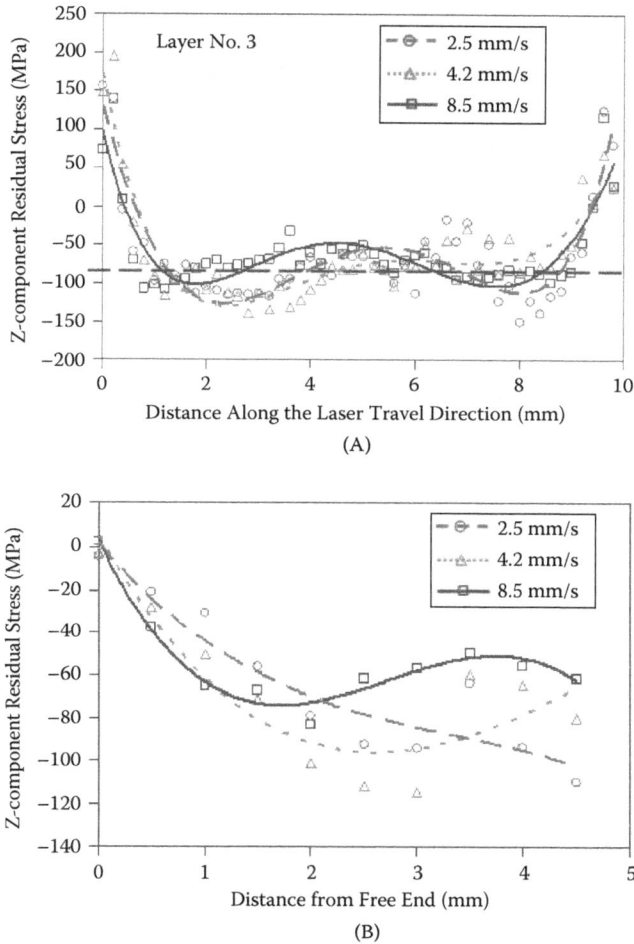

FIGURE 9.7 Z-component residual stress along (A) the travel direction and (B) the vertical direction as a function of laser travel speed.[37]

9.5 POROSITY IN LASER-DEPOSITED MATERIALS

Although the laser deposition process yields a small heat-affected zone and high cooling rates, significant porosity may form during the process, either at the interface of two layers or within the bulk material.[40] Process parameters can be adjusted to minimize the pore formation. The interlayer pores usually have an irregular shape that elongates along the layer boundary. As one of the possible sources, non-optimum combinations of process parameters can lead to lack of fusion in the deposited parts and consequently form these pores.[40] By careful adjustment of process parameters such as laser power and travel speed the tendency to form these pores can be reduced.[32,41] Another possible contributing factor is the mechanical properties of the underlying layer. Studies[42] have shown that low ductility of the deposited material leads to more interlayer porosity. The mechanism for formation of interlayer pores is not understood yet and it is very difficult to control them.[15]

Enclosed gas is believed to be another main source of porosity. Enclosed gas may come from contamination, entrainment during turbulent impact of particles into the molten pool, contamination by powder-feed gases, gases contained within the powder particles, or vaporization of high vapor pressure alloy constituents. Due to the rapid solidification rates, the dissolved gases may not

TABLE 9.3
Process Parameters for AISI 410 Single-Walled Deposits

Sample No.	Laser Power (W)	Laser Speed (mm/s)	Layer Thickness (mm)	Powder Flow Rate (g/min)	No. of Layers	Sample Length (mm)
S1	600	2.5	0.5	0.6	25	25.4
S2	450	2.5	0.5	0.6	25	25.4
S3	450	4.2	0.5	0.8	25	25.4
S4	450	8.4	0.5	1.3	50	38.1

Source: Wang, L. et al., *J. Manuf. Sci. Eng.*, 131(5), 051008, 2009.

have enough time to reach the surface of the molten pool and are trapped within the solid structure. Studies[40] have shown that initial porosity and gas content of the powders have a significant influence on formation of pores within the layers. Statistical analysis[43] suggests that the mass flow rate of the powder plays an important role in porosity formation.

The effect of process parameters and type of materials on porosity formation was studied in Wang et al.[44] AISI 410 martensitic stainless steel was used for the study. The powders were produced with gas atomization with a nominal size of 53 to 173 μm in diameter. Deposits were made for single-wall geometries in atmospheres with an oxygen level between 3 and 15 ppm in the chamber. The process parameters for samples are listed in Table 9.3. A LENS 750 machine was used to deposit all samples. Laser depositions were performed at Optomec, Inc. (Albuquerque, NM). To measure the porosity distributions, a Phoenix x-ray tomography system was used. The single-wall laser deposits were all made of AISI 410 stainless steel powders. The deposited walls were 22 to 38 mm wide, 15 mm long, and 1 to 3 mm thick, depending on the applied laser energy. As listed in Table 9.3, three process parameters were changed for deposition of these samples: laser power, travel speed, and powder flow rate.

The cross-sectional optical micrographs of samples S1 to S4 are shown in Figures 9.8 to 9.11, respectively. Spherical gas porosity with a random distribution and a size range of 50 to 130 μm is observed in samples S1, S2, and S3. However, pores bigger than 200 μm can be observed in S4. The laser travel speed for S4 was much higher in comparison with the other three samples, limiting the time available for the gas pores to move to the free surface. However, the pores may merge inside the bulk material and form larger pores with irregular shapes, as can be seen in Figure 9.11. This is also related to the large fraction of pores that are present, because only near-surface pores can escape the bulk material. This agrees with other studies using titanium-based powders.[42] It seems that laser power does not have a significant effect on the pore size and density, which might be due to the large values of laser power employed in this study.

9.6 SOLIDIFICATION MICROSTRUCTURE IN LASER-DEPOSITED MATERIALS

In order to understand the microstructure and properties of layer-additive deposited materials, it is essential to investigate the solidification and transport phenomena that take place in the molten pool during deposition. This is the first step to properly address the study of the complex phase transformations occurring during cooling in the solid state.[45] How can we predict the final microstructure if we do not know the starting condition right after solidification? It is clear that the continuum-type model is not adequate for this situation if we want to learn about the interaction of the pool with the forming solid structure. The growth of the solid phases, with dendritic, cellular, lamellar, or other structures, is significantly affected by the transport phenomena in the molten pool.[46–54]

Thermal history determines the microstructure, which in turn determines the mechanical properties. Many experimental studies have been performed to analyze the morphology of the microstructures. The experimental results show that fine microstructures form during the laser deposition because of the high cooling rates. Ghosh and Choi[55] came up with an equation that relates the

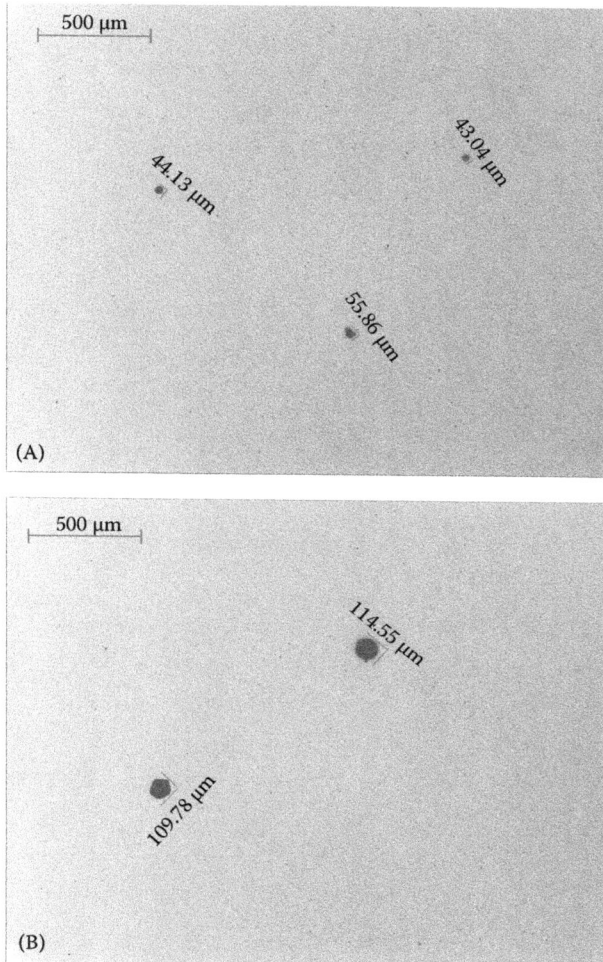

FIGURE 9.8 Optical photography of AISI 410 thin plate for sample S1 showing gas porosity in the transverse direction: (A) gas pores about 50 μm in diameter, and (B) gas pores about 100 μm in diameter.[44]

dendrite arm spacing (DAS) to the thermal behavior. There are very few studies[56,57] on numerical modeling of the solidification microstructure during the laser deposition process, none of which provides much detail about the dendrite growth phase. Yin and Felicelli[58] studied the simulation of the solidification microstructure during the laser deposition process. Dendritic morphologies, DAS, and the effect of process parameters were also studied. A finite element (FE)–cellular automaton (CA) model was employed to solve for transport equations and track the solidification interface.

By controlling the temperature gradient and cooling rate, different morphologies can be obtained.[59–61] The morphology that results during the LENS process can be equiaxed, columnar, or a mixture of both. Because the material experiences many thermal cycles with high cooling rates, it may undergo many phase transformations and the resulting microstructure is very complex. The last layer in the laser-deposited material usually has a different microstructure. The microstructure in the last layer is usually dendritic, whereas it can be cellular/dendritic structure in the other layers.[5,8,16,33,41,62–66] The DAS is greatly affected by cooling rate and influences the mechanical properties of the laser-deposited parts. The relationship between DAS and cooling rate has been studied experimentally. For 310 stainless steel in laser welding, a general relationship was found as $\lambda_2 = 25\varepsilon^{-0.28}$ between the cooling rate, ε, and the secondary DAS (SDAS), λ_2.[67]

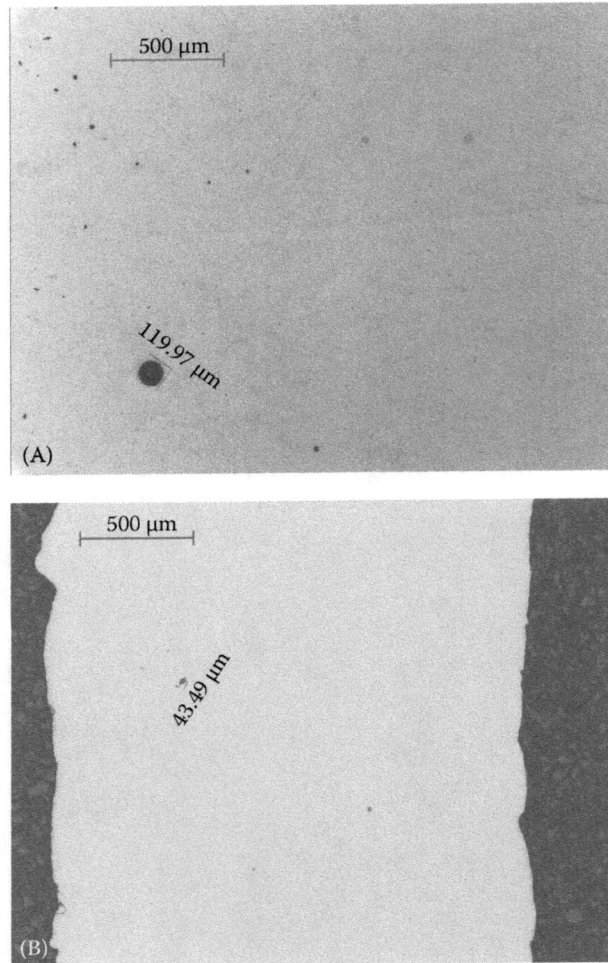

FIGURE 9.9   Optical photography of AISI 410 thin plate for sample S2 showing gas porosity: (A) transverse direction, and (B) longitude direction.[44]

A two-dimensional FE–CA model was developed to simulate dendritic solidification in the molten pool of Fe–0.13wt%C LENS samples.[58] Dendrite growth simulations must be done in microscale which is much smaller than the thermal simulations in macroscale; therefore, two length scales are used in the model: one for thermal simulations and one for dendrite growth simulations. Temperature calculations were done in a domain surrounding the molten pool area, as shown in Figure 9.12A. The dendrite growth simulations were performed in small square domains with 100-mm side lengths, and the dendritic microstructure was studied at two different locations at the top and bottom of the pool, as shown in Figure 9.12B.

The SDAS and primary DAS (PDAS) were studied for different cooling rates. As shown in Figure 9.13, both PDAS and SDAS decrease with increasing cooling rate. The experimental relationship of Katayama and Matsunawa[67] and the experimental results for Fe–Ni–Cr ternary alloys with 59% Fe produced by electron beam surface melting[68] are also included in Figure 9.13A and show good agreement with the simulation results with a comparable trend. The discrepancy in results can be attributed to differences between the chemical compositions of the samples. By curve fitting, a new equation is obtained for the relationship between SDAS and cooling rate in SS310 laser deposits, as presented in Figure 9.13A.

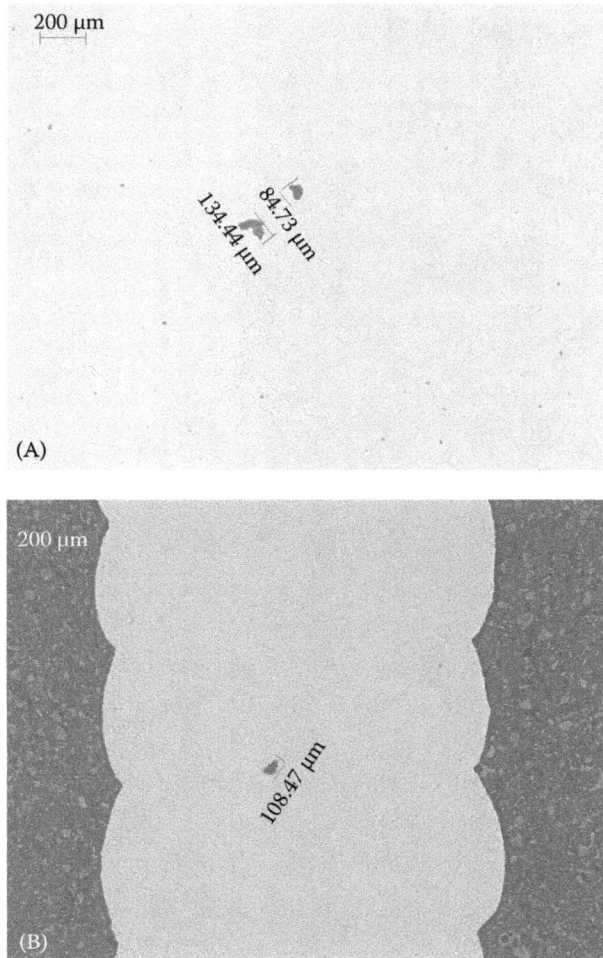

FIGURE 9.10 Optical photography of AISI 410 thin plate for sample S3 showing gas porosity: (A) transverse direction, and (B) longitude direction.[44]

The predicted values for PDAS are much higher than SDAS values and reach 20 μm for lower cooling rates. An FE model was developed by Choi and Mazumder[69] to simulate thermal history and PDAS in welding. Their results showed that PDAS may vary between 7.5 and 20 μm at a cooling rate around 400 K/s for AISI 304 stainless steel, depending on welding speed. The calculated values are also consistent with experimental results,[68] as shown in Figure 9.13B. The predicted cooling rate was found to be about 10^4 K/s, which is in agreement with the experimental results that report cooling rates in the range of 10^3 to 10^4 K/s.[70,71] The experimental results for different materials show that the arm spacing can be just a few microns. Smugeresky et al.[72] reported SDAS values in the range of 2 to 15 μm. Hofmeister et al.[11] found that when the laser power is increased for deposition of SS316 alloy the average dendrite mean intercept length increases from 3 to 9 μm. A PDAS of about 5 μm and a SDAS in the range of 1.5 to 2.5 μm was reported for laser deposition of Ni-based alloys.[65] Table 9.4 lists the reported microstructure sizes for various materials and processes. The difference in the dendrite size for similar materials may be due to the difference in the process parameters used for deposition.

The resultant microstructure in the square domain located in the lower region of the pool (Figure 9.12B) is shown in Figure 9.14 for different travel speeds. The substrate and deposited layer are assumed to be 0.25 mm and 5 mm thick, respectively. The molten pool is slightly deeper than

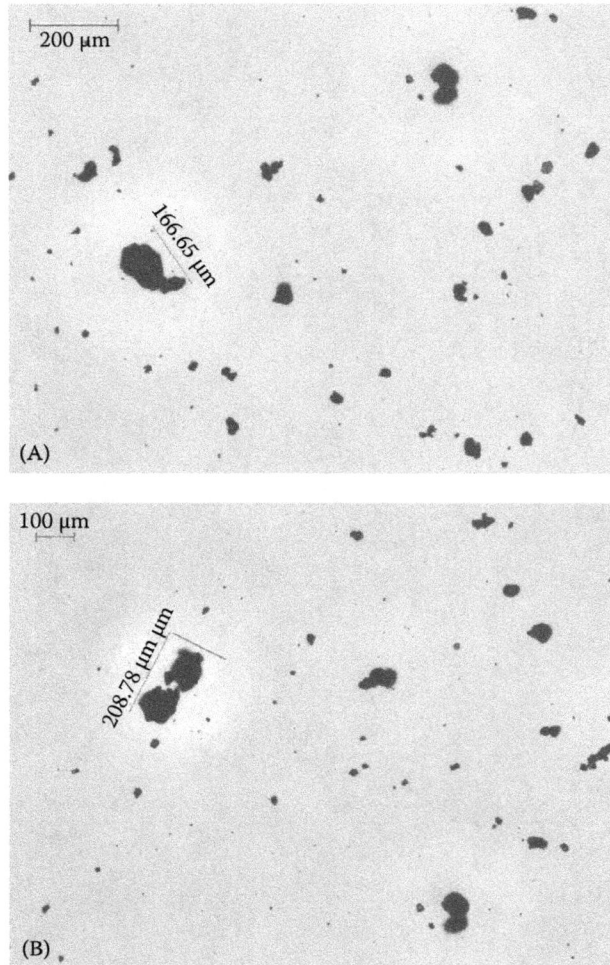

FIGURE 9.11 Optical photography of AISI 410 thin plate for sample S4 showing gas porosity.[44]

0.25 mm, showing an overlap with the substrate. The color legends show the carbon content levels in wt%. The calculated cooling rates are also presented. Increasing laser travel speed results in the increase of cooling rate and solidification velocity. The laser power is adjusted for different travel speeds, so the molten pool size remains constant. Increasing the travel speed from 2 mm/s to 20 mm/s leads to an increase in the cooling rate from 1050 K/s to 9000 K/s. Columnar dendritic microstructure can be observed in all figures, but DAS varies with the cooling rate. Both PDAS and SDAS decrease by increasing travel speed. Because the cooling rate is very high at high travel speeds, PDAS is very small and the formation of secondary and tertiary branches is not possible, leading to a transition from dendritic to cellular microstructure (Figure 9.14C). Experimental studies[62,70,71] also confirm this observation.

The simulated dendritic structures for two different layer thicknesses of 0.25 mm and 0.5 mm are shown in Figures 9.15A and 9.15B, respectively. The laser travel speed was 10 mm/s for both cases, but a higher laser power was used for the case with the thicker layer to be able to melt larger amounts of powder. The molten pool is bigger in the case with a thicker layer, which results in a lower cooling rate. Therefore, the PDAS and SDAS values are larger for the case with larger layer thickness. For this case, because the material is kept at high temperatures for a longer time, the

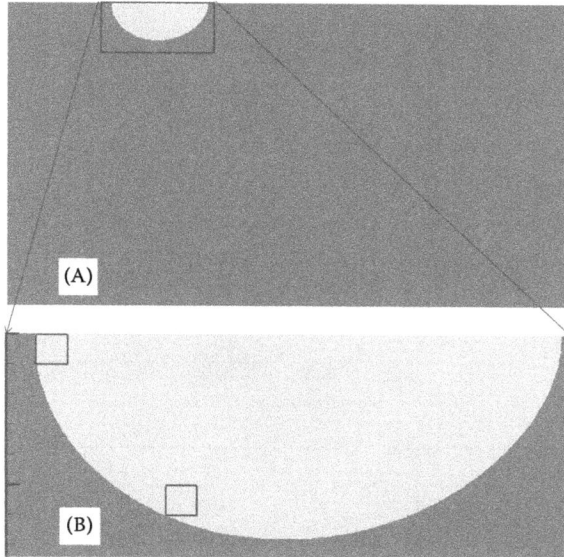

FIGURE 9.12 (A) Thermal simulation domain, showing the molten pool. (B) Small square domains at the top and bottom of the molten pool used for dendrite growth simulations.[58]

temperature gradients are smaller and grains have more time to grow. Experimental results also confirm that a coarser microstructure can be obtained when a higher laser power is used.[11,76,77] Figure 9.16 compares the predicted PDAS values with the result of an experimental study,[76] in which two samples with different layer thicknesses were deposited. Other experiments[11] also suggest that for lower laser powers the cooling rates are significantly higher. Their results show that in SS316 LENS deposits the microstructure is finer for lower laser powers, as shown in Table 9.5. In another study,[77] grain size was measured in H13 steel materials deposited using different laser powers, showing finer grains for lower laser powers. These results are presented in Table 9.6 and support the simulation results.

TABLE 9.4
Reported Microstructure for Various Deposition Materials and Processes

Material	Dimension (μm)	Process	Ref.
SS316	DAS: 1.31–3.0	LENS	Song et al.[73]
308L	PDAS: 4	LENS	Griffith et al.[74]
308L	PDAS: 4	LENS	Griffith et al.[75]
H13	Grain width: 4–20	DMD	Mazumder et al.[76]
H13	SDAS: 2	LENS	Choi and Chang[43]
SS316	Mean intercept length: 3.25–8.68	LENS	Hofmeister et al.[11]
H13	PDAS: 1.5–4; SDAS: 2–5.5	DMD	Ghosh and Choi[55]
304L	Grain width: ~10	LENS	Brooks et al.[5]
SS316	SDAS: <5	LENS	Syed et al.[64]
Ni-based alloy	PDAS: 5; SDAS: 1.5–2.5	DMD	Dinda et al.[65]
SS316	PDAS: 8–20	LCF	Zheng et al.[71]
SS316	SDAS: 2–15	LENS	Smugeresky et al.[72]
H13	Grain width: 6.4–12.2	DMD	Sparks et al.[77]

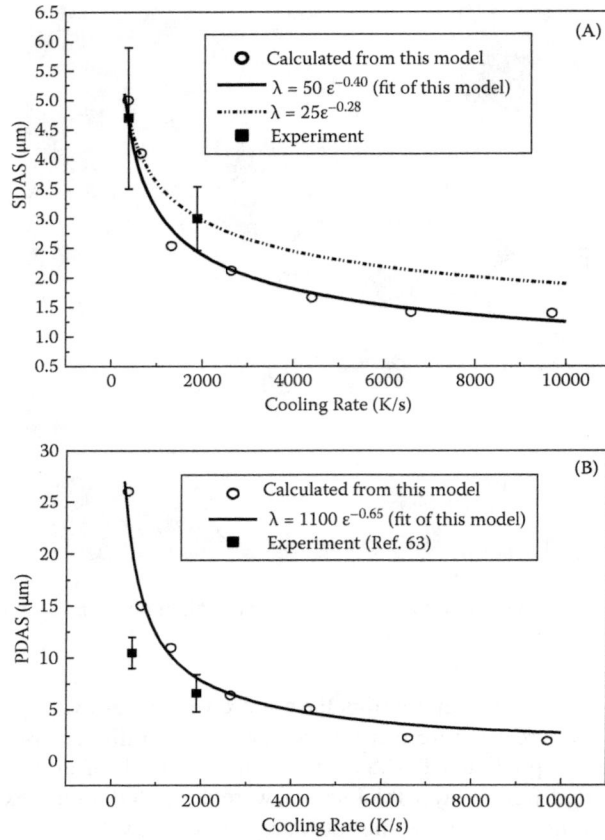

FIGURE 9.13 Predicted and measured DAS for Fe–0.13wt%C as a function of cooling rate: (A) SDAS, and (B) PDAS.[58]

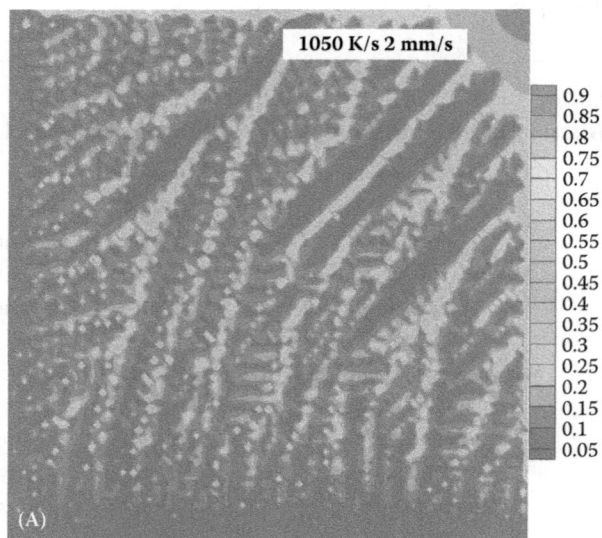

FIGURE 9.14 Solidification microstructure for different travel speeds: (A) 2mm/s, (B) 10mm/s, and (C) 20mm/s. The legend shows carbon concentration in wt%.[58]   *(Continued)*

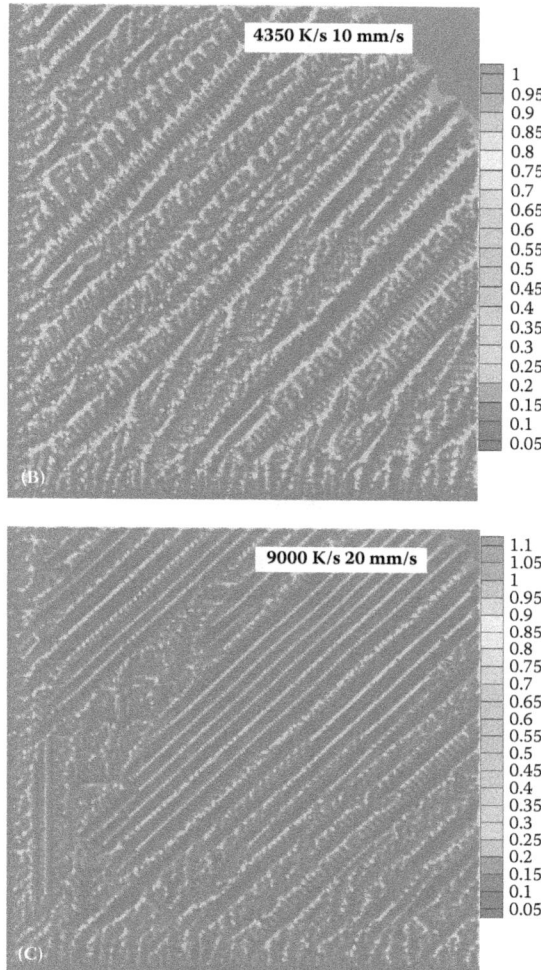

FIGURE 9.14 (Continued) Solidification microstructure for different travel speeds: (A) 2mm/s, (B) 10mm/s, and (C) 20mm/s. The legend shows carbon concentration in wt%.[58]

The dendritic microstructures are compared for two different locations at the top and bottom of the molten pool in Figures 9.17A and 9.17C. The substrate was considered to be 1.5 mm thick. The temperature contours for those two locations are depicted in Figures 9.17B, and 9.17D, respectively. A competitive growth can be observed with co lumnar dendrites growing in the direction of the highest temperature gradient. The dendrites parallel to the highest temperature gradient direction block the other dendrites. Therefore, at the surface of the pool, the dendrites tend to grow parallel to the direction of the laser movement. However, at the bottom of the pool, the laser effect is combined with the heat conduction effect from the substrate and the maximum temperature gradient, and the dominant growth direction is almost upright. The simulation results are consistent with the experimental observations,[43,71] which show that dendrite orientations change with the location in the deposited parts.

9.6.1 ADVANCES IN MODELING SOLIDIFICATION MICROSTRUCTURE

The microscopic cellular automaton (MCA) has already demonstrated great potential to simulate several dendrites with modest hardware resources and little parallelization efforts. Cellular automaton algorithms are particularly suitable for massively parallel supercomputers because

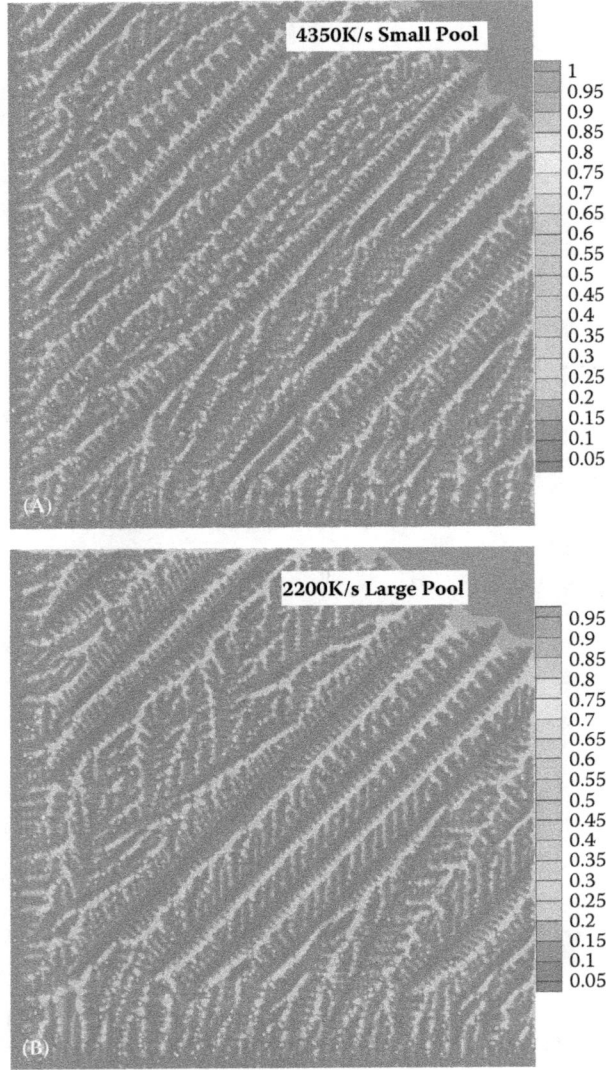

FIGURE 9.15 Solidification microstructure for different layer thickness values: (A) 0.25 mm, and (B) 0.5 mm at a travel speed of 10 mm/s. The legend shows carbon concentration in wt%.[58]

they involve simple local operations at the cell level which can be distributed independently. The computation of the temperature and solute fields by solving the continuum conservation equations is, however, non-local (it requires information from all the mesh sites) and less amenable to parallelization. If fluid flow is also considered, the solution of the flow equations becomes the most demanding computational task, and resorting to high-scale parallelization is imperative. In this respect, the lattice Boltzmann method (LBM) is an attractive alternative. The lattice Boltzmann method is a well-established tool to simulate transport phenomena. With MCA, LBM calculations are also local (no assembling and solving of large systems of equations is required), involving only neighboring lattice sites, modest memory requirements, and excellent scalability on parallel computers.[78] The basic structure of the LBM resembles that of a cellular automaton, making it an ideal platform for combining with the MCA. 2D models are usually unable to capture all

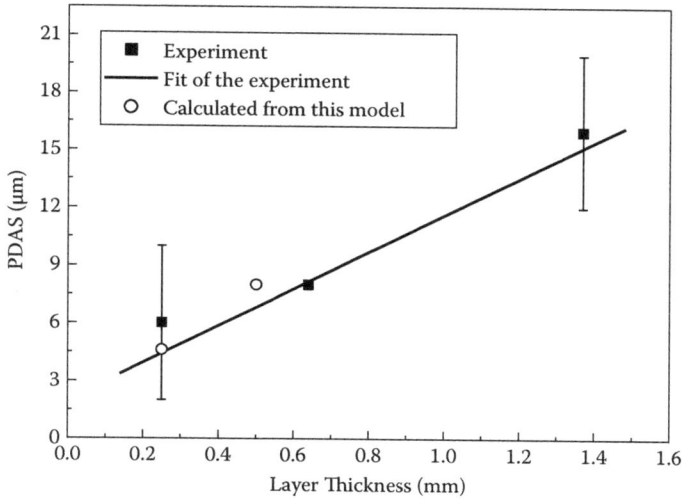

FIGURE 9.16 Predicted and measured[76] PDAS vs. layer thickness.

features of microstructures which are determinative in many materials properties. Our studies[79] have shown that the growth of dendrites in 3D is considerably different than in 2D. Therefore, in order to obtain correct physical results, it is necessary to perform the simulations in 3D. The simulation of dendrite growth in this domain in three dimensions is computationally unaffordable for a FE–CA scheme.

TABLE 9.5

Grain Mean Intercept Length for Different Laser Powers in SS316 LENS Deposits

	Laser Power (W)					
	410	345	275	200	165	115
Mean intercept length (μm)	8.68	8.55	7.12	6.46	4.63	3.25

Source: Hofmeister, W. et al., *JOM*, 51(7), 1999.

TABLE 9.6

Grain Size for Different Laser Powers and Powder Flow Rates in H13 Steel Deposits

			Grain Size (μm)	
Run Order	Power (W)	Powder Flow Rate (grams/min)	Mean	Standard Deviation
1	750	6	6.43	2.86
4	1000	8	12.19	5.20

Source: Sparks, T. et al., in *Solid Freeform Fabrication Symposium Proceedings*, 2006, pp. 261–267.

In Yin et al.,[80] a 2D LB-CA model was developed for dendritic solidification. LBM was used to solve for transport phenomena, and CA was employed to capture the new interface cells. The velocity of the interface was determined from the solution of the transport equations and boundary conditions at the interface. The above model was parallelized to perform very large scale 2D simulations.[81,82] The parallelized solidification model showed a very good scalability up to centimeter-size domains, including more than ten million dendrites. Motivated by this success, the authors developed a 3D LB-CA model for simulating solute-driven dendrite growth.[83,84] The model successfully captured the morphology of dendritic microstructure in three dimensions with a good computational efficiency and parallel scalability. The local nature of the model made it possible to perform dendrite growth simulations in macroscale 3D domains as depicted in Figure 9.18.[85] The developed model can be employed for simulation of solidification microstructure in additive manufacturing

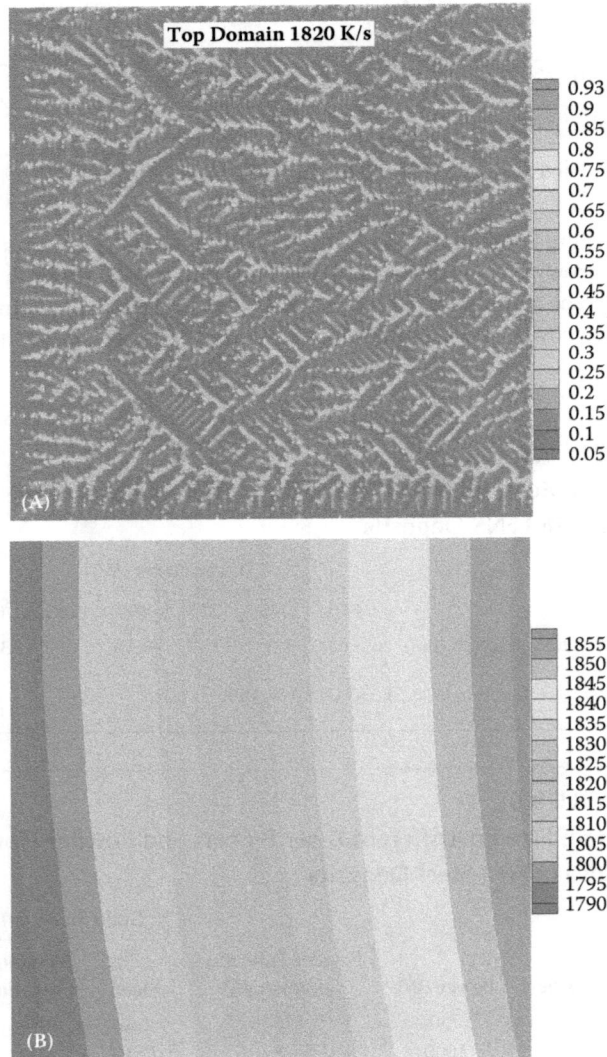

FIGURE 9.17 (A) Solidification microstructure and (B) temperature contour at the upper domain. (C) Solidification microstructure and (D) temperature contour at the bottom domain. The travel speed was 5 mm/s for both cases.[58]   *(Continued)*

processes. The developed innovative all-local LB–CA strategy is highly scalable and can efficiently exploit tera- and peta-scale computing power to simulate growth of large 3D dendritic structures. These simulations will provide the grain structures and distribution of alloying elements.

9.7 SUMMARY

Some of the recent studies on laser deposition of metallic materials were discussed in this chapter. The main attention was focused on the effect of process parameters on the thermal behavior, residual stresses, porosity, microstructure, and properties of the laser deposits. The experimental results for study of the molten pool size and thermal behavior for SS410 laser deposits were presented. The results show that cooling rate and size of the molten zone strongly depend on the laser power and

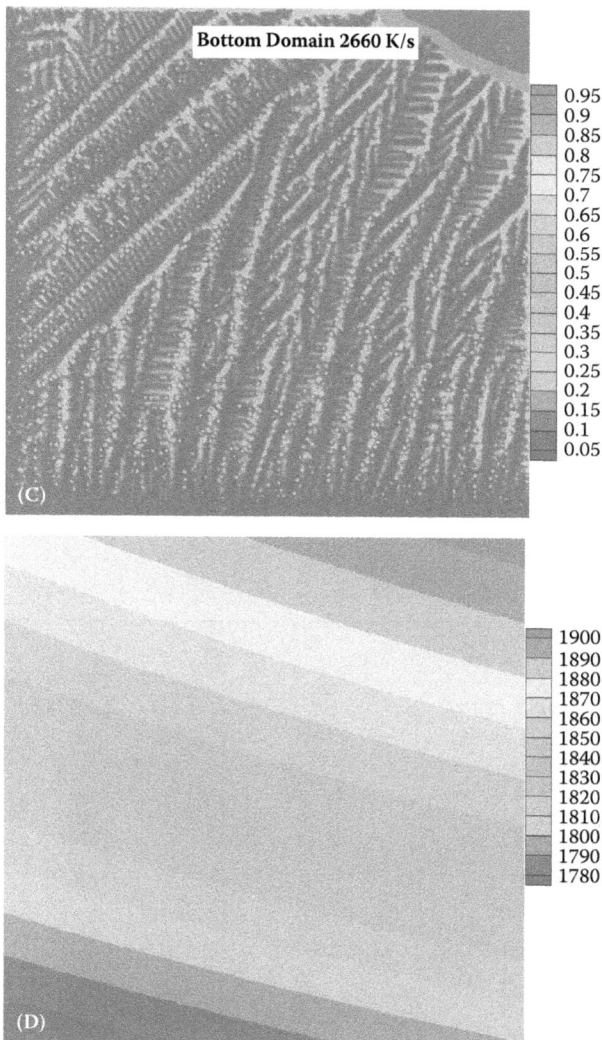

FIGURE 9.17 (Continued) (A) Solidification microstructure and (B) temperature contour at the upper domain; (C) solidification microstructure and (D) temperature contour at the bottom domain. The travel speed was 5 mm/s for both cases.[58]

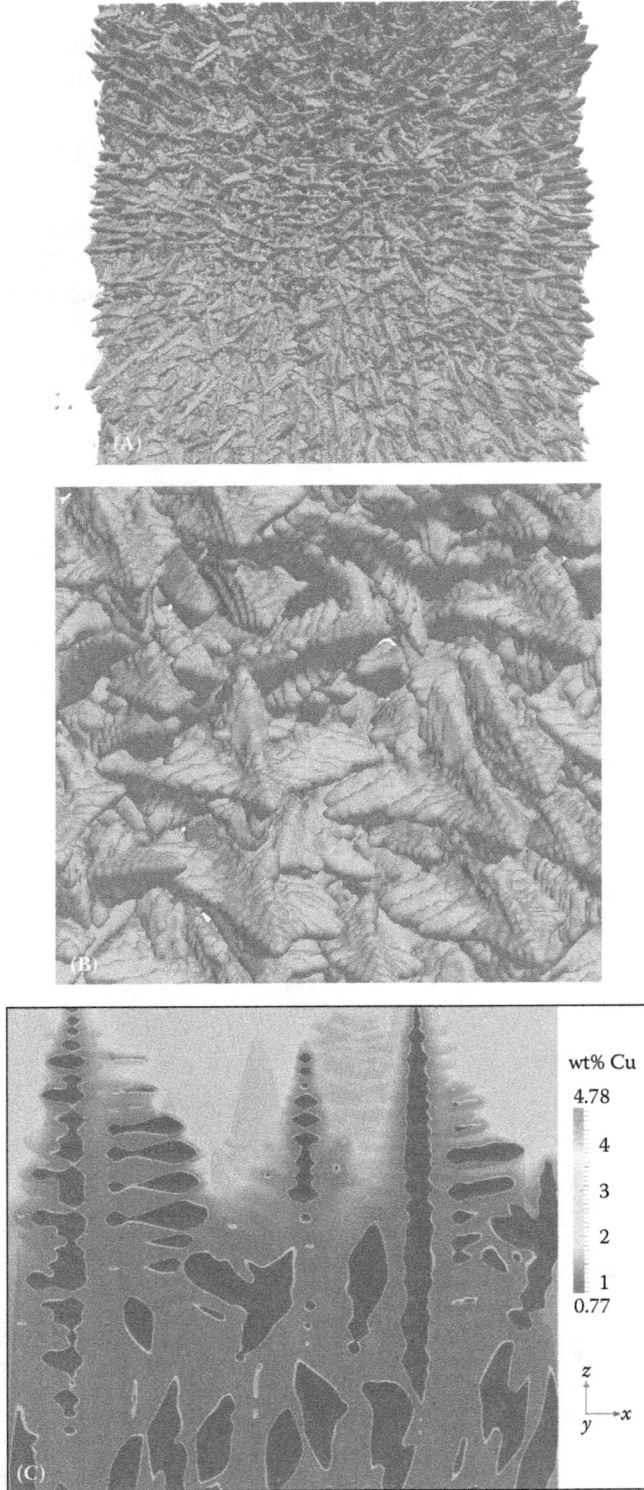

FIGURE 9.18   (A) Large-scale 3D simulation of dendrite growth in a 1-mm³ domain; (B) dendrite features in a higher resolution; and (C) distribution of alloying elements around the solidified dendrites.[85]

travel speed. FE numerical simulations of the LENS process were also presented. Numerical models can help better understand the thermal behavior and evaluate the effect of process parameters on the molten pool size and properties of the laser deposits. The simulation results show that, once the initial layers are deposited, the laser power should decrease with the number of layers to keep a constant molten pool size. Also, in order to maintain the molten pool size, the laser power should increase with increasing travel speed. The results can be employed to maintain a steady molten pool size and temperature distribution and consequently obtain a more uniform microstructure in the laser-deposited parts.

The residual stress measurement results for SS410 thin-wall laser deposits were discussed. The measurements were performed using the neutron diffraction method. The residual stress component along the growth direction was mostly compressive and dominant compared to the other components, which is consistent with the previous studies. No meaningful relationship could be found between process parameters and the residual stresses. However, increasing the laser power seems to cause a slight increase in the magnitude of the residual stresses. The simulation results for residual stresses were also consistent with the experimental results.

The porosity formation in SS410 steel laser deposits was discussed. The studies were performed using x-ray computed tomography and optical microscopy. It was found that the intralayer pores can be affected by laser travel speed, but not by laser power, at least in the range considered for the study. Also, the intralayer pores seem to be sensitive to powder flow.

The simulation results for microstructural evolution during solidification of laser-deposited materials were discussed. The simulations were performed using a multi-scale FE–CA model. A dendritic microstructure was observed even at very high cooling rates of the laser deposition process. The DAS was found to be on the order of a few microns. Columnar growth was also observed at the bottom of the pool. Secondary and tertiary branches did not form, and a transition from dendritic to cellular structure was observed, when a high travel speed was considered. Using a computationally more efficient LB–CA scheme can help to conduct large 3D simulations and include the convection effects in the molten pool.

When combined with experimental validation, the physics-based computational modeling of AM processes will allow a more faithful prediction of part microstructure and properties, as well as the proper optimization of process parameters for the desired specifications. The successful development and commercial implementation of the additive manufacturing technique will represent a game-changing advance in the way manufacturing components are designed, built, repaired, and maintained. The potential impacts include but are not limited to aerospace, automotive, medical, military, and commercial industrial markets.

REFERENCES

1. GE Aviation. (2010). Private communication.
2. Atwood, C.L., Griffith, M.L., Schlienger, M.E., Harwell, L.D., Ensz, M.T., Keicher, D.M., Schlienger, M.E., Romero, J.A., and Smugeresky, J.E. (1998). Laser Engineered Net Shaping (LENS®): a tool for direct fabrication of metal parts. In: *Proceedings of ICALEO '98*, pp. E-1–E-7.
3. Mazumder, J. and Qi, H. (2004). Fabrication of 3D components by laser-aided direct metal deposition. In: *Proceedings of SPIE 5706: Critical Review: Industrial Lasers and Applications*, pp. 38–59.
4. Keicher, D.M., Miller, W.D., Smugeresky, J.E., and Romero, J.A. (1998). Laser Engineered Net Shaping (LENS™): beyond rapid prototyping to direct fabrication. In: *Proceedings of the 1998 TMS Annual Meeting*, pp. 369–377.
5. Brooks, J.A., Headley, T.J., and Robino, C.V. (2000). Microstructures of laser deposited 304L austenitic stainless steel. In: *Materials Research Society Symposium—Proceedings*, Vol. 625, pp. 21–30.
6. Brooks, J.A., Robino, C.V., Headley, T.J., and Michael, J.R. (2003). Weld solidification and cracking behavior of free-machining stainless steel. *Weld. J.*, 82: 51S–64S.
7. Griffith, M.L., Ensz, M.T., Puskar, J.D., Robino, C.V., Brooks, J.A. et al. (2000). Understanding the microstructure and properties of components fabricated by Laser Engineered Net Shaping (LENS). In: *Materials Research Society Symposium—Proceedings*, Vol. 625, pp. 9–20.

8. Liu, W. and Dupont. J.N. (2000). *In-situ* reactive processing of nickel aluminides by Laser Engineered Net Shaping. *Metall. Mater. Trans. A*, 34A: 2633–2641.

9. Kelly, S.M., Kampe, S.L., and Crowe, C.R. (2000). Microstructural study of laser formed Ti-6Al-4V. In: *Materials Research Society Symposium—Proceedings*, Vol. 625, pp. 3–8.

10. Kelly, S.M. and Kampe, S.L. (2004). Microstructural evolution in laser-deposited multilayer Ti-6Al-4V builds. Part I. Thermal modeling. *Metall. Mater. Trans. A*, 35: 1869–1879.

11. Hofmeister, W., Wert, M., Smugeresky, J., Philliber, J.A., Griffith, M., and Ensz, M. (1999). Investigating solidification with the Laser Engineered Net Shaping (LENS™) process. *JOM*, 51(7).

12. Grylls, R. (2003). Laser Engineered Net Shapes. *Adv. Mater. Process.*, 161: 45–46.

13. National Research Council. (2008). *Integrated Computational Materials Engineering: A Transformational Discipline for Improved Competitiveness and National Security*. National Academies Press, Washington, DC.

14. Horstemeyer, M.F. (2009). Multiscale modeling: a review. In: *Practical Aspects of Computational Chemistry* (Leszczynski, J. and Shukla, M.K., Eds.), pp. 87–135. Springer, New York.

15. Smugeresky, J.E., Gill, D.D., and Atwood, C.J. (2007). *New Low Cost Material Development Technique for Advancing Rapid Prototyping Manufacturing Technology*, Sandia Report, SAND2007-7832. Sandia National Laboratory, Albuquerque, NM.

16. Bi, G., Gasser, A., Wissenbach, K., Drenker, A., and Poprawe, R. (2006). Characterization of the process control for the direct laser metallic powder deposition. *Surf. Coat. Technol.*, 201: 2676–2683.

17. Hu, D., Mei, H., and Kovacevic. R. (2002). Improving solid freeform fabrication by laser based additive manufacturing. *Proc. Inst. Mech. Eng. Part B Eng. Manuf.*, 216: 1253–1264.

18. Hu, D. and Kovacevic, R. (2003). Modeling and measuring thermal behavior of molten pool in closed-loop controlled laser-based manufacturing. *Proc. Inst. Mech. Eng. Part B J. Eng. Manuf.*, 217: 441–452.

19. Griffith, M.L., Schlienger, M.E., Harwell, L.D., Oliver, M.S., Baldwin, M.D. et al. (1998). Thermal behavior in the LENS process. In: *Solid Freeform Fabrication Symposium Proceedings*, pp. 89–97.

20. Griffith, M.L., Schlienger, M.E., Harwell, L.D., Oliver, M.S., Baldwin, M.D. et al. (1999). Understanding thermal behavior in the LENS process. *Mater. Des.*, 20: 107–113.

21. Hofmeister, W., MacCallum, D., and Knorovsky. G. (1999). Video monitoring and control of the LENS process. In: *Proceedings of the Ninth International Conference on Computer Technology in Welding*, pp. 187–196.

22. Hofmeister, W., Griffith, M., Ensz, M., and Smugeresky, J. (2001). Solidification in direct metal deposition by LENS processing. *JOM*, 53(9): 30–34.

23. Wei, W., Zhou, Y., Ye, R., Lee, D., Craig, J.E., Smugeresky, J.E., and Lavernia, E.J. (2002). Investigation of the thermal behavior during the LENS process. In: *Proceedings of the International Conference on Metal Powder Deposition for Rapid Manufacturing*, pp. 128–135.

24. Wang, L., Felicelli, S.D., and Craig, J.E. (2009). Experimental and numerical study of the LENS rapid fabrication process. *J. Manuf. Sci. Eng.*, 131(4): 041019.

25. Ye, R., Smugeresky, J.E., Zheng, B., Zhou, Y., and Lavernia, E.J. (2006). Numerical modeling of the thermal behavior during the LENS process. *Mater. Sci. Eng. A*, 428: 47–53.

26. Wang, L. and Felicelli, S. (2006). Analysis of thermal phenomena in LENS deposition. *Mater. Sci. Eng. A*, 435–436: 625–631.

27. Labudovic, M., Hu, D., and Kovacevic, R. (2003). A three dimensional model for direct laser metal powder deposition and rapid prototyping. *J. Mater. Sci.*, 38: 35–49.

28. Beuth, J., Vasinonta, A., and Griffith. M. (1999). Process maps for laser deposition of thin-walled structures. In: *Solid Freeform Fabrication Symposium Proceedings*, pp. 383–391.

29. Vasinonta, A., Beuth, J.L., and Griffith. M.L. (2001). A process map for consistent build conditions in the solid freeform fabrication of thin-walled structures. *J. Manuf. Sci. Eng.*, 123: 615–622.

30. Costa, L., Vilar, R., Reti, T., and Deus, A.M. (2005). Rapid tooling by laser powder deposition: process simulation using finite element analysis. *Acta Mater.*, 53: 3987–3999.

31. Wang, L., Felicelli, S.D., Gooroochurn, Y., Wang, P.T., and Horstemeyer, M.F. (2008). Optimization of the LENS process for steady molten pool size. *Mater. Sci. Eng. A*, 474: 148–156.

32. Wang, L. and Felicelli, S.D. (2007). Process modeling in laser deposition of multilayer SS410 steel. *J. Manuf. Sci. Eng.*, 129: 1028–1034.

33. Lewis, G.K. and Schlienger, E. (2000). Practical considerations and capabilities for laser assisted direct metal deposition. *Mater. Des.*, 21: 417–423.

34. Unocic, R.R. and DuPont, J.N. (2004). Process efficiency measurements in the Laser Engineered Net Shaping process. *Metall. Mater. Trans. B*, 35: 143–152.

35. Olson, D.L. et al., Eds. (2005). *ASM Handbook*. Vol. 6. *Welding, Brazing, and Soldering*. ASM International, Material Park, OH.

36. Rangaswamy, P., Griffith, M.L., Prime, M.B., Holden, T.M., Rogge, R.B., Edwards, J.M., and Sebring, R.J. (2005). Residual stresses in LENS® components using neutron diffraction and contour method. *Mater. Sci. Eng. A*, 399: 72–83.

37. Pratt, P., Felicelli, S.D., Wang, L., and Hubbard, C.R. (2008). Residual stress measurement of Laser-Engineered Net Shaping AISI 410 thin plates using neutron diffraction. *Metall. Mater. Trans. A*, 39(13): 3155–3163.

38. Holden, T.M., Suzuki, H., Carr, D.G., Ripley, M.I., and Clausen, B. (2006). Stress measurements in welds: problem areas. *Mater. Sci. Eng. A*, 437: 33–37.

39. Wang, L., Felicelli, S.D., and Pratt. P. (2008). Residual stresses in LENS-deposited AISI 410 stainless steel plates. *Mater. Sci. Eng. A*, 496: 234–241.

40. Susan, D.F., Puskar, J.D., Brooks, J.A., and Robino, C.V. (2006). Quantitative characterization of porosity in stainless steel LENS powders and deposits. *Mater. Charact.*, 57(1): 36–43.

41. Kobryn, P.A., Moore, E.H., and Semiatin, S.L. (2000). The effect of laser power and traverse speed on microstructure, porosity, and build height in laser-deposited Ti-6Al-4V. *Scr. Mater.*, 43(4): 299–305.

42. Groh, H.C. (2006). *Development of Laser Fabricated Ti-6Al-4V*, NASA Report No. NASA/TM-2006-214256. National Aeronautics and Space Administration, Washington, DC.

43. Choi, J. and Chang, Y. (2005). Characteristics of laser aided direct metal/material deposition process for tool steel. *Int. J. Mach. Tools Manuf.* 45: 597–607.

44. Wang, L., Pratt, P., Felicelli, S.D., El Kadiri, H., Berry, J.T., Wang, P.T., and Horstemeyer, M.F. (2009). Pore formation in laser-assisted powder deposition process. *J. Manuf. Sci. Eng.*, 131(5): 051008.

45. El Kadiri, H., Wang, L., Horstemeyer, M.F., Yassar, R., Shahbazian, Y., Felicelli, S.D., and Wang, P.T. (2007). Phase transformations in low alloy steel laser deposits. *Mater. Sci. Eng. A*, 494: 10–20.

46. Felicelli, S.D., Heinrich, J.C., and Poirier, D.R. (1993). Numerical models for dendritic solidification of binary alloys. *Numer. Heat Trans. B*, 23: 461–481.

47. Felicelli, S.D., Heinrich, J.C., and Poirier, D.R. (1991). Simulation of freckles during vertical solidification of binary alloy. *Metall. Trans. B*, 22: 847–859.

48. Felicelli, S.D., Poirier, D.R., Giamei, A.F., and Heinrich, J.C. (1997). Simulating convection and macrosegregation in superalloys. *JOM*, 49: 21–25.

49. Felicelli, S.D., Poirier, D.R., and Heinrich, J.C. (1998). Modeling freckle formation in three dimensions during solidification of multicomponent alloys. *Metall. Mater. Trans. B*, 29: 847–855.

50. Felicelli, S.D., Poirier, D.R., and Sung, P.K. (2000). A finite element model for prediction of pressure and redistribution of gas forming elements in multicomponent casting alloys. *Metall. Mater. Trans. B*, 31: 1283–1292.

51. Felicelli, S.D. and Escobar de Obaldia, E. (2007). Quantitative prediction of microporosity in aluminum alloys. *J. Mater. Process. Technol.*, 191: 265–269.

52. Felicelli, S.D., Escobar de Obaldia, E., and Pita, C.M. (2007). Simulation of hydrogen porosity during solidification. *Am. Foundry Soc. Trans.*, 115: 1–13.

53. Felicelli, S.D., Pita, C.M., and Escobar de Obaldia, E. (2007). Modeling the onset and evolution of hydrogen pores during solidification. In: *Shape Casting: The 2nd International Symposium* (Crepeau, P. et al., Eds.), pp. 201–208. Minerals, Metals & Materials Society, Warrendale, PA.

54. Felicelli, S.D., Wang L., Pita, C.M., and Escobar de Obaldia, E. (2009). A model for gas microporosity in aluminum and magnesium alloys. *Metall. Mater. Trans. B*, 40: 169–181.

55. Ghosh, S. and Choi, J. (2006). Modeling and experimental verification of transient/residual stresses and microstructure formation in multi-layer laser aided DMD process. *J. Heat Trans.*, 128: 662–679.

56. Grujicic, M., Gao, G., and Figliola, R.S. (2001). Computer simulations of the evolution of solidification microstructure in the LENS™ rapid fabrication process. *Appl. Surf. Sci.*, 183: 43–57.

57. Miller, R.S., Cao, G., and Grujicic, M.J. (2001). Monte Carlo simulation of three-dimensional non-isothermal grain-microstructure evolution: application to the LENS™ process. *J. Mater. Synth. Process.*, 9: 329.

58. Yin, H. and Felicelli, S.D. (2010). Dendrite growth simulation during solidification in the LENS process. *Acta Mater.*, 58: 1455–1465.

59. Gaumann, M., Bezencon, C., Canalis, P., and Kurz, W. (2001). Single-crystal laser deposition of superalloys: processing–microstructure maps. *Acta Mater.*, 49: 1051–1062.

60. Liu, W. and DuPont, J.N. (2004). Effects of melt-pool geometry on crystal growth and microstructure development in laser surface-melted superalloy single crystals: mathematical modeling of single-crystal growth in a melt pool, part I. *Acta Mater.*, 52: 4833–4847.
61. Pinkerton, A., Karadge, M., Syed, W., and Li, L. (2006). Thermal and microstructural aspects of the laser direct metal deposition of waspaloy. *J. Laser Appl.*, 18: 216–226.
62. Collins, P.C., Banerjee, R., Banerjee, S., and Fraser, H.L. (2003). Laser deposition of compositionally graded titanium-vanadium and titanium-molybdenum alloys. *Mater. Sci. Eng. A*, 352: 118–128.
63. Bao, Y., Ruan, J., Sparks, T.E., Anand, J., Newkirk, J., and Liou, F. (2006). Evaluation of mechanical properties and microstructure for laser deposition process and welding process. In: *Solid Freeform Fabrication Symposium Proceedings*, pp. 280–289.
64. Syed, W., Pinkerton, A.J., and Lin, L. (2005). A comparative study of wire feeding and powder feeding in direct diode laser deposition for rapid prototyping. *Appl. Surf. Sci.*, 247: 268–276.
65. Dinda, G.P., Dasgupta, A.K., and Mazumder, J. (2009). Laser aided directed metal deposition of Inconel 625 superalloy: microstructural evolution and thermal stability. *Mater. Sci. Eng. A*, 509: 98–104.
66. Bi, G., Gasser, A., Wissenbach, K., Drenker, A., and Poprawe, R. (2006). Investigation on the direct laser metallic powder deposition process via temperature measurement. *Appl. Surf. Sci.*, 253: 1411–1416.
67. Katayama, S. and Matsunawa, A. (1984). Solidification microstructure of laser welded stainless steels. In: *Proceedings of the Materials Processing Symposium (ICALEO '84)*, p. 60.
68. Elmer, J.W., Allen, S.M., and Eager. T.W. (1989). Microstructural development during solidification of stainless steel alloys. *Metall. Trans. A*, 20: 2117–2131.
69. Choi, J. and Mazumder, J. (2002). Numerical and experimental analysis for solidification and residual stress in the GMAW process for AISI 304 stainless steel. *J. Mater. Sci.*, 37: 2143–2158.
70. Zheng, B., Zhou, Y., Smugeresky, J.E., Schoenung, J.M., and Lavernia, E.J. (2008). Thermal behavior and microstructural evolution during laser deposition with laser-engineered net shaping. Part I. Numerical calculations. *Metall. Mater. Trans. A*, 39: 2228–2236.
71. Zheng, B., Zhou, Y., Smugeresky, J.E., Schoenung, J.M., and Lavernia, E.J. (2008). Thermal behavior and microstructure evolution during laser deposition with LENS. Part II. Experimental investigation and discussion. *Metall. Mater. Trans. A*, 39: 2237–2245.
72. Smugeresky, J.E., Keicher, D.M., Romero, J.A., Griffith, M.L., and Harwell, L.D. (1997). *Laser Engineered Net Shaping (LEMS™) Process: Optimization of Surface Finish and Microstructural Properties*. Sandia National Laboratory, Albuquerque, NM.
73. Song, J., Deng, Q., Chen, C., Hu, D., and Li, Y. (2006). Rebuilding of metal components with laser cladding forming. *Appl. Surf. Sci.*, 252: 7934–7940.
74. Griffith, M.L., Ensz, M.T., Greene, D.L., Reckaway, D.E., Romero, J.A. et al. (1999). *Solid Freeform Fabrication Using the Wirefeed Process*. Sandia National Laboratory, Albuquerque, NM.
75. Griffith, M.L., Ensz, M.T., Greene, D.L., Reckaway, D.E., Morin, J.A. et al. (1999). *Laser Wire Deposition (WireFeed) for Fully Dense Shapes LDRD*. Sandia National Laboratory, Albuquerque, NM.
76. Mazumder, J., Schifferer, A., and Choi, J. (1999). Direct materials deposition: designed macro and microstructure. *Mater. Res. Innov.*, 3: 118–131.
77. Sparks, T., Ruan, J., Fan, Z., Bao, Y., and Liou, F. (2006). Effect of structured laser pulses on grain growth in H13 tool steel. In: *Solid Freeform Fabrication Symposium Proceedings*, pp. 261–267.
78. Wellein, G., Lammers, P., Hager, G., Donath, S., and Zeiser, T. (2006). Towards optimal performance for lattice Boltzmann applications on terascale computers. In: *Parallel Computational Fluid Dynamics: Theory and Applications, Proceedings of the 2005 International Conference on Parallel Computational Fluid Dynamics* (Deane, A. et al., Eds.), pp. 31–40. Elsevier, Amsterdam.
79. Eshraghi, M., Jelinek, B., and Felicelli. S.D. (2013). A three-dimensional lattice Boltzmann-cellular automaton model for dendritic solidification under convection. In: *Proceedings of the TMS Annual Meeting & Exhibition*, pp. 493–499.
80. Yin, H., Felicelli, S.D., and Wang, L. (2011). Simulation of a dendritic microstructure with the lattice Boltzmann and cellular automaton methods. *Acta Mater.*, 59: 3124–3136.
81. Jelinek, B., Eshraghi, M., and Felicelli. S.D. (2013). Large scale parallel lattice Boltzmann model of dendritic growth. In: *Proceedings of TMS Annual Meeting & Exhibition*, pp. 39–46.
82. Jelinek, B., Eshraghi, M., and Felicelli, S.D. (2014). Large-scale parallel lattice Boltzmann–cellular automaton model of two-dimensional dendritic growth. *Comput. Phys. Commun.*, 185(3): 939–947.
83. Eshraghi, M., Jelinek, B., and Felicelli, S.D. (2012). Three dimensional simulation of solutal dendrite growth using lattice Boltzmann and cellular automaton methods. *J. Cryst. Growth*, 354(1): 129–134.

84. Felicelli, S.D., Eshraghi, M., and Jelinek, B. (2013). Large-scale simulation of dendritic solidification. In: *Proceedings of the 8th Pacific Rim International Conference on Advanced Materials and Processing*, pp. 2931–2939.

85. Eshraghi, M. and Felicelli, S.D. (2015). Large-scale three-dimensional simulation of dendritic solidification using lattice Boltzmann method. *J. Minerals Metals Mater.*, DOI: 10.1007/s11837-015-1446-0.

10 Integration of Gas-Permeable Structures in Laser Additive Manufactured Products

Christoph Klahn and Mirko Meboldt

CONTENTS

ABSTRACT

Gas-permeable sections are required in a variety of industrial and end-user applications to perform different functions. Usually the integration of these sections requires additional components made from gas-permeable materials. Laser additive manufacturing is a manufacturing technology to produce three-dimensional parts by metal powder bed fusion based on a 3D-CAD model. The large freedom in design of laser additive manufacturing offers the possibility to integrate gas-permeable structures into solid parts. Different strategies to laser additive manufacture gas-permeable material are presented here. The strategy with the most flexibility to design the appearance and permeability to a specific application is presented in detail. This chapter covers manufacturing process parameters and the resulting geometry, mechanical properties, and flow characteristics. The potential to create new solutions to challenges of industrial applications by integrating gas-permeable structures is demonstrated with two cases for injection molding tools.

10.1 INDUSTRIAL APPLICATIONS FOR GAS-PERMEABLE MATERIALS

Gas-permeable materials are solid materials with an internal structure that makes them permeable for air and other gases. This internal structure distinguishes gas-permeable material from bulk material of loose powder where gas may flow around unconnected, impermeable particles. Many industrial and consumer products require gas-permeable sections to allow a flow of gas in or out of the device. The function of gas permeability in the flow of gas varies among the different applications.

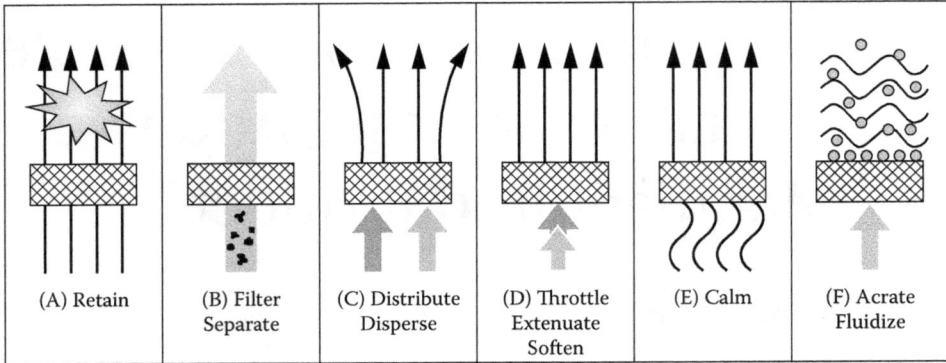

FIGURE 10.1   Applications for gas-permeable material.[1]

A selection of applications is shown in Figure 10.1. The one presented in Figure 10.1A is a safety function for handling flammable or oxidizing gases or fumes. The gas-permeable material is used as a flame arrestor and prevents an intended or unintended fire from spreading in the system. The second application uses permeable material to filter the gas and separate particles of various sizes from the flow. In this application, the size of the internal structure is important for the function of the part. The turbulences in a gas-permeable material are used in the application shown in Figure 10.1C to mix and distribute different gases. In the next two applications, gas-permeable materials are used to throttle and extenuate pulsations and pressure peaks as well as calm flow turbulence. The application shown in Figure 10.1F allows blowing gas into a container with liquid or solid particles without contaminating the supply line of the gas. During operation the constant direction of the flow keeps particles from entering the gas-permeable material. Nevertheless, requirements on the material are similar to the application of filters with regard to the gap size of the internal structure, because operating conditions without a flow have to be covered by the material.

The different applications shown in Figure 10.1 have already been realized in commercially available products. This is usually done either by woven or non-woven filters, by inserts made out of permeable material (e.g., sintered metals, ceramics, fabrics), or by utilizing the gaps between parts, as has been done for venting injection molds through the parting line.[2,3] These conventional methods lead to additional parts in the product. This increases manufacturing and assembly costs and could compromise overall quality due to additional tolerances. To overcome these issues, the freedom of design provided by laser additive manufacturing is used to integrate gas-permeable sections into solid parts.

10.2   LASER ADDITIVE MANUFACTURING

Additive manufacturing, as defined by the standard ISO/ASTM 52921,[4] refers to manufacturing processes to build three-dimensional (3D) objects from a three-dimensional computer-aided design (CAD) model by adding material. This is usually done layer by layer in a cyclic process. The variety of processes for various materials and different joining mechanisms are grouped under the term *additive manufacturing*.[4] An additive manufacturing technology suited for the production of complex metal parts for end users and industrial applications is laser additive manufacturing. The process is based on powder bed fusion and uses a laser to selectively solidify powder by fully melting its particles. All powder particles are from the same alloy. A variety of alloys, including steel, titanium, aluminum, and Inconel, are commercially available. Other process names used by companies for marketing their machines and powders are selective laser melting (SLM), direct metal laser sintering (DMLS), and LaserCUSING®.[5]

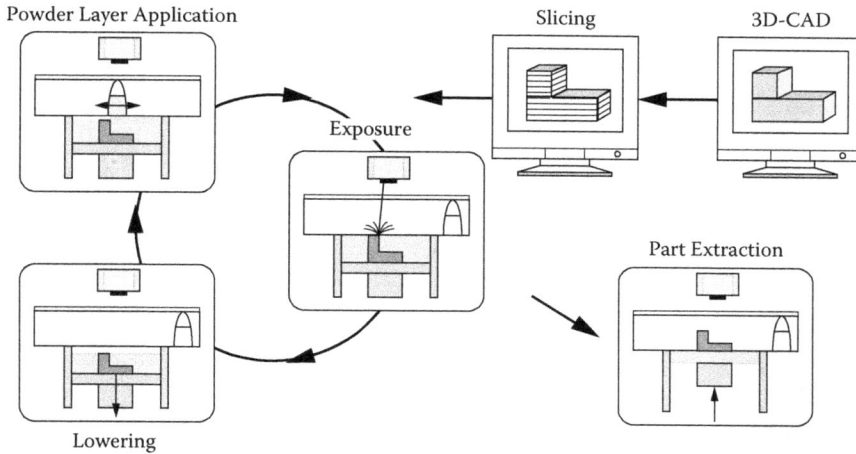

FIGURE 10.2 Process of laser additive manufacturing.[7]

The process cycle of laser additive manufacturing is shown in Figure 10.2. The 3D-CAD model of the part is sliced in layers of constant thickness. Typical layer thicknesses range from 30 to 50 μm. The file containing the layer information is transferred to the additive manufacturing machine. The manufacturing process itself consists of three steps. In the first step of the cycle a layer of powder is applied by a coating device. Next a laser beam is directed onto the powder bed and melts the powder according to the shape of the current slice. After the exposure of the slice, the platform with the part and the powder bed is lowered by the layer thickness and the next cycle begins.[6]

The exposure process is a laser welding process using continuous wave (CW) Nd:YAG lasers with a wavelength of 1064 nm and up to 1000 W of laser power.[5] A melt pool forms at the place where the energy of the laser beam enters the powder bed. By moving the laser focus over the powder bed, an area of the current layer is solidified. The linear energy density, E_s, influences the process stability and the density of the material.[8,9] The energy density is defined by Equation 10.1 as a function of laser power (P_L), scan speed (v_s), and layer thickness (s):[10]

$$E_s = \frac{P_L}{v_s \times s} \tag{10.1}$$

The appropriate energy density for processing a material is an important factor for ensuring the stability of the manufacturing process and the resulting quality of the part. An insufficient energy input does not fully melt the powder, and pores are formed between the partially melted particles. Too much energy leads an increased amount of sparks and splashes and reduces the stability of the melt pool due to the higher temperature and reduced viscosity of the melt. The continuous melt pool changes its shape and eventually falls apart into droplets. This so-called *balling* effect makes it impossible to produce fully dense parts in a stable process.[9,11–13]

To stabilize the melt pool and ensure a good quality of the produced parts, the melt tracks are placed on the powder bed following certain strategies. In single-melt tracks without neighboring structures, the melt pool tends to take a cylindrical form due to the surface tension of the melt.[14] This is depicted in Figure 10.3A. To stabilize the melt of the following tracks, each melt track overlaps the neighboring tracks. The melt is drawn to the neighboring track by the surface tension, as shown in Figure 10.3B. The larger contact area between the melt pool and the solid structure stabilizes the melt pool. The manufacturing process becomes more stable and the number and size of material

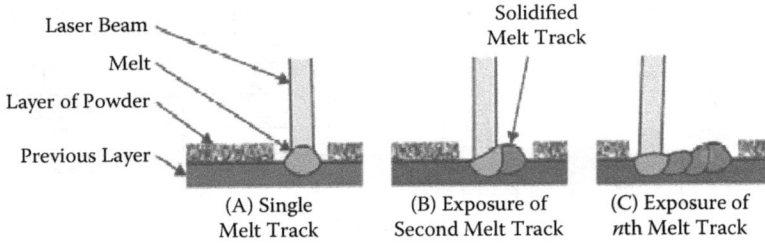

FIGURE 10.3 Powder consolidation during additive manufacturing.[17]

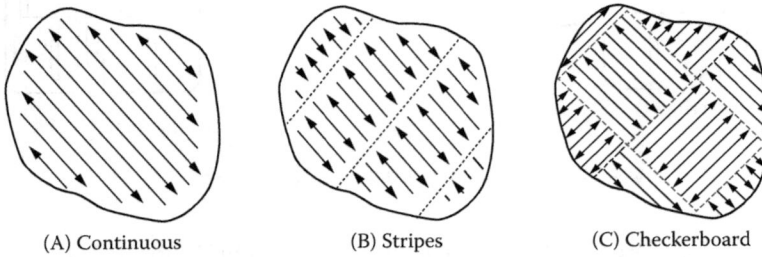

FIGURE 10.4 Exposure strategies used in laser additive manufacturing.[21]

defects are reduced.[15,16] Therefore, the parallel melt tracks of a layer are placed with an overlap as shown in Figures 10.3B and 10.3C. A hatch distance (h_s) of 0.7 of the melt track width b empirically shows the best results regarding process stability and part density:[11,18]

$$h_s = 0.7 \times b \tag{10.2}$$

In the laser welding process, the focused energy input into the melt pool leads to thermal stresses. The mechanical properties and thermal stresses along a melt track are different than those perpendicular to the track or in the build direction. Different scan strategies are used to fill the area with scan vectors. Figure 10.4 presents three scan strategies to fill an area with melt tracks. The continuous strategy on the left is commonly not used, because long scan vectors lead to higher thermal stresses and material defects.[8,19,20] Machine manufacturers recommend either the stripes scan strategy shown in Figure 10.4B or the checkerboard strategy depicted in Figure 10.4C. A heat treatment after the manufacturing process eliminates the thermal stresses.

To further reduce mechanical anisotropy, the scan strategy alters the position and orientation of the scan vectors from one layer to the next.[20,22,23] Figure 10.5 depicts how the checkerboard exposure strategy is altered on a Concept Laser machine. The orientation of the hatching of each square is turned by 90° and the position is shifted in the x and y directions.

In addition to creating a more isotropic material, the alternation between the layers also reduces material defects. The starting and end points of the melt tracks are no longer located at the same position, and there is a chance that a void formed at these locations will be closed by the melt track of the next layer. The scan strategy can be found in the microstructure of laser additive material. Figure 10.6 shows a metallographic sample of laser additive manufactured CoCrMo. The photograph was taken with the build direction pointing to the top of image and view direction along one of the two perpendicular scan directions. The layer-wise alternating direction of the melt tracks is clearly visible.

Using the strategies above with overlapping melt tracks and alternating scan strategies a relative density of over 99% is achieved.[9] The static mechanical properties of laser additive manufactured material are similar to those of conventional material, as can be seen in Table 10.1 for 316L stainless

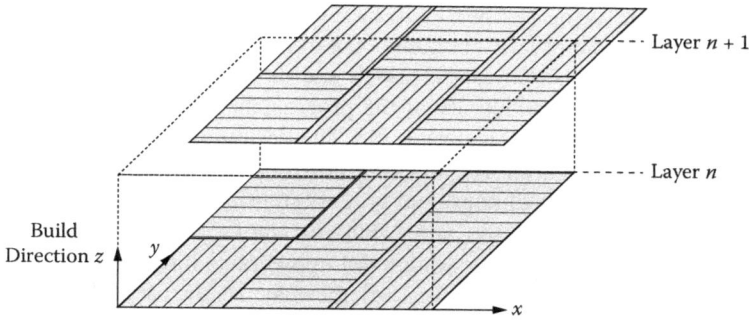

FIGURE 10.5   Shift of pattern from one layer to the next.

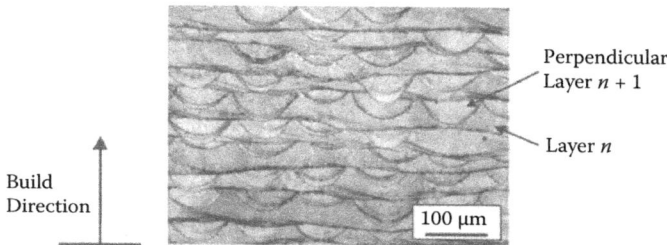

FIGURE 10.6   Microstructure of laser additive manufactured CoCrMo.[24]

steel and in Table 10.2 for Maraging 300 steel. Dynamic mechanical properties of laser additive manufactured materials depend on the surface quality. Samples of laser additive and conventional metals with a machined surface have similar dynamic properties.[25]

TABLE 10.1
Tensile Strength of Conventional and Laser Additive Manufactured 316L Stainless Steel

Material	Yield Strength (MPa)	Ultimate Tensile Strength (MPa)	Elongation (%)
Conventional[26]	310	620	30
Laser additive manufactured[27]	640	760	30

Source: Spierings, A.B. et al., Rapid Prototyping J., 19, 88–94, 2013.

TABLE 10.2
Tensile Strength of Conventional and Laser Additive Manufactured Maraging 300 Steel

Material	Yield Strength (MPa)	Ultimate Tensile Strength (MPa)	Elongation (%)
Conventional annealed[28]	900	950–1100	10
Conventional hardened (490°C)[28]	1800	1900–2100	9
Laser additive manufactured as- build[29]	1214	1290	13.3
Laser additive manufactured hardened (480°C)[29]	1998	2217	1.6

Sources: Wegst, C. and Wegst, M., Stahlschlüssel, 23rd ed., Verlag Stahlschlüssel Wegst, Marbach, 2013; Kempen, K. et al., Phys. Proc., 12, 255–263, 2011.

Laser additive manufacturing has proven to be a useful manufacturing technology for ensuring the industrial and end-user quality of complex parts. The material properties already match those of conventional materials. Current research and development activities on materials and processes aim to increase the productivity of the process, as well as the size and quality of the parts, in addition to developing new materials.[30,31] The design of additive manufactured parts is a second emerging field of research. By breaking down the manufacturing of a complex three-dimensional part into simple two-dimensional process steps the manufacturing costs are no longer determined by the complexity of the part. Additive manufacturing offers a greater freedom of design, thus changing how products are designed. Engineers no longer have to focus on creating simple parts that are easy to manufacture. Instead, they can use the capabilities of additive manufacturing to increase the performance characteristics of the products beyond the limits of design for the manufacture of conventional processes.[32–34] This liberty to create new geometrical shapes applies not only to the macroscopic shape of the part but also to smaller scales. A result of these activities has been the development of metal parts with gas-permeable sections. The characteristics of laser additive manufacturing enable the design and production of such mesostructured materials for industrial applications.

10.3 STRATEGIES FOR GAS-PERMEABLE STRUCTURES

The high degree of freedom in design offered by laser additive manufacturing can be utilized to create gas-permeable structures. Different strategies are presented here to create a gas-permeable material and to adjust the permeability to match the requirements of the specific application. The ability to adapt a structure allows the use of laser additive manufactured, gas-permeable materials in very different applications without developing a new type of material. As a general rule a material is gas permeable if channels or connected voids allow air or other gas to flow through the material.[3,35] A variety of techniques are available to create these connections in a permeable structure. The choice of strategy depends on the requirements of the specific application. A generic requirement of all applications is a low degree of unconnected or blind pores (Figure 10.7), which have no or only one connection to the environment. They deteriorate the thermal and mechanical properties of the material without contributing to the permeability and should therefore be avoided.

The distinction between through pores and other pores with less than two connections to the environment shows clearly that it is not sufficient to increase the porosity of the material by reducing the energy density, for example. Research has shown that reducing the linear energy density increases the porosity of the material, but those pores are randomly distributed in the part and have few connections. In general, this is not desired, because the higher porosity reduces the mechanical properties of the material and a sufficient permeability is observed only at high porosities.[13,22,36,37] A more targeted approach is necessary to create a permeable material with mostly through pores. Such an approach would also allow better control over size and the number of connections, which determines the degree of permeability. Different industrial and end-user applications require different permeabilities, thus influencing the choice of strategy for additive manufacturing the parts.

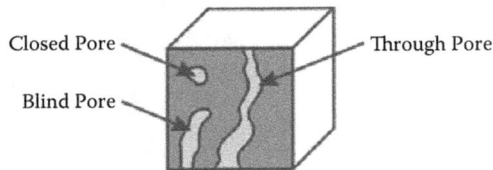

FIGURE 10.7 Classification of pores.[35]

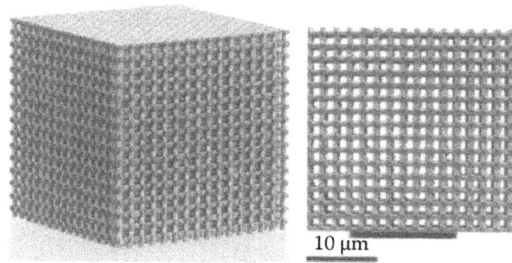

FIGURE 10.8 Lattice spatial structure fabricated from 316L stainless steel.[38]

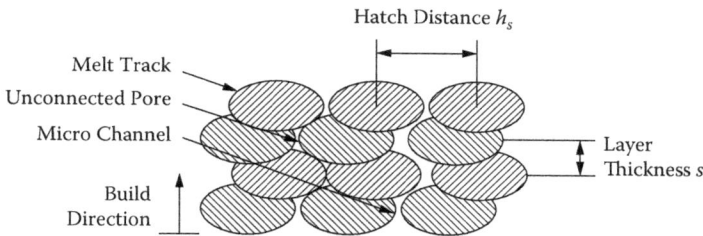

FIGURE 10.9 Low permeability due to an increased porosity.[41]

Additive manufactured lattice structures allow a high gas flow. The air flows around the rods of the structure in any direction. Lattices like the one presented in Figure 10.8 and other cellular mesostructures can be designed in CAD. This straightforward approach takes advantage of the geometric freedom in the design of additive manufacturing and works well on lattices with large rod diameters.[38,39] Small rod diameters exceed the capabilities of the process chain to prepare the data for additive manufacturing. With proper software it is possible to reduce the rod diameter to the diameter of the solidified melt pool.[40]

For applications with only one direction of flow and a need for smaller passages, other structures are more suited than lattices. To produce these structures, the user has to manipulate the exposure strategy and process parameters. A more promising strategy than introducing a random porosity is to increase the hatch distance above the melt track width and maintain a constant direction of scan vectors between layers. Doing so without controlling the position of the scan vectors in the following layers results in a structure like the one shown in Figure 10.9. The scan vectors in Figure 10.9 are parallel and the melt tracks do not overlap. Pores form between the melt tracks. They are located in lines parallel to the melt tracks within a layer. Along these lines it is possible for the pores to connect and create through pores. In the build direction, through pores are created only in locations where the microchannels are not blocked by a melt track. The material is gas permeable in the direction of the scan vectors and has little permeability in the build direction.[37] Because the length of a scan vector is limited by thermal stress, the thickness of the permeable material is limited.

To optimize permeability in the build direction, the structure in Figure 10.9 can be further improved. The approach to create gas-permeable material by increasing the hatch distance is maintained, but in addition the position of each scan vector is controlled with respect to the melt tracks in the other layers. The melt tracks are placed on top of each other as shown in Figure 10.10. The stacked tracks form a lamellar structure with walls and continuous gaps in the build direction. Just as for the structure displayed in Figure 10.9, the material is permeable in the direction of the scan vectors, but the permeability in the build direction is largely improved.

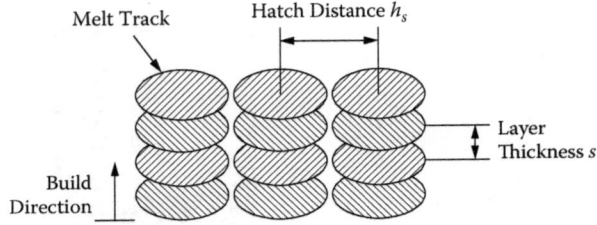

FIGURE 10.10 Stacked melt tracks for high air flow in the build direction.[41]

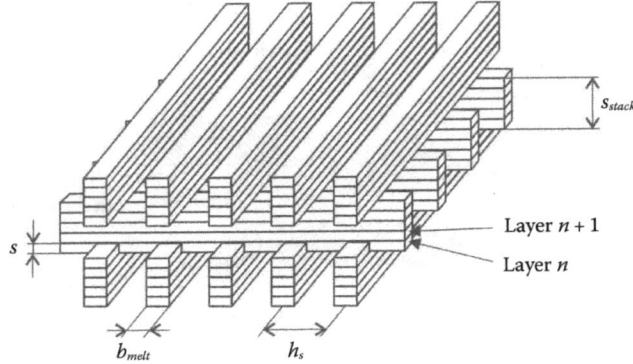

FIGURE 10.11 Scanning strategy for manufacturing porous material for medical applications.[42]

The walls of the lamellar structure are as thick as the solidified melt track, and the length and height can be up to a few millimeters. The large ratio between the size of the wall and its thickness makes the structure vulnerable to lateral forces. To produce parts for industrial applications the stability of the walls must be increased. Connecting the walls with perpendicular melt tracks, as shown in Figure 10.11, adds additional stability to the structure. To create this structure, the orientation of the scan vectors is turned after a predefined number of layers. The scanning strategy in Figure 10.11 was developed for porous biomaterials on medical implants[42] but can also be used for gas-permeable materials.

Most applications demand post-machining the laser additive manufactured parts by grinding, milling, electrical discharge machining (EDM), or other subtractive manufacturing processes to meet the dimension and surface quality requirements. Machining removes material from the surface of the part without respect to the structure of the permeable material. A slope machined into the grid depicted in Figure 10.11 would show different orientations of gaps in the surface. This results in an irregular structuring of the surface, which is not acceptable in many end-user and industrial products. To meet the technical and aesthetic requirements, a structure has to be robust against mechanical loads and invariant in the building direction of additive manufacturing. The mechanical stability of the grid in Figure 10.11 is achieved by the intersections of the perpendicular melt tracks.

An alternative approach is to fixate the ends of the walls. The exposure strategy shown in Figure 10.12 does this by increasing the hatch distance and adapting the checkerboard strategy. The area of the current slice of the part is separated in smaller areas, as is done in laser additive manufacturing of solid material. To produce permeable material, the position of the fields of the checkerboards no longer varies between the layers, and the position of the melt tracks is constant.[41] The design is not limited to checkerboards. Tilings with other geometric elements are possible, such as with triangle- or hexagon-shaped tiles as shown in Figure 10.13. These tilings or any other combination can be used to fill the gas-permeable section of a part. At points where the end of a wall is not fixed by an intersecting wall an additional melt track is required.

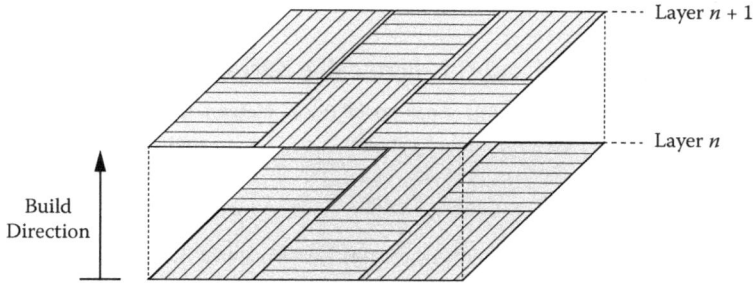

FIGURE 10.12 Structure of additive manufactured material for high air flows.[41]

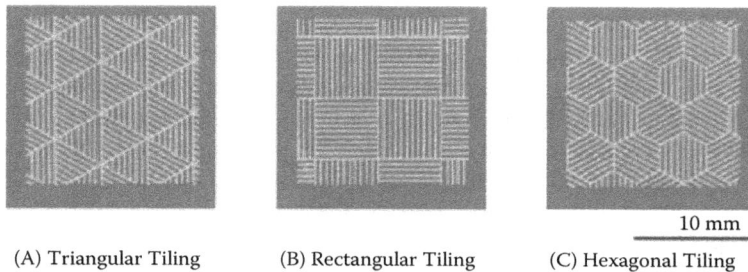

(A) Triangular Tiling   (B) Rectangular Tiling   (C) Hexagonal Tiling

FIGURE 10.13 Different tilings to create gas-permeable structures.

10.4 LASER ADDITIVE MANUFACTURING OF PERMEABLE MATERIAL

Further investigations are necessary for the successful use of gas-permeable structures in an industrial or end-user product. Current research on manufacturability and material properties is presented in the following. In a structure like the one shown in Figure 10.12 melt tracks are stacked on top of each other. The closest control over the position of each track is achieved by defining the scan vectors and entering the list of scan vectors into the laser additive manufacturing machine. Other approaches, such as designing the structure in CAD or manipulating the process parameters, are less reliable in creating a structure like that shown in Figure 10.12.

The stacked melt tracks affect the laser additive manufacturing process. The laser energy entering into the melt pool has to be dissipated in order for the melt to solidify. The thermal conductivity of the powder bed is significantly lower than that of the solid material;[43] therefore, the majority of heat is conducted into the already solidified material of the part. In laser additive manufacturing of solid materials, the heat is conducted into the surrounding material in all directions. The stacked melt tracks of gas-permeable material are separated by gaps that isolate the walls from each other. Heat can only be transported within the walls. Due to this limitation in space, less heat is removed from the melt pool at a given time step. The temperature of the pool rises and the cooling rate decreases. The melt pool becomes longer and wider and the melt itself is less viscous. This has two effects that both diminish the process stability of laser additive manufacturing. A longer melt pool is less stable and eventually fragments into drops. This is referred to as *balling*.[15] The second effect results from the wider melt pool. Generally, the surface tension of the molten metal forms the melt pool into a cylinder. This occurs in all laser processes with significant amounts of melt.[44] A wider melt pool contains a larger melt volume, and this increases the height of the solidified melt track. If the diameter exceeds the layer thickness the coating device comes in contact with the solidified melt tracks. Excessive wear of machinery and possible blockage of the coater are the undesired results of these elevated melt tracks.[16]

TABLE 10.3

Process Parameters for Solid and Gas-Permeable X3NiCrMoTi Tooling Steel

Material	Laser Power (P_L) (W)	Scan Speed (v_s) (mm/s)	Hatch Distance (h_s) (μm)	Layer Thickness (s) (μm)
Solid material	180	600	105	30
Gas-permeable material	100	800	120–200	30

Source: Klahn, C., Laseradditiv gefertigte, luftdurchlässige Mesostrukturen: Herstellung und Eigenschaften für die Anwendung, PhD dissertation, Hamburg, 2014.

Laser additive manufacturing of solid parts from X3NiCrMoTi powder on a Concept Laser M2 cusing machine uses a laser power of $P_L = 180$ W, a scan speed of $v_s = 600$ mm/s, and a hatch distance of $h_s = 105$ μm. With these process parameters, the mechanical properties shown in Table 10.2 are achieved. To compensate for less conducted heat in gas-permeable material, the energy density needs to be reduced. According to Equation 10.1, this can be done by reducing the laser power (P_L), increasing the scan speed (v_s), or adjusting both parameters. The layer thickness (s) affects all parts of the build job and remains unchanged. The hatch distance (h_s) is a parameter that adjusts the permeability. The support structure is a structure also built by adding single tracks on top of each other, and it connects the part to the build platform during laser additive manufacturing. The development of parameters for gas-permeable structures made from a specific material can be based on the support parameters, as recommended by the manufacturer of the laser additive manufacturing system. On the M2 cusing machine, the tooling steel supports are built with a laser power of $P_L = 120$ W and a scan speed of $v_s = 800$ mm/s. To reduce elevated melt tracks the laser power can be further decreased to $P_L = 100$ W. All samples from the tooling steel were manufactured with the process parameters provided in Table 10.3.

Different products require different permeabilities. To identify a suitable hatch distance (h_s) for specific applications, specimens with hatch distances between 120 and 200 μm are manufactured out of X3NiCrMoTi tooling steel using the parameters shown in Table 10.3. A selection of metallographic samples prepared from these specimens is presented in Figure 10.14. Microscopy inspection of the samples reveals different properties of the gaps between the melt track walls.

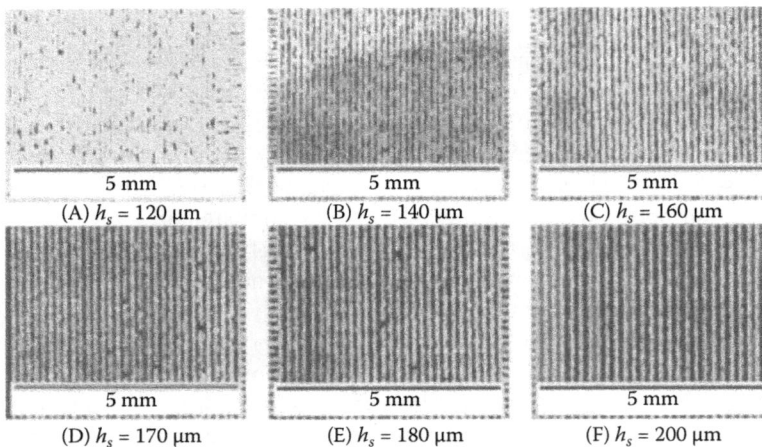

FIGURE 10.14 Metallographic samples of gas-permeable tooling steel with various hatch distances h_s: (A) $h_s = 120$ μm, (B) $h_s = 140$ μm, (C) $h_s = 160$ μm, (D) $h_s = 170$ μm, (E) $h_s = 180$ μm, and (F) $h_s = 200$ μm.[21]

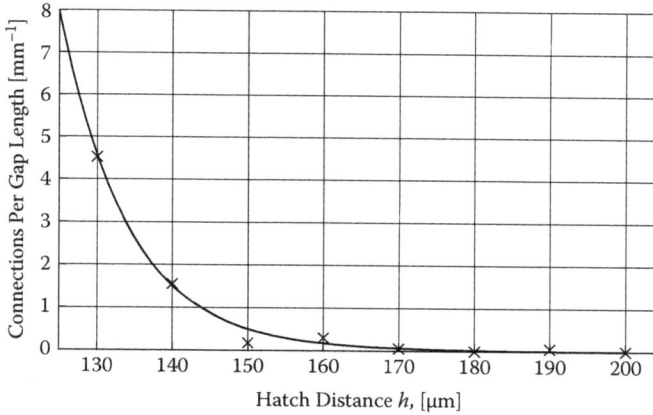

FIGURE 10.15   Number of connections between walls for different hatch distances.[21]

The sample with a hatch distance of $h_s = 120$ µm (Figure 10.14A) shows a mostly solid structure with systematic pores. The walls are not separated by a gap. Instead, rows of large pores with an irregular shape indicate the position of the space between melt tracks. A hatch distance of $h_s = 120$ µm is not wide enough to create the desired gas-permeable structure. Samples with hatch distances greater than 120 µm show continuous gaps between melt track walls. Smaller hatch distances show many connections between the walls, while there are fewer connections at larger hatch distances. Those random connections are the result of melt pool dynamics and partially molten powder particles. The wider the gaps between the melt tracks become, the less likely are these connections. The number of connections in an area is normalized to the average number of connections in a gap length. Figure 10.15 shows the decrease in connections resulting from increasing hatch distance.

The previously described influence of heat conduction on the shape and size of the melt pool indicates that the melt track width (*b*) is not constant. To confirm this, the melt track width of samples is measured close to the build platform and at a distance of 25 mm above the base plate. The build platform is a 250×250-mm plate of 25- to 45-mm thick steel and represents a large heat sink. Figure 10.16 depicts the melt track width of tooling steel with hatch distances ranging from 120 to 200 µm at the described distances. At the upper position, the melt pool is larger due to the low heat conductivity of the structure underneath. The gaps between the melt tracks isolate the walls from

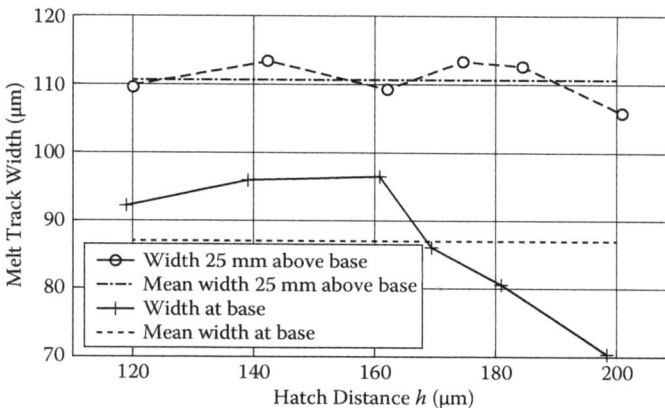

FIGURE 10.16   Melt track width of laser additive manufactured tooling steel at various hatch distances.[21]

each other, and the hatch distance has no effect on the width. Close to the base plate the melt tracks are narrower because of the higher heat flow into the platform and the resulting lower temperature of the melt. A second effect of the building platform is that it acts as a connection between the walls at small hatch distances. This can be obtained with wider tracks for hatch distances that are ≤160 μm at the lower position.

In an industrial or end-user application, the gas-permeable material is integrated into a solid part, with gas distribution channels underneath. It is assumed that the upper position of the samples better represents the situation during the manufacturing of such a part. Therefore, a mean melt track width (b) of 110.9 μm at a distance of 25 mm above the base plate is considered to be the melt track width of gas-permeable material made from X3NiCrMoTi. The evaluation of the sample also confirmed that the measured hatch distances agreed with the distance defined in the scan vector list. The mean gap width (w) of the laser additive manufactured, gas-permeable material is defined by Equation 10.3 as the difference between the hatch distance and mean melt track width:

$$w = h_s - b \qquad\qquad (10.3)$$

Laser additive manufactured parts are known for their rough surface. This roughness is caused by two mechanisms. The first one is inherent to all additive manufacturing processes and causes an increased roughness on sloped surfaces. This is due to the layer-wise manufacturing process, which requires a discrete approximation of the geometry in the build direction. The step size of the discretization equals the layer thickness (s) of the manufacturing process. Figure 10.17 shows this so-called stair-step effect on a simplified geometry. Because the walls of the gas-permeable structures are perpendicular to the base plate, there are no sloped surfaces inside the gaps and the stair-step effect is avoided.

The second mechanism results from the laser welding process in laser additive manufacturing. The melt pool dynamic causes variations in the shape of the melt pool, and because the melt pool solidifies rapidly those variations are preserved in the solid melt track. The melt also comes into contact with powder particles. The heat of the molten metal partially melts the particles and welds them to the surface. The variations of both the melt track and the attached particles increase the roughness of a laser additive manufactured part. Normally, the part is sandblasted after removal from the process chamber. This simple post-processing step removes attached powder particles and smoothes out the surface.[45] Figure 10.18 shows that variations in the melt pool and attached powder particles can also be observed on the walls of the gas-permeable structure. The inside surfaces of the gaps are not accessible for any kind of post-processing, so the surface remains as built. The average surface roughness (R_{zDIN}) is 164.3 μm, and the arithmetical mean roughness (R_a) was measured as 13.9 μm on the walls of gas-permeable samples made from tooling steel with the parameters in Table 10.3. Comparing these roughness values to the gap width (w) at different hatch distances (h_s) explains the observed large number of interconnections between walls at smaller hatch distances, as shown in Figure 10.15.

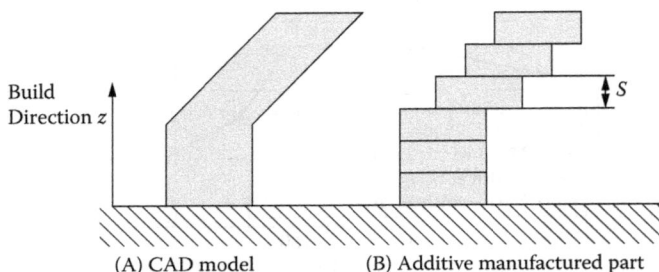

(A) CAD model   (B) Additive manufactured part

FIGURE 10.17   Stair-step effect on sloped surfaces.

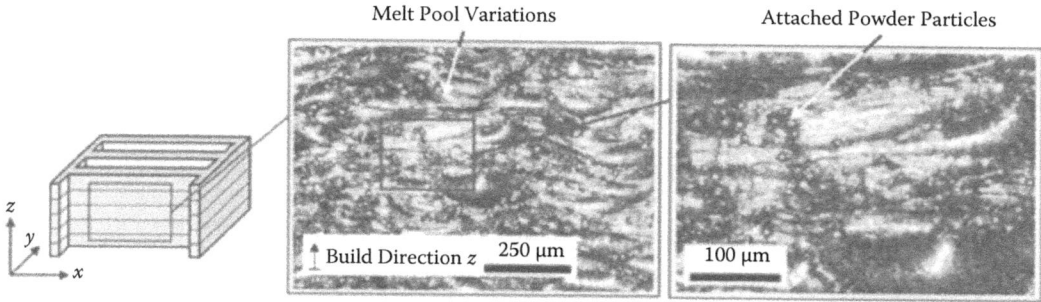

FIGURE 10.18   Surface of the walls inside the gas-permeable structure.

10.5   PROPERTIES OF GAS-PERMEABLE MATERIAL

The strategy to create gas-permeable material by stacking melt tracks offers a variety of possibilities to design structures for specific applications. The structure has to meet different requirements with regard to surface appearance, mechanical stability, and permeability. The appearance is determined by choosing the shape and the size of the tiling.

10.5.1   Mechanical Properties of Gas-Permeable Material

The mechanical properties describe how a structure behaves under mechanical loads. For industrial applications, a robust structure is necessary; it must have the ability to withstand mechanical loads without deteriorating in performance. Two factors determine the mechanical properties of gas-permeable materials: (1) the macroscopic structure of the material (e.g., long walls have a lower bending stiffness and are therefore more vulnerable to lateral forces), and (2) the properties of the walls on a microscopic level. It is not possible to fully describe the effect of the first factor due to the large freedom in design. For each application, the properties of the macroscopic structure have to be determined by experiments or by simulations that reflect the expected load cases. For simulating the gas-permeable structure, the microscopic properties of the material are needed.

Empirical correlations are known between the hardness of a material and other mechanical properties. These correlations are used to identify possible differences between solid and gas-permeable materials and to test for the influence of hatch distance. Microhardness measures are a suitable method to test the Vickers hardness of laser additive manufactured walls. The indentations are small enough to place them along the melt track (Figure 10.19).

Figure 10.20 shows the obtained Vickers hardness HV0.1 of solid and gas-permeable tooling steel that was not heat treated after additive manufacturing using the parameters in Table 10.3. The samples were tested on three walls with at least nine measurements on each wall. The average hardness, standard deviation, and minimum and maximum values are marked in Figure 10.20. It can clearly be seen that neither the hatch distance nor the gas permeability itself has a significant influence on the microhardness of the additive manufactured tooling steel.

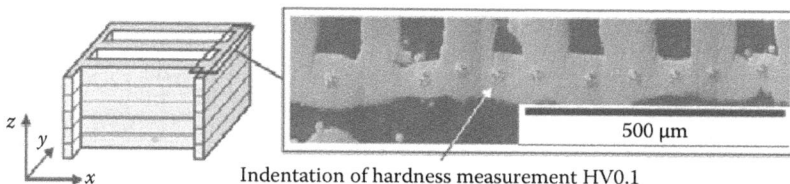

FIGURE 10.19   Indentations of microhardness measurement on a melt track.[21]

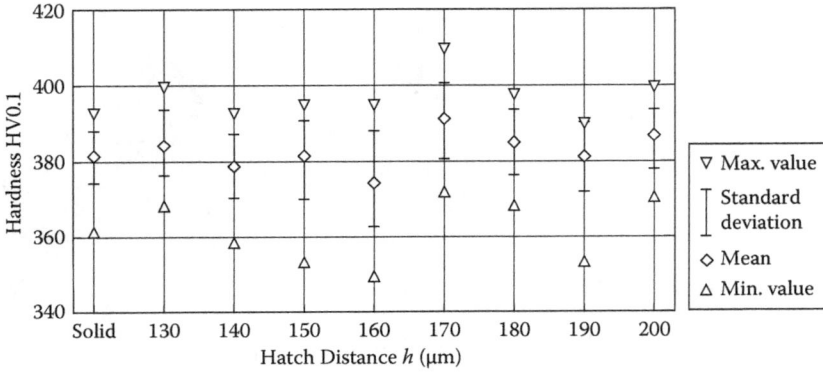

FIGURE 10.20 Vickers hardness HV0.1 of solid and gas-permeable material before heat treatment.[21]

The results of the Vickers hardness indicate no difference between the mechanical properties within the walls and solid material in the as-built state, although different parameters were used for solid and gas-permeable materials, as shown in Table 10.3. The mechanical properties of the tooling steel used can be further improved upon by precipitation hardening by following the heat treatment process recommended by the powder supplier.[46] The samples were heated up at a rate of 100°C/hr to a temperature of 480°C. After precipitation hardening for 6 hours, the temperature was lowered at a rate of −100°C/hr back to room temperature. The hardness measurement was repeated to identify possible differences in the hardening process of solid and gas-permeable materials. Figure 10.21 shows an increase of Vickers hardness HV0.1 between 70% and 80% compared to the values of the as-built samples. As before, no significant difference can be observed between the solid and the gas-permeable samples or between the samples with different hatch distances.

The hardness measurements indicate no difference between the mechanical properties of the tooling steel in solid material or in the walls of the gas-permeable structure. Therefore, it is reasonable to apply the mechanical properties of solid additive manufactured tooling steel to the technical design and simulation of walls of a specific gas-permeable structure. Nevertheless one should be aware that the gaps of the permeable structure reduce the overall mechanical properties of laser additive manufactured, gas-permeable materials.[21]

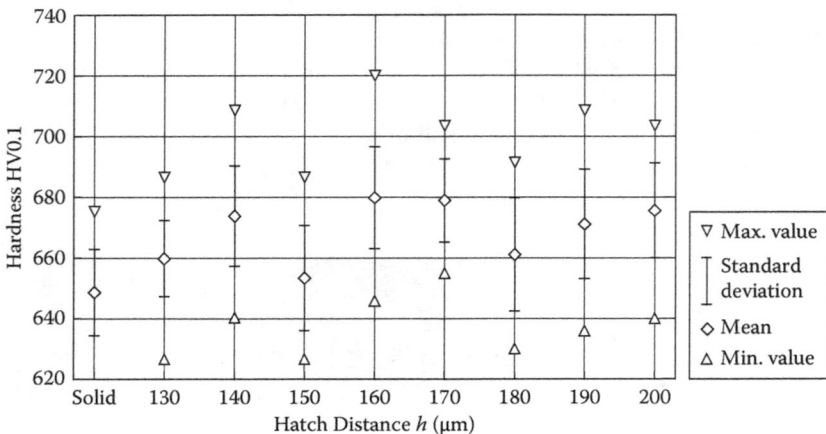

FIGURE 10.21 Vickers hardness HV0.1 of solid and gas-permeable material after heat treatment.[21]

10.5.2 Gas Permeability of Laser Additive Manufactured Material

Different industrial and end-user applications require various levels of permeability for air and other gases. The designer can tune the permeability to meet the requirements by selecting a suitable gap size of the structure. The depth (s) of the gap equals the thickness of the permeable layer and is usually determined by the shape of the part and the location of channels for gas transportation inside the part. Therefore, the design space to adjust permeability by selecting a gap depth is limited. The two remaining parameters of the gap size are the gap width (w), defined by Equation 10.3 as a result of the hatch distance (h_s), the melt track width (b), and the gap length (l), which is usually a result of the size of the tiling. To select these two parameters in order to create a gas-permeable part with a predefined permeability, the designer needs to know the relationship between the gap size and the permeability.

A metric for the permeability is the volume flow or the mean speed of fluid as a function of the pressure difference. In theory, the pressure difference, Δp, in a given channel is described by Equation 10.4 as a function of the mean speed (u) of a fluid with density ρ. The geometry of the channel in the permeable material is described by the gap depth (s) and the hydraulic diameter (d_h). The friction factor (f) describes the physics of pipe friction:

$$\Delta\rho = f\times\frac{s}{d_h}\times\frac{\rho}{2}\times u^2 \tag{10.4}$$

The hydraulic diameter (d_h) is used to transform noncircular channels into circular pipes of equivalent diameter. According to Equation 10.5, the hydraulic diameter is calculated from the perimeter (P) and the area (A) of the channel's cross-section:[47]

$$d_h = \frac{4A}{P} \tag{10.5}$$

Examining Equations 10.4 and 10.5 reveals the available parameters for the designer to adjust the permeability of the structure to meet the requirements of a specific industrial or end-user application. A certain pressure difference Δp at a given air flow or *vice versa* is the aim of the adjustment. The friction factor (f) is a result of flow conditions, Reynolds number, and surface roughness and cannot be controlled by the designer. As described earlier, the thickness (s) of the gas-permeable material is often determined by the overall design of the part and therefore is not a tuning parameter for gas permeability. To control the flow of gas and the pressure drop the designer can vary the hydraulic diameter by choosing an appropriate gap width (w) and length (l) for the gas-permeable structure. Because the gaps in the gas-permeable material are of rectangular shape, Equation 10.5 can be converted into Equation 10.6:

$$d_h = \frac{2\times w\times l}{w+l} \tag{10.6}$$

The two parameters of the gap size have a different impact on the hydraulic diameter (d_h). Table 10.4 demonstrates this by comparing the hydraulic diameter of different gap sizes. An engineer can change the hydraulic diameter significantly by selecting a different hatch distance, but doubling the gap length from 5 mm to 10 mm has little effect. Therefore, the tuning parameter of choice for adjusting permeability is the hatch distance (h_s).

In fluid dynamics, different friction laws are used to calculate the friction factor (f) for different flow conditions of a continuous fluid. The question arises as to whether the flow through the small gaps of the gas-permeable material can be treated as a continuous fluid or if its behavior changes toward a molecular flow. Fluid channels with diameters between 3 mm and 0.2 mm are called

TABLE 10.4

Hydraulic Diameters of Gas-Permeable Materials with Various Gap Lengths and Gap Widths

Hatch Distance (h_s) (µm)	Gap Width (w) (µm)	Hydraulic Diameter (d_h) (µm)	
		Gap Length l = 5 mm	Gap Length l = 10 mm
130	19.1	38.1	38.1
140	29.1	57.9	58.0
160	49.1	97.2	97.7
170	59.1	116.8	117.5
180	69.1	136.3	137.3
200	89.1	175.1	176.6

minichannels, and microchannels are in a range from 0.2 to 0.01 mm.[48] Based on the dimensions in Table 10.4 the gaps in the laser additive manufactured, gas-permeable material are microchannels. The Knudsen number (Kn) indicates if the assumption of a continuous fluid is applicable. It is the dimensionless ratio between the mean free path of the gas molecules (λ) and a characteristic length of the fluid dynamic problem. The mean free path of the gas molecules in air at T = 300 K is λ = 68 nm.[49,50] The characteristic length of gas-permeable material is the gap width (w), because it is the smallest dimension of the gap. The Knudsen number of a gas-permeable material is calculated according to Equation 10.7:

$$Kn = \lambda/w \qquad (10.7)$$

According to Kandlikar et al.,[48] a gas can be treated as a continuous fluid at Knudsen numbers < 10^{-3}. In the range $10^{-3} < Kn < 10^{-1}$, the gas still behaves like a continuous fluid, but discontinuities may occur at the boundaries. Herwig[51] observed that Knudsen numbers < 10^{-2} are necessary for a continuous gas. Table 10.5 shows the gap widths and calculated Knudsen numbers of gas-permeable materials. The Knudsen numbers are in a range close to the described empirical limits of a continuous fluid. Therefore, one can assume a continuous flow but must also be aware that there might be a different behavior due to discontinuities in the boundary layer.

A second point of uncertainty is the effect of the surface on the flow of air through the laser additive manufactured gaps. The high roughness and the connections between the walls will influence friction factor f. There is ongoing discussion in fluid mechanics as to whether the friction laws of macroscopic channels with rough walls also apply to microchannels with the same relative roughness ε/d_h, where ε quantifies the roughness.[49,52–55] The so-called roughness height (ε) has the dimension of length and is derived from Nikuradse's experiments with sand-covered pipes.[56] The roughness height is a phenomenological description of the effect of roughness on the flow compared

TABLE 10.5

Gap Width and Knudsen Numbers of Gas-Permeable Samples

Hatch Distance (h_s) (µm)	Gap Width (w) (µm)	Knudsen Number (Kn)
130	19.1	3.4×10^{-3}
140	29.1	2.2×10^{-3}
160	49.1	1.3×10^{-3}
170	59.1	1.1×10^{-3}
180	69.1	0.9×10^{-3}
200	89.1	0.72×10^{-3}

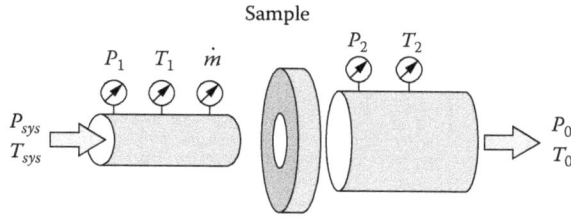

FIGURE 10.22 Test setup for measuring the gas permeability of laser additive manufactured samples.

to the grain size of a sand cover on the walls. There is no generally applicable rule to convert the roughness values of engineering, like the arithmetical mean roughness (R_a), into fluid dynamic's roughness ε. One can estimate from the ratio of the measured surface roughness values R_a or R_{zDIN} and the gap width (w) that the relative roughness (ε/d_h) of the gas-permeable material will be much higher than is usually the case in fluid dynamic applications.

To test the permeability of the structure presented in Figure 10.12, samples with a checkerboard tiling are manufactured on a Concept Laser M2 cusing machine. The square size of ~5 mm is adjusted to fill the area with an integer number of melt tracks at a defined hatch distance ranging from 120 to 200 μm. The total permeable area of the samples is a 19-mm × 38-mm rectangle with a sample thickness of 10 mm. The permeable section is surrounded by a 5-mm-wide solid rim to allow proper sealing between the test bench and the sample. The parameters in Table 10.3 were used to manufacture solid and gas-permeable sections of the samples in one build job.[41]

The permeability of these samples is tested on the test bench sketched in Figure 10.22. A supply system provides dry and unlubricated compressed air at temperature T_{sys} and an adjustable system pressure (p_{sys}) in a range of 0 to 8 bar. At the end of the test bench, the air is vented into ambient atmosphere. The test setup records pressures p_1 and p_2 before and after the sample as well as the temperatures T_1 and T_2. The volume flow ($\dot{V}$) at standard conditions is measured only once at the entry of the test bench and is converted into a mass flow ($\dot{m}$) that is constant along the entire setup.

Figure 10.23 shows the gas flow measured and difference pressure of a set of six samples. The influence of the various hatch distances is clearly demonstrated. Although the samples with a hatch distance of 140 μm and above show good permeability, no significant air flow is observed at the 120-μm hatch distance. This confirms the result of the visual inspection of Figure 10.14.[41] Figure 10.23 indicates that the end of the measurement range of the flow meter at $\dot{V}_{max}$ = 200 L/min was reached at very low pressure differences for most hatch distances. This is due to the large gas-permeable section of the samples, and it was assumed that the results only cover a laminar flow because of the low speed of air in the gaps. To increase the speed and to achieve a turbulent flow a second set of samples was produced with a smaller permeable section. Those samples had 10-mm-long parallel gaps, hatch distances of 130 to 200 μm, and thicknesses of s_{short} = 5 mm and s_{long} = 10 mm. The samples were tested on the test bench, and the volume flow at different pressure differences was measured. For each of these datasets, the friction factor (f) was calculated using Equation 10.4. The samples were manufactured in two different thicknesses to eliminate the pressure drop at the entry and exit of the sample by comparing the friction factors of the two samples. Equation 10.8 allows calculation of the friction factor of the fully developed flow inside the samples:[52]

$$f = \frac{f_{long} \times s_{long} - f_{short} \times s_{short}}{s_{long} - s_{short}} \tag{10.8}$$

Figure 10.24 plots the friction factors for hatch distances from 130 to 200 μm and compares them to the friction laws. There is a significant difference between the friction factors of hatch distances ≤ 140 μm and those ≥ 150 μm. The plots of all friction factors all have a similar shape, but at

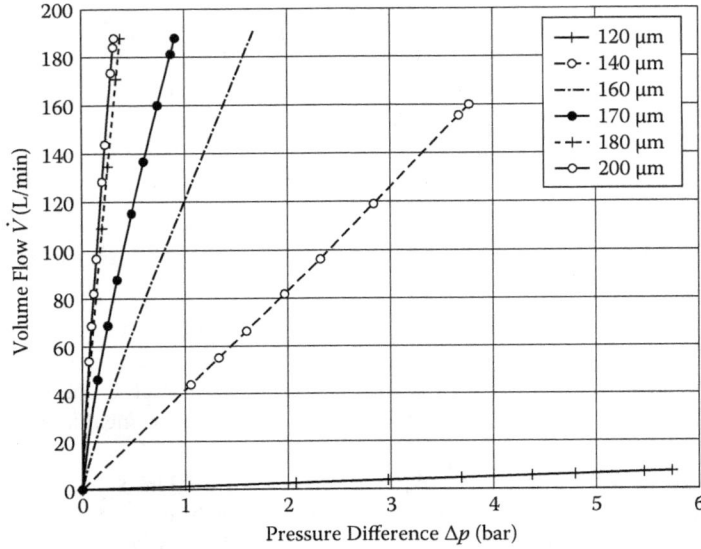

FIGURE 10.23   Gas permeability at hatch distances of 120 to 200 μm.[41]

hatch distances ≤ 140 μm the friction factors are approximately ten times larger. This correlates to the number of connections across the gaps in Figure 10.15, which decreases rapidly above $h_s = 140$ μm. It is likely that these connections disturb the flow through the gaps and increase the pressure drop. The change of slope in the plots indicates a change in the flow conditions between $Re = 100$ and $Re = 300$. Below this transition, the friction factors for hatch distances ≥ 150 μm are similar in size to the ones calculated with the friction law for laminar flows through rectangular channels in Equation 10.9.[57] The factor φ is calculated by Equations 10.10 and 10.11 from the width (w) and height (l) of the channel.[57]

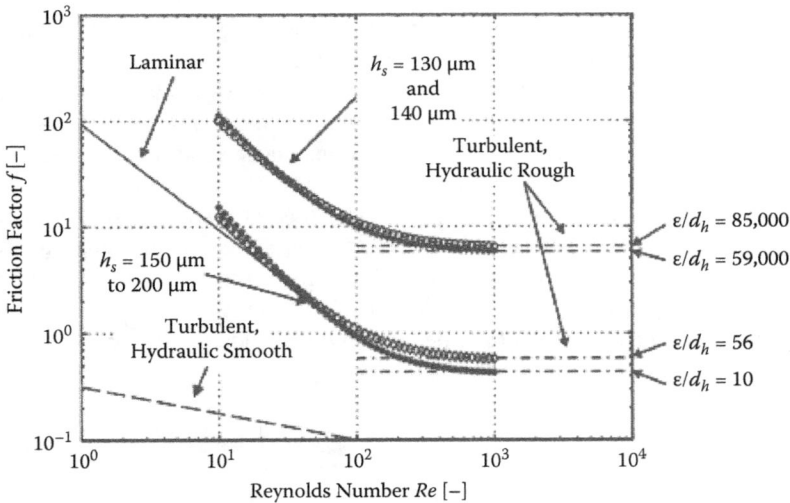

FIGURE 10.24   Friction factors f of laser additive manufactured, gas-permeable material and friction laws for laminar and turbulent flows.

$$f = \varphi(64/Re) \tag{10.9}$$

$$\varphi = 0.878 + 0.0566\varepsilon + 0.758\varepsilon^2 + 0.193\varepsilon^3 \tag{10.10}$$

$$\varepsilon = (w - 1)/(w + 1) \tag{10.11}$$

The alignment of the measured friction factors with the laminar friction laws is not perfect, but from a mechanical engineer's point of view it is sufficient to use laminar friction laws to estimate the pressure loss across the laser additive manufactured, gas-permeable material. The transition from laminar to turbulent flow conditions usually occurs around $Re \approx 2300$. Experiments by Dean[58] and Wibel[52] have shown that the critical Reynolds number for this transition decreases as the relative roughness (ε/d_h) increases.

Beyond the transition, the turbulent flow can follow three different friction laws. As long as the roughness profile of the walls is covered by a laminar boundary layer, friction factor f is independent of the surface roughness. This flow condition is considered to be hydraulically smooth, and the friction is a function of the Reynolds number as shown in Equation 10.12.[57] Figure 10.24 shows no agreement between the measured friction factors and this friction law.

$$f = 0.3164/Re^{1/4} \tag{10.12}$$

The second friction law describes the flow condition when the laminar boundary layer does not cover the entire surface roughness, and individual peaks disturb the turbulent flow in the channel. Because there was no agreement with the smooth pipe law there is also no agreement with this law.

When the surface roughness is no longer covered with a laminar boundary layer the third friction law for rough pipes applies. As shown in Equation 10.13, the friction factor is independent of the Reynolds number and only determined by the relative roughness (ε/d_h). In a Moody diagram, the plots of the rough pipe law resemble parallel horizontal lines for the different relative roughness values.[59,60]

$$f = 0.0055 + 0.15(\varepsilon/d_h)^{1/3} \tag{10.13}$$

Equation 10.13 is used to determine the relative roughness (ε/d_h) from the measured friction factors in Figure 10.24. The relative roughness for hatch distances ≤ 140 µm varies between 59,000 and 85,000, and the roughness height varies between 2.3 and 4.9 m. A roughness in the range of a few meters highlights the large disturbance of the gas flow by the connections between the walls. At hatch distances ≥ 150 µm, the relative roughness varies from 10 to 56 with roughness heights between 1.1 and 4.6 mm. A relative roughness ≥ 10 seems illogical, because it says that the roughness height is considerably larger than the available gap width. At this point one has to keep in mind the phenomenological nature of the roughness height. The roughness of the laser additive manufactured walls has the same effect on the flow as a sanded wall with a homogeneous grain size of ε.

To validate these roughness heights, they are compared to the range of values given as guidance; for example, for rusty cast iron $\varepsilon = 1.0$ to 1.5 mm and for encrusted steel $\varepsilon = 0.5$ to 2.0 mm.[62] The measured average surface roughness (R_{zDIN}) of 164.3 µm of the walls inside the additive manufactured, gas-permeable material is comparable to the values given for sand casting; therefore, the roughness height (ε) is assumed to be a realistic assessment of the effect of the wall's surface on the flow.[63]

Based on the results presented in Figure 10.23, laser additive manufactured parts with a hatch distance of 130 µm and above are suitable for gas-permeable applications in industrial and end-user applications. For applications requiring a high-volume flow, a hatch distance of at least $h_s = 150$ µm should be selected.

Figure 10.15 suggests a higher mechanical strength of permeable material with a hatch distance of 140 µm or less due to the connections between the walls. When an additional mechanical load is placed on the gas-permeable material, the engineer has to balance the hatch distance between a high gas permeability and the additional strength given to the material by the connections.

10.6 APPLICATIONS IN PLASTIC INJECTION MOLDING

Plastic injection molding was selected to demonstrate the benefits of laser additive manufactured, gas-permeable materials in industrial applications. Injection molding was chosen, because it is a challenging process with great relevance to the production of industrial and end-user goods. Polymers are the material of mass production. Most products for industry and end-users contain polymer parts in their assembly or as part of the packaging. This makes polymer processing industries an important economic branch. Almost 85% of the processed materials are thermoplastics.[61] The majority of thermoplastics are processed by extrusion, which is a continuous process to manufacture simple, elongated shapes like tubes, cable ducts, sheets, and films. Second comes plastic injection molding. Injection molding is capable of mass producing complex parts at high quality and low cost.[2,64] In a cyclic process, thermoplastic parts are produced on injection molding machines using molds designed and manufactured individually for each product. The machine injects plastic melt into a closed mold consisting of a movable half and a fixed half. In the mold, the melt is cooled down until it solidifies. The mold opens and ejects the plastic part once it is solid enough to maintain its shape. Two applications for gas-permeable material in injection molding tools have been developed and tested.

10.6.1 VENTING OF THE MOLD

In the injection phase of the injection molding process, the mold is filled with melt in a very short time. As the melt enters the cavity the air inside has to leave the mold. Usually most air escapes through the parting plane between the movable half and the fixed half and through small gaps at inserts, sliders, and ejector pins. Insufficient venting of the cavity results in quality issues and defects. Possible quality issues are differences in surface appearance and gloss due to small amounts of air between melt and mold surface. Situations like the one depicted in Figure 10.25 result in large amounts of air being trapped in the mold. The residue air hinders complete filling of the mold, and the combustion of vapors due to the microdiesel effect is possible.[2,64]

Different options to release the trapped air are available to mold designers. Separating the mold half into two or more parts creates additional gaps in the mold. The air can vent through these gaps. The venting is further improved by grinding the parting lines with a coarse paper and introducing venting channels between the parts at a certain distance to the surface of the cavity. Another option is adding gas-permeable inserts to the mold design. Standardized inserts made of sintered metal are commercially available in a limited number of shapes. The inserts have to be customized to fit into the mold design at the desired location.[64]

Standard inserts are not available for all possible locations in a mold, and the customization of inserts is costly. Laser additive manufactured gas-permeable material offers the possibility to integrate air vents at arbitrary positions. Even more interesting is the option to integrate the gas-permeable material into an existing additive manufactured insert. As a result, no additional manufacturing steps are required.[37]

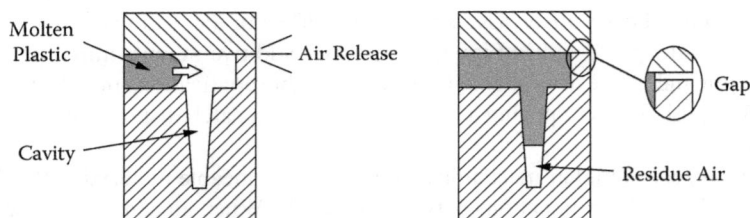

FIGURE 10.25   Insufficient venting of injection molds.[65]

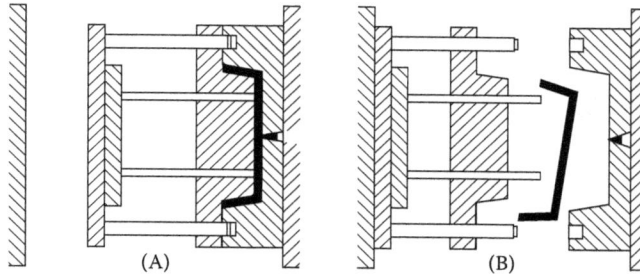

FIGURE 10.26 Simplified injection mold with mechanical ejectors during (A) cooling phase and (B) ejection process.[41]

10.6.2 Gas-Permeable Materials in Pneumatic Ejector Systems

Venting air from the cavity is not the only application for gas-permeable material in injection molding tools. The material allows the flow of air in and out of the mold but blocks the melt. This is used to create a pneumatic ejector system capable of fully replacing mechanical ejectors. Ejectors are an essential part of an injection molding tool. During the cooling phase, the volume of the thermoplastic decreases and the part shrinks onto the movable half of the mold. In classic mold designs, the finished part is removed from the mold by a mechanical ejector system. Figure 10.26 shows a simplified injection mold closed during the cooling phase and while mechanically ejecting the part. Each ejector pin is connected to a base plate. When the movable half opens, the ejector base plate hits an end stop while the half still moves. The ejection pins apply a force on the part to overcome the breakout and friction forces between part and movable half.

Although the working principle of ejectors is quite simple, the system becomes very complex in reality. The flexibility and mechanical strength of the plastic part require many ejector pins to avoid deformations and punch marks. These pins compete with other systems, such as cooling channels, on the limited space inside the tool. The dimensions of the pins have to be within tight tolerances for the production of high-quality parts. If a pin is too long or too short, it will leave marks on the plastic parts. The same applies to the diameters of pins and holes. If the gap between pin and hole is too wide, then melt enters and the quality of the part is reduced. If the gap is too tight, then the pin gets stuck and damages the mold. Therefore, the design, manufacturing, and assembly of the ejector system have a high impact on manufacturing costs and lead time of the mold.

Maintenance and repairs of the ejector system are another disadvantage of mechanical systems. A study recorded maintenance and repairs on 16 molds over 9 years.[66] During this time, ejector systems were responsible for 50% of repairs on injection molds, and 20% of the maintenance hours were invested in the ejector system.[66] To overcome these disadvantages, a pneumatic ejector system was developed. The system is integrated into a laser additive manufactured insert. Figure 10.27 presents the schematic setup of cooling channels and pneumatic ejectors in an additive manufactured

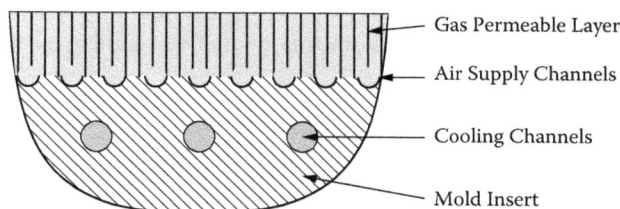

Gas Permeable Layer
Air Supply Channels
Cooling Channels
Mold Insert

FIGURE 10.27 Mold insert with cooling channels and a pneumatic ejector system.[41]

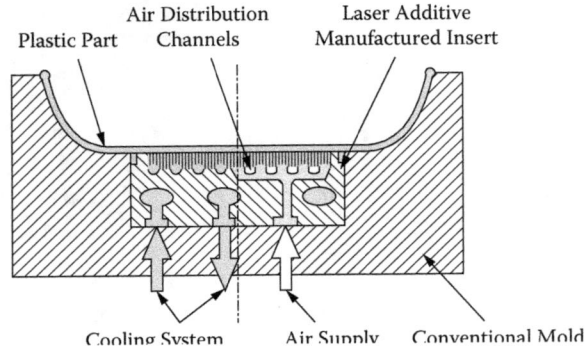

FIGURE 10.28 Movable half with a pneumatic ejector system for a dish.

mold. Within the insert, pressurized air is transported close to the surface of the mold by air supply channels. A top layer of about 5 mm is made of gas-permeable material. A hatch distance of 170 μm was used to create a material with openings wide enough to allow a sufficient air flow from the supply channels to the surface, but narrow enough to keep the melt from entering.[41]

To validate the system, an injection molding tool for plastic dishes was equipped with a pneumatic ejector. As shown in Figure 10.28, the part of the mold forming the lower surface of the dish was chosen as the ejector surface for two reasons. First, this surface of the product is not visible to the user. Second, the elevated rim around the lower surface acts as a seal and forms a pneumatic cylinder with a large piston area during ejection. This gives the part additional acceleration and ensures a reliable removal of the part from the mold.

Injection molding trials were performed with polypropylene (PP) and a blend of polycarbonate and acrylonitrile butadiene styrene (PC+ABS). Two inserts with different scan vector lengths were manufactured and tested. The insert with 10-mm-long scan vectors failed due to melt entering into the air supply channels during the injection phase. This happened during the first shots and indicates a limitation of the scan vector length in the injection molding process. The other insert had 5-mm-long scan vectors. About 60 parts of each of the two different plastics were produced and ejected with this insert. This proves the concept of the pneumatic ejector system.[41]

The trials showed the reliability of the system as well. 90% of the parts made out of polypropylene were successfully ejected. For the blend of PC+ABS a success rate of 100% was achieved. All parts were successfully removed from the mold by the pneumatic ejector system.[41] After the trials, the mold inserts were inspected under a microscope. The insert with 10-mm scan vectors failed due to the lateral force of the injected melt on the walls of the material. The visual inspection of the second insert showed no traces of plastic within the gaps. Inspection of the surface of the plastic parts revealed a light structuring of the surface but no defects that would be considered a quality issue. This is a big advantage compared to mechanical ejectors, which require tight dimensional tolerances in order to leave no marks on the surface of the plastic part. Manufacturing and adjusting such mechanical systems require skilled labor and time. The advantage of the pneumatic system is the integration of an insert with a continuous surface. Machining this surface to give it a smooth look is a lot easier than manufacturing and adjusting a set of mechanical ejectors.[41]

10.7 CONCLUSIONS

Laser additive manufacturing is capable of producing gas-permeable material. For applications in challenging environments not only is a high permeability required but also sufficient strength. Pores with no or only one connection to the environment weaken the material without contributing to the flow through the material. From the strategies discussed here, the one offering the most through channels and the least number of closed or blind pores was chosen for use in an industrial application. By placing melt tracks on top of each other, straight walls with the width of one track are created. Because the walls are vulnerable to lateral forces, a scan strategy is necessary to support the ends of the walls. To reduce the deformation of the walls by lateral forces, the length of the walls must be limited. In injection molding tools, a length of 5 mm proved to be of sufficient stiffness to withstand the forces of the injection molding cycle. The size of the gap between two walls can be adjusted by choosing a suitable hatch distance.

Hatch distances of 130 μm and above show continuous gaps between the walls with sufficient permeability. The permeability is demonstrated by measuring the flow of pressurized air through samples at different pressure differences. Based on these experiments the friction factors of the flow through the gaps of the material were calculated. The values indicate that the critical Reynolds numbers for the transition from laminar to turbulent flow are between $Re = 100$ and 300. At hatch distances ≤ 140 μm, the friction factors are very large and show no alignment with the pipe friction laws. This is due to the number of interconnections across the gaps of the gas-permeable material. The friction factors of hatch distances ≥ 150 μm agree with the friction laws for laminar flows and show a transition directly into a turbulent, hydraulic rough flow condition. The roughness height (ε) was calculated based on the friction factors to vary from 1.1 to 4.6 mm. This is comparable to the roughness height of similar surfaces produced with other manufacturing processes. The acquired data on permeability can be used to select a suitable hatch distance for an industrial application.

To demonstrate that the presented gas-permeable material is suitable for the rough conditions in industrial processes, two applications were identified for injection molding tools. Injection molding is an important industrial manufacturing process with challenging conditions regarding the efficiency and reliability of the process and strength of the used materials. Therefore, the successful use of gas-permeable material in venting and in a pneumatic ejection system verifies the capabilities of laser additive manufactured gas-permeable material in industrial applications.

REFERENCES

1. Sehrt, J.T. (2010). *Möglichkeiten und Grenzen bei der generativen Herstellung metallischer Bauteile durch das Strahlschmelzverfahren*. Shaker, Aachen.
2. Johannaber, F. (2008). *Injection Molding Machines: A User's Guide*, 4th ed. Carl Hanser Publishers, Cincinnati, OH.
3. Purchas, D.B. and Sutherland, K. (2002). *Handbook of Filter Media*, 2nd ed. Elsevier, Oxford.
4. ASTM. (2013). *Terminology for Additive Manufacturing—Coordinate Systems and Test Methodologies*, ISO/ASTM52921:2013. American Society for Testing and Materials, West Conshohocken, PA.
5. Wohlers, T.T. (2014). *3D Printing and Additive Manufacturing State of the Industry, Annual Worldwide Progress Report*. Wohlers Associates, Inc., Fort Collins, CO.
6. Gebhardt, A. (2011). *Understanding Additive Manufacturing: Rapid Prototyping, Rapid Tooling, Rapid Manufacturing*. Hanser, Munich.
7. Poprawe, R. (2005). *Lasertechnik für die Fertigung: Grundlagen, Perspektiven und Beispiele für den innovativen Ingenieur*. Springer, New York.
8. Kruth, J., Froyen, L., Van Vaerenbergh, J., Mercelis, P., Rombouts, M., and Lauwers, B. (2004). Selective laser melting of iron-based powder. *J. Mater. Process. Technol.*, 149: 616–622.
9. Yasa, E., Kempen, K., Kruth, J., Thijs, L., and Van Humbeeck, J. (2010). Microstructure and mechanical properties of maraging steel 300 after selective laser melting. In: *Solid Freeform Fabrication Symposium Proceedings*, pp. 383–396.
10. Lü, L., Fuh, J., and Wong, Y.S. (2001). *Laser-Induced Materials and Processes for Rapid Prototyping*. Kluwer Academic, Boston, MA.
11. Meiners, W. (1999). *Direktes Selektives Laser Sintern einkomponentiger metallischer Werkstoffe*. Shaker, Aachen.
12. Eisen, M.A. (2010). *Optimierte Parameterfindung und prozessorientiertes Qualitätsmanagement für das Selective Laser Melting Verfahren*. Shaker, Aachen.
13. Thijs, L., Verhaeghe, F., Craeghs, T., Van Humbeeck, J., and Kruth, J. (2010). A study of the microstructural evolution during selective laser melting of Ti-6Al-4V. *Acta Mater.*, 58: 3303–3312.

14. Gusarov, A.V. and Smurov, I. (2010). Modeling the interaction of laser radiation with powder bed at selective laser melting. *Phys. Proc.*, 5(B): 381–394.

15. Yadroitsev, I., Gusarov, A.V., Yadroitseva, I., and Smurov, I. (2010). Single track formation in selective laser melting of metal powders. *J. Mater. Process. Technol.*, 210: 1624–1631.

16. Yasa, E., Deckers, J., Craeghs, T., Badrossamay, M., and Kruth, J. (2009). Investigation on occurrence of elevated edges in selective laser melting. In: *Solid Freeform Fabrication Symposium Proceedings*, pp. 180–192.

17. Yadroitsev, I. and Smurov, I. (2011). Surface morphology in selective laser melting of metal powders. *Phys. Proc.*, 12: 264–270.

18. Over, C. (2003). *Generative Fertigung von Bauteilen aus Werkzeugstahl X38CrMoV5-1 und Titan TiAl6V4 mit Selective Laser Melting*. Shaker, Aachen.

19. Munsch, M. (2013). *Reduzierung von Eigenspannungen und Verzug in der laseradditiven Fertigung*. Cuvillier, Göttingen.

20. Kruth, J., Mercelis, P., Van Vaerenbergh, J., Froyen, L., and Rombouts, M. (2005). Binding mechanisms in selective laser sintering and selective laser melting. *Rapid Prototyping J.*, 11: 26–36.

21. Klahn, C. (2014). Laseradditiv gefertigte, luftdurchlässige Mesostrukturen: Herstellung und Eigenschaften für die Anwendung, doctoral dissertation, Hamburg.

22. Meier, H. and Haberland, C. (2008). Experimentelle Untersuchungen zum Laserstrahlgenerieren metallischer Bauteile [experimental studies on selective laser melting of metallic parts]. *Materialwiss. Werkstofftech.*, 39: 665–670.

23. Tolosa, I., Garciandía, F., Zubiri, F., Zapirain, F., and Esnaola, A. (2010). Study of mechanical properties of AISI 316 stainless steel processed by "selective laser melting," following different manufacturing strategies. *Int. J. Adv. Manuf. Technol.*, 51: 639–647.

24. Gibson, I., Rosen, D.W., and Stucker, B. (2010). *Additive Manufacturing Technologies*. Springer, New York.

25. Spierings, A.B., Starr, T.L., and Wegener, K. (2013). Fatigue performance of additive manufactured metallic parts. *Rapid Prototyping J.*, 19: 88–94.

26. ASM Handbook Committee. (1995). *Properties and Selection: Irons, Steels, and High-Performance Alloys*, 10th ed. American Society for Metals, Metals Park, OH.

27. Spierings, A.B., Herres, N., and Levy, G. (2011). Influence of particle size distribution on surface quality and mechanical properties in AM steel parts. *Rapid Prototyping J.*, 17: 195–202.

28. Wegst, C. and Wegst, M. (2013). *Stahlschlüssel*, 23rd ed. Verlag Stahlschlüssel Wegst, Marbach.

29. Kempen, K., Yasa, E., Thijs, L., Kruth, J., and van Humbeeck, J. (2011). Microstructure and mechanical properties of selective laser melted 18Ni-300 steel. *Phys. Proc.*, 12: 255–263.

30. Gausemeier, J., Echterhoff, N., and Wall, M. (2013). *Thinking Ahead the Future of Additive Manufacturing: Innovation Roadmapping of Required Advancements*. Universität Paderborn.

31. Gausemeier, J. and Wall, M. (2013). *Thinking Ahead the Future of Additive Manufacturing: Exploring the Research Landscape*. Universität Paderborn.

32. Klahn, C., Leutenecker, B., and Meboldt, M. (2014). Design for additive manufacturing—supporting the substitution of components in series products. *Proc. CIRP*, 21: 138–143.

33. Hopkinson, N., Hague, R.J., and Dickens, P.M., Eds. (2006). *Rapid Manufacturing: An Industrial Revolution for the Digital Age*. John Wiley & Sons, Chichester.

34. Emmelmann, C., Herzog, D., Kranz, J., Klahn, C., and Munsch, M. (2013). Manufacturing for Design—Laseradditive Fertigung ermöglicht neuartige Funktionsbauteile. *Indust. Manage.*, 29: 58–62.

35. Stoffregen, H., Fischer, J., Siedelhofer, C., and Abele, E. (2011). Selective laser melting of porous structures. In: *Solid Freeform Fabrication Symposium Proceedings*, pp. 680–695.

36. Li, R., Liu, J., Shi, Y., Du, M., and Xie, Z. (2010). 316L stainless steel with gradient porosity fabricated by selective laser melting. *J. Mater. Eng. Perform.*, 19: 666–671.

37. Trenke, D. (2006). *Selektives Lasersintern von porösen Entlüftungsstrukturen am Beispiel des Formenbaus*. Papierflieger, Clausthal-Zellerfeld.

38. Yadroitsev, I., Shishkovsky, I., Bertrand, P., and Smurov, I. (2009). Manufacturing of fine-structured 3D porous filter elements by selective laser melting. *Appl. Surf. Sci.*, 255: 5523–5527.

39. Williams, C.B. (2008). Design and Development of a Layer-based Additive Manufacturing Process for the Realization of Metal Parts of Designed Mesostructure, PhD dissertation, Georgia Institute of Technology, Atlanta.

40. Rehme, O. (2009). *Cellular Design for Laser Free Form Fabrication*. Cuvillier, Göttingen.

41. Klahn, C., Bechmann, F., Hofmann, S., Dinkel, M., and Emmelmann, C. (2013). Laser Additive Manufacturing of Gas Permeable Structures. *Phys. Proc.*, 41: 873–880.

42. Stamp, R., Fox, P., O'Neill, W., Jones, E., and Sutcliffe, C. (2009). The development of a scanning strategy for the manufacture of porous biomaterials by selective laser melting. *J. Mater. Sci. Mater. Med.*, 20: 1839–1848.

43. Rombouts, M., Froyen, L., Gusarov, A.V., Bentefour, E.H., and Glorieux, C. (2005). Photopyroelectric measurement of thermal conductivity of metallic powders. *J. Appl. Phys.*, 97: 024905.

44. Mazumder, J. (1991). Overview of melt dynamics in laser processing. *Opt. Eng.*, 30: 1208–1219.

45. Gibson, I., Rosen, D.W., and Stucker, B. (2010). *Additive Manufacturing Technologies*. Springer, New York.

46. Concept Laser. (2010). *Heat Treatment CL 50WS*. Concept Laser GmbH, Lichtenfels.

47. Kleinstreuer, C. (2008). *Modern Fluid Mechanics: Intermediate Theory and Applications*. Springer, Berlin.

48. Kandlikar, S.G., Garimella, S., Li, D., Colin, S., and King, M.R. (2006). *Heat Transfer and Fluid Flow in Minichannels and Microchannels*. Elsevier, Amsterdam.

49. Kandlikar, S.G. and Grande, W.J. (2003). Evolution of microchannel flow passages: thermohydraulic performance and fabrication technology. *Heat Transfer Eng.*, 24: 3–17.

50. Möller, D. (2003). *Luft: Chemie, Physik, Biologie, Reinhaltung, Recht*. Walter de Gruyter, Berlin.

51. Herwig, H. (2006). *Strömungsmechanik: Eine Einführung in die Physik und die mathematische Modellierung von Strömungen*. Springer-Verlag, Berlin.

52. Wibel, W. (2009). *Untersuchungen zu laminarer, transitioneller und turbulenter Strömung in rechteckigen Mikrokanälen*. Forschungszentrum, Karlsruhe.

53. Gloss, D., Herwig, H., and Peters, F. (2009). *Der Einfluss von Wandrauheiten auf laminare Strömungen: Untersuchungen in Mikrokanälen. Techn. Univ., Institut für Thermofluiddynamik, Diss. Hamburg-Harburg*. Cuvillier Verlag, Göttingen.

54. Kandlikar, S.G., Schmitt, D., Carrano, A.L., and Taylor, J.B. (2005). Characterization of surface roughness effects on pressure drop in single-phase flow in minichannels. *Phys. Fluids*, 17: 100606.

55. Croce, G. and D'Agaro, P. (2005). Numerical simulation of roughness effects on microchannel heat transfers and pressure drop in laminar flow. *J. Phys. D Appl. Phys.*, 38: 1518–1530.

56. Nikuradse, J. (1933). *Strömungsgesetze in rauhen Rohren*. VDI Verlag, Berlin.

57. Brauer, H. (1971). *Grundlagen der Einphasen- und Mehrphasenströmungen*. Sauerländer, Frankfurt.

58. Dean, R.B. (1978). Reynolds number dependence of skin friction and other bulk flow variables in two-dimensional rectangular duct flow. *J. Fluids Eng.*, 100: 215.

59. Moody, L.F. (1944). Friction factors of pipe flow. *Trans. ASME*, 66: 671–684.

60. Bohl, W. and Elmendorf, W. (2005). *Technische Strömungslehre*, 13th ed. Vogel, Würzburg.

61. Xanthos, M. and Todd, D.B. (2004). Plastics processing. In: *Encyclopedia of Polymer Science and Technology*, 3rd ed. (Mark, H.F., Ed.), pp. 1–29. Wiley-VCH, Weinheim.

62. ISO. (2003). *Measurement of Fluid Flow by Means of Pressure Differential Devices Inserted in Circular Cross-Section Conduits Running Full. Part 1. General Principles and Requirements*, ISO 5167-1:2003. International Organization for Standardization, Geneva, Switzerland.

63. Fischer, U., Kilgus, R., Paetzold, H., Schilling, K., Heinzler, M., Näher, F. et al. (1999). *Tabellenbuch Metall*, 41st ed. Europa Lehrmittel, Haan-Gruiten.

64. Pötsch, G. and Michaeli, W. (2008). *Injection Molding: An Introduction*, 2nd ed. Carl Hanser Verlag, Munich.

65. Kojima, M., Narahara, H., Nakao, Y., Fukumaru, H., Koresawa, H. et al. (2008). Permeability characteristics and applications of plastic injection molding fabricated by metal laser sintering combined with high speed milling. *Int. J. Autom. Technol.*, 2: 175–181.

66. Feldhaus, A. (1993). *Instandhaltung von Spritzgießwerkzeugen Analyse des Ausfallverhaltens und Entwicklung angepaßter Maßnahmen zur Steigerung der Anlagenverfügbarkeit*. Verlag der Augustinus Buchhandlung, Aachen.

11 Additive Manufacturing of Components from Engineering Ceramics

James D. McGuffin-Cawley

CONTENTS

ABSTRACT

Description is given of two technical approaches to the exploitation of additive manufacturing in the forming of powder compacts for the production of engineering ceramic components. Both strategies permit the development of complex, finely resolved internal features. This is a notable development for powder processing technologies. A green-tape lamination method is preferred for platelike geometries. Tape allows ready formation of defect-free layers from finely divided sinterable ceramic powders. An effort is made to describe how additive manufacturing is similar to and different from well-established green-tape lamination applications. A casting method is described that takes advantage of a combination of additive manufacturing methodologies, or technologies, to produce permanent and soluble tooling. Some examples of parts being evaluated for performance are presented.

11.1 INTRODUCTION

Additive manufacturing (AM) is a point on a continuum of advances in manufacturing. Bourell et al.,[1] for example, traced its history back to the mid-1800s. In the modern era (i.e., since the mid-1980s), the terminology associated with AM has evolved from rapid prototyping to solid freeform fabrication to either additive manufacturing or three-dimensional (3D) printing. This latter term, 3D printing, was coined by Sachs et al.[2] to refer specifically to the process of using inkjet printing to locally apply a binder to sequential layers of a powder bed, but it has now been long used as a colloquial term for the entire field independent of the technology employed. It is notable that an early application of 3D printing was to create "green" (in the sense of unfired) ceramic parts compatible with conventional firing.[3] In fact, the general strategy of AM for ceramics continues to be the production of green parts that are later fired conventionally, as is evident in several comprehensive reviews.[4–7]

In principle, direct fabrication of ceramic parts by a process such as selective laser sintering/melting is possible. However, there are two problems that are fundamental. The first is that thermal stresses typically lead to cracking. The preheating temperatures necessary to avoid thermal shock are extreme (>1600°C). Also, the requirement of free-flowing dry powders requires coarse powders (in excess of tens of microns), and this creates engineering limits. These constraints are evident in the results obtained, for example, by Betrand et al.[8] and Hagedorn et al.[9]

The process development described herein was intentionally constrained to employ conventional ceramic powders to create green powder compacts that are compatible with conventional binder burnout and sintering. This reduces the burden of developing special powders (that might restrict available compositions and be associated with high cost) or the need for specialized capital equipment, or cycles, for thermal processing. It also means that prior work done to optimize raw-material/thermal-processing to achieve particular properties can be readily adapted to AM.

Geometrically, the process development goal was to achieve equal flexibility with respect to internal and external features, the aspirational analogy being to develop a process or processes having the geometric flexibility normally associated with cored investment casting of metal alloys. Both direct production of fully shaped green parts (i.e., freeform fabrication) and indirect production of tooling (both permanent and fugitive) are explored. In this context, two well-established techniques of ceramic processing are leveraged. The first is lamination of green tape and the second is gelcasting.

The use of multilayer green tape lamination dates back to the 1960s.[10,11] A number of workers have recognized the potential of laminating tape to effect freeform fabrication of a variety of ceramic materials to form an array of components.[12–15] Similarly, gelcasting is a widely used process dating from the early 1990s[16,17] that is ideally suited for complex shaping. It also is increasingly widely exploited in concert with tooling generated using additive manufacturing.[18–21]

The following review is intended to demonstrate the importance of keeping focus on the ultimate application of the intended parts or components, how depth of knowledge about conventional processing of engineering ceramics assists in optimized additive processing, and the quality of parts that can be so produced.

11.2   GREEN-TAPE-BASED ADDITIVE MANUFACTURING

Qualitatively, green tape lamination to produce ceramic substrates for high-density electronics anticipates additive manufacturing. From the beginning, it was the use of patterned thin layers that were first independently contoured, subsequently assembled in order, and fused into a block to create a 3D structure (of dissimilar materials and the 3D information was internal). To be generalized to true additive manufacturing, however, several qualitative accomplishments were required. The result is what has been termed computer-aided manufacturing of laminated engineering materials (CAM-LEM).[22] In essence, the lamination of tape is a powder bed process, the difference being that each layer of the powder bed is formed separately via tape casting. For a review of the fundamentals of tape casting, see Moreno[23,24] and Mistler and Twiname.[25]

The advantages of using tape casting are several. The first is that it is compatible with the creation of a homogeneous layer of the very fine particle sizes associated with sinterable ceramics. Typically, liquid-phase sintered materials need to be only several microns in diameter and solid-state sintered materials, especially covalent materials, require an order of magnitude smaller characteristic size. Such fine powders are inherently unflowable in the dry state. Fine powders spontaneously agglomerate and therefore do not flow in a predictable or uniform manner.[26,27] The standard solution to this problem in conventional ceramic processing is to intentionally agglomerate (i.e., either granulate or spray dry) the powder to form flow units containing several thousands of particles held together by an organic binder that is removed during the initial stages of firing. This approach is also used in AM methods relying on flowability.[28,29]

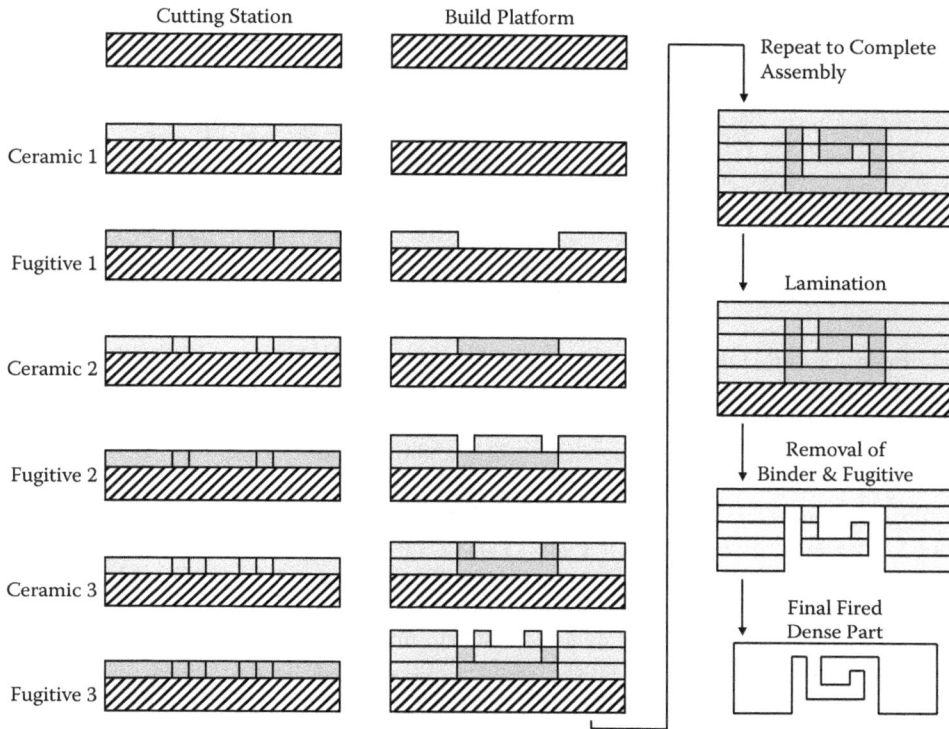

FIGURE 11.1 Schematic illustration of a cut-then-stack approach using two green tapes (one carrying the desired ceramic in particulate form and the other carrying a combustible particulate). The two columns on the left are meant to represent a time series in stacking. The cutting is done in a separate location than that of the stacking. The left-most column represents the cutting station and next to that is the location at which the stack is created. (In actual implementation, the locations may be physically separated.) The process of building each layer requires two time steps: one to cut and stack the selected portions of the ceramic tape and one to cut and stack the complementary pieces from the fugitive. The strategy and mechanisms for selectively gripping the appropriate portion of each are given in Newman et al.[70] The process is repeated as necessary to build the desired complete object. The right-hand-most column indicates the sequence of steps associated with post-assembly processing. The stack is converted to a green-state monolith; that is, the lamination step erases the interfaces between the layers so that the powder compact within the tapes is continuous across the entire part. This takes place at temperatures that are low compared to the binder burnout temperature. The binder burnout step includes a variety of phenomena. In addition to thermal decomposition of the binder, any particulate in the fugitive tape must be converted to gases and vented from the interior. Finally, the ceramic powder is sintered to the desired state of densification. Some details of each step are described in the text.

A second important advantage is that contoured or cut tape is readily manipulated, permitting the use of multiple materials in a given layer. An illustration of one use of this is given in Figure 11.1, in which a fugitive tape is used to occupy space during the build which becomes an interior channel after firing. Properly designed systems can form interior cavities that are disconnected from the free surface after firing.

The first thing to consider is the process of making ceramic tape. It is a well-established process in which a slurry (or slip) that contains insoluble powders suspended in a solution of an organic binder in a mixture of one or more solvents is spread on a belt. Due to forced convection and heating, the slurry dries quickly to form a "tape" that is stripped off the belt and coiled up. The formed tapes are stable, with a consistency similar to that of leather, and can be stored in contact with each other for extended periods of time without degradation of properties or sticking together. Some

FIGURE 11.2　A three-layer part made of two tapes. The ceramic tape is that of a glass-containing alumina and the fugitive tape has particulates that completely burn away during firing. (A) After assembly and lamination; (B) fired to full density. The part is approximately 25 mm in diameter after firing.

typical dimensions are a thickness measured in 10s to 100s of microns, a width of 10s of cm, and an arbitrary length. A number of manufacturers sell unfired tape as a product, toll vendors provide tape-casting services for development work, and a wide-ranging set of electronic devices contain laminated ceramic tape-based components. It is straightforward to set up tape casting in a laboratory setting.

To illustrate the observed behavior of a well-designed fugitive, Figure 11.2A presents a partial assembly (three layers of a normally five-layer part to leave an exposed area for viewing) and Figure 11.2B shows a final fired part. The fired part is ash free and flat and has retained layer adherence. The net firing shrinkage for this system is visible by comparing the respective diameters.

Prior to describing the processing controls that determine outcomes, the nature of green ceramic tapes will be briefly reviewed. Green tapes contain multiple phases: the particulate, the organic binder (which may be multiphase), and porosity. They are typically compressible and the compressibility changes as a function of temperature. The characteristics of several tapes used for tape-based AM are given in Table 11.1. Key aspects are the fineness of the powders, the high-volume organic fraction, and the large amount of porosity. The in-house tapes were produced in a laboratory setting. The industrial tape is known as a good performer with respect to laser cutting and lamination. In-house tapes were developed to emulate the properties of the industrial tape. The formulations of the slurries used to prepare these tapes are variants on classic non-aqueous formulations[30] with polyvinyl butyral as the binder and dioctyl phthalate and/or polyethylene glycol (PEG) as plasticizers. A typical sequence for slurry preparation is milling of the ceramic powder with the appropriate dispersant in the solvent for several hours to mechanically disrupt agglomerates and generate a stable dispersion. At this point the organics are added and low-intensity milling is continued for several more hours (often overnight). This achieves mixing without inducing foaming or polymer degradation. An illustrative set of slurry formulations used with silicon nitride is provided in Table 11.2.

TABLE 11.1
Characteristics of Green Ceramic Tape

Characteristic	Industrial[a] Alumina Tape	In-House Alumina Tape	In-House Silicon Nitride Tape
Ceramic powder	Al_2O_3 + glass	Al_2O_3[b]	Si_3N_4 + glass[c]
Particle size (m)	3–4	~0.5	~0.5
Packing factor (%)	55	57	50
Binder content (vol%)	20	31	31
Porosity (vol%)	25	12	19

[a]　Courtesy of K.R. McNerney, CoorsTek, Golden, CO.
[b]　Alcoa, AKP SG-16.
[c]　GS-44 formulation of AlliedSignal.

TABLE 11.2
Silicon Nitride Slurry Formulations
in Weight Percent (wt%)

System 1		System 2		System 3	
GS-44	52.7	GS-44	50.0	GS-44	63.8
DM-55[a]	1.1	B-73305[e]	40.0	PAA[f]	0.1
PVB[b]	5.3	Toluene	2.0	B-1000	17.0
PEG[c]	2.1	Ethanol	8.0	H_2O	19.1
DOP[d]	2.1				
Toluene	26.4				
Butanol	10.3				

[a] Dispersant; Rohm and Haas, Philadelphia, PA.
[b] Butvar-98 (polyvinyl butyral); Monsanto, St. Louis, MO.
[c] Polyethylene glycol; Aldrich Chemical, Milwaukee, WI.
[d] Dioctylphthlate; Aldrich Chemical, Milwaukee, WI.
[e] Vinyl binder solution; Ferro Corporation, Santa Barbara CA.
[f] Polyacrylic acid; Polysciences, Warrington, PA.

The fugitive is constrained beyond just burning out cleanly. As will be discussed below, the elastic modulus of the fugitive tape and the ceramic tape must be matched over at least the strain encountered during lamination, which includes a temperature cycle. Second, either the thermal expansion must be matched or the tape must lose mechanical integrity during the higher temperature excursion of the binder removal. Finally, it is an advantage if the tape composition is such that the formation of gases during decomposition is spread over a wide range in temperatures so the associated rise in internal pressure is minimized. Initial experiments using just graphite powder (8 to 10 μm; Superior Graphite, Chicago, IL) gave poor results. The tape was too stiff and the burnout happened over a very narrow range in temperature. The solution was to admix corn starch to the graphite. The corn starch has a particle size[31] similar to that of the graphite so they will be part of the same particle–particle network. It is also of practical utility that the corn starch has a glass transition temperature between 110 and 200°C.[32] This means it will be a solid during lamination (see below) but will melt early in the heat-up cycle (eliminating back stresses due to differential thermal expansion of the two tapes) and prior to the onset of the production of gaseous decomposition products. The properties lend themselves to empirical control by altering the ratio of the two particulates; an optimized formulation for one system is given in Table 11.3.

TABLE 11.3
Optimized Fugitive (Corn Starch/Graphite) Tape Formulations

Formulation	Corn Starch (wt%)	Graphite (wt%)	PS-21A[a] (wt%)	Solisperse 24000[b] (wt%)	EtOH (wt%)	Toluene (wt%)	Polyvinyl Butyral (PVB) (wt%)	Dioctylphthlate (DOP) (wt%)
CG-7-2	38	32	0.46	0.6	42	57	13	6
CG-8	38	32	0.46	0.6	35	64	13	6

[a] Dispersant for corn starch.
[b] Dispersant for graphite.

Given a set of feedstocks, the first task evident in the schematic that is Figure 11.1 is the cutting of parting lines so the appropriate portions of the ceramic and the fugitive tapes can be selectively stacked. The approach used was laser cutting using a low-wattage (25-W) CO_2 laser with a concentric gas jet. The laser melts the binder and the gas jet blows the powder and molten polymer from the kerf. This approach has the great advantage of reducing capital costs and minimizes (but does not eliminate) the need for laser safety equipment and procedures. It also means that the cutting efficiency is to first order independent of the particulate, and the ceramic and fugitive tapes cut well under essentially identical laser settings.

The interaction of laser cutting introduces at least three aspects that require control: width, taper, and roughness.[33] First, it is important to control the kerf width so the contours can be appropriately offset such that the pieces stacked form the correct fired part. The ceramic and the fugitive contours should be offset so that they fit closely together, as this is an advantage for force distribution in subsequent lamination. Contours also need to be offset to account for the different shrinkage that is observed in tape cast ceramics. The casting direction, the transverse direction, and the thickness all shrink different amounts. This is due to two effects. The irregular particles will orient under the shear field in the tape-casting process. This will cause the shrinkage in the plane of the tape to be smaller in the casting direction than in the transverse, as the long axes of the particle end up co-aligned with the casting direction. Second, there is some segregation of binder to the top and bottom surface of the tape so that the through thickness shrinkage is typically significantly larger than in either of the other two directions. As long as it is reproducible, and the lamination step does not change it, then it can be accounted for by pre-distorting the computer-aided design (CAD) information by appropriate scalars in the three normal directions.

For obvious reasons, it is also desirable for the kerf edge to be normal to the tape face and for the kerf to be smooth. The standard approach to help determine the process window is to determine a *P/V* curve. That is the ratio of the laser power scaled with the travel speed so that the unit energy deposited per unit line length travel is held constant. One important measure is the critical energy per unit line length needed to cut through a tape; an example is shown in Figure 11.3. Above a critical value, in this case ≈5 to 10 J/mm, the critical cut-through energy is independent of travel speed. The cutoff identifies the speed at which lateral conduction of heat becomes competitive with the energy. Above this energy, the machine speed can be controlled for other reasons (e.g., slowing at

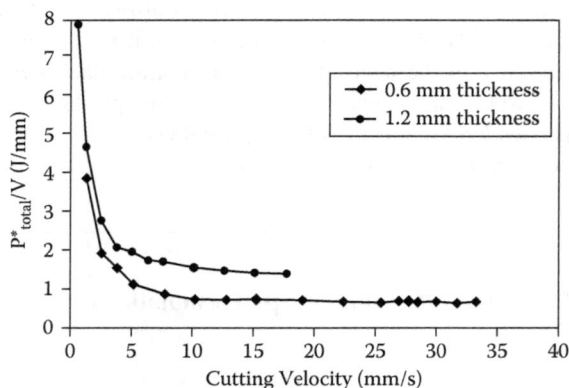

FIGURE 11.3 The critical power-to-travel speed ratio to cut through two alumina tapes that differ in thickness by a factor of two (0.6 mm and 1.2 mm). Above a critical ratio, the cut-through threshold becomes constant for each and scales directly with the tape thickness. This indicates that under these conditions all the energy is consumed in melting the material in the kerf. At lower speeds, lateral heat conduction bleeds away heat, which causes other problems, such as increased kerf width and roughness.

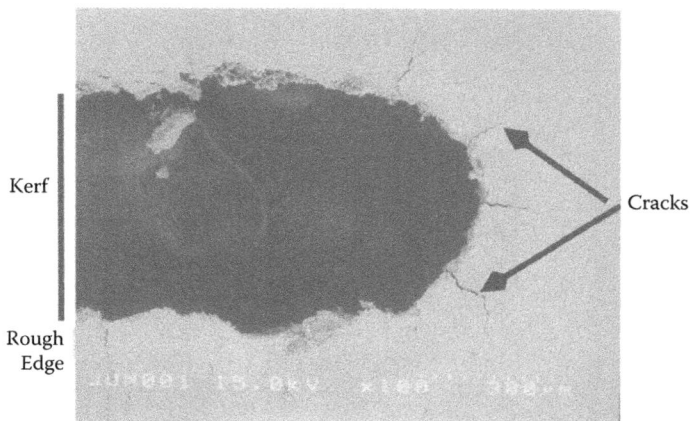

FIGURE 11.4 Scanning electron microscope image revealing the phenomenology of near-kerf crack during low-speed laser cutting due to lateral heat conduction causing a heat-affected zone in which the plasticizer is preferentially evaporated from the binder system.

corners to lower inertial forces). The lateral heat conduction that occurs at slow travel speeds gives rise to an increase in kerf width and a higher degree of kerf taper (kerfs are much wider on the top that bottom of the green tape) and leads to increased roughness due to transverse cracking that occurs as the laser moves.

This latter effect, the increase in roughness, appears to be caused by a special form of heat-affected zone, the phenomenology of which is shown in Figure 11.4. The lateral heat flow appears to boil the plasticizer preferentially from the binder adjacent to and ahead of the spot at low speeds. This results in volumetric shrinkage at the same time that the modulus is increasing, leading to cracks. The cracked material is blown away by the gas jet, and the kerf is both ragged and wide. The engineering solution to this involved switching to a plasticizer with a lower vapor pressure, decreasing the overall binder content, and avoiding low speeds whenever possible.[33] The combination proved very effective.

Two other effects of the laser cutting proved to be best addressed through shielding. The first was the fact that laser beams typically have an intensity that is nominally Gaussian, with the intensity highest at the center and then dropping off as a function of radial distance. After some efforts to filter and threshold the beam using optics, a simple empirical solution was found. A sheet of low-ash filter paper, found in every chemistry lab, was simply placed on top of the tape prior to cutting. This served as a *de facto* threshold as the paper was not cut but only charred below a certain laser intensity, giving clean cuts; that is, the region adjacent to the kerf was no longer softened by the beam and subject to deformation under the gas jet.

The second effect was reflections off of the top of the aluminum honeycomb that supported the tape during cutting. The cell sizes of the honeycombs usually were 6 mm with aluminum wall thicknesses of 200 μm. Thus, the reflections were spaced relatively far apart and were of modest size to the unaided eye. However, this meant there was a high probability of several hundred micron flaws in every layer. Such flaws do not get removed by lamination, nor do they close up during sintering. The solution to this was to use an aluminized polymer sheet under the tape, with the reflective side down. The primary laser then encounters the polymer before the aluminum, absorbs the light, and melts, presenting no obstacle to the normal cutting action of the laser. The reflections, however, hit the reflective side and do not go into the backside of the green tape. This empirical solution has proven to be very successful. Clean, smooth, and relatively narrow kerfs were obtained with all of these strategies used in combination, as seen in Figure 11.5.

FIGURE 11.5 An end-on view of a laser-cut kerf in a silicon nitride tape with optimized lower binder concentration and low volatility plasticizer, and sandwiched between paper and aluminized polymer sheeting. The cut is nearly perpendicular, of nearly constant width, and very smooth.

Feedstocks that are cleanly cut and stacked with a high degree of registry must now be successfully laminated. Initially, this seemed to be a solved problem as thermocompression is the basis of decades' worth of well-established multilayer ceramic substrate production; however, AM presents challenges not routinely encountered in that industry. First, the thermocompression typically involves intentional resolved shear at the interfaces between the tapes and macroscopic densification of the tapes.[34] This is helpful in closing gaps and ensuring a high degree of lamination efficiency, but thermocompression is not well suited to AM, where the desired shape will not, in general, be platelike. With arbitrary shapes, the resolved loads will not be uniform and thermocompression will lead to variations in green density that in turn cause warpage during firing. Also, when high-aspect-ratio parts are built, the interior of the part will be shielded from the load during post-assembly compression. And, finally, parts of complex curved external surfaces do not present a flat surface against which to push.

The solution is to chemically modify the surfaces of the tapes so they can be caused to migrate together through the application of low pressures at low temperatures, so the body of the tape remains in the elastic region. This was done is two ways. The first was through the use of a spray application of a mixture of solvents and a plasticizer.[35] Alternatively, it was determined that a dry tack-free coating derived from a polymer emulsion yielded a feedstock that was well suited to these goals.[33,36] In the time since, a number of studies have reported extensions or independent generation of similar strategies.[37–41] The behavior of the dry tack-free system will be described here.

Figure 11.6 shows a cross-section of a laboratory alumina tape coated using a doctor blade system with a water-based ethylene acrylic acid (EAA) dispersion (Michem® Prime 4983R; Michelman Chemical, Cincinnati, OH). This forms a thin discrete coating of a material that behaves like a heat-sealable coating. As assembled, the microstructure clearly shows that the green tapes are separated by a discrete particle-free polymer layer. If a normal heating rate is used to reach the binder burnout temperature, this layer vaporizes and the assembly delaminates. An intermediate temperature hold is required to permit the EAA to migrate into the pore space of the tape immediately adjacent to the joint. Figure 11.7 shows the results of 1-hour holds at several temperatures. At 75°C the polymer layer is unchanged. At 125°C and 150°C, the layer is thinner (and appears phase separated). After 1 hour at 175°C, the adhesive has migrated sufficiently to allow the ceramic particles to form an uninterrupted homogeneous network across the interface. For processing parts, it was found that inclusion of a short, 1-hour hold at 175°C is readily accommodated and results in defect-free microstructures.

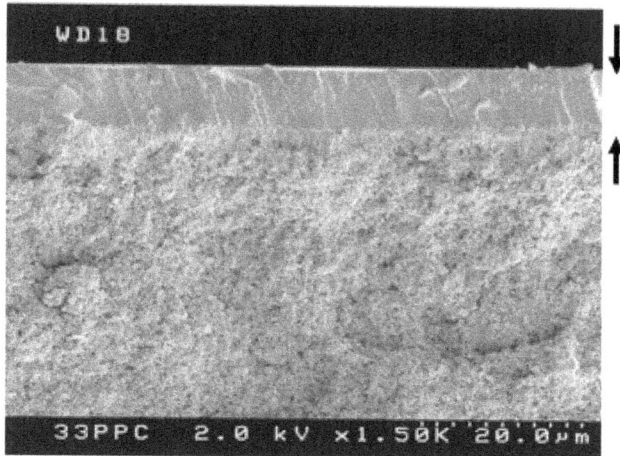

FIGURE 11.6 Scanning electron microscope image of a fracture surface (obtained by dipping the tape in liquid nitrogen and snapping it) showing the uniform dense layer of heat-sensitive adhesive prepared by doctor blading an aqueous dispersion and drying.

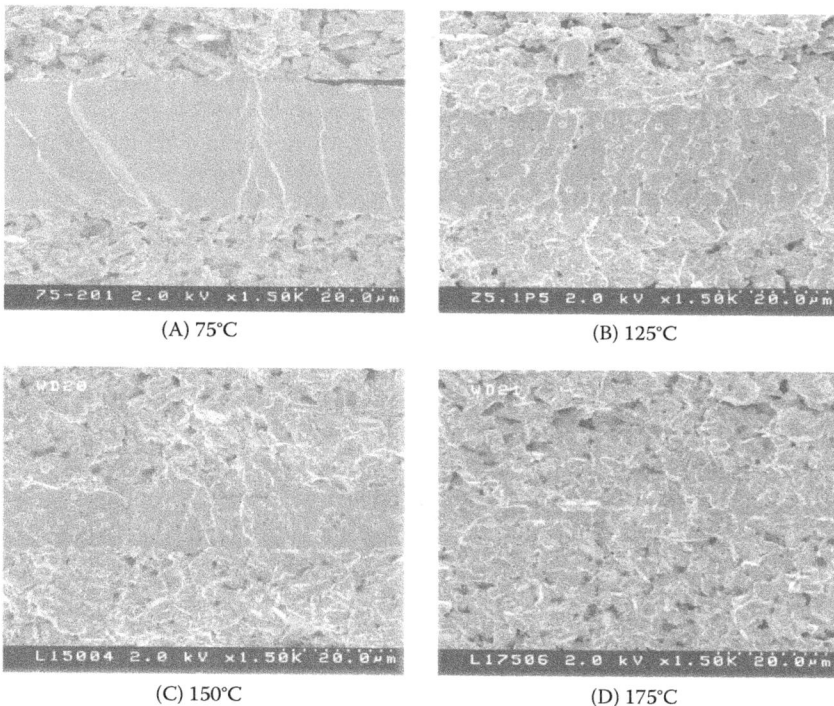

(A) 75°C

(B) 125°C

(C) 150°C

(D) 175°C

FIGURE 11.7 The evolution of the EAA adhesive layer used to join the alumina tape with a polyvinyl butyral (PVB) binder system. The EAA goes into solution in PVB and the mass transport is accommodated by the existence of substantial porosity in the tape; no swelling or distortion is observed. When the migration of the EAA is complete (D), the ceramic particulates form a continuous network across the interface and sinter homogeneously to a defect-free final state. Typically, a small load, <0.8 MPa, is applied to the stack of tapes. The tapes remain in the elastic regime. Macroscopic strain measurements confirm that the only change in dimension is in the stacking direction. Furthermore, measurements over time confirm that the strain ceases when it reaches a magnitude corresponding to the thickness of the original adhesive layer.

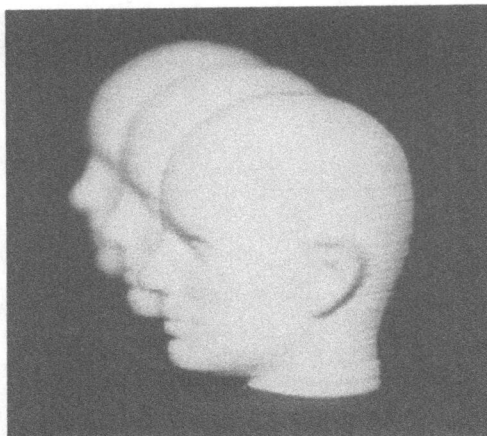

FIGURE 11.8 Stylized human head forms from a public-domain CAD file. The heads stand nearly 4 cm high and are solid 96% Al_2O_3. The lamination area is substantial, on the order of 600 cm^2 in total. The feedstock was an industrial green tape coated in-house with adhesive. Fugitive can used to "picture frame" each layer so there is medium to transmit force during lamination.

Shown in Figures 11.8 and 11.9 are two example parts that demonstrate two limiting types of parts. Figure 11.8 shows a series of solid heads, derived from a CAD file, that are roughly 4 cm high and were built, laminated, and fired. These parts represent "blocky" volumes in which the challenge is to achieve 100% lamination efficiency across a large number of layers, each of which is of substantial area. These parts also demonstrate that with the appropriate heating cycle it is possible to successfully thermally degrade and vent the gaseous content from the (relatively) high binder content and the adhesive. Registration errors layer to layer are visible but generally less that 0.1 mm.

Figure 11.9 shows a part that has a simple exterior platelike shape but a complex internal geometry that requires precision. It is a fluidic amplifier designed to operate in dusty gas streams; therefore, the erosion resistance of a ceramic is of value. Figures 11.9A shows the cut, assembled, and laminated green part with the fugitive in place. Figure 11.9B shows a close-up highlighting the cleanness of the edge and an emergent fugitive from the midst of the part. Figures 11.9C–E show three orthogonal views of the fired part. The part is flat and well bonded (as was determined later through sectioning), and the geometric dimensions were within specification.

11.2.1 SUMMARY: GREEN-TAPE-BASED ADDITIVE MANUFACTURING

Qualitatively, green-tape lamination to additively manufactured ceramic components is a straightforward extension of multilayer ceramic substrate production, which is a very well-established process. Both internal and external geometrical features are readily achieved. The key to optimizing the process is to focus on the behavior of the organic binder phase throughout each unit operation associated with AM. The primary literature on ceramic processing with broad contributions from multiple academic, industrial, and governmental researchers offers a tremendous resource for interested process engineers.

11.3 ADDITIVELY MANUFACTURED MOLDS WITH SOLUBLE CORES FOR GELCASTING

Gravity-driven casting is generally regarded as the technology with the greatest geometrical flexibility at the lowest cost. In concept, the technique is the essence of simplicity. A liquid is poured into a shaped container and allowed, or induced, to solidify. The solid part is then extracted from

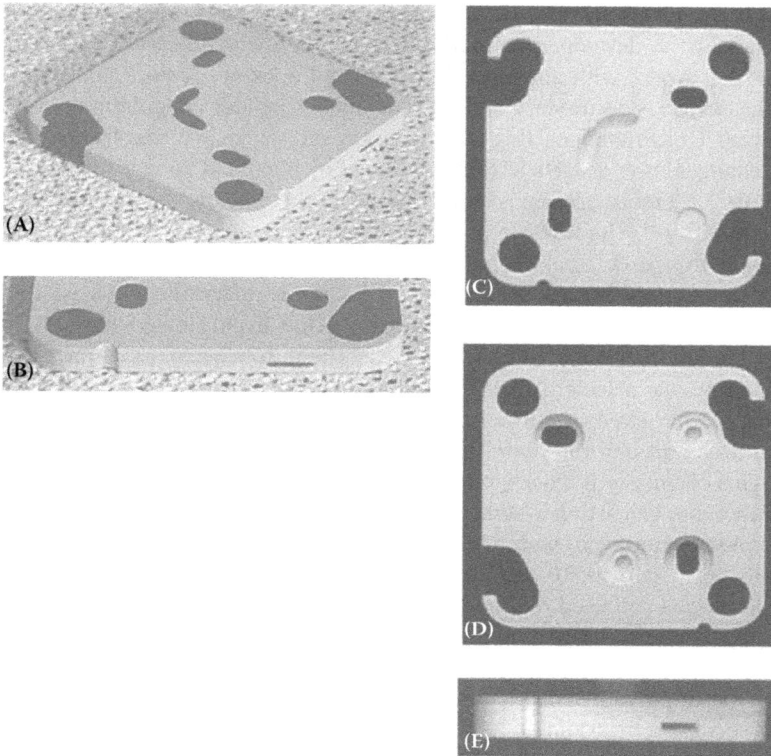

FIGURE 11.9 (A,B) Two photographs showing green state laminates. These tapes were cut using paper shielding between the laser and the top surface of the tape as well as aluminized polymer sheeting between the tape and the supporting aluminum honeycomb. The parts were fired on a zirconia open cell foam (Selee Corp.), which assists in venting gases from the bottom of the part during binder/adhesive removal. (C–E) Three views of the fired part. It shows good edge retention, planarity of features at all levels in the stack, and no evidence of de-cohesion.

the cavity. In some cases, the extraction requires the destruction of the tooling (e.g., sand casting of metals) whereas in other cases the tooling is reused and is considered permanent. As in any forming operation, however, the details ultimately determine performance characteristics of the produced parts. In fusion casting of metals (e.g., see Campbell[42]), the alloy is taken to a temperature in excess of the melting point, caused to flow into the mold (with or without coring to define internal passages or cavities), and allowed to cool to solidify. Fusion casting of ceramics is done, but it is a niche technology associated with particular engineering requirements. One example is the production of fusion cast alumina–zirconia refractories, which exhibit special resistance to corrosion in contact with molten glass and are used in its production.[43,44]

However, the general approach to casting in ceramics is the production of green powder compacts that are subsequently thermally processed (i.e., fired) to produce dense parts with desired density, grain structure, and concomitant properties. The most venerated of these techniques is slip casting, which is widely used in the production of both artistic and engineering components. In slip casting, the system is isothermal and solidification is accomplished by selective extraction of fluid; that is, a cake develops on the surface of a porous mold (e.g., plaster) due to capillary-driven flow of the suspension medium (e.g., water). In the realm of engineering components, the technique has proven to be highly effective, relevant to a broad array of materials, and applicable to very different application areas.[45–50]

Producing porous molds by additive manufacturing, however, is inherently challenging. Most AM methods have been developed to produce dense polymeric materials. Therefore, a ceramic casting process that works with an impervious tool and does not require applied force (i.e., uses gravity to fill the cavity) is desirable. Candidate techniques include coagulation casting, freeze casting, and gelcasting. Coagulation casting (i.e., changing interparticle forces from repulsive to attractive) relies on changes in interparticle forces induced by chemical changes that alter either pH or the ionic strength.[51–53] Freeze casting is simply reducing the ambient temperature to cause the fluid of the slurry to solidify.[54] Although, in principle, both are compatible with AM-produced tooling, neither was used in this work. Coagulation casting typically offers narrow processing windows as it is strongly dependent on the surface chemistry of the ceramic material being used. Freeze casting in polymer-based tooling is hampered by both the high thermal expansion coefficients of the polymers and their relatively low thermal conductivity. These affect dimensional tolerances and cycle time. Therefore, gelcasting was selected.

In gelcasting, the particulate is dispersed in a polymer solution. In the mold, the polymer solution is induced to gel around the particles. In this work, the standard acrylamide chemistry was employed.[16,17] This chemistry is closely related to that used in the production of electrophoresis gels for DNA,[55] as are other gelcasting systems based on agarose.[56]

Two distinct AM technologies were used to prepare the tooling.[57] One technique was used to prepare functionally permanent molds and a different technique to generate soluble cores. The processing sequence developed and used required tooling to contact three liquids. First, all tooling had to be stable in contact with the water-based gelcasting slurry during casting. Subsequently, immersion in alcohol was used to selectively dissolve cores, while the permanent tooling remained unaltered. Finally, immersion in polyethylene glycol was used for chemical drying, during which tooling is unaffected and so should be available for subsequent reuse.

In this work, stereolithography was used to produce epoxy parts that acted as the permanent tooling, and the droplet-on-demand technology of Solidscape (formerly Sanders Prototyping) was used to create alcohol-soluble cores. The selective dissolution of this material has been used by others in preparing ceramic parts.[58]

Three types of parts are described here, all of which were produced from a silicon nitride formulation capable of rapid changes in temperature without experiencing thermal shock: (1) converging–diverging rocket nozzle, (2) burner rig standard specimen, and (3) a thin-walled cylinder with continuous internal channels. The rocket nozzle required only permanent tooling, the desired geometry being a body of revolution that could be formed with a two-part pullable core (see Figure 11.10). Casting required only quiescent pouring of a de-aired gelcasting slurry into the space between the core and mold. Gelation occurred in under an hour, and the wet gel strength was sufficient that the part could be de-molded with appropriate care. After drying, the part was subjected to standard binder burnout and sintering with an overpressure of nitrogen following the standard protocol for this silicon nitride. As is characteristic of gelcasting, firing shrinkage was substantial (see Figure 11.11) but isotropic.

The rocket nozzle was subjected to oxygen–hydrogen firing in a test stand at NASA Glenn Research Center (Figure 11.12). This constitutes a severe thermal shock as the system goes from flowing cryogenic gases to high temperatures in moments during ignition. As is visible in Figure 11.12B, even at steady state the spatial thermal gradients are substantial. This is related to one of the advantages of producing powder compacts rather than trying to do direct fabrication (which would not be possible with silicon nitride anyway as it sublimates rather than melts at normal pressures). The powder compacts are fired using a heating schedule that leads to both densification and the development of a whisker-like microstructure that offers high fracture toughness.

The ability to pull the core from the wet gel in the case of the rocket thrustor is a special case. Both of the sections being pulled are conical, and small relative motion causes full separation. Trying to extract a long cylindrical core, especially without draft, is likely to distort the gelled part

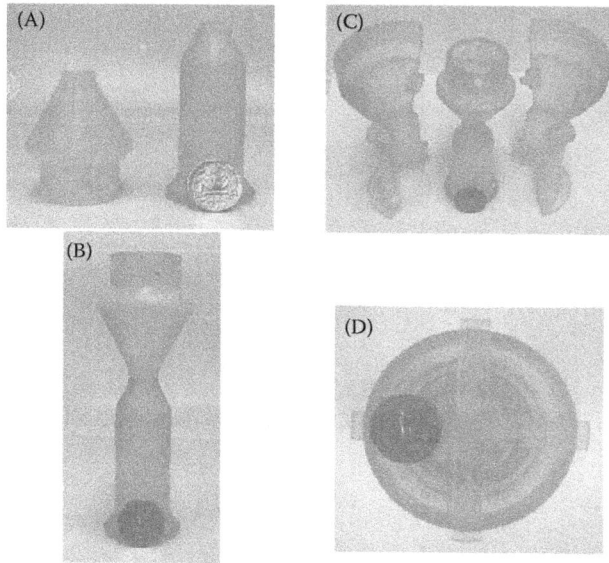

FIGURE 11.10 Epoxy tooling produced using stereolithography to form a converging–diverging rocket nozzle. The symmetry of the desired part is such that the core (A,B) could be segmented perpendicular to the axis of the part and it would be pullable. The outer mold (C) was a simple clam-shell design, and the core was registered to the mold (D) with a slotted cross.

or cause fracture. One solution to this is to use a soluble mold liner. (Often grease or mold release, which might be another strategy, interferes with the chemistry leading to gel formation.) This was used to produce a ceramic equivalent of the test part originally designed for casting superalloy parts in the development of thermal barrier coatings.[59] The characteristic geometrical features of the specimen are shown in Figure 11.13.

The tooling created to reproduce this part in silicon nitride was designed to include appropriate scaling to account for drying and firing shrinkage. A soluble core was produced from an alcohol-soluble polymer using the Solidscape droplet-on-demand technology. In addition, a thin mold liner was created to define the external surface of the part. This was held in place by epoxy parts

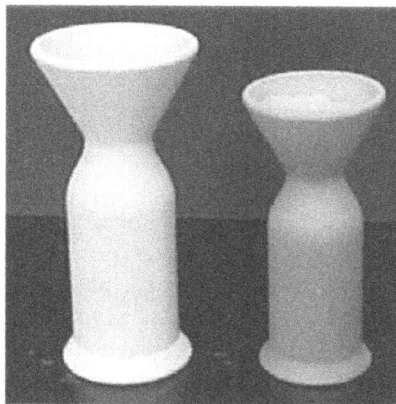

FIGURE 11.11 Silicon nitride rocket thrustors manufactured using gelcasting into the mold shown in Figure 11.10. The large, but isotropic, shrinkage exhibited during firing is evident.

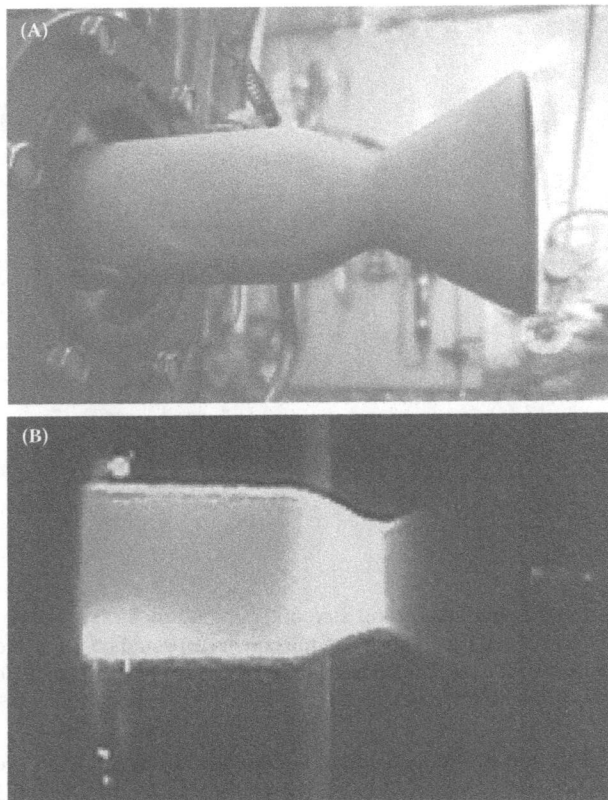

FIGURE 11.12   Rocket thrustor (A) mounted in the NASA Glenn test stand, and (B) during testing with O_2/N_2. In (B), the room lights are off and the color (grayscale) is due to incandescence.

produced using stereolithography. The epoxy tooling also included the runner system to feed the slurry into the assembled mold. The design of the parts, the assembled system, and the cast part are shown in Figure 11.14.

Drying of gelcast parts is often a very slow process requiring time on the order of days.[60] This is due to the small effective capillary size of the polymer gel. Attempts to rapidly dry in low humidity lead to steep moisture gradients, high stresses, and cracking. It is possible to use elevated temperatures in a high-humidity environment, but that requires close process control and capital equipment. In this work, it is was suggested to immerse the wet gels in polyethylene glycol (PEG),[61] which proved remarkably successful. A part such as that depicted in Figure 11.14 was simply immersed in PEG for 30 minutes and then dried in room air, typically overnight. It is difficult to overstate the practical significance of this process, as a perceived road block was effectively removed. (Interestingly, subsequent review of the literature revealed that this PEG immersion is a well-established technique used by archeologists interested in preserving water-logged wood[62] or shale.[63] Furthermore, in the botany literature, PEG is routinely used to simulate drought conditions in soils for laboratory experiments with plants.[64,65]) The process became akin to old-style darkroom operations in which the part was moved from bath to bath and then allowed to dry. An advantage of using the mold liner is that the entire free surface of the cast part is accessible to the PEG.

Systematic studies of liquid desiccants to dry gelcast parts have recently been published.[66–69] The use of liquid desiccants will allow tighter definition of process windows and allow broader deployment of this method of gelcasting in AM tooling. The result of this process was a testable part. Figure 11.15 shows a burner 0.3 mach rig test carried out at NASA Glenn Research Center.

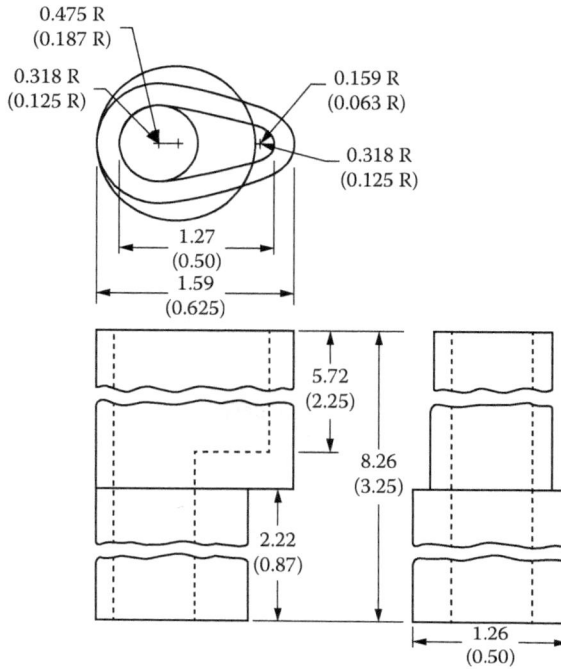

0.475 R
(0.187 R)

0.318 R
(0.125 R)

0.159 R
(0.063 R)

0.318 R
(0.125 R)

1.27
(0.50)
1.59
(0.625)

5.72
(2.25)

8.26
(3.25)

2.22
(0.87)

1.26
(0.50)

FIGURE 11.13 Design of cast superalloy test specimen reproduced in silicon nitride.[59]

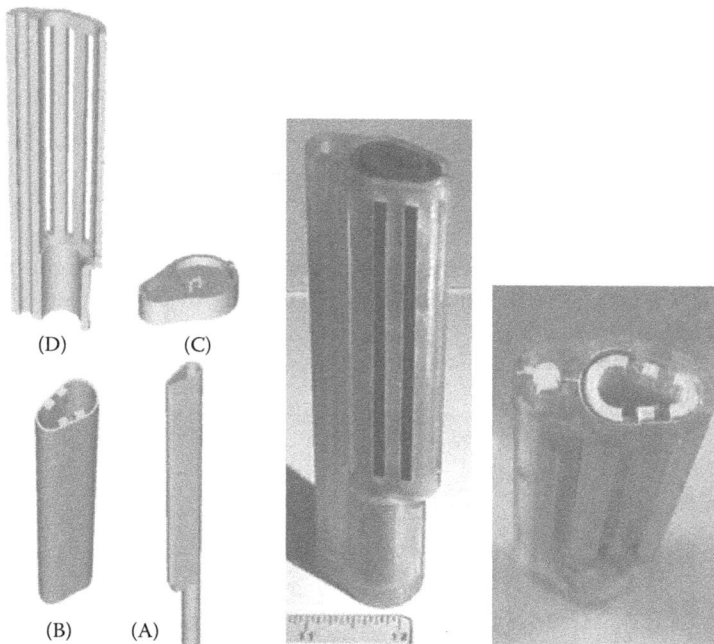

(D)

(C)

(B)

(A)

FIGURE 11.14 The four figures at the left are renderings of CAD files used to make (A,B) epoxy permanent parts, (C) a soluble mold liner, and (D) a soluble core. The tabs at the top of the liner are used to register the core. The center figure shows the assembly ready for slurry, and the figure on the right shows the cast parts after soaking in alcohol to remove the soluble parts and drying. The gap between the cast part and the epoxy at the top is produced by leaching the liner.

FIGURE 11.15 Burner rig test of silicon nitride part manufactured to the dimensions of a standard specimen (developed for superalloys). Comparison of the two figures shows the impact of internal cooling air flow on surface temperature as evidenced by the brightness of the heated surface.

The final part described here was a test specimen designed to probe the feasibility of creating cooling channels with dimensions on the order of those used in investment casting of superalloy metal parts. The specimen design was a short thin-walled cylinder. The inner diameter of the cylinder was 25 mm, the length 38 mm, and the wall thickness 1 mm. The channels were designed as 300-μm squares with semicircular ends that were then transformed to an array conforming to a circle in the middle of the wall. The part itself is shown in the series of images in Figure 11.16. The details of the molding system produced through AM and shown in Figure 11.17 are renderings of the CAD files.

The key advantage of using a combination of AM methods for this part was the ability to design, refine, and redesign the combination of permanent and fugitive parts that permitted good mold filling, access to the liquids for both dissolution of the soluble elements and PEG drying, and, finally, ease of demolding. Some of the details are highlighted in the CAD renderings shown in Figure 11.17. It is difficult to imagine non-AM methods, even in combination, being able to provide the required complexity and detail independent of cost and time.

11.3.1 SUMMARY: ADDITIVE MANUFACTURED MOLD AND SOLUBLE-CORE SETS

A combination of AM methods that offer different materials and resolutions has been demonstrated to be most useful when creating tooling for a particular ceramic casting process. It is to be expected that such continued open-mindedness will generally offer benefits, especially when it results in making the least amount of required changes to the process using the tooling (i.e., casting process). The presented examples included duplication of preexisting metal parts in engineering ceramics in a short time and at low cost, as well as the creation of a new highly complex set of tooling to assess the feasibility of offering something previously not considered—in this case, internally cooled ceramic components for aerospace applications.

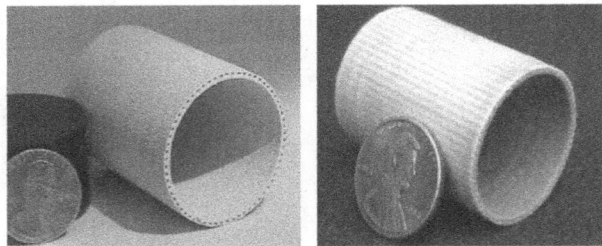

FIGURE 11.16 A thin-walled, 1-mm cylinder designed to assess feasibility of producing fine internal features using soluble cores produced via AM. (Left) The end-on view of the unfired part shows the location within the wall and the shape of the cross-section of the channel. (Right) In the fired part, the length of the cooling channel is visible due to the extreme thinness of the silicon nitride over the channel, just 200 μm. Testing confirmed that the channels were hermetic.

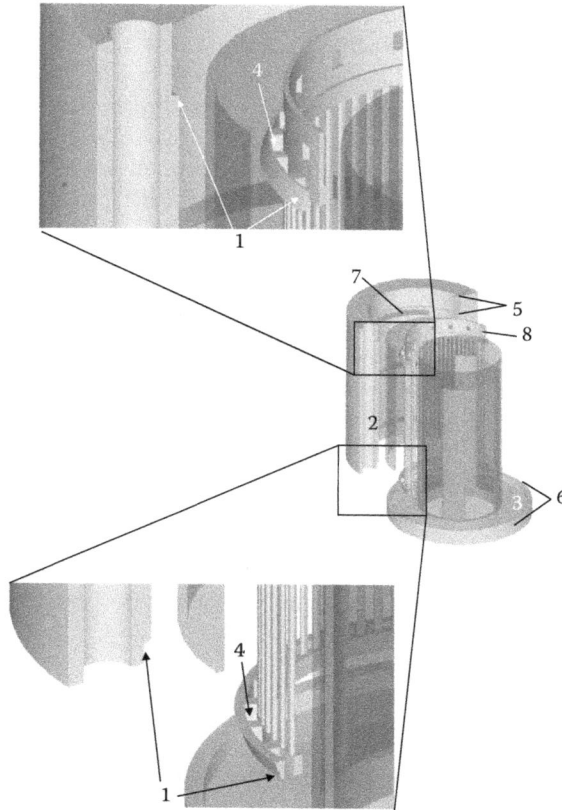

FIGURE 11.17   A demonstration of the complexity that is readily realized using AM, this CAD rendering shows the individual parts in the center overview from right to left: inner gating, inner lining, core half, outer lining half, and outer gating half. Design features that are highlighted numerically include (1) registration features, (2) the slurry casting tube, (3) a bottom filling tray that guides flow, (4) flow path slats in core to give uniform filling, (5) overflow space, (6) plane provided for gate removal through cutting, (7) uppermost of three slots for dissolution of outer lining, and (8) leachable ethanol path for core dissolution.

11.4   OVERALL SUMMARY

Additive manufacturing methods are rooted in the unit operations of conventional manufacturing but offer particular flexibility. Whether used directly to create parts or indirectly through AM tooling, the ultimate goal is the same—to create high-integrity components with microstructures and properties equivalent to the best of those of conventional methods. A fundamental advantage of AM in ceramics (and other particulate-based systems such as powder metallurgy) is that it can be used to create complex internal geometrical features. This opens up a design space to engineers wishing to exploit the often unique properties of engineering ceramics. Finally, AM is likely best exploited by individuals with a depth of knowledge in the processing strategies and constraints associated with the material class of interest.

ACKNOWLEDGMENTS

The work reviewed here is the result of a number of different studies funded from a variety of sources over a 10-year period. I was first introduced to additive manufacturing in the early days of rapid prototyping by Prof. A. Heuer, who had learned of it from his service on the Defense Sciences

Research Council. Much of the work on tape lamination involved a collaboration with him as well as Prof. W. Newman, an expert in mechatronics. This work eventually led to a start-up company named CAM-LEM, Inc. The chief technology engineer was Mr. B. Mathewson, who was responsible for designing and realizing the CL-1, a fully robotic system for lamination of green tape. The company continues to this day. Recently, Dr. H. Shulman of Ceralink, Inc., in Troy, NY, has acquired much of the CAM-LEM technology and is the recommended contact for individuals seeking commercial applications. Funding included grants from DARPA, NSF, and ONR. The company was supported through Army Research Office STTR funding.

The gelcasting work was carried out under sponsorship from a cooperative agreement with NASA to support ceramic processing work through support of the Great Lakes Professorship for Ceramic Processing. Beyond critical financial support the professorship offered access the facilities at NASA Glenn and direct collaboration with NASA engineers and technical staff. Key figures in these collaborations were Drs. A. Eckel, R. Bhatt, and S.R. Levine. This provided a rich experience and furthered the work in a manner that would not have been otherwise possible.

As is always the case, university research relies on talented graduate students. In this case, key students on the materials side were Paul Wei, Jun Mou, Jon Best, Nitinat Suppakarn, and Brian Kernan. Jon Best was responsible for the design and execution of the gelcasting molds, and he produced the internally cooled ceramic parts. A visiting scholar, Dr. Zhien Liu is a wizard in the laboratory. His openness to new ideas, ability to translate concepts into practice, and work ethic were instrumental to his own success and that of many others. The dry-sheet warm lamination process was reduced to practice in his skilled hands.

REFERENCES

1. Bourell, D. et al. (2009). A brief history of additive manufacturing and the 2009 roadmap for additive manufacturing: looking back and looking ahead. In: *Proceedings of the U.S.–Turkey Workshop on Rapid Technologies*, pp. 5–11.
2. Sachs, E. et al. (1992). Three dimensional printing: rapid tooling and prototypes directly from a CAD model. *J. Manuf. Sci. Eng.*, 114(4): 481–488.
3. Yoo, J. et al. (1993). Structural ceramic components by 3D printing. In: *Solid Freeform Fabrication Symposium Proceedings*, pp. 43–50.
4. Halloran, J.W. (1999). Freeform fabrication of ceramics. *Br. Ceram. Trans.*, 98(6): 299–303.
5. Cawley, J.D. (1999). Solid freeform fabrication of ceramics. *Curr. Opin. Solid State Mater. Sci.*, 4(5): 483–489.
6. Tay, B.Y., Evans, J.R.G., and Edirisinghe, M.J. (2003). Solid freeform fabrication of ceramics. *Int. Mater. Rev.*, 48(6): 341–370.
7. McGuffin-Cawley, J.D. (2012). Manufacturing technology: rapid prototyping. In: *Ceramics Science and Technology*. Vol. 3. *Synthesis and Processing* (Riedel, R. and Chen, I.-W., Eds.), pp. 415–433. Wiley, New York.
8. Bertrand, Ph. et al. (2007). Ceramic components manufacturing by selective laser sintering. *Appl. Surf. Sci.*, 254(4): 989–992. .
9. Hagedorn, Y.-C. et al. (2010). Net shaped high performance oxide ceramic parts by selective laser melting. *Phys. Proc.*, 5(B): 587–594.
10. Blodgett, A.J. and Barbour, D.R. (1982). Thermal conduction module: a high-performance multilayer ceramic package. *IBM J. Res. Dev.*, 26(1): 30–36.
11. Greig, W.J. (2007). *Integrated Circuit Packaging, Assembly and Interconnections*. Springer, New York, pp. 233–244.
12. Li, J. and Ananthasuresh, G.K. (2002). Three-dimensional low-temperature co-fired ceramic shells for miniature systems applications. *J. Micromech. Microeng.*, 12(3): 198–203.
13. Das, A. et al. (2003). Binder removal studies in ceramic thick shapes made by laminated object manufacturing. *J. Euro. Ceram. Soc.*, 23(7): 1013–1017.
14. Weisensel, L. et al. (2004). Laminated object manufacturing (LOM) of SiSiC composites. *Adv. Eng. Mater.*, 6(11): 899–903.

15. Schindler K. and Roosen, A. 2009 Manufacture of 3D structures by cold low pressure lamination of ceramic green tapes. *J. Eur. Ceram. Soc.*, 29(5): 899–904.
16. Young, A.C. et al. (1991) Gelcasting of alumina. *J. Am. Ceram. Soc.*, 74(3): 612–618.
17. Omatete, O.O., Janney, M.A., and Nunn, S.D. (1997). Gelcasting: from laboratory development toward industrial production. *J. Eur. Ceram. Soc.*, 17(2): 407–413.
18. Guo, D. et al. (2002). Gelcasting based solid freeform fabrication of piezoelectric ceramic objects. *Scripta Mater.*, 47(6): 383–387.
19. Cai K. et al. (2003). Solid freeform fabrication of alumina ceramic parts through a lost mould method. *J. Eur. Ceram. Soc.*, 23(6): 921–925.
20. Gyger, Jr., L.S. et al. (2005). Porous gelcast ceramics for bone repair implants. In: *Proceedings of the 2005 SEM Annual Conference & Exposition on Experimental and Applied Mechanics—Experimental Mechanics Applied to Advanced Materials Systems*. The Printing House, Inc., Stoughton, WI, pp. 251–257.
21. Hanemann, T., Honnef, K., and Hausselt, J. (2007). Process chain development for the rapid prototyping of microstructured polymer, ceramic and metal parts: composite flow behaviour optimisation, replication via reaction moulding and thermal postprocessing. *Int. J. Adv. Manuf. Technol.*, 33(1–2): 167–175.
22. Liu, Z.E. et al. (1997). CAM-LEM processing: materials flexibility. In: *Solid Freeform Fabrication Symposium Proceedings*, pp. 379–382.
23. Moreno, R. (1992). The role of slip additives in tape-casting technology. I. Solvents and dispersants. *Am. Ceram. Soc. Bull.*, 71(10): 1521–1531.
24. Moreno, R. (1992). The role of slip additives in tape casting technology. II. Binders and plasticizers. *Am. Ceram. Soc. Bull.*, 71(11): 1647–1657.
25. Mistler, R.E. and Twiname, E.R. (2000). *Tape Casting: Theory and Practice*. Wiley-American Ceramic Society, Westerville OH.
26. McColm, I.J. and Clark, N.J. (1988). *Forming, Shaping and Working of High Performance Ceramics*. Blackie, London, pp. 118–120.
27. Tomas, J. and Kleinschmidt, S. (2009). Review: improvement of flowability of fine cohesive powders by flow additives. *Chem. Eng. Technol.*, 32(10): 1470–1483.
28. Butscher, A. et al. (2011). Review: Structural and material approaches to bone tissue engineering in powder-based three-dimensional printing. *Acta Biomater.*, 7(3): 907–920.
29. Zocca, A. et al. (2014). Powder-bed stabilization for powder-based additive manufacturing. *Adv. Mech. Engr.*, 2014: 491581.
30. Mistler, R.E., Shanefield, D.J., and Runk, R.B. (1978). Tape casting of ceramics. In: *Ceramic Processing Before Firing* (Onoda, G.Y. and Hench, L.L., Eds.), pp. 411–448. Wiley, New York.
31. Jane, J. et al. (1992). Preparation and properties of small-particle corn starch. *Cereal Chem.*, 69(3): 280–283.
32. Shogren, R.L. (1992). Effect of moisture content on the melting and subsequent physical aging of corn-starch. *Carbohydr. Polym.*, 19(2): 83–90.
33. Cawley, J.D. et al. (1998). Materials issues in laminated object manufacturing of powder-based systems. *Solid Freeform Fabrication Symposium Proceedings*, pp. 503–510.
34. Plucknett, K.P. et al. (1994). Processing of tape–cast laminates prepared from fine alumina/zirconia powders. *J. Am. Ceram. Soc.*, 77(8): 2145–2153.
35. Suppakarn, N., Ishida, H., and Cawley J.D. (2001). Roles of poly(propylene glycol) during solvent-based lamination of ceramic green tapes. *J. Am. Ceram. Soc.*, 84(2): 289–294.
36. Liu, Z., Suppakarn, N., and Cawley, J.D. (1999). Coated feedstock for fabrication of ceramic parts by CAM-LEM. In: *Solid Freeform Fabrication Symposium Proceedings*, pp. 393–401.
37. Roosen, A. (2001). New lamination technique to join ceramic green tapes for the manufacturing of multilayer devices. *J. Euro. Ceram. Soc.*, 21(10): 1993–1996.
38. Nair, B. et al. (2003). Ceramic microfabrication techniques for microdevices with three-dimensional architecture. *MRS Proc.*, 782: A5–A30.
39. Gurauskis, J., Sanchez-Herencia, A.J., and Baudin, C. (2005). Joining green ceramic tapes made from water-based slurries by applying low pressures at ambient temperature. *J. Eur. Ceram. Soc.*, 25(15): 3403–3411.
40. Malecha, K., Jurków, D., and Golonka, L.J. (2009). Comparison of solvent and sacrificial volume-material-based lamination processes of low-temperature co-fired ceramics tapes. *J. Micromech. Microeng.*, 19(6): 065022.
41. Jurków, D. and Golonka, L. (2009). Novel cold chemical lamination bonding technique—a simple LTCC thermistor-based flow sensor. *J. Euro. Ceram. Soc.*, 29(10): 1971–1976.

42. Campbell, J. (2003). *Castings*, 2nd ed. Elsevier, New York.
43. Asokan, T. (1994). Microstructural features of fusion cast Al_2O_3–ZrO_2–SiO_2 refractories. *J. Mater. Sci. Lett.*, 13(5): 343–345.
44. Lataste, E. et al. (2009). Microstructural and mechanical consequences of thermal cycles on a high zirconia fuse-cast refractory. *J. Euro. Ceram. Soc.*, 29(4): 587–594.
45. Hayashi, K. et al. (1988). NMR imaging of advanced ceramics during the slip casting process. *J. Phys. D Appl. Phys.*, 21(6): 1037.
46. Pröbster, L. and Diehl, J. (1992). Slip-casting alumina ceramics for crown and bridge restorations. *Quintessence Int.*, 23(1): 25–31.
47. Oliveira, M.L., Chen, K., and Ferreira, J.F. (2002). Influence of the deagglomeration procedure on aqueous dispersion, slip casting and sintering of Si_3N_4-based ceramics. *J. Eur. Ceram. Soc.*, 22(9): 1601–1607.
48. Suzuki, T.S. and Sakka, Y. (2002). Fabrication of textured titania by slip casting in a high magnetic field followed by heating. *Jpn. J. Appl. Phys.*, 41(11A): L1272.
49. Hotta, Y. et al. (2007). Slip casting using wet-jet milled slurry. *J. Eur. Ceram. Soc.*, 27(2): 753–757.
50. Zhang, L. et al. (2009). Fabrication and characterization of anode-supported tubular solid-oxide fuel cells by slip casting and dip coating techniques. *J. Am. Ceram. Soc.*, 92(2): 302–310.
51. Graule, T.J., Gauckler, L.J., and Baader, F.H. (1996). Direct coagulation casting—a new green shaping technique. Part 1. Processing principles. *Indust. Ceram.*, 16(1): 31–34.
52. Yin, L. et al. (2008). Fabrication of three-dimensional inter-connective porous ceramics via ceramic green machining and bonding. *J. Eur. Ceram. Soc.*, 28(3): 531–537.
53. Binner, J.G.P. et al. (2006). *In situ* coagulation moulding: a new route for high quality, net-shape ceramics. *Ceram. Int.*, 32(1): 29–35.
54. Sofie, S.W. and Dogan, F. (2001). Freeze casting of aqueous alumina slurries with glycerol. *J. Am. Ceram. Soc.*, 84(7): 1459–1464.
55. Stellwagen, N.C. (2009). Electrophoresis of DNA in agarose gels, polyacrylamide gels and in free solution. *Electrophoresis*, 30(S1): S188–S195.
56. Santacruz, I., Nieto, M.I., and Moreno, R. (2005). Alumina bodies with near-to-theoretical density by aqueous gelcasting using concentrated agarose solutions. *Ceram. Int.*, 31(3): 439–445.
57. Cawley J.D. et al. (1999). Production strategies for production-quality parts for aerospace applications. In: *Solid Freeform Fabrication Symposium Proceedings*, pp. 833–840.
58. Reis, N., Ainsley, C., and Derby, B. (2003). Digital microfabrication of ceramic components. *Am. Ceram. Soc. Bull.*, 82(9): 9102–9106.
59. Hodge, P.E. et al. (1980). Thermal barrier coatings: burner rig hot corrosion test results. *J. Mater. Energy Sys.*, 1(4): 47–58.
60. Ghosal, S. et al. (1999). A physical model for the drying of gelcast ceramics. *J. Am. Ceram. Soc.*, 82(3): 513–520.
61. Janney, M. (2000). Personal communication.
62. Oddy, W.A. (1975). *Problems in the Conservation of Waterlogged Wood*. National Maritime Museum, London.
63. Oddy, W.A. and Lane, H. (1976). The conservation of waterlogged shale. *Stud. Conserv.*, 21(2) 63–66.
64. Krizek, D.T. (1985). Methods of inducing water stress in plants. *Hort. Sci.*, 20: 1028–1038.
65. Zwiazek, J.J. and Blake, T.J. (1989). Effects of preconditioning on subsequent water relations, stomatal sensitivity, and photosynthesis in osmotically stressed black spruce. *Can. J. Bot.*, 67: 2240–2244.
66. Barati, A., Kokabi, M., and Famili, M.H.N. (2003). Drying of gelcast ceramic parts via the liquid desiccant method. *J. Eur. Ceram. Soc.*, 23(13): 2265–2272.
67. Barati, A., Kokabi, M., and Famili, N. (2003). Modeling of liquid desiccant drying method for gelcast ceramic parts. *Ceram. Int.*, 29(2): 199–207.
68. Zheng, Z., Zhou, D., and Gong, S. (2008). Studies of drying and sintering characteristics of gelcast $BaTiO_3$-based ceramic parts. *Ceram. Int.*, 34(3): 551–555.
69. Trunec, M. (2011). Osmotic drying of gelcast bodies in liquid desiccant. *J. Eur. Ceram. Soc.*, 31(14): 2519–2524.
70. Newman, W.S. et al. (1996). A novel selective-area gripper for layered assembly of laminated objects. *Robot. Comput. Integr. Manuf.*, 12(4): 293–302.

12 Reactive Inkjet Printing of Nylon Materials for Additive Manufacturing Applications

Saeed Fathi

CONTENTS

ABSTRACT

The existing additive manufacturing processes for plastic parts have a number of limitations, as they were originally conceived as methods to make prototypes. The plastics involved in existing processes are generally limited to photosensitive polymers or powdered nylons. High resolution could be achieved via inkjet printing of low-viscosity photopolymers; however, they are usually based on epoxy, tend to be brittle, and suffer from absorption of moisture. Higher viscosity photopolymers used in stereolithography also have limited mechanical properties. With lower resolution processes such as laser sintering, the porosity of sintered nylon parts also reduces mechanical properties. Although inkjet-based processes may have low throughput compared with the other techniques, they have the potential to be much faster as many thousands of jets could be used via multiple print-heads. On the other hand, engineering plastics such as nylon 6 have too high a viscosity and melting point for commercially available printheads. Such polymers however, could be formed during layer fabrication in an *in situ* polymerization in an additive manner. This was the basis of a novel research for reactive inkjet printing of nylon 6. The idea was to combine inkjet printing and anionic polymerization of caprolactam by depositing mixtures of caprolactam with activator and catalyst on top of each other under the appropriate conditions.

12.1 INTRODUCTION

12.1.1 Nylon

Anionic polymerization of caprolactam was patented by DuPont in 1939 as a process for producing high monomer conversion in a short reaction time.[1] The commercialized polymer, referred to as *nylon* by DuPont in 1948, was used for molding and extrusion processes; the material later became known as nylon 6 or polyamide 6 (PA6). Shortly after, the introduction of a nylon suitable for spinning and drawing to produce nylon fiber led to fast growth in its production.[3] Nylon 6 is made from caprolactam ($C_6H_{11}NO$), which has six carbon atoms as a ring. It is white and solid at room temperature and melts at about 68°C. When molten, it is transparent and has a dynamic viscosity of less than 10 mPa·s at 80°C.[2] Various methods have been described for the polymerization of caprolactam, among which anionic polymerization provides the possibility of producing polymers with a well-defined structure and the desired molecular weight distribution.[3] Nylon 6 produced by anionic polymerization is a semicrystalline polymer suitable for many engineering applications, as it provides a combination of high mechanical strength, heat resistance, and chemical stability.[3] It is processed by reactive extrusion and molding techniques for a wide variety of applications such as molded automotive components, extruded tubing and wire coating, and rotationally molded tanks.

12.1.2 Inkjet Additive Manufacturing of Polymeric Materials

Inkjet technology has been successfully employed in additive manufacturing. Commercialized processes include 3D Systems' Multi-Jet Modeler (MJM) technology for three-dimensional (3D) printing of waxes which was introduced in 1997 and Stratasys' PolyJet™ technology for 3D printing of ultraviolet (UV)-curable resins which was introduced in 2001.[4] However, research into additive manufacturing is ongoing, and there is a demand for processes to produce parts from functional polymers with high resolution and good surface finish. Although inkjet technology could fulfill these requirements, a challenge is that polymers with reasonable mechanical properties for structural applications have too high melt temperatures and viscosities to be used in commercial print-heads. In spite of this, research into inkjet additive manufacturing has been reported using a wide range of materials, including polymers, ceramics, metals, and even living cells in the form of diluted solutions, suspensions, and melts.[5–10]

Two main modes of inkjet technology have been used for deposition: continuous and drop-on-demand (DOD).[11] Piezoelectric DOD printheads have attracted more interest for manufacturing use because of their compatibility with various materials and their ability to reliably deposit them onto a substrate with higher throughput. In a DOD inkjet printhead, the liquid material should have a viscosity and surface tension of typically less than 20 mPa·s and between 20 and 70 mN/m, respectively.[6] With particle sizes larger than 5% of the nozzle size, droplet formation instabilities can occur due to partial blocking of the nozzle.[12,13] Further references on research into various aspects of inkjet technology include Gao and Sonin,[14] Beulen et al.,[15] and Wang et al.[16]

The concept of printing reactive polymers was reported in patents as early as 1986 by Hewlett-Packard (HP) for graphical printing applications.[17] The technique was to print two water-based inks containing a low concentration of reactants via two separate printheads onto each other. Upon deposition onto the paper, this would produce cross-linked polymeric lattices encapsulating the inks for increased print quality. The idea of printing multiple reactive epoxy-based materials via separate printheads was first published as a patent by Johnson et al.[18] Voxeljet Technology GmbH filed a similar patent on the concept in 2005[19] and reported the printing of lines and multiple layers with this approach.[20] The concept was developed for 3D printing of epoxy-based materials. Three printheads were employed to deposit three reactive components onto each other. By selective deposition of different compositions and mixing ratios of the components, multi-graded structures with the desired graduated mechanical properties could be printed.[20] Fabrication of micron-size 3D features using reactive printing of polyurethane (PU) was also reported.[21] The reaction was based on cross-linking of printed initiator and catalyst agents deposited as micron-size lines (beads). Solid thermoset polyurethane structures were obtained within 3 minutes on a glass substrate at 90°C.

Very limited reports have been published on the concept of employing polymerization of a functional polymer during an additive manufacturing approach. The concept of extrusion freeforming of nylon 6 was employed for *in situ* polymerization of nylon 6.[22] The approach had limitations, however, such as low resolution and poor surface finish. In addition, premixing of the two reactive mixtures to extrude via a single nozzle could result in nozzle clogging, and the exothermic nature of the reaction could diminish melt flow control during extrusion.

12.1.3 Reactive Inkjet Printing of Nylon 6 for Additive Manufacturing

Reaction injection molding (RIM) is a family of processes developed for two-part reactive polymers such as polyurethanes, polyesters, epoxies, and nylons. The RIM process to produce molded nylon 6 parts via anionic polymerization is known as *cast nylon*, which gives the highest mechanical properties in the group of reactive polymers used for RIM.[3] The process involves supplying two reactive mixtures in the molten state to a mixer just before injection into a mold, as shown in Figure 12.1. Polymerization occurs inside the mold, which is preheated to the required reaction temperature. This solidifies the part before de-molding after about 10 minutes.[23]

There is a need for an additive technique to produce parts from a functional polymer with high resolution and good surface finish. Inkjet technology could fill this gap with nylon 6; however, nylon 6 in the melt state has such a high viscosity that it cannot be jetted. Because it can be polymerized

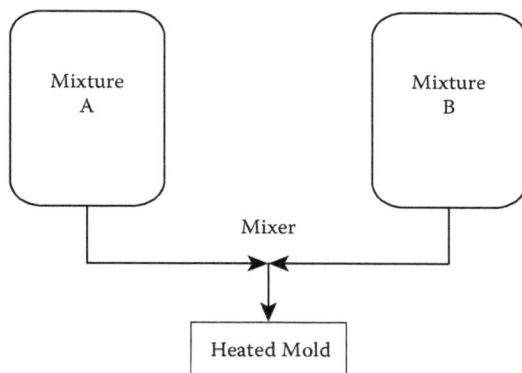

FIGURE 12.1 Schematic of reaction injection molding process.

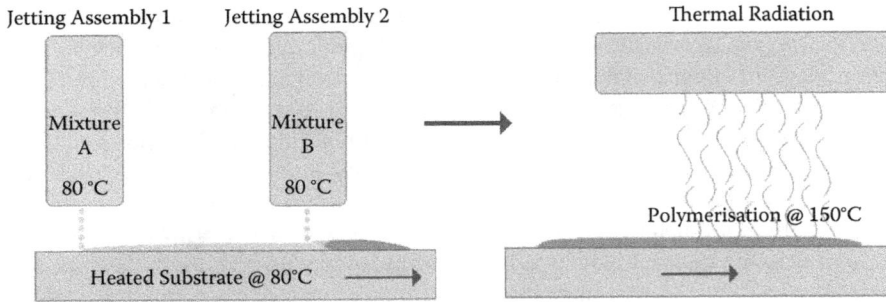

FIGURE 12.2   Concept of reactive inkjet printing of nylon 6.

from mixtures of low-viscosity caprolactam at elevated temperatures, the idea of jetting nylon could become a possibility. The polymerization concept would be similar to the cast nylon process, but instead of premixing the two reactive compounds, consisting of caprolactam plus catalyst and caprolactam plus activator, they could be deposited via inkjet printing. By depositing a layer of the two reactive mixtures on top of each other onto a substrate, the two mixtures could be expected to mix and start the reaction under appropriate conditions (Figure 12.2). This could produce nylon 6 as a solid layer before fabricating the next layer in an additive approach, making it possible to produce solid nylon parts with fine resolution and acceptable build speed.

12.1.4   Research into the Concept of Jetting of Nylon

Research into the concept of jetting nylon was undertaken at Loughborough University. Caprolactam and a number of reactive mixtures were characterized for jetting. The physical properties of the materials—namely, the surface tension, dynamic viscosity, and particle content—were investigated. In addition, by developing an experimental setup for jetting of nylon, the behavior of the materials in the melt supply unit was investigated. Jetting of molten caprolactam and the reactive mixtures began with pure caprolactam to obtain an understanding of the role and the range of parameters affecting the jetting behavior. The process window for jet stability was the main focus. Investigation of the instabilities and their origin were undertaken. Different phenomena associated with jet instability and failure of jets were investigated at this stage.

Formation of droplets was characterized to achieve the highest consistency and optimal droplet condition for deposition. The focus was to study the droplet formation process including the meniscus oscillation, droplet shape evolution after ejection, and kinetics of droplets for pure caprolactam and the reactive mixtures within the process parameter window. High-speed imaging with microscopy was used to observe droplet/nozzle plate interaction and droplet characteristics. Image analysis was undertaken to extract information on different phenomena associated with droplet formation. This stage was intended to help further narrow down the jetting parameters for the deposition stage, where the interaction between the droplets and the surface during deposition was studied with high-speed imaging and image analysis.

Mixing of droplets was another aspect of the deposition stage. This could define whether the droplets of the two mixtures were mixed upon deposition to start the reaction with the appropriate thermal and environmental conditions. Dye tracing and *in situ* fluorescent microscopy were used to observe this. The reaction of nylon was investigated when droplets of the two reactive mixtures were deposited on top of each other. The heating approach was developed and assessed for reaction temperature and also evaporation. After DOD deposition of the samples, they were exposed to different heating conditions. Using thermal analysis, the outcome of the reaction was analyzed.

12.1.5 Design of Experimental Setup

The physical properties of caprolactam were considered as the starting point to find the appropriate printhead and design of the experimental setup. The caprolactam was supplied by Sigma Aldrich GmbH as granules. The dynamic viscosity of molten caprolactam was measured over a range of temperatures, and it was found that caprolactam had a viscosity below 10 mPa·s at temperatures higher than 75°C and shear rates well below those experienced in jetting, which are on the order of 10^4.[6] The surface tension was also found to be 35 mN/m measured at 80°C using the pendant drop method. Based on these findings, a Xaar 126/80 DOD printhead was chosen. The printhead had 126 nozzles (50-μm diameter) in an array 17.2 mm wide and was capable of jetting with a maximum frequency of 5.2 kHz. A jetting assembly was dedicated to each catalyst and activator mixture to avoid cross-contamination and possible pre-jetting reaction and clogging of the printhead. Each assembly consisted of a melt supply unit with thermal and pneumatic control, a printhead with attached resistors as heaters, mount plates, fixtures, and pneumatic connections. Further details of the jetting assemblies can be found in Fathi et al.[24]

A digital microscope camera was used to monitor the nozzle plate for jetting stability. To characterize the droplet formation and deposition, a high-speed camera with a long lens was used in combination with an intensive light source, placed opposite the camera lens for backlight imaging. A heated substrate set at 80°C was considered for deposition. Positioning of the substrate was achieved by a 3D motion system. A second digital microscope camera with fluorescent imaging capability was also installed in line with the deposition setup. It was switchable between normal (white light) and fluorescent imaging to study the molten materials mixing after DOD deposition.

The deposited mixtures required enough heat to initiate the nylon polymerization. The appropriate temperature for the reaction was about 150°C.[25] Deposition of samples in small volumes onto the hot substrate at this temperature gave a high rate of evaporation; therefore, an alternative heating mechanism was developed using radiation to heat up the deposited samples from 80°C to about 150°C. A surface radiation heating system was then used based on linear halogen lamps inside a cast aluminum box. By setting the required distance to the deposition surface, different power settings of the radiation heating could be set. The heating time for the samples could be controlled by programming the substrate motion. More details of the setup can be found in Fathi and Dickens.[26]

12.2 JETTING OF CAPROLACTAM

12.2.1 Process Parameters for Jetting of Caprolactam

Controllable variables for the jetting were the melt temperature, voltage signal for droplet generation, and vacuum level. The melt temperature was set at 80°C in the jetting assembly. For the voltage signal, the amplitude and frequency could be varied from the printhead electronic peripherals (from 0 to 40.0 V and 0 to 5.2 kHz, respectively), whereas the waveform and pulse width were fixed by the printhead manufacturer. The vacuum level on the melt supply to the printhead could affect the meniscus shape and therefore the droplet formation process, so it was varied from 5 to 50 mbar. A start-up sequence ensured a supply of the melt to the printhead before nozzle actuation, as shown in Figure 12.3.

12.2.2 Process Parameters and Jetting Stability

An important research objective was to investigate whether molten caprolactam could be jetted with a normal graphics industry printhead. Figure 12.4 shows a jet array of molten caprolactam from 126 nozzles at 17.5 V, 5 kHz, and 25 mbar, where all jets were stable.[24] Within the

~ Melt under atmospheric pressure

~ Pressure purging of melt
(more than 80 mbar)

~ Under atmospheric pressure
~ Head pressure by gravity

State 1: Dry nozzle plate

State 2: Dripping under pressure

State 3: Dripping under melt gravity
(head pressure)

2 mm

2 mm

2 mm

~ Under increasing vacuum

~ Under vacuum
(10 ~ 100 mbar)

State 4: Retracting meniscus
under vacuum

State 5: Wet surface
(nozzle plate)

Wetting layer

2 mm

2 mm

Big melt meniscus being retracted

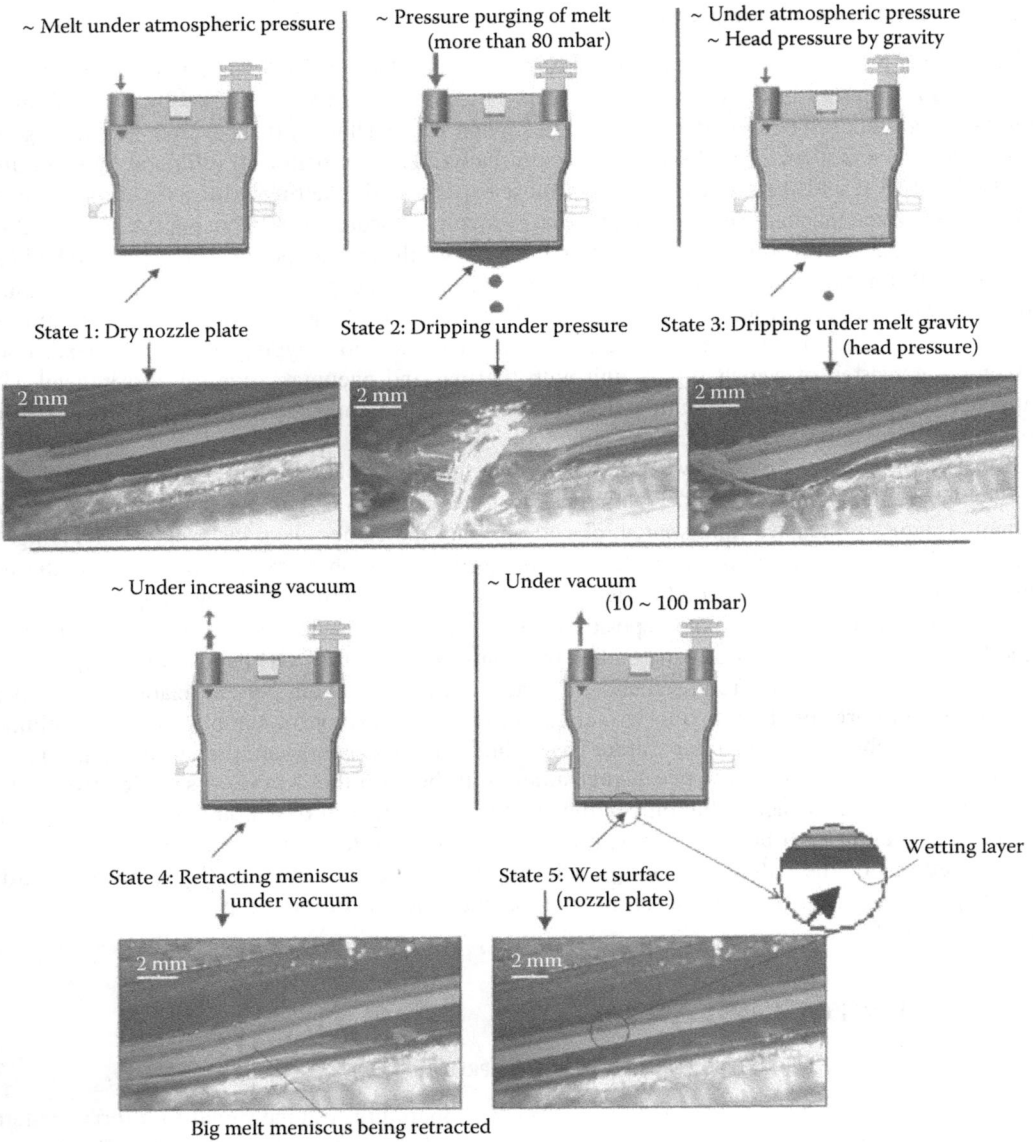

FIGURE 12.3   Sequence of the start-up for jetting trials.

1 mm

FIGURE 12.4   Jet array of melt caprolactam (17.5 V, 5 kHz, 25 mbar).

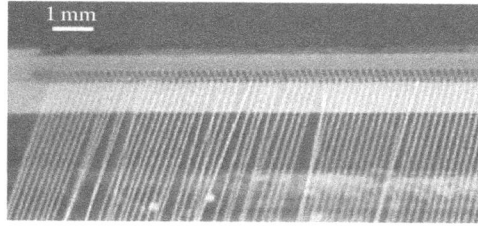

FIGURE 12.5 Jet instabilities during trials with molten caprolactam.

trials, however, some instability occurred, especially when jetting an array as seen in Figure 12.5. Frequency did not have an effect on the jet array stability. Because the instability could have originated externally due to contamination or air motion, the trials with instability were repeated after a purging period (for a further three times). This was to check if the instability occurred due to setting the parameters improperly.

No jet occurred with jetting voltage below 10.0 V and vacuum level below 5 mbar. In such a case, the pressure wave generated by the nozzle actuation inside the melt channel was dissipated when propagating toward the meniscus on the nozzle. A combination of high values of jetting voltage and vacuum level resulted in unstable jets. Some jets failed in the jet array with the vacuum set at 50 mbar at all jetting voltages at the very start of the trial. All of the remaining jets failed within seconds which indicated a time-dependent phenomenon responsible for the jet failures. With the jetting start-up, the risk of air entrapment in the melt supply was eliminated, but there were still instabilities in individual jets or in the array of jets in some experiments. Jet instability was observed when the jetting voltage and vacuum level were low. A combination of low voltage and vacuum resulted in insufficient droplet kinetics and hence jet instability. In such a situation, the jet trajectory error could lead to a jet failure as recorded by the high-speed camera (Figure 12.6). From the start, there was a trajectory error that increased over time until the jet of droplets finally touched the nozzle plate, leading to a failure within about 1 second. This shows how the individual jets could become unstable, probably due to the effect of air motion due to heat convection when using inadequate jetting voltage and vacuum level.

Jet instability was also observed at high vacuum levels. Individual nozzles failed during the trials, typically with a vacuum level at 50 mbar. When next applying a pressure pulse in the melt supply, molten caprolactam was purged through all of the nozzles along with air bubbles from the failed nozzles. In trials with voltage and vacuum levels equal to or below 25.0 V and 30 mbar, no nozzle blocking by air bubbles was observed. It was likely that with high vacuum levels air ingestion occurred during the meniscus oscillation. A similar situation was reported by de Jong et al.[13] for single nozzle jetting with a piezoelectric DOD printhead as shown in Figure 12.7.

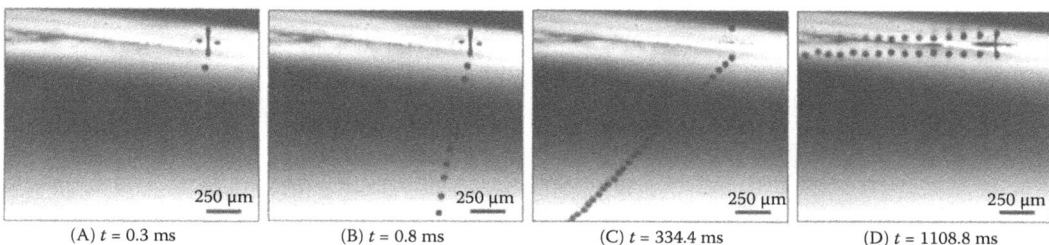

(A) t = 0.3 ms   (B) t = 0.8 ms   (C) t = 334.4 ms   (D) t = 1108.8 ms

FIGURE 12.6 Jet failure due to low jetting voltage and air motion (13.0 V, 5 kHz, 25 mbar).

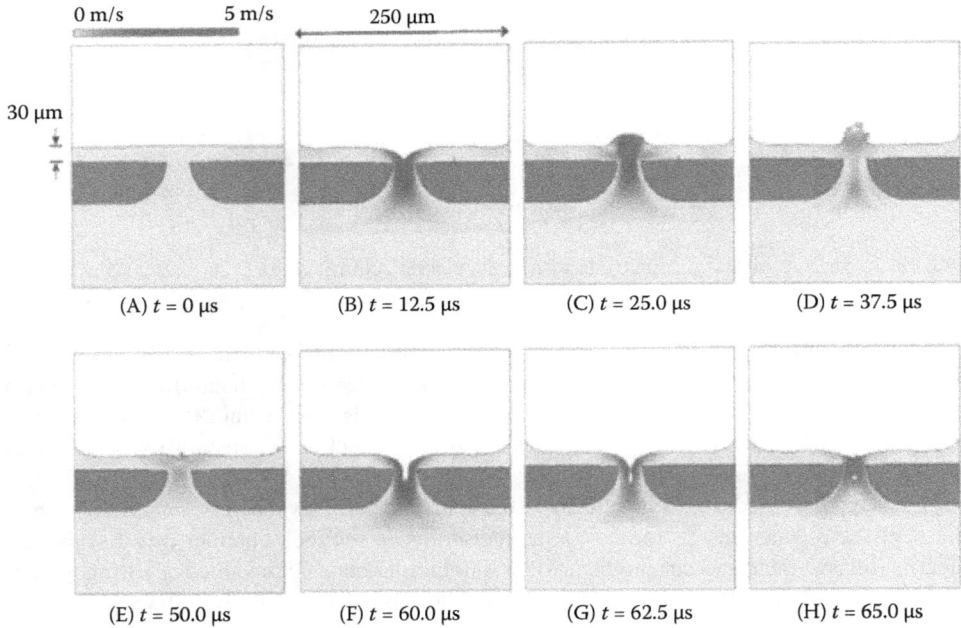

FIGURE 12.7   Modeling of air ingestion during meniscus oscillation in an inkjet printhead.[13]

12.2.3   RECOMMENDED WINDOW FOR JET ARRAY STABILITY

Figure 12.8 gives a general guideline for setting the two parameters of voltage and vacuum in relation to each other for stable jetting of molten caprolactam. Jetting stability was sensitive to jetting voltage and vacuum level whereas jetting frequency did not play a significant role. Because a combination of high voltage and high vacuum level led to air ingestion and consequent nozzle blocking, it is recommended that at higher jetting voltages the vacuum level should be decreased, and *vice versa*.

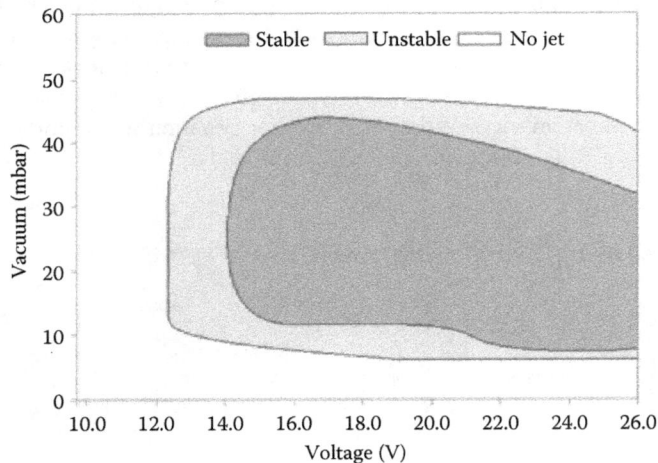

FIGURE 12.8   Guideline jetting parameters for caprolactam at 80°C.[24]

12.3 JETTING OF REACTIVE MIXTURES

12.3.1 REACTIVE MIXTURES

The catalyst and activator mixtures (known here as reactive mixtures) were based on commercial concentrates supplied from Brueggemann Chemical GmbH. C10 as the catalyst and its matching activator, C20P, were diluted in molten caprolactam at 80°C to form C10-CL and C20P-CL, respectively. The C10 concentrate consisted of 17 to 19% of sodium caprolactamate (NaCL) in caprolactam. C20P was made of blocked diisocyanate in caprolactam with approximately 17% free isocyanate content as reported by the supplier. Earlier research suggested the use of 20% concentration of the concentrate.[25] A similar jetting temperature to caprolactam was chosen for jetting reactive mixtures due to limitations of the printhead. Both of the mixtures (C20P-CL and C10-CL) at 20% concentration behaved similarly and had a viscosity of about 15 mPa·s at 80°C, which was about 50% higher than the pure molten caprolactam.

12.3.2 MICROCRYSTAL CONTENT OF REACTIVE MIXTURES AND MELT SUPPLY BEHAVIOR

Due to the higher viscosity of the reactive mixtures, a particle characterization was performed to study the behavior. Hot-stage light microscopy was performed and showed that the pure caprolactam and activator mixture had no particles in the molten state. Figure 12.9, however, shows an image of the catalyst at 20% concentration and after 2 mL of the melt purged through the printhead. Comparing images of brightfield and polarized light microscopy indicated agglomeration of undissolved salt of caprolactam magnesium bromide (CLMgBr) in molten caprolactam as seen in the form of microcrystals. An elevated temperature of 150°C was found to dissolve the microcrystals; however, in addition to the limitation of the printhead's operating temperature, this could result in very high evaporation in the system. As it was found that there was a consistent and sufficient concentration of microcrystals through the system, it was decided to continue the research in jetting the catalyst mixture. More details can be found in Fathi et al.[27]

Melt supply behavior of the activator mixture was similar to the molten caprolactam. With the catalyst mixture, however, an initial pressure of 150 mbar did not provide dripping through the printhead. Increasing the pressure to 750 mbar resulted in dripping of the melt at a rate below 0.6 mL/min at the start of purging. This confirmed that the agglomeration of catalyst microcrystals resulted in partial blocking of the filter (5-μm pore size) in the jetting assembly. The test was repeated several times, each time with a new filter membrane. A vacuum of 10 mbar for up to 5 seconds could retract the meniscus into the printhead completely. The results of repeated tests showed that the purging could provide 6 to 8 mL before blocking the filter membrane, and this was sufficient for researching the drop-on-drop reaction of nylon. It was decided not to search for a new catalyst mixture.

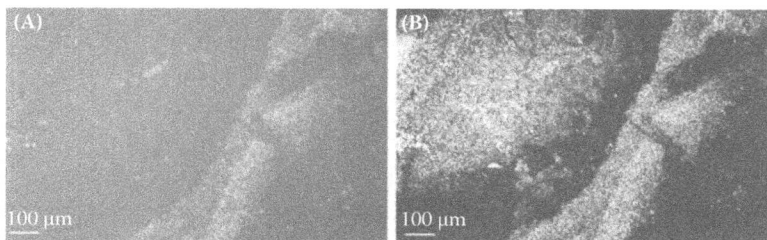

FIGURE 12.9 (A) Polarized light microscopic image of the catalyst mixture, and (B) detection of highly agglomerated particles by image processing.[27]

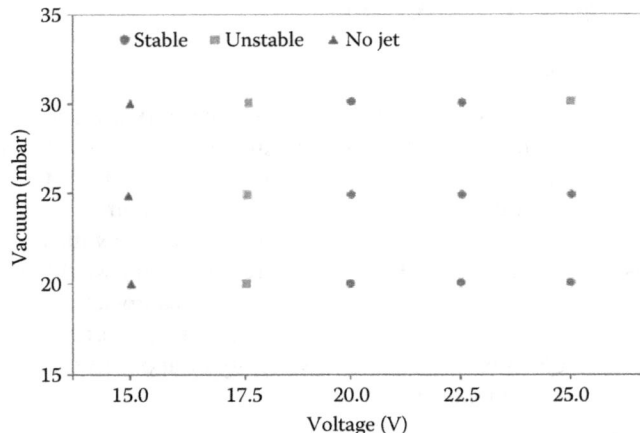

FIGURE 12.10   Stability status in single jet trials with the activator and catalyst mixtures.

12.3.3   JETTING STABILITY OF REACTIVE MIXTURES

Although caprolactam dominated the reactive mixtures, systematic jetting trials with the reactive mixtures showed a different stability behavior. Due to concern over the effect of filter blocking on the concentration of the catalyst mixture, jetting trials with this mixture were carried out after 1, 2, 3, and 4 mL of melt were purged through the printhead, referred to here as melt supply levels. Figure 12.10 shows the results for the jet stability of the reactive mixtures which indicate that the caprolactam and the reactive mixtures did not have the same jetting voltage range for jet stability. The range was narrower for the reactive mixtures (20.0 ~ 22.5 V). A higher jetting voltage was required to develop a stable jet with the reactive mixtures. This means that greater pressure wave dissipation occurred with the mixtures compared with the molten caprolactam. This could be due to the higher viscosity of the mixtures (50% higher).

With both the activator and the catalyst mixtures at different melt supply levels, similar stable ranges of parameters were achieved for a single jet. However, with the catalyst mixture, there were instabilities in the form of jet disturbance (temporary trajectory variations) in a few trials within the stable range. On such occasions, further trials with the same parameter settings showed stability with the single jet. It was reported that contamination within an ink could result in jet failure.[13] The instability could be due to microcrystal agglomerations and the way they interacted with the jetting nozzles. This possibility was studied via particle tracking of melt flow behavior on the wetting layer on the nozzle plate to show how particles can disturb actuating nozzles and result in jet array failure.[28]

12.4   DROPLET FORMATION AND DEPOSITION

12.4.1   DROPLET CHARACTERIZATION OF CAPROLACTAM

High-speed imaging revealed that jetting voltages equal to or higher than 15.0 V formed a train of droplets at all jetting frequencies. Figure 12.11 shows a droplet of the molten caprolactam being ejected. Before the droplet separated from the nozzle, a tail formed which was attached to and traveled with the droplet. This was the same for all jetting parameters that could generate droplets. The tail attached to the droplet disintegrated into small satellite droplets at voltages higher than 17.5 V. Deposition of droplets without satellites is preferable to help with maintaining the uniformity of layer deposition. With a voltage signal, the channel walls of an actuating nozzle

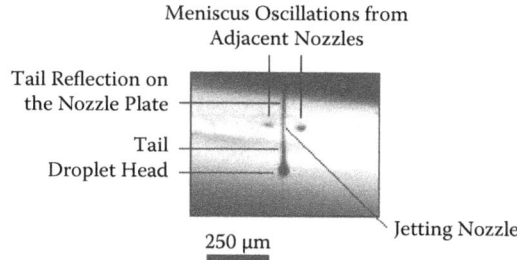

FIGURE 12.11 High-speed imaging of a droplet being ejected from a nozzle (15.0 V, 3 kHz).

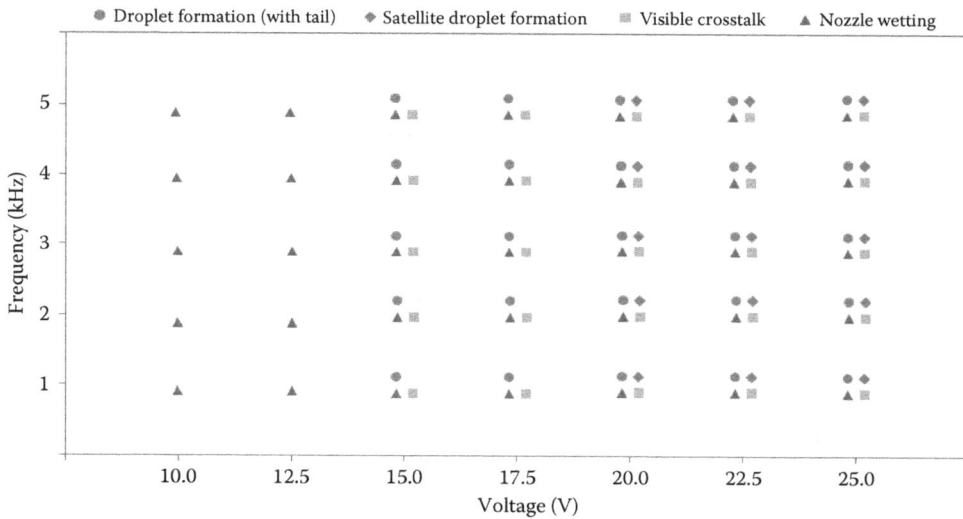

FIGURE 12.12 Summary of observations in droplet formation experiments.[30]

vibrated to expand and contract the melt channel to generate a droplet. With Xaar's printhead shared-wall technology, the adjacent channels were also affected, resulting in a partial expansion and contraction as seen in the form of dark spots (known as *crosstalk*[29]) around the jetting nozzle in Figure 12.11. With the dry nozzle plate, a wetting area developed over time around the actuating nozzle. This phenomenon was observed in all the parameter settings. The nozzle wetting during the droplet formation process is important, as it could interact with the droplet formation stability (e.g., by asymmetric wetting). Figure 12.12 summarizes the observations made in droplet formation experiments.

Figure 12.13 shows the variation of caprolactam droplet size with jetting voltage and frequency. It was almost constant at around 50 μm, which was also the nozzle size. This shows that the droplet size was not affected by the jetting parameters and was due to the nozzle size. The variation of the droplet kinetics with jetting voltage and frequency was quantified. Figure 12.14 shows that the range of frequencies used did not have an effect on the droplet velocity. With a higher jetting voltage, a higher amplitude pressure wave was formed and propagated toward the meniscus, giving the droplet a higher velocity. The velocity fluctuations and linear regression of the curve show that the droplet kinetic behavior was repeatable within the range. The range of jetting frequencies used in the experiments was not observed to provide a droplet inconsistency or fluctuations in the droplet velocity, as shown in Figure 12.14.

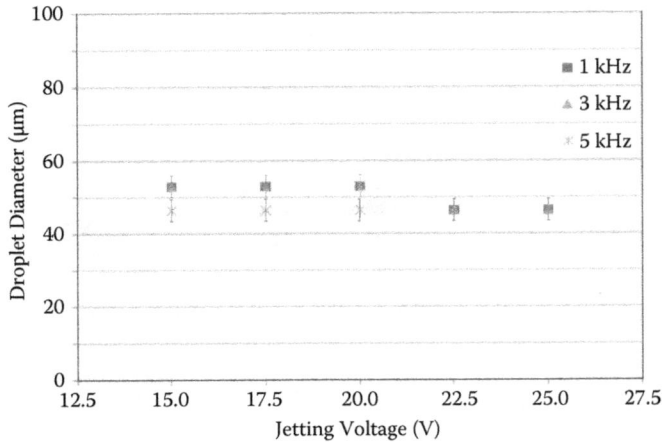

FIGURE 12.13 Droplet size vs. voltage at different frequencies.

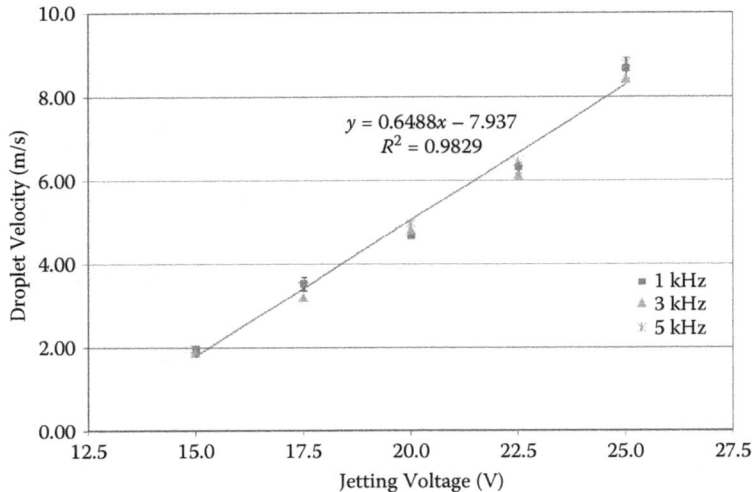

FIGURE 12.14 Droplet velocity vs. voltage at different frequencies.

12.4.2 Droplet Characterization of Reactive Mixtures

Oscillation of the melt meniscus was observed at the actuating nozzle with all jetting voltages and frequencies; however, droplets were formed with 17.5 V or higher at all frequencies. As with caprolactam, jetting frequency was found to have no significant effect on the droplet formation and its reliability. With voltages higher than 17.5 V, sustainable formation of droplets was observed with both the molten mixtures. Formation of a wetting area around the actuating nozzle was also seen. This was due to the higher surface energy of the nozzle plate material (40 mN/m) compared with the jetting materials. The droplet size of both the mixtures was similar to that of molten caprolactam irrespective of the melt supply level, which suggests that the droplet size was mainly dependent on the nozzle geometry and that the physical properties and the jetting parameters range reported had no significant effect. The droplet velocity, however, was lower with the mixtures mainly due to the higher viscosity of the mixtures than that of caprolactam. Figure 12.15 shows the droplet velocity of the catalyst mixture at different melt supply levels. The droplet velocity increased with the jetting voltage almost linearly.

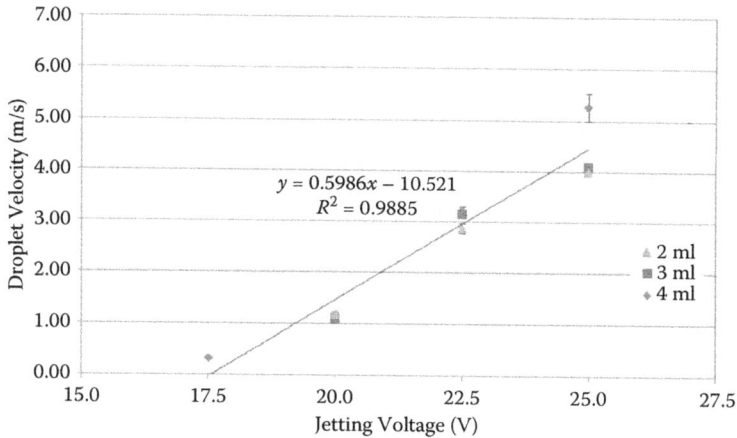

FIGURE 12.15 Droplet velocity vs. jetting voltage at different catalyst mixture melt supply levels (3 kHz, 25 mbar).

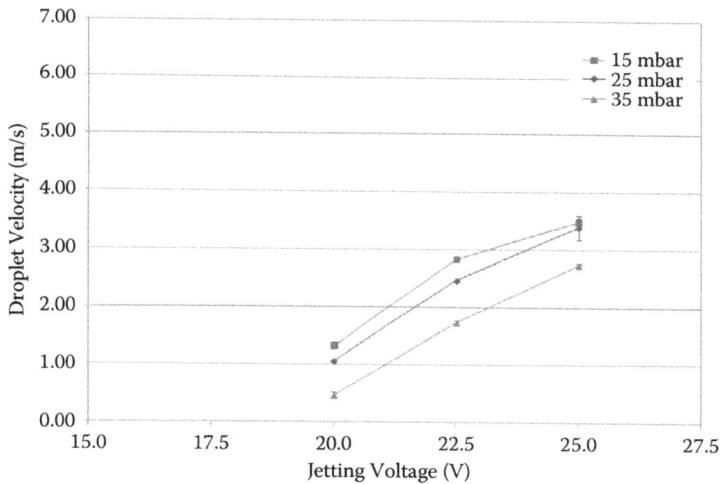

FIGURE 12.16 Droplet velocity vs. voltage with different vacuum levels when jetting the activator mixture.

Figure 12.16 compares droplet velocity against jetting voltage for the activator mixture. The effect of vacuum level on the velocity is clear. Increasing the vacuum level from 15 mbar to 35 mbar decreased the droplet velocity by almost 1 m/s for all of the jetting voltages. The vacuum level affects the meniscus shape and therefore the interaction between the pressure wave and the meniscus surface tension. Due to the higher surface tension of the nozzle plate than the jetting materials, the vacuum applied should form a concave meniscus inside the nozzle. Increasing the vacuum level could increase the surface area of the meniscus.[31] This may also suggest why air ingestion occurred in the jetting trials with high vacuum levels where a highly concaved meniscus had more potential to entrap an air bubble.

12.4.3 DROPLET IMPACT ON SURFACE AND DEPOSITION STABILITY

Figure 12.17 shows the start of a bead formation on the moving heated glass surface in a lateral high-speed image. Splashing was not observed upon droplet impact. Due to the droplet impact kinetics, the surface energy, and the chosen droplet spacing, coalescence of the droplets occurred

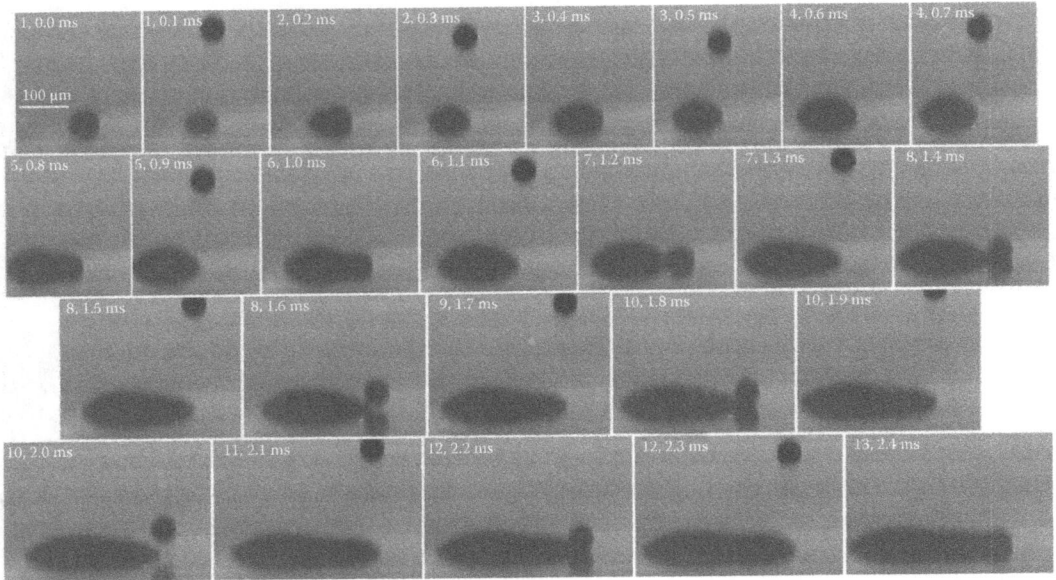

FIGURE 12.17 Spreading of molten caprolactam droplets onto the heated moving surface forming a bead.

to form a bead. However, for the first few droplets, the coalescence resulted in the bead having a greater height at the start (left side) compared with the advancing front where new droplets were impinging. This formed a bulge at the start of the bead and is a form of bead instability, according to Derby.[10] When a pattern is being deposited, uniform beads are preferred and the bulge formation at the start of the bead could affect the accuracy of the layer's edge. As shown in Figure 12.17, following the initial bulge formation, the bead was stable and uniform after the initial 2.0 ms. However, because bead stability is required until solidification by polymerization (seconds after the deposition), it was necessary to monitor the bead stability for a longer period. Figure 12.18 shows bead-on-bead deposition onto a cold glass surface after a period of 1 second. The bead and its reflection on the surface can be seen. Contamination on the surface demonstrates the substrate movement in

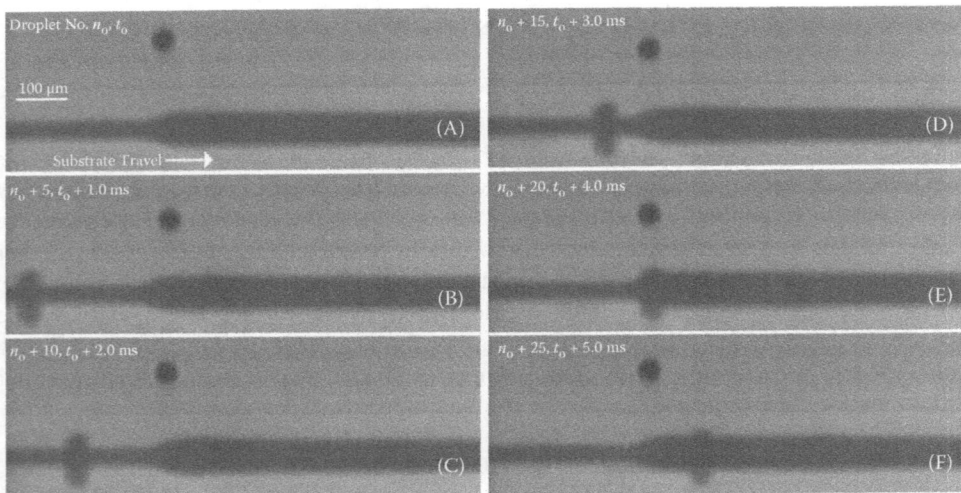

FIGURE 12.18 Monitoring the stability of a deposited bead of caprolactam on the cold surface (bead-on-bead deposition).

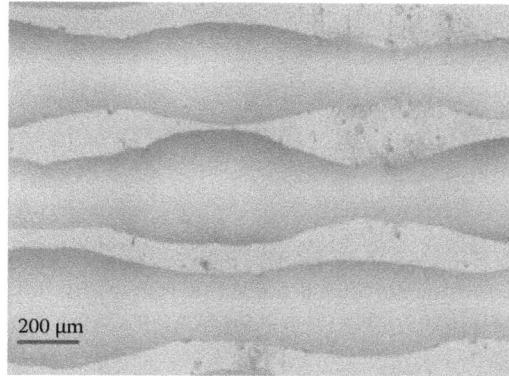

FIGURE 12.19 Formation of bulges and ridges in deposition of molten caprolactam beads onto the heated glass substrate.

the sequence shown. The first deposited bead remained stable and had a uniform height of 25 μm, and the bead-on-bead deposition produced a uniform total height of 40 μm. Figure 12.19, however, shows a top-view image of multiple beads deposited side by side onto the heated surface and better demonstrates the instability of the beads due to formation of bulges and ridge features when a heated substrate is used. More details can be found in Fathi and Dickens.[26]

12.5 DROP-ON-DROP REACTION

For drop-on-drop deposition of mixtures, stable and repeatable droplet placement of the catalyst mixture is necessary. The microcrystals must properly dissolve or sustainably disperse within the molten caprolactam. Research into better dispersion could resolve issues with this process; however, drop-on-drop mixing and the reaction of nylon 6 after exposure to top-down heating would have to be proven for such research to be useful. This possibility could be demonstrated by the reaction of nylon upon accumulation of a large number of droplets of the two mixtures on top of each other as detailed in this section.

12.5.1 DROP-ON-DROP MIXING UPON ACCUMULATIVE DEPOSITION

Drop-on-drop mixing was investigated with the reactive mixtures deposited accumulatively on top of each other onto a glass slide set at 80°C. Fluorescent dye was stirred well into the molten activator mixture at a ratio of 0.1% (volumetric), whereas the catalyst mixture was used without a dye. Single-nozzle jetting was used with the same jetting parameters chosen in deposition trials. In each trial, the transparent catalyst mixture was deposited first via a single nozzle for 10 seconds. Then, the same amount of dyed activator mixture was deposited on top. The in-line microscope camera with a fluorescent light source observed the mixed areas. The trials were repeated three times and image processing was undertaken.

Figure 12.20A shows the transparent catalyst mixture deposited onto the glass slide (placed on the heated aluminum substrate with a 6-mm hole underneath the slide for scaling). Figure 12.20B, on the other hand, shows the dyed activator mixture of the same amount deposited onto the catalyst. Fluorescent imaging of the drop-on-drop-deposited mixtures is shown in Figure 12.20C. Image processing was used to evaluate the mixing outcome by extracting the contours of the accumulated materials before and after the dyed activator mixture deposition and also the dyed area. The contours were then superimposed as shown in Figure 12.20D. Then, areas 1, 2, and 3 (catalyst mixture only, dyed, and transparent areas after the activator deposition, respectively) were measured using the image processing software.

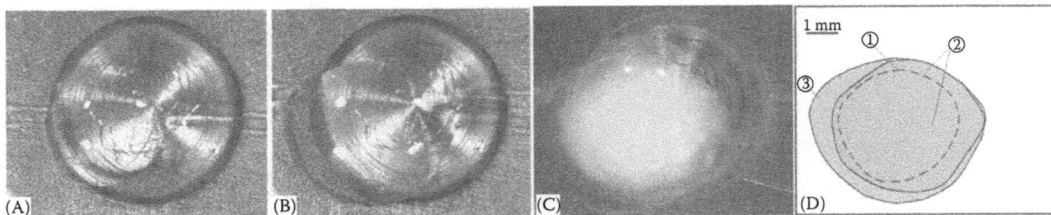

FIGURE 12.20 Drop-on-drop mixing of the reactive mixtures onto a heated glass slide: (A) transparent catalyst mixture only, (B) dyed activator mixture deposited onto the catalyst mixture, (C) mixing areas traced by the fluorescent imaging, and (D) superimposition of the contours extracted by processing of images.[26]

Assuming that the two molten materials mixed upon droplet impact in the dyed area, there were some limited areas where they did not mix. The impact kinetics of the dyed activator mixture droplets could have pushed away the transparent catalyst mixture. Restriction of the catalyst mixture for further spreading upon consecutive impingement of the activator mixture droplets could resolve this problem and any drop-on-drop placement inaccuracies. In fact, in order to investigate *in situ* polymerization of nylon 6, a small differential scanning calorimetry (DSC) pan was used as the receiving medium. The samples that would receive thermal radiation heating upon accumulative drop-on-drop deposition into the pan would be evaluated for the monomer to polymer conversion rate using DSC analysis. This could show that, first, polymerization would be achieved using the inkjet printing concept and, second, how control over the thermal radiation conditions would affect the reaction outcome.

12.5.2 Heating to Obtain the Drop-on-Drop Reaction

An elevated substrate temperature of 150°C was initially used for the heat-up process but due to very high level of evaporation a surface radiation heating was then considered. The plan was to deposit the mixtures into DSC pans on the substrate set at 80°C followed immediately by moving the substrate underneath a unit to expose the mixtures to radiation heating. By adjusting the power of the unit, the substrate/reaction unit gap, and the period of time that the deposited samples were exposed to the radiation, the heating conditions could be varied. The heating outcome was assessed by monitoring a DSC pan temperature. As Figure 12.21 shows, a sharp temperature increase was

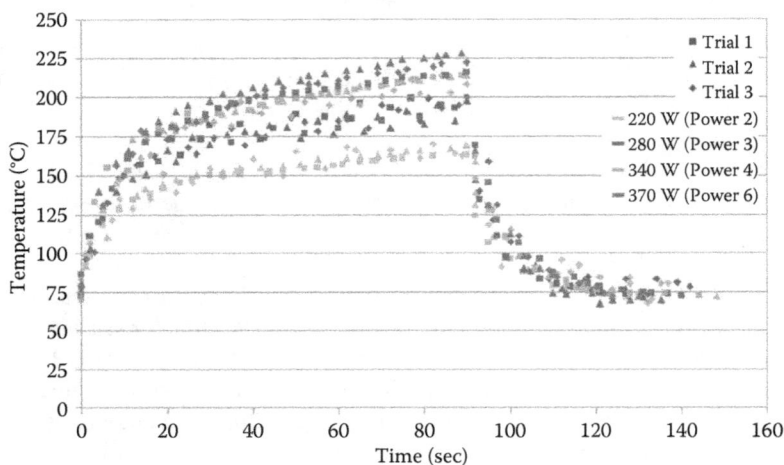

FIGURE 12.21 Temperature monitoring trials within 90 seconds of radiation heating at different power settings.

recorded within the first 20 seconds which then decreased rapidly by moving the sample away from the radiation source. This showed that reaction temperatures would be provided using 1-minute radiation with all power settings (labeled as Power 1 to 4). The sample weight was measured after the test and showed some level of evaporation during the thermal radiation. From an additive manufacturing perspective, evaporation during radiation heating would induce inaccuracies in the amount of material deposition for each layer. A compensation factor could be anticipated for an accurate final volume of drop-on-drop deposition. This, however, would not be an easy task as evaporation and polymerization could be occurring simultaneously during layer fabrication; however, with a fast reaction, the evaporation of caprolactam would be minimized. This emphasizes the importance of rapid reactions in the possible inkjet additive manufacturing of nylon.

12.5.3 Drop-on-Drop Reaction of Mixtures

The mixtures were jetted at 20.0 V, 5 kHz, 25 mbar, and 4 mL through a single nozzle into a DSC pan at a set melt temperature of 80°C. The sample deposition for drop-on-drop reaction followed a procedure similar to the drop-on-drop mixing. DSC was used (DSC-60; Shimadzu Corp.) for the thermal analysis of caprolactam, the mixtures, and the drop-on-drop reaction samples. This technique monitors the flux of heat absorption or release against a reference (empty pan) during heating or cooling at constant rate. Figure 12.22 shows the typical thermal behavior of the samples. The thermogram indicates that the sample contained a significant amount of monomer, as a large caprolactam melting peak is observed according to a report by Khodabakhshi et al.[25] The broad endothermic peak that occurred after the caprolactam was melted was typical of the caprolactam evaporation also seen with the mixtures. Measuring the sample weight after the DSC test confirmed that a considerable amount of evaporation (typically 70%) had occurred.

It was not clear whether reaction occurred during the DSC run or during radiation heating, as the endothermic heat of evaporation of unpolymerized monomer could have dominated any heat released by a possible reaction during the DSC run. Therefore, a second DSC run was produced as shown in Figure 12.23. There was no weight loss after the second run. Therefore, the endothermic and exothermic peaks in Figure 12.22 are assumed to be respectively due to the melting and crystallization of the polymer. However, the melting point recorded was about 195°C, which was about 20 to 30°C lower than that produced with bulk polymerization.[25] This could have been due to the lower

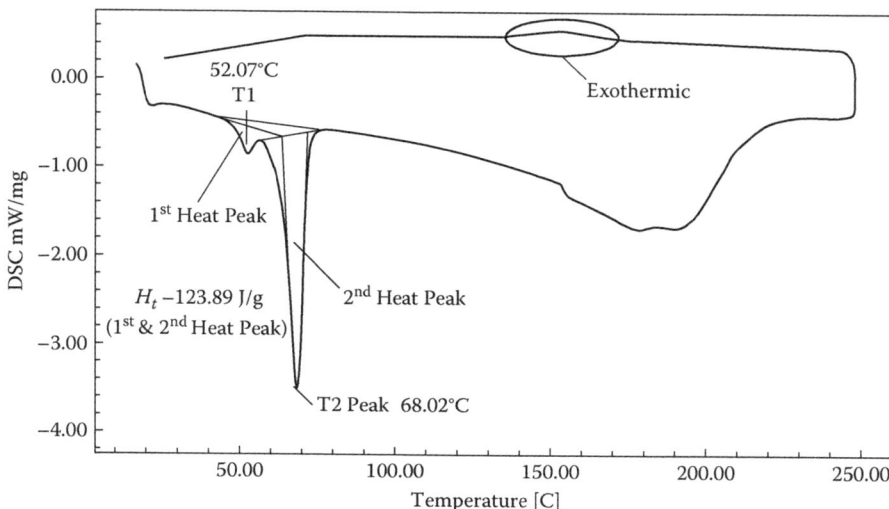

FIGURE 12.22 DSC thermogram of a sample (Powder 3 exposed to sample for 7 minutes).

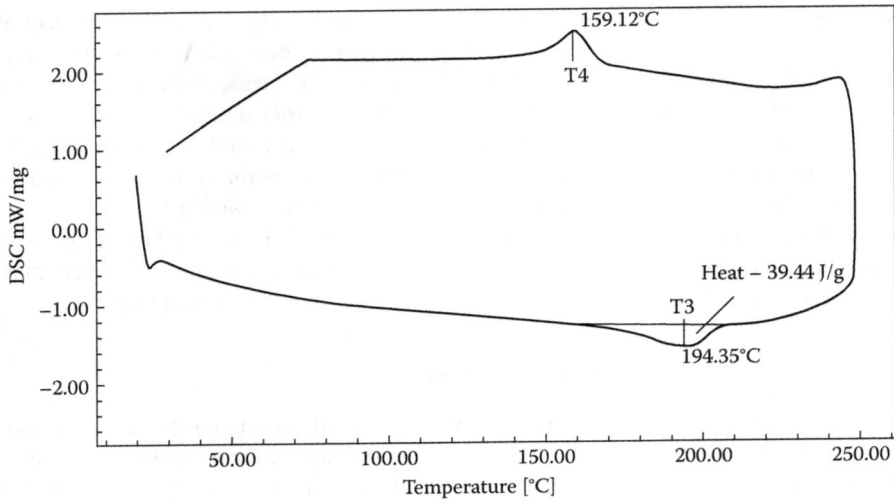

FIGURE 12.23 Thermogram from second DSC run of the sample corresponding to Figure 12.22.

molecular weight of the nylon made in the drop-on-drop reaction samples. Dividing the sample weight after the first DSC run by the original drop-on-drop reaction sample weights was used to calculate the monomer conversion rates. Figure 12.24 shows the results of the percent monomer conversion for different radiation heating conditions. The range of monomer conversion was 10 to 30%. The results showed no particular effect due to the radiation heating conditions.

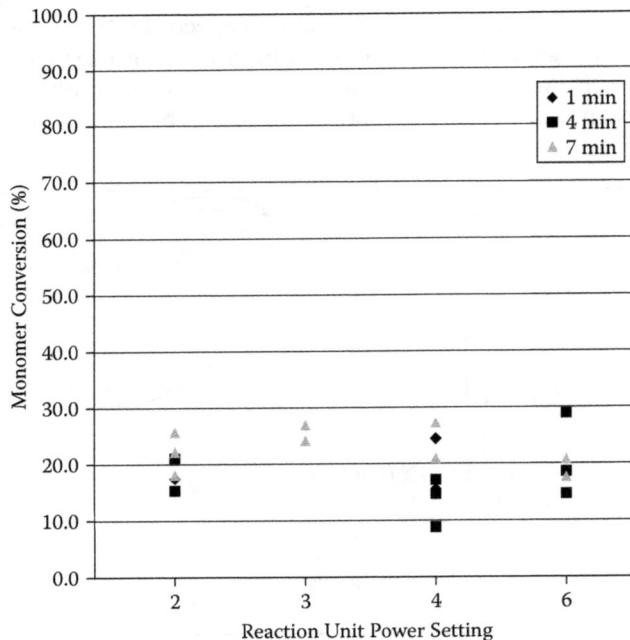

FIGURE 12.24 Monomer conversion of drop-on-drop reaction samples.

12.5.4   OTHER IMPORTANT PARAMETERS IN DROP-ON-DROP REACTION

The above results confirmed that samples had conversion rates up to 30%, much lower than what was needed for a feasible additive manufacturing process. Considering the range of heating parameters used, other factors had a major effect on the results. The results for the research into the drop-on-drop reaction to date did not produce adequate conversion rates for the concept, which would require a high conversion in a short time to solidify a layer before the next layer was deposited. It needs to be established why the monomer conversion in the drop-on-drop reaction was not as high as the bulk polymerization method. The environment was assumed to be a likely reason, although the atmosphere in the glove box used for the drop-on-drop reaction was the same as for the bulk polymerization using normal atmosphere conditions. For this reason, the influence of environment on the samples for these two methods should be considered. A main difference was the total surface of molten mixture exposed to air in the two approaches. Droplets of the reactive mixtures were hundreds of millions of times smaller than the bulk polymerization samples (30 mL). The high surface-to-volume ratio of droplets gave them added exposure to oxygen and moisture. Upon deposition, droplets in flight were exposed to the environment in a similar way as blowing air with 1 m/s speed for about 1 ms (based on the droplet velocity and the nozzle-to-surface distance of 1 mm). This could give the droplet surface molecules added opportunity for contact with oxygen and moisture. All of these aspects of the drop-on-drop reaction could have deactivated the molten mixtures on the surface of the droplets before the reaction stage and therefore limited the polymerization outcome. This suggests the importance of researching the effect of environment on the drop-on-drop reaction.

12.6   SUMMARY AND CONCLUSIONS

This chapter focused on research into a novel additive manufacturing concept that utilizes inkjet deposition of reactive materials to produce nylon 6 by *in situ* anionic polymerization. The research required integration of different technologies and an understanding of several aspects of material processing, including melt supply behavior, jetting stability, droplet formation, mixing, and the formation of nylon upon radiation heating. The surface tension and viscosity of the materials tested were found to be within a range suitable for inkjet technology. Jetting frequency had no significant effect on the jetting stability; however, the jets were sensitive to the voltage and vacuum level. Too low or too high values of these parameters induced instability. It was also found that, due to their higher viscosity, jetting the reactive mixtures required higher voltages than did caprolactam to obtain stable jets. Jet array stability was found to be considerably more sensitive with the catalyst mixture than with the activator mixture within the normal stable range of parameters which was assumed to be due to the agglomeration of microcrystals of the catalyst complex.

The investigation showed that, with the chosen jetting parameters, spreading occurred upon droplet impact. With a moving surface, a stable bead was obtained with a cold surface whereas with the heated surface instability was observed in the form of formation of bulges. Maintaining a stable and uniform bead for bead-on-bead deposition of the mixtures would have required further investigations on the droplet spacing and surface temperature, so it was decided to deposit multiple droplets into a DSC pan to pursue research on the drop-on-drop reaction. Dye tracing showed that the use of a DSC pan for the accumulative drop-on-drop deposition approach gave good mixing.

With regard to initiating the reaction of drop-on-drop deposited mixtures, an elevated substrate temperature at 150°C was found to result in rapid evaporation and unbalancing the mixtures ratio, which could affect the reaction. Therefore, surface radiation was used to heat up the samples after the two mixtures were deposited onto each other. The monomer conversion rates for samples generated under various radiation heating conditions were found to be less than half the results for the bulk polymerization approach and not adequate for a deposited-material phase change by solidification due to polymerization. The biggest difference between the two approaches is that in the deposition method jetting thousands of tiny droplets in air instead of a large volume of sample could have resulted in a very high monomer deactivation before the start of polymerization cycle. Therefore, the effect of environment on the jetting of nylon was thought to be significant.

One of the most important aspects of future research will be developing an understanding of the catalyst mixture microcrystals and their interaction with the jetting system. The main objective will be to obtain stable jetting with high reliability. Research into the agglomeration mechanism to develop a methodology for dispersing the microcrystals in molten caprolactam is suggested. Possibilities could include charging microcrystals electrostatically or applying a vibration mechanism similar to ultrasonic. Another possibility would be the use of chemical surfactants that would not affect the reaction. For bead stability, the main parameters are the droplet spacing and surface energy of the substrate. The latter could be controlled by melt and surface temperatures and possibly by chemical surfactants. Research into the deactivation mechanism of monomers during flight may be required if polymerization in normal air environment is to be pursued. Otherwise, inert nitrogen chambers may have to be considered which would increase the cost and complexity of future processes.

To develop the concept into a layer fabrication process, several materials and process considerations should be taken into account. This research explored some of the main aspects to provide a fundamental understanding of the parameters involved, and the results point toward a very challenging future for the concept of jetting nylon. Successful development of a reliable and economical process would have a huge commercial impact because of the on-demand deposition and additive manufacturing features in addition to the demand of nylon 6 due to its functionality.

REFERENCES

1. Hanford, W.E. and Joyce, R.M. (1948). Polymeric amides from epsilon-caprolactam. *J. Polym. Sci.*, 3: 167–172.
2. Ritz, J., Fuchs, H., Kieczka, H., and Moran, W.C. (2005). Caprolactam. In: *Ullmann's Encyclopedia of Industrial Chemistry*. Wiley-VCH, Weinheim, Germany.
3. Kohan, M.I. (1995). *Nylon Plastics Handbook*. Hanser, New York.
4. Gibson, I., Rosen, W.D., and Stucker, B. (2009). *Additive Manufacturing Technologies: Rapid Prototyping to Direct Digital Manufacturing*. Springer, New York.
5. Orme, M. (1991). On the genesis of droplet stream microspeed dispersions. *Phys. Fluids*, 3: 2936–2947.
6. de Gans, B.J., Duineveld, P.C., and Schubert, U.S. (2004). Inkjet printing of polymers: state of the art and future developments. *Adv. Mater.*, 16: 203–213.
7. Mironov, V., Reis, N., and Derby, B. (2006). Organ printing: computer-aided jet-based 3D tissue engineering. *Tissue Eng.*, 12: 631–634.
8. Hon, K.K.B., Li, L., and Hutchings, I.M. (2008). Direct writing technology: advances and developments. *CIRP Ann. Manuf. Technol.*, 57: 601–620.
9. Singh, M., Haverinen, H.M., Dhagat, P., and Jabbour, G.E. (2010). Inkjet printing: process and its applications. *J. Adv. Mater.*, 22: 673–685.
10. Derby, B. (2011). Inkjet printing ceramics: from drops to solid. *J. Eur. Ceram. Soc.*, 31: 2543–2550.
11. Le, H.P. (1998). Progress and trends in inkjet printing technology. *J. Imaging Sci. Technol.*, 42: 42–69.
12. Pique, A. and Chrisey, B. (2002). *Direct-Write Technologies for Rapid Prototyping Applications*. Academic Press, San Diego CA, p. 184.
13. de Jong, J., de Bruin, G., Reinten, H., van den Berg, M., Wijshoff, H., Versluis, M., and Lohse, D. (2006). Air entrapment in piezo-driven inkjet printheads. *J. Acoust. Soc. Am.*, 120: 1257–1265.
14. Gao, F. and Sonin, A.A. (1994). Precise deposition of molten microdrops: the physics of digital microfabrication. *Proc. R. Soc. Lond. Philos. Trans. Ser. A Math. Phys. Eng. Sci.*, 444: 533–554.
15. Beulen, B., de Jong, J., Reinten, H., van den Berg, M., Wijshoff, H., and van Dongen, M.E.H. (2007). Flow on the nozzle plate of an inkjet printhead. *Exp. Fluids*, 42: 217–224.
16. Wang, T., Hall, D., and Derby, B. (2004). Inkjet printing of wax-based PZT suspensions. *Key Eng. Mater.*, 264/268: 697–700.
17. Hackleman, D.E. and Pawlowski, N.E. (1987). Reactive Ink-jet Printing, U.S. Patent 4,694,302A.
18. Johnson, D.R., Kynaston-Pearson, A.W., and Damarell, W.N. (2003). Inkjet Printer Which Deposits at Least Two Fluids on a Substrate Such That the Fluids React Chemically to Form a Product Thereon, U.K. Patent GB2382798(A).

19. Elsner, P., Dreher, S., Ederer, I., Voit, B., Gudrun, J., and Stephan, M. (2010). Method and Device for Production of a Three-Dimensional Article, U.S. Patent 7,767,130.
20. Uhlmann, E. and Elsner, P.C. (2005). New printing technology for fully graduated material properties. In: *Solid Freeform Fabrication Symposium Proceedings*, pp. 13–18.
21. Kröber, P., Delaney, J.T., Perelaer, J., and Schubert, U.S. (2009). Reactive inkjet printing of polyurethanes. *J. Mater. Chem.*, 19: 5234–5238.
22. Lombardi, J.L. and Calvert, P. (1999). Extrusion freeforming of nylon 6 materials. *Polymer*, 40: 1775–1779.
23. Crawford, R.J. (1998). *Plastics Engineering*. Elsevier, Amsterdam.
24. Fathi, S., Dickens, P.M., and Hague, R.J. (2012). Jetting stability of molten caprolactam in an additive inkjet manufacturing process. *Int. J. Adv. Manuf. Technol.*, 59: 201–212.
25. Khodabakhshi, K., Gilbert, M., Fathi, S., and Dickens, P.M. (2014). Anionic polymerisation of caprolactam at the small-scale via DSC investigations. *J. Therm. Anal. Calorim.*, 115: 383–391.
26. Fathi, S. and Dickens, P.M. (2012). Challenges in drop-on-drop deposition of reactive molten nylon materials for additive manufacturing. *J. Mater. Process. Technol.*, 213: 84–93.
27. Fathi, S., Dickens, P.M., Khodabakhshi, K., and Gilbert, M. (2013). Microcrystal particle characterisation in inkjet printing of reactive nylon materials for additive manufacturing. *ASME J. Manuf. Sci. E*, 135: 011013.
28. Fathi, S. and Dickens, P.M. (2013). Jet array driven flow on the nozzle plate of an inkjet printhead. *J. Mater. Process. Technol.*, 213: 283–291.
29. Raman, G. (1999). Reduced Crosstalk Inkjet Printer Printhead, U.S. Patent 5,912,685.
30. Fathi, S. and Dickens, P.M. (2012). Droplet characterisation of molten caprolactam for a additive manufacturing applications. *Proc. IMechE Part B: J. Eng. Manufact.*, 226: 1052–1060.
31. Fathi, S. and Dickens, P.M. (2012). Nozzle wetting and instabilities during droplet formation of molten nylon materials in an inkjet printhead. *ASME J. Manuf. Sci. E*, 134: 041008.

13 Comparison of Additive Manufacturing Materials and Human Tissues in Computed Tomography Scanning

John Winder, Darren Thompson, and Richard Bibb

CONTENTS

ABSTRACT

Additive manufacturing (AM), covering processes frequently referred to as rapid prototyping, rapid manufacturing, or three-dimensional (3D) printing, provides new opportunities in the manufacture of highly complex and custom-fitting medical devices and products. Although many medical applications of AM have been developed and physical properties of the resulting parts have been studied, the characterization of AM materials using computed tomography (CT) has not been explored. The aim of this study was to determine the CT number of commonly used AM materials. There are many potential applications of the information resulting from this study in the design and manufacture of wearable medical devices, implants, prostheses, and medical imaging test phantoms. Twenty-nine AM material samples were CT scanned and the resultant images analyzed to ascertain the materials' CT numbers and appearance in the images. It was found that some AM materials have CT numbers very similar to human tissues; that fused deposition modeling, stereolithography, and selective laser sintering produce samples that appear uniform on CT images; and that 3D printed materials show a variation in internal structure. AM materials may be suitable for the development of anatomically accurate phantoms for image quality and radiation dose tests.

13.1 INTRODUCTION*

Additive manufacturing (AM) is increasingly used to refer to a variety of technologies that are used to manufacture physical models, prototypes, or functional components directly from three-dimensional (3D) computer-aided design (CAD) data. In AM, physical objects are constructed in a layer-by-layer manner. AM covers all applications and encompasses previous commonly used terms, including 3D printing, rapid prototyping (RP), and rapid manufacturing (RM). It is not appropriate to describe each available AM process here. Full descriptions and technical details are available from the respective manufacturers (a list of contact details and websites is provided below), and several texts are available that provide comprehensive overviews of the processes.[1–4] Continuous development means that AM technologies can now produce objects in a wide variety of materials ranging from soft, flexible polymers to high-performance metal alloys.

Additive manufacturing technologies have been successfully applied in medicine since the early 1990s.[5] Initially, RP processes, such as stereolithography, were used to make highly accurate models of skeletal anatomy directly from three-dimensional computed tomography (CT) data. Typically referred to as medical modeling or biomodeling, this has now become widely accepted as good practice, and many papers have been published reporting cases and the benefits achieved, particularly in craniomaxillofacial surgery. Medical models have typically been used to plan and rehearse surgery and in the design and manufacture of custom-fitting prostheses. The use of medical models has become commonplace, and a number of texts and review papers are available that describe a wide range of medical applications and their principal advantages.[6–11]

More recently, AM technologies have been used to directly manufacture custom-fitting medical devices, such as facial prosthetics, removable partial denture frameworks, surgical guides, and even implants directly from 3D CAD data.[12–14] AM principles are also being exploited in tissue engineering, where the advantages of layer additive manufacture are being utilized to build highly complex porous scaffolds that can support the growth of living cells.[15,16] To date, no polymer-based AM materials have been specifically developed or approved for implantation, and most of the materials tested in this research are not considered biocompatible. The limitations of currently available materials means that this research will be used to provide an indicator of material types that may be further developed for medical applications in the future.

Some assessment of the physical properties of AM materials has been carried out; for example, the dimensional accuracy, roughness of surface, and mechanical properties have been established for ZPrinter® 310 Plus and the Objet Eden 330.[17] In addition, much research has been conducted on the utilization of CT data in building objects and models using AM technologies; however, the CT properties of AM materials have not been investigated. The characteristics of AM materials under radiological conditions will become important in the future, as a variety of medical devices and custom-fitting patient products may be manufactured using AM and subsequently scanned using CT for design, testing, or treatment purposes. Therefore, the aim of this work was to determine the CT number (also known as the Hounsfield unit) of a selection of AM materials and establish their appearance in CT images. A CT scanner measures the spatial distribution of the linear attenuation coefficient or amount of absorption of x-rays. To allow this measure to be compared among scanners, a CT number range was developed based on the linear attenuation to x-rays of water. This scale or CT number range is typically from –1024 to 3072 (10-bit storage required).[18] Table 13.1 indicates the CT numbers of a range of human tissues.

---

* This chapter is based on work previously published and is reproduced here with the permission of the copyright holders: Bibb, R., Thompson, D., and Winder, J., Computed tomography characterisation of additive manufacturing materials, *Med. Eng. Phys.*, 33(5), 590–596, 2010. With permission from Elsevier. Additional data have been added from Winder, R.J., Thompson, D., and Bibb, R.J., Comparison of additive manufacturing materials and human tissues in computed tomography scanning, in *Proceedings of the 12th National Conference on Rapid Design, Prototyping and Manufacture*, Bocking, C.E. and Rennie, A.E.W., Eds., CRDM, Ltd., Buckinghamshire, U.K., 2011, pp. 79–86. With kind permission of the editors.

TABLE 13.1
CT Numbers of Selected
Human Tissues

Tissue	CT Number
Air	−1005 to −995
Lungs	−950 to −550
Fat	−100 to −80
Water	−4 to 4
Kidney	20 to 40
Pancreas	30 to 50
Blood	50 to 60
Liver	50 to 70
Spongious bone	50 to 300
Compact bone	300+

Source: Adapted from Kalender, W.A., *Computed Tomography: Fundamentals, System Technology, Image Quality, Applications*, Publicis MCD, Erlangen, 2000, p. 30.

It should be noted that since this work was carried out there has been some consolidation through mergers and acquisitions in the AM industry. Z Corp. was acquired by 3D Systems, and the technology is now marketed under their ProJet label. Stratasys merged with Objet, but the technology remains labeled Objet and Connex. The Huntsman materials business has also been acquired by 3D Systems. These developments have not altered the basic materials or process capabilities of the machines used and these results remain valid.

13.2   BACKGROUND ON COMPUTED TOMOGRAPHY

Computed tomography (CT) works by passing focused x-rays through an object and measuring the amount of the x-ray energy absorbed. The amount of x-ray energy absorbed is the proportional linear x-ray attenuation coefficient, which is related to the density of the object. By taking multiple measurements from many angles, the density measurements can be reconstructed into a cross-sectional image using a computer. This produces a gray-scale image where the density at a given position within the object is indicated by shades of gray ranging from black, indicating the lowest density (i.e., that of air), to white, representing the highest density (which in medical CT corresponds to the hardest bone or teeth, for example). The fundamental principles and practice of medical CT are described in medical training and research literature.[8,19,20]

Computed tomography is used in industry and medicine to ascertain internal features. For example, castings may be CT scanned to reveal cracks in engines or pipes. In medicine, CT scanning is used to investigate internal abnormalities or injuries and provide superior detail compared to plane x-rays images. As bones are much denser than surrounding soft tissues they show up very clearly in CT images, which make CT an important tool when investigating skeletal anatomy. Similarly, the density difference between soft tissues and air is great, allowing, for example, the nasal airways to be clearly seen. However, the density difference between different adjacent soft tissues is small and therefore it may be difficult to distinguish between different adjacent organs in a CT image.

Computed tomography uses ionizing radiation (x-rays) so exposure should be minimized, particularly to sensitive organs. Within the CT scanner, the x-ray tube and detectors are rotated around the patient in a continuous manner. Typically, the patient lies on their back and is passed through the scanner on the moving table. The detector array acquires profiles of x-ray attenuation through the patient. Most modern scanners perform a continuous spiral of exposure enabling three-dimensional CT scanning to be performed. Modern CT scanners employ multiple arrays to speed up three-dimensional data capture.

Computed tomography images are generated as a gray-scale pixel image (i.e., a bitmap computer image). If the distance between a series of axial images, called the slice thickness, is known, they can be interpolated from one image to the next to form voxels. Software can be used to re-slice the voxel data, enabling cross-sectional images from any orientation to be generated from the original data.

The radiographers who conduct medical CT scans have specific parameters and settings for different types of scan. These are standardized and referred to as protocols. When embarking on using CT data for medical modeling using additive manufacturing, it is helpful to discuss it first with the radiographers, who may well develop a protocol specifically for medical modeling.[21] CT scans are time consuming, expensive, and potentially harmful, so every care must be taken to ensure that the scan is conducted correctly the first and only time.

13.3 MATERIALS AND METHODS

13.3.1 MATERIALS

A total of 29 AM materials were constructed from a CAD-generated stereolithography (STL) file defining a rectangular block of material with dimensions 40 mm × 20 mm × 10 mm. The samples represented a variety of commonly used materials from the most popular AM processes. However, the sample set was not intended to be comprehensive, as there are potentially hundreds of process and material combinations that could have been used. The STL data were sent to a range of AM service providers across the United Kingdom for manufacture (see Acknowledgments section for details). Table 13.2 provides the AM machines, materials, and physical descriptions of the sample blocks. Twenty-four of the blocks were solid and five "sparse," or quasi-hollow. In industry, quasi-hollow parts are typically built to reduce material consumption, build time, and therefore cost in some AM processes. In this study, for each quasi-hollow sample, there was an equivalent solid sample. The solid samples were used to ascertain CT number ranges, and the quasi-hollow samples were included only to investigate their appearance in CT images. The density of each sample (excluding the "sparse" samples) was calculated by measuring the sample weight in grams using a Sartorius precision balance and the sample volume in cubic centimeters using a digital Vernier caliper (g/cm³). Objet "DM" materials are made from combinations of the rigid VeroWhite and soft, flexible Tango+ materials. For the range from DM9740 to DM9795, the final two digits relate to Shore hardness (i.e., from 40 to 95 Shore). DM8410 and DM8430 approach the rigidity of some common thermoplastics.

13.3.2 CT SCANNING

The AM blocks underwent two computed tomography scans. First, they were scanned suspended in air. This was facilitated by using a low-density expanded polystyrene foam support (CT number = –963), as shown in Figure 13.1. This low-density foam material has a CT number very similar to air (CT air = –1000) and was selected to minimize any beam hardening effects of the support.[22,23] Second, the blocks were scanned in contact with a tissue-equivalent head phantom, as shown in Figure 13.2 (phantom supplied by Imaging Equipment, Ltd., Bristol, U.K.). This was to mimic the situation of materials being adjacent to the body, as they might be in the case of an AM prosthesis or wearable medical device, and to determine any shift to the absolute CT numbers compared to those measured when scanned in air.

TABLE 13.2
Description of AM Material Samples

Sample No.	AM Machine	AM Material	Physical Description
1	Z Corp. 450	Z Bond (cyanoacrylate)	White/gray, opaque, solid
2	Z Corp. 450	ZP130 (wax)	White/gray, opaque, solid
3	EOS P100 Formiga	Nylon 12 (polyamide)	White, opaque, solid
4	3D Systems 250	ProtoCast AF19120, DSM Somos	Orange/red, translucent, solid
5	3D Systems 250	Watershed XC11122, DSM Somos	Clear/blue, translucent, solid
6	3D Systems 250	9420 EP (white), DSM Somos	White/cream, opaque, solid
7	3D Systems 250	RenShape SL Y-C 9300, Huntsman	Pink, translucent, solid
8	3D Systems InVision SD	VisiJet SR	Clear, translucent, solid
9	Dimension 1200 SST	ABS	White, opaque, solid
10	Fortus 400mc	ABS (solid)	White, opaque, solid
11	Fortus 400mc	ABS (sparse)	White, opaque, quasi-hollow
12	Fortus 400mc	ABS+ (solid)	White, opaque, solid
13	Fortus 400mc	ABS+ (sparse)	White, opaque, quasi-hollow
14	Fortus 400mc	PPSF (solid)	Light brown, opaque, solid
15	Fortus 400mc	PPSF (sparse)	Light brown, opaque, quasi-hollow
16	Fortus 400mc	PC (solid)	White, opaque, solid
17	Fortus 400mc	PC (sparse)	White, opaque, quasi-hollow
18	Fortus 400mc	PC/ABS (solid)	Black, opaque, solid
19	Fortus 400mc	PC/ABS (sparse)	Black, opaque, quasi-hollow
20	Objet500 Connex	VeroWhite	White, opaque, solid
21	Objet500 Connex	Tango+	Clear, translucent, soft
22	Objet500 Connex	DM 9740	Off-white, translucent, flexible
23	Objet500 Connex	DM 9750	Off-white, translucent, flexible
24	Objet500 Connex	DM 9760	Off-white, opaque, flexible
25	Objet500 Connex	DM 9770	Off-white, opaque, flexible
26	Objet500 Connex	DM 9785	Off-white, opaque, flexible
27	Objet500 Connex	DM 9795	Off-white, opaque, rigid
28	Objet500 Connex	DM 8410	Off-white, opaque, rigid
29	Objet500 Connex	DM 8430	Off-white, opaque, rigid

Note: ABS, acrylonitrile butadiene styrene; PC, polycarbonate; PPSF, polyphenylsulfone.

Computed tomography scanning was performed using a Philips Brilliance 10 multislice system (www.medical.philips.com) and a sinus/facial/head CT protocol (exposure of 67 mAs, peak voltage 120 kV, slice thickness 2 mm, rotation time 1 second, and convolution kernel type "D"), software version 1.2.0. Although a small field of view would have resulted in a smaller pixel size and therefore more pixels within the sample images to analyze, a field of view of 27.9 cm was specified to replicate the typical field of view and, therefore, pixel size encountered in a wide range of clinical CT applications. CT images were stored in DICOM format and imported into image analysis software (AnalyzeAVW V9.0, Lenexa, KS) for CT number measurement. Visual inspection and analysis of the images was also performed using another software package (Mimics version 13, Materialise NV, Leuven, Belgium).

FIGURE 13.1　Samples suspended in low-density polystyrene foam.

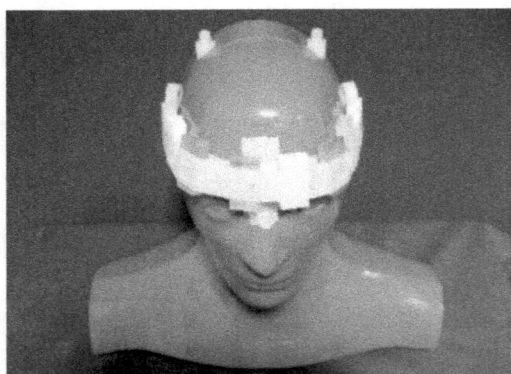

FIGURE 13.2　Samples attached to head phantom.

13.3.3　Data Analysis

The mean and standard deviation (SD) of the CT numbers for each sample were recorded and averaged over the volume of the material to ensure it was representative of the whole sample. The mean CT numbers measured from the samples scanned in air and positioned on the phantom were compared to determine any shift in CT number due to the presence of the tissue-equivalent phantom. Pixels at the edges of the sample images were not used in the analysis to avoid the partial volume effect, which may lower the average CT number as indicated in Figure 13.3. CT images of the sample cross-sections were visually inspected to ascertain material structure and any effects of cross-sectional variation. Cross-sections were visually inspected to determine whether expected material densities were present, noting any unexpected features such as voids, porosity, or cracking. CT number profiles were also generated to illustrate variation in density.

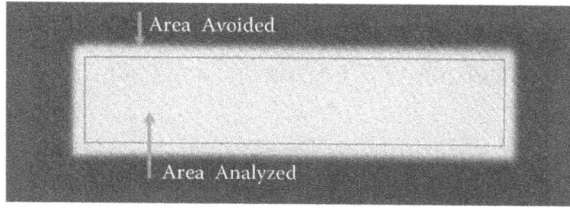

FIGURE 13.3 The image areas analyzed and avoided in determining the CT number.

13.4 RESULTS

Figure 13.4 shows the relationship between the sample density and CT number. The pseudo-hollow sparse samples would obviously have an average sample density much lower than the actual material density due to the presence of air and therefore were not included. The result presents the average density for each sample, and it should be noted that some samples are not homogeneous and their density varies considerably across their sections (especially the Z Corp. samples). The differences in CT numbers reflect the differences in sample densities. As might be expected, the relationship between CT number and average sample density is essentially linear. It is well known that the CT number of a material is dependent on a range of properties including density, x-ray beam energy, and sample thickness. As x-ray beam energy and section thickness were constant, the variations in CT numbers can be attributed to the differences in material and is related to their density. It can be seen that there is a cluster of samples around the density of 1.0 to 1.2 g/cm^3, which is typical for polymers, and the CT numbers are clustered, suggesting that the CT number for these polymers is also similar. The two denser materials are from the Z Corp. process and are not polymers, but it is interesting to note that their CT numbers are also proportional to their density.

Table 13.3 shows the sample name, material, mean CT number, standard deviation of pixel values, and density of each solid sample. The Objet material Vero White is rigid white acrylate-based material and the Tango+ material is a soft rubber-like material. The Objet digital materials (DM) are composite materials made from a selective mixture of Vero White and Tango+. This produces a range of physical properties that can replicate the stiffness of a variety of thermoplastics. All materials are proprietary and specific to the relevant AM process. The table presents the average density for each sample.

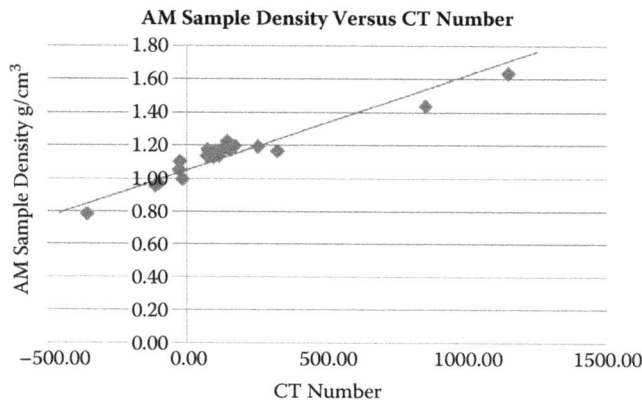

FIGURE 13.4 Linear relationship between CT number and average sample density.

TABLE 13.3
CT Number, Standard Deviation of CT Numbers, and Density

Sample No.	Manufacturer and Machine	Material	CT Number	Standard Deviation	Density (g/cm^3)
1	Z Corp. 450	Z Bond (cyanacrylate)	850.17	51.28	1.44
2	Z Corp. 450	ZP130 (wax)	1146.41	71.72	1.64
3	EOS P100	Nylon 12 (polyamide)	−17.80	29.88	1.00
4	3D Systems 250	ProtoCast AF19120, DSM Somos	168.50	28.57	1.20
5	3D Systems 250	Watershed XC11122, DSM Somos	320.82	27.62	1.17
6	3D Systems 250	9420 EP (white), DSM Somos	251.57	26.35	1.19
7	3D Systems 250	RenShape SL Y-C 9300, Huntsman	142.43	28.67	1.22
8	3D Systems InVision SD	VisiJet SR	126.44	26.05	1.18
9	Dimension 1200 SST	ABS	−115.74	34.43	0.96
10	Fortus 400mc	ABS	−102.86	32.61	0.97
12	Fortus 400mc	ABS+	−358.93	31.08	0.79
14	Fortus 400mc	PPSF	151.60	46.01	1.17
16	Fortus 400mc	PC	−26.37	29.88	1.11
18	Fortus 400mc	PC/ABS	−30.21	26.34	1.05
19	Fortus 400mc	PC/ABS	110.36	8.86	1.17
20	Objet500 Connex	Vero White	99.75	5.06	1.17
21	Objet500 Connex	Tango+	118.28	6.00	1.17
22	Objet500 Connex	DM 9740	111.96	5.99	1.14
23	Objet500 Connex	DM 9750	93.13	5.54	1.13
24	Objet500 Connex	DM 9760	75.09	5.54	1.14
25	Objet500 Connex	DM 9770	72.60	5.70	1.13
26	Objet500 Connex	DM 9785	69.65	6.17	1.14
27	Objet500 Connex	DM 9795	75.96	5.08	1.16
28	Objet500 Connex	DM 8410	71.55	6.16	1.18
29	Objet500 Connex	DM 8430	119.82	6.16	1.17

Note: ABS, acrylonitrile butadiene styrene; PC, polycarbonate; PPSF, polyphenylsulfone.

The mean CT number for the solid samples ranged from a minimum of −359 to a maximum of 1146. It is interesting to note that many of the AM sample CT number ranges coincide with or are similar to those of the human tissues as shown in Table 13.1. For example, samples 1 and 2 have CT number ranges that are similar to cortical bone, which may range from 200 to 1200. Samples 4, 6, 7, and 8 are similar to cancellous bone with a CT number range of 50 to 300. Both samples 9 and 10 have mean CT numbers that are very similar to the range found for fat tissue in the body at approximately −100. Sample 16 has a value similar to water when scanned in air (CT number = 0) but shifted by approximately 20 CT numbers when adjacent to the phantom. The standard deviations of the sample CT numbers range from approximately 5.0 to over 90.0, with the larger deviations being measured in samples that had a higher CT number. This is in keeping with CT scans of human tissue where bone (CT number > 300) has the highest standard deviation due to increased noise present in that tissue type, while air (CT number = −1024) had the lowest standard deviation. The standard deviation of the measurements within the AM samples was due to two factors: inherent noise due to the CT imaging system and any material density variation within the structure of the AM sample. Sample numbers 21 to 29 were all constructed using the Objet500 Connex system, and they demonstrate a limited range of CT numbers, from 70 to 120. This CT number range is similar in CT scanning to contrast enhanced blood. It is interesting to note that the difference in physical properties of the Objet samples is not detectable in the CT images.

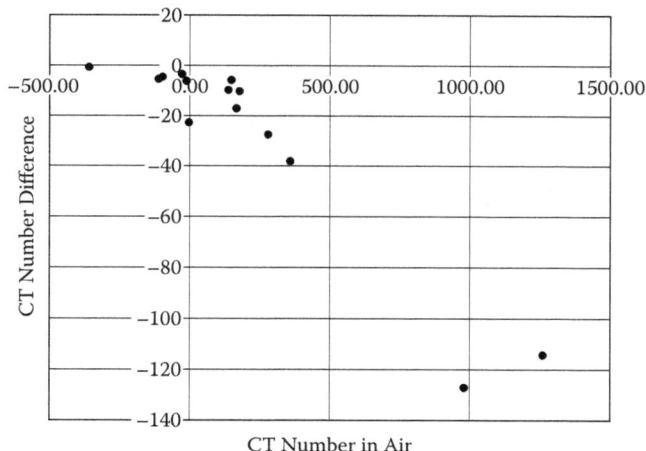

FIGURE 13.5 The shift in CT numbers due to the presence of a tissue equivalent phantom.

The differences in the CT numbers of the samples when scanned in air compared to the CT number when attached to a soft tissue equivalent head phantom are shown in Figure 13.5. Note that the differences were small (<20) for samples within the CT number range −360 to 320, while there are more significant deviations for samples with CT numbers 850 and 1150. In general, the CT numbers measured from the samples scanned in air were greater than those recorded from the samples scanned against the tissue phantom.

Generally, there was no internal structural variation visible in the CT images. There were also no signs of voids or cracking in any of the samples scanned. As would be expected from an understanding of the respective RP processes, the fused deposition modeling (FDM), stereolithography (SLA), and selective laser sintering (SLS) samples appeared to have uniform density throughout. This is shown by samples 3, 5, and 18 in Figure 13.6 and can also be seen in the generally flat CT number profile taken lengthwise through sample 2, as shown in Figure 13.8. However, sample 1 shows a variation in pixel value across the sample due to the manufacturing process.

Figure 13.6 shows the internal structure of AM samples (3, 5, and 18). The images show an overall uniform internal structure, although there is some noise present within the image. This is typical of CT scanning, where the x-ray photons are detected in a random manner, creating small variations in pixel values for a constant material. Although it is known that certain AM processes produce inherently porous parts, the porosity is not apparent in all of the CT images.

FIGURE 13.6 Selected sample images: (A) sample 3, EOS P100 Formiga; (B) sample 5, SLA DSM XC11122; and (C) sample 18, FDM PC-ABS solid.

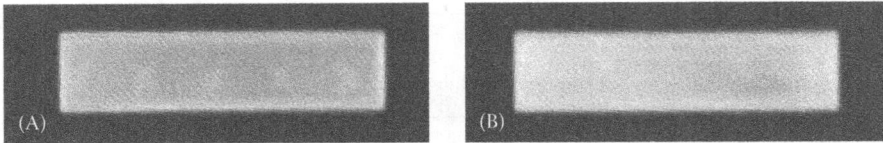

FIGURE 13.7 3D printing sample images: (A) sample 1, ZCorp. Z450 ZBond; and (B) sample 2, ZCorp. Z450 ZP130.

This is because the porosity is at a very small scale compared to the resolution of the CT scanner (high-contrast objects less than 0.5 mm can be visualized) and appears uniform throughout the parts. If we consider the example of SLS, the process works by sintering together thermoplastic particles with a typical average particle size of around 60 μm (PA2200 Material Data Sheet, EOS GmbH, Munich, Germany). The particles do not fully melt but fuse together to form a sintered, porous structure. Therefore, it is reasonable to assume that this results in a slight lowering of the CT number compared to fully dense nylon produced by injection molding, extrusion, or casting. Further work will be conducted to ascertain whether the differences between SLS nylon and solid nylon can be detected in CT images.

A distinct variation of internal density was visible in the images of the powder bed 3D printed samples, 1 and 2, as shown in Figure 13.7. A variation in density can be seen in the CT images, which show a higher density around the periphery. This is a result of the powder bed 3D printing process, whereby the manufactured part is initially very fragile. The parts are therefore subjected to infiltration of a liquid hardener, typically a cyanoacrylate resin (as in sample 1) or wax (as in sample 2). It is known that the hardeners penetrate into the part through capillary action but that this penetration is limited to a few millimeters. This leads to a higher density "skin" or "shell" that is clearly visible in the CT image and CT number profile. The variation can also be clearly seen in the peaks in the CT number profile for sample 1, shown in Figure 13.8, and can be compared to the much flatter profile for sample 5.

Each of the AM materials was inspected for the presence of the beam hardening artifact. This is visible in CT scans covering areas of the body where dense material is adjacent to much less dense material. It is particularly noticeable where dense or thick bone is near soft tissue. On inspection of the CT scans, the beam hardening artifact could be detected near samples 1 and 2. These are the materials with the highest CT numbers, and no other artifact was detected for any of the other samples. Figure 13.9 shows the slight but visible beam hardening effect (a darker region) near samples 1 and 2 as indicated by the arrows.

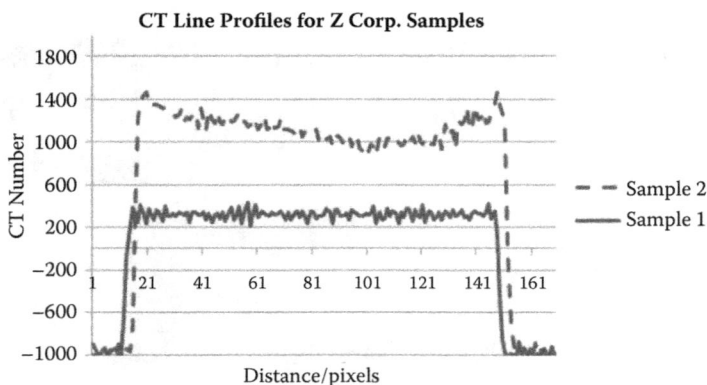

FIGURE 13.8 CT number profiles for sample 1 and 2.

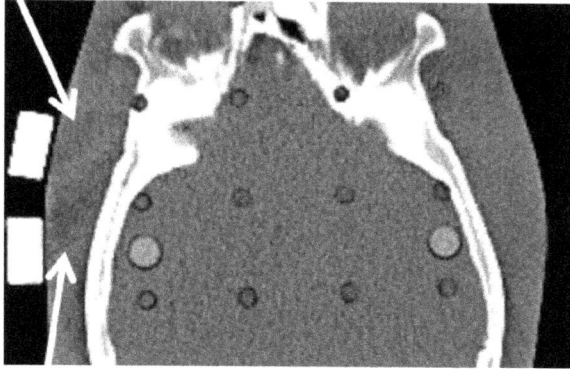

FIGURE 13.9   Beam hardening adjacent to samples 1 and 2 (contrast enhanced).

13.5 DISCUSSION

There are many potential applications of the information resulting from this study. Knowledge of the CT number for a particular material may enable medical devices or rehabilitation products to be designed to exhibit a specific appearance in CT images. For example, the ability to predict the appearance of a particular material in CT images could be used to either match a desired CT number in order to replicate another material or to provide a known image contrast with other adjacent materials. Knowledge of the CT number may simply be useful when distinguishing between a medical device and adjacent tissues in subsequent diagnoses. As the number and variety of medical applications of AM grows both inside and outside the human body it will be increasingly important to be able to predict the appearance of AM materials in CT images.

If particular AM materials are found to be equivalent to those of specific human tissues it would allow accurate and highly detailed anatomical models to be constructed which would have CT imaging properties similar to those of the relevant tissue. Table 13.1 shows CT number ranges for specific human tissues.[18] Such models could be used in the design and manufacture of anatomical phantoms or test objects that can be used for research, training, and testing in medical imaging. The information may also find application in the design and manufacture of immobilization devices that can secure a patient in a given position during scanning and yet be easily distinguished from other materials or human tissues in the subsequent CT images. This would enable efficient isolation and removal of an artificial object from CT images through segmentation of the object according to its CT number range. It could also enable the production of custom-fitting support devices, orthotics, or prostheses that display particular characteristics in CT images. Such devices may be appropriate for patients requiring a long-term, wearable device or prosthesis and who might be expected to undergo further or repetitive radiological procedures.

These findings demonstrate that there is significant potential to use AM materials for sophisticated test objects for use in medical image modality testing, including image quality and radiation dose. Some AM materials have CT numbers very similar to those of human tissues, as summarized in Table 13.4 and therefore may be used to develop anatomically accurate phantoms produced from CT scans using AM. Phantoms designed using these materials may have the added advantage of having CT numbers corresponding to real tissues. Anatomically complex, multi-tissue phantoms could be developed from existing patient CT scan data using well-established image segmentation techniques to provide more accurate phantoms for test purposes. Anthropomorphic phantoms have been developed for use in radiation dose studies for diagnostic radiology and therapeutic radiology. Soft-tissue, lung, and bone equivalent tissue substitutes (at the diagnostic x-ray energy range of 80 to 120 kVp) were created from urethane-based compounds mixed with other materials.[24] This particular phantom suffered from manufacturing difficulties in that molds would display variations

TABLE 13.4
Potential AM Materials That Mimic Human Tissues in CT Images

Sample No.	AM Machine	AM Material	CT Image Equivalent Tissue
1	Z Corp. 450	Z Bond (cyanoacrylate)	Cortical bone
2	Z Corp. 450	ZP130 (wax)	Cortical bone
4	3D Systems 250	ProtoCast AF19120, DSM Somos	Spongious bone
6	3D Systems 250	9420 EP (white), DSM Somos	Spongious bone
7	3D Systems 250	RenShape SL Y-C 9300, Huntsman	Spongious bone
8	3D Systems InVision SD	VisiJet SR	Spongious bone
9	Dimension 1200 SST	ABS	Fat
10	Fortus 400mc	ABS (solid)	Fat
16	Fortus 400mc	PC (solid)	Water

Note: ABS, acrylonitrile butadiene styrene; PC, polycarbonate.

in depth or suffer from physical distortion. AM has the capability to provide accurate anatomical definition, geometrical shape, and the appropriate x-ray attenuation. The potential to expand the application to radiation dosimetry for diagnostic and therapeutic procedures is obvious.

The similarity between the CT numbers for some of the samples and human tissues can also be visually illustrated by utilizing preset thresholds for specific tissues available in CT image analysis software. For example, the preset threshold for bone in Mimics software (226 and higher) perfectly segments sample 1. Similarly, the preset thresholds for spongious bone (148 to 661) provide very good segmentation of sample 6, and the preset thresholds for fat (−205 to −51) provide a good segmentation of sample 10.

Figure 13.10A shows a normal CT scan of a human spine vertebra and Figure 13.10B shows a CT scan of a spine model manufactured using a powder bed 3D printing system (the same process as used to produce sample 1). The CT number range for the cortical bone is 1000 to 1300, and the range within the cancellous bone is 490 to 815. These ranges mimic the CT number ranges for human cortical and cancellous bone very closely. As described previously, this is due to the AM process, which hardens the outer few millimeters of the model, resulting in an elevated CT number at the periphery of the sample. This relatively simple example demonstrates the potential to manufacture anatomically correct, sophisticated test objects with a mixture of hard and soft tissue materials that would be useful in radiation therapy dosimetry experiments where test objects may be created from a combined approach to model creation.

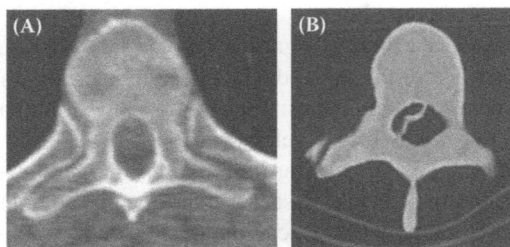

FIGURE 13.10 (A) CT of human vertebra. (Reproduced with permission of National Library of Medicine Visible Human Project®; Bethesda, MD.) (B) CT scan of a Z Corp. lumbar spine model.

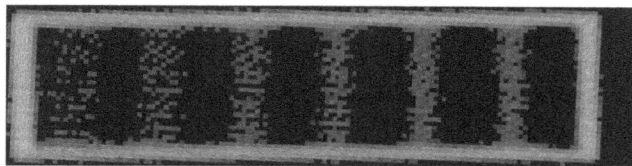

FIGURE 13.11   CT image of a quasi-hollow FDM part (sample 19).

13.6   FUTURE WORK

The authors plan to repeat the experiments with more AM materials and develop a comprehensive database of CT numbers for a wider selection of AM materials. For the benefit of comparison, the authors also plan to include samples of well-known conventional materials such as ultra-high molecular weight polyethylene and silicone. This work set out to demonstrate the appearance of AM materials in CT imaging rather than fully characterize all AM sample properties. However, further physical characterisation would complement this work and may enable identifying relationships between physical properties and CT number. For example, nano- or micro-indentation techniques could be useful to demonstrate other properties of the samples. A porosity measure would also be useful, as this will have an effect on sample density. However, the particle size typically encountered in AM powder materials is an order of magnitude smaller than the pixel size of a typical CT image. Therefore, it is likely that CT image noise would dominate any variation in image appearance rather than porosity of the samples, which would occur at a much smaller physical scale and not show directly on the images.

As mentioned earlier, many AM processes can produce quasi-hollow structures, as shown in Figure 13.11. These structures can be varied in section to simulate different or varying densities throughout the volume of a part. These volumes could be filled with fluids such as water, fat, or oil in order to simulate different tissues or even whole organs. A more recent and developing advantage of AM is the ability to build objects with multiple materials and graded mixtures of materials. It is anticipated that further work in this area will prove particularly interesting when investigating objects made using AM machines that are capable of depositing multiple materials simultaneously such as the Objet Connex and 3D Systems ProJet machines. This will be particularly interesting, as the machines are capable of producing graded structures where material composition can be graded to produce areas of differing physical properties within a single object manufactured in a single-step process. Therefore, material compositions could be adjusted to replicate a combination of specific tissues or whole sections of anatomy potentially including hard and soft tissues.

13.7   CONCLUSIONS

This study has revealed several interesting facts relating to the appearance of AM materials in CT images. First, the images provide an indication of material uniformity of density at the macroscale. This analysis can be used to corroborate other observations from visual analysis and mechanical testing. Second, the actual CT numbers of a number of commonly used AM materials have been established. This may enable the specification of AM materials for specific medical devices that are required to present a specific CT number or characteristic in CT images. Further work is required to analyze a greater variety of AM materials and in particular samples from AM processes that produce mixed, graded, and multiple material parts.

ACKNOWLEDGMENTS

The authors thank the following for donation of the samples: Phil Dixon, Loughborough Design School, and Dr. Russ Harris, Wolfson School of Mechanical and Manufacturing Engineering, Loughborough University, United Kingdom; Dr. Dominic Eggbeer, National Centre for Product Design and Development Research, University of Wales Institute, Cardiff; and Jeremy Slater, Technical Sales Engineer, Design Engineering Group, Laser Lines, Ltd., Banbury United Kingdom.

MANUFACTURER DETAILS

AnalyzeAVW V9.0, Lenexa, KS (www.analyzedirect.com)

DSM Somos, 1122 St. Charles Street, Elgin, IL 60120 (www.dsm.com/en_US/html/dsms/home_dsmsomos.htm)

EOS GmbH, Robert-Stirling-Ring 1, D-82152 Krailling, Munich, Germany (www.eos.info/en/home.html)

Materialise NV, Technologielaan 15, 3001 Leuven, Belgium (www.materialise.com/mimics)

Stratasys, Inc., 7655 Commerce Way, Eden Prairie, MN 55344 (www.stratasys.com)

3D Systems Corp., 333 Three D Systems Circle, Rock Hill, SC 29730 (www.3dsystems.com)

REFERENCES

1. Chua, C.K., Leong, K.F., and Lim, C.S. (2010). *Rapid Prototyping: Principles and Applications*, 3rd ed. World Scientific, Singapore.
2. Gibson, I., Rosen, D.W., and Stucker, B. (2009). *Additive Manufacturing Technologies: Rapid Prototyping to Direct Digital Manufacturing*. Springer, New York.
3. Noorani, R.I. (2005). *Rapid Prototyping: Principles and Applications*. John Wiley & Sons, New York.
4. Hopkinson, N., Hague, R., and Dickens, P., Eds. (2006). *Rapid Manufacturing: An Industrial Revolution for a Digital Age*. John Wiley & Sons, New York.
5. Arvier, J.F., Barker, T.M., Yau, Y.Y., D'Urso, P.S., Atkinson, R.L., and McDermant, G.R. (1994). Maxillofacial biomodelling. *Br. J. Oral Maxillofac. Surg.*, 32(5): 276–283.
6. Giannatsis, J. and Dedoussis, V. (2009). Additive fabrication technologies applied to medicine and health care: a review. *Int. J. Adv. Manuf. Technol.*, 40(1–2): 116–127.
7. Azari, A. and Nikzad, S. (2009). The evolution of rapid prototyping in dentistry: a review, *Rapid Prototyping J.*, 15(3): 216–225.
8. Bibb, R. (2006). *Medical Modelling: The Application of Advanced Design and Development Technologies in Medicine*. Woodhead Publishing, Cambridge.
9. Gibson, I., Ed. (2005). *Advanced Manufacturing Technology for Medical Applications: Reverse Engineering, Software Conversion, and Rapid Prototyping*. John Wiley & Sons, Chichester.
10. Petzold, R., Zeilhofer, H., and Kalender, W. (1999). Rapid prototyping technology in medicine: basics and applications. *Comput. Med. Imaging Graph.*, 23: 277–284.
11. Webb, P.A. (2000). A review of rapid prototyping (RP) techniques in the medical and biomedical sector. *J. Med. Eng. Technol.*, 24(4): 149–153.
12. Bibb, R., Eggbeer, D., and Evans, P. (2010). Rapid prototyping technologies in soft tissue facial prosthetics: current state of the art. *Rapid Prototyping J.*, 16(2): 130–137.
13. Bibb, R., Eggbeer, D., Evans, P., Bocca, A., and Sugar, A.W. (2009). Rapid manufacture of custom fitting surgical guides. *Rapid Prototyping J.*, 15(5): 346–354.
14. Bibb, R., Eggbeer, D., and Williams, R. (2006). Rapid manufacture of removable partial denture frameworks. *Rapid Prototyping J.*, 12(2): 95–99.
15. Peltola, S.M., Melchels, F.P., Grijpma, D.W., and Kellomäki, M. (2008). A review of rapid prototyping techniques for tissue engineering purposes. *Ann. Med.*, 40(4): 268–280.
16. Leong, K.F., Chua, C.K., Sudarmadji, N., and Yeong, W.Y. (2008). Engineering functionally graded tissue engineering scaffolds. *J. Mech. Behav. Biomed. Mater.*, 1(2): 140–152.
17. Pilipovic, A., Raos, P., and Sercer, M. (2009). Experimental analysis of properties of materials for rapid prototyping. *Int. J. Adv. Manuf. Technol.*, 40: 105–115.

18. Kalender, W.A. (2000). *Computed Tomography: Fundamentals, System Technology, Image Quality, Applications*. Publicis MCD, Erlangen, p. 101.
19. Hofer, M. (2000). *CT Teaching Manual*. Thieme-Stratton Corp., Stuttgart.
20. Henwood, S. (1999). *Clinical CT: Techniques and Practice*. Greenwich Medical Media, Valley Stream, NY.
21. Bibb, R. and Winder, J. (2010). A review of the issues surrounding three-dimensional computed tomography for medical modelling using rapid prototyping techniques. *Radiography*, 16: 78–83.
22. Shikhaliev, P.M. (2005). Beam hardening artefacts in computed tomography with photon counting, charge integrating and energy weighting detectors: a simulation study. *Phys. Med. Biol.*, 250(24): 5813–5827.
23. Brooks, R.A. and Di Chiro, G. (1976). Beam hardening in x-ray reconstructive tomography. *Phys. Med. Biol.*, 21(3): 390–398.
24. Winslow, J.F., Hyer, D.E., Fisher, R.F., Tien, C.J., and Hintenlang, D.E. (2009). Construction of anthropomorphic phantoms for use in dosimetry studies. *J. Appl. Clin. Med. Phys.*, 10(3): 195–204.

14 Additive Manufacturing of Medical Devices

Jayanthi Parthasarathy

CONTENTS

ABSTRACT

Recent advances in medical imaging and image data processing, combined with advances in additive manufacturing technologies and the introduction of a wider spectrum of biocompatible materials, are changing the medical device world. Patient-specific devices for use as surgical planning models, surgical guides, and implantable devices are being manufactured using additive manufacturing. The geometric design freedom offered by additive manufacturing has made possible the creation of complex implantable devices that can be adapted to individual requirements and the region of implantation. Surgeons can now choose from an array of patient-specific devices promising better performance, precise fit, and a reduction in precious surgical time. Processes such as stereolithography (SLA) and fused deposition modeling (FDM) are used in the production of patient-specific models and guides. Electron beam melting (EBM) and direct metal laser sintering (DMLS) are used for the fabrication of implantable devices using titanium and its alloys, stainless steel, and chrome cobalt, which have been used for decades. Selective laser sintering (SLS) is being used for the production of non-implantable guides for use during surgery and implantable devices. This chapter discusses the manufacturing processes and devices that are currently available for routine clinical use.

14.1   INTRODUCTION

A medical device is an instrument, apparatus, implant, *in vitro* reagent, or similar or related article that is used to diagnose, prevent, or treat disease or other conditions and is intended to affect the structure or any function of the body of humans or other animals. It does not achieve its purposes through chemical action within or on the body and is not dependent upon being metabolized for the achievement of any of its primary intended purpose like drugs.[1] Medical devices vary widely in complexity, ranging from simple tongue depressors and thermometers to very complex neuro and cardiac stimulators. Most are off-the-shelf products but some are patient specific or custom. Several challenges exist in the design, material requirements, and manufacturing of a medical device. This chapter provides an overview of the product lifecycle management of devices as it relates to additive manufacturing (AM).

Custom implants for the reconstruction of defects offer better performance over their generic counterparts, which can be attributed to their precise adaptation to the region of implantation, reduced surgical times, and improved cosmesis. The recent introduction of direct (AM) technologies that enable the fabrication of implants from patient-specific data has opened up a new horizon for the next generation of customized implants.

In any reconstructive surgery (e.g., secondary to trauma or ablative tumor resection, to address infection or congenital/developmental deformities), the restoration of aesthetics and function is the primary goal and calls for precise presurgical planning and execution of the plan. Surgeons have adapted to enhanced visualization techniques for close to two decades, and even today this is an advancing field. Such virtual reality is useful only when transferred to the clinical scenario—the operatory—to achieve the desired results. The development of computer-aided design (CAD) and computer-aided manufacturing (CAM) systems that adapt to a surgeon's needs has resulted in a growing armamentarium for computer-assisted surgery (CAS). Such systems specifically focus on enhanced visualization tools; for example, three-dimensional (3D) modeling, or virtual reality, allows for precise preoperative planning, such as performing virtual osteotomy resections, or for the production of patient-specific implants. These virtual models can be imported into an intraoperative navigation system for precise placement of bone segments, implants, and hardware. Advances in manufacturing technology and materials science have turned such virtual models into reality as physical replica models, patient-specific implants, surgical guides, and jigs or splints for intraoperative use.

Additive manufacturing technologies such as stereolithography (SLA), PolyJet™, fused deposition modeling, 3D printing, selective laser melting (SLM), and electron beam melting (EBM) lend themselves to the manufacture of complex anatomic parts without any barriers of design constraints. SLM and EBM use biocompatible implantable materials such as titanium and its alloys (e.g., Ti6Al4V), chrome cobalt, polyetheretherketone (PEEK), and polyetherketoneketone (PEKK) to facilitate the direct production of implants with engineered properties that match the properties of the tissues at the region of implantation. Surgeons can now have access to the facility's service providers. Additive manufacturing can be used to produce patient-specific models to be used as templates for simulation surgery, guides to be used for appropriate positioning of bones, next-generation implants with engineered mechanical properties equivalent to the region of implantation, and off-the-shelf products such as tissue-engineered burr hole covers and hip implants with enhanced functionality.

FIGURE 14.1 EBM-printed Ti6Al4V extra-low interstitial (ELI)-grade parts showing the interconnecting repeated pores.

14.1.1 Why Additive Manufacturing?

Computed tomography (CT) and AM are layer-based technologies, and the transfer of data from CT scans for rapid prototyping (RP) manufacturing has been achieved with good precision, thus it has become the chosen method for fabricating direct custom titanium implants. AM has no limitations with regard to the shape of the product, which makes it very suitable for manufacturing organic shapes. Conventional manufacturing processes fail to produce the repeatable desired porosity in the structure. AM is the only manufacturing process offering the capability of manufacturing net-shaped parts with interconnecting pores similar to trabecular bone and with the mechanical properties desired (see Figure 14.1). AM utilizes a range of biocompatible materials for manufacturing guides and implants. Producing any design of lattice structures is a unique capability of AM that no other manufacturing process can offer today. Direct fabrication of

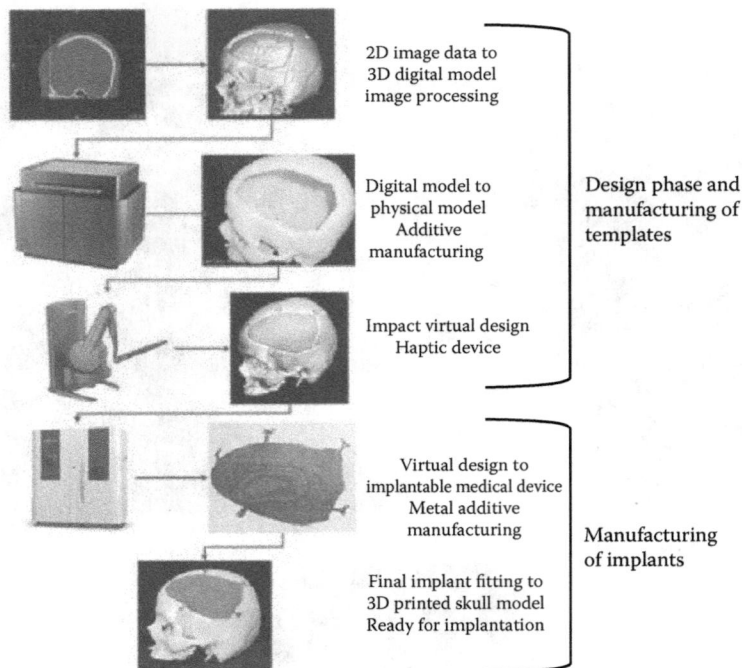

2D image data to
3D digital model
image processing

Digital model to
physical model
Additive
manufacturing

Impact virtual design
Haptic device

Design phase and
manufacturing of
templates

Virtual design to
implantable medical device
Metal additive
manufacturing

Final implant fitting to
3D printed skull model
Ready for implantation

Manufacturing
of implants

FIGURE 14.2   Art to part process.

implants from patient-specific data with controlled mechanical properties and precise adaptation
to the region of implantation is made possible with EBM, direct metal laser sintering (DMLS), and
selective laser sintering (SLS), thus eliminating expensive secondary processing such as machining, forging, swaging, or forming and their related lead times.

14.2   METHODOLOGY FOR MANUFACTURING PATIENT-SPECIFIC DEVICES

The process of creating patient-specific models and implants is shown in Figure 14.2. It entails the
complete transfer of scans (medical data) to CAD (engineering data) and the generation of appropriate file formats for the manufacturing process.

14.2.1   PRODUCT REALIZATION PROCESS

The product realization process (PRP) for creating patient-specific models and implants is two
phased. The first phase involves the conversion of medical data derived from CT scans, magnetic
resonance imaging (MRI), and ultrasound images to a 3D model for enhanced visualization of
the surgical procedure required. This phase is when patient-specific implants are designed with
engineered properties that are akin to the region of implantation, in addition to the generation of
patient-specific models and surgical guides. The second phase is the actual manufacturing process.
A suitable additive manufacturing technique is chosen based on the material and biological and
performance requirements of the product. The process of digital modeling can be divided into data
acquisition and data processing, which is the confluence of medicine and engineering to produce a
CAD file format from the medical images obtained.

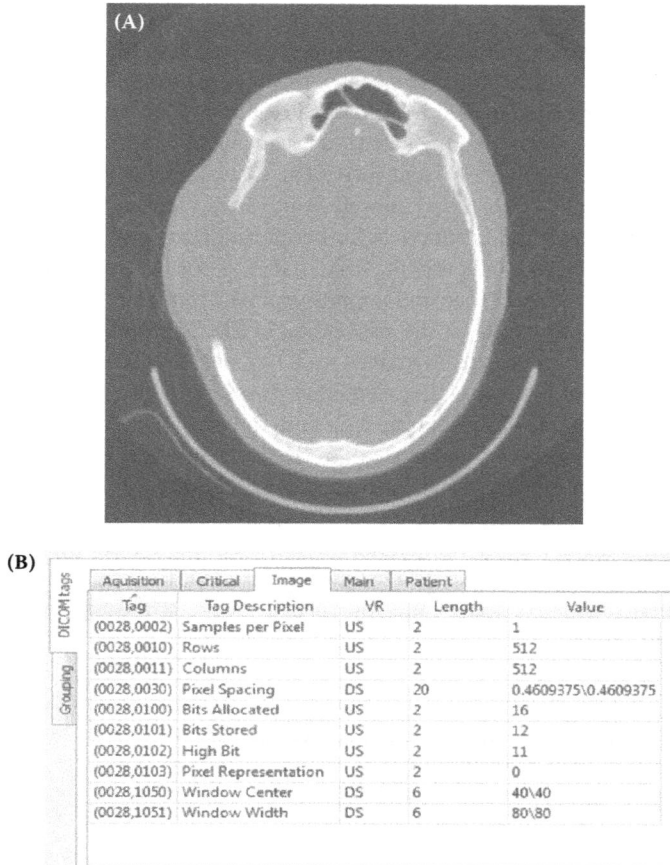

FIGURE 14.3 (A) CT scan image; (B) DICOM tags and information.

14.2.1.1 Data Acquisition

Input data are compiled from CT, MRI, or ultrasound images. Stacked slice images of the region of interest are acquired in a Digital Imaging and Communication in Medicine (DICOM)-compatible format. The image formats are specified by the National Electrical Manufacturers Association (NEMA) for the distribution and viewing of images obtained by CT, MRI, or ultrasound. This standard describes a file format for the distribution of radiological images. A single DICOM file contains a header (which stores information about the patient's name, type of scan, image dimensions, etc.) and image data and 3D information. Figure 14.3 shows a CT scan image and the DICOM tag information as read by the imaging software MIMICS. The input data can be used to acquire two-dimensional (2D) images of the region of interest. The quality and precision of the final 3D model are directly related to the slice thickness. The greater the number of slices, the more accurate the results will be. From our experience, a range of 0.5 to 1.25 mm is optimum to develop a good 3D model that can be used for designing an implant, performing simulation studies, or manufacturing the device itself. Images are taken with no gantry tilt. The region to be scanned is fixed so no movement occurs during the scanning process. State-of-the-art helical scans are capable of continuously acquiring data with short exposure times. The image data are then transferred to a compact disc for further work-up or are transferred to the end user via a File Transfer Protocol (FTP) site.

14.2.1.2   Data Processing

Advances in computer graphics and image processing have revolutionized imaging. 3D objects are being visualized at different angles and distances, with varying colors, lighting, and surface properties. The amount of data provided by CT, MRI, and ultrasound is most often excessive compared to the way they are utilized. Medical imaging software is capable of handling DICOM and other image formats, such as TIFF, JPEG, PNG, GIF, and BMP. The software converts 2D image data into a 3D model that can be rotated, moved, cut, and viewed from various angles under varying lighting conditions, thus providing virtual real-life models for better diagnosis and surgical planning. Medical imaging software output is in the form of STL, OBJ, IGES, STEP, VRML, and DXF. The STL files can be sent directly to any AM machine and the physical part realized.

Popular software commonly used for the conversion of 2D CT, MRI, ultrasound, positron emission tomography (PET), and microscopic images to 3D models are 3D Doctor, MMICS, Biobuild, Amira, and Analyze, to name a few. The essential features of medical imaging software include brightness/contrast adjustment, thresholding/segmentation, rotation, translation scale reslicing, measuring, editing, examination of volumetric data, assignment of different object names for various tissues, visualization of objects either individually or together, visualization of internal structures by assigning transparency feature to external objects, and exporting 3D digital objects in a compatible format for AM machines to manufacture physical parts or to other CAD programs for further analysis before the final product design is ready for manufacture.

Three-dimensional reconstruction of the region of interest of the patient-specific anatomy is done in the following steps:

1. Creation of an image dataset
2. Thresholding
3. Region growing and 3D reconstruction
4. Reconstruction of the implant's external geometry

The process of deriving a digital model for a patient model or implant is shown in Figure 14.4. Thresholding or segmentation of tissues in the region of interest results in the creation of a 3D model that defines the defect, creating an implant design that can be manufactured using the appropriate materials and process. The CAD STL file is then generated and sent to appropriate equipment for realizing the final anatomical model.

14.2.2   APPLICATIONS OF AM TECHNOLOGIES

14.2.2.1   Anatomical Model Manufacturing

Anatomical models and templates are used in the early stages of surgical planning and the implant design verification process; they are classified as Class 1 exempt devices by the U.S. Food and Drug Administration (FDA). They must be a 1:1 replica of the human anatomy and must be present in the operatory to serve as a reference model during the surgery. The model must also be cut using tools similar to those that will be used in surgery. Therefore, they must be made of a sturdy material that can withstand vigorous handling, that can be cut, and that can be sterilized to be allowed in the operatory. These models, however, do not come in direct contact with the surgical site.

The STL file generated earlier can be utilized by any of the AM technologies, including PolyJet and fused deposition modeling (Stratasys), stereolithography (3D Systems), or starch models (Z Corp.) so patient-specific physical models can be fabricated. The limitations are only the size of the printing bed and the choice of material.

PolyJet printers (the Connex family in particular) have multi-material printing capabilities and are a great benefit to the medical community. Models can be printed with a transparent outer region and an opaque region that defines the pathology and the path of the nerves that can be printed. This

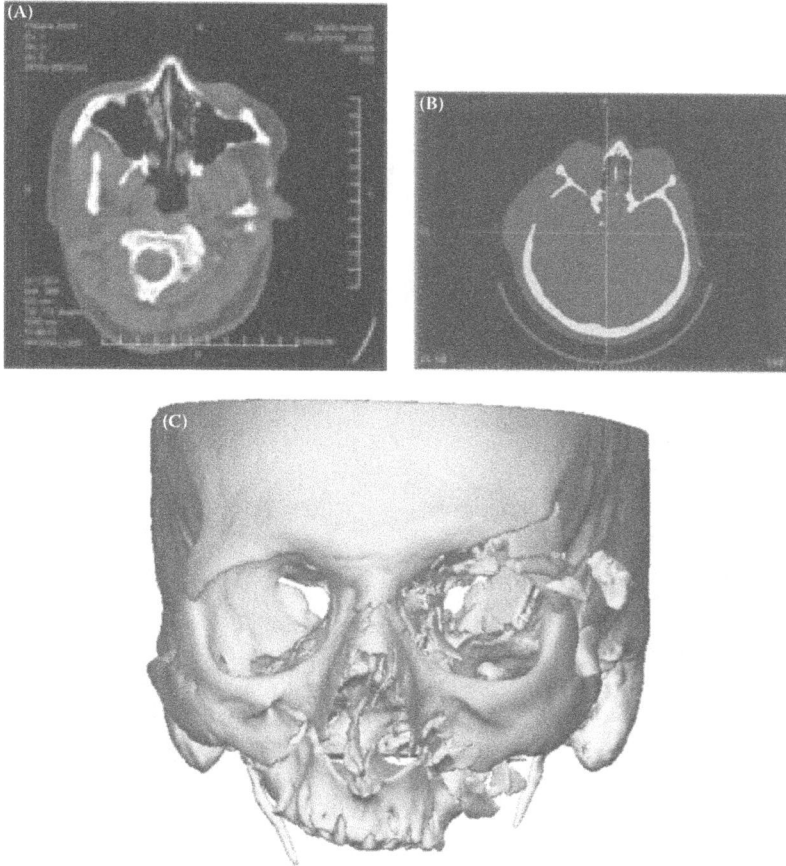

FIGURE 14.4 Process flow for 2D to 3D model creation: (A) CT scan image, (B) region growing, and (C) reconstructed 3D model of the defect

unique feature gives the surgeon the precise definition of the tumor or the pathology and its relation to the adjoining normal structures. The course of the nerve and vasculature can be very well defined and the surgery planned accordingly. Figure 14.5 shows a transparent PolyJet skull model with the pathology defined in an opaque white color. The models are highly precise, as the machine prints 16- to 30-μm layers. These models are quite suitable for printing such complex thin-walled structures as the nasal septum and internal structures of the cranium. Post-processing is very simple, as the wax-like support material can be washed away with a water jet. Single-color transparent models can be printed using the Connex or Eden family of printers. OrthoDesk printers from Stratasys are specifically designed for the fabrication of dental crown and bridge models. PolyJet models can be sterilized using gluteraldehyde.

The fused deposition modeling (FDM) process uses ABS-M30i thermoplastic (acrylonitrile butadiene styrene), which is a biocompatible ISO 10993 certified and USP Class VI material; it is an ideal material for the medical, pharmaceutical, and food packaging industries. Models are sterilizable using gamma radiation or ethylene oxide (EtO) sterilization methods. Soluble supports can be removed using a proprietary ultrasonic process. An FDM mandible model is shown in Figure 14.6. Models are dimensionally stable at room environment and do not absorb water. PolyJet and FDM processes do not produce any fumes and can be placed in any room, including hospitals and doctors' offices. Post-processing is simple and does not require any additional oven-curing process. Models do not change color over time on exposure to atmospheric conditions.

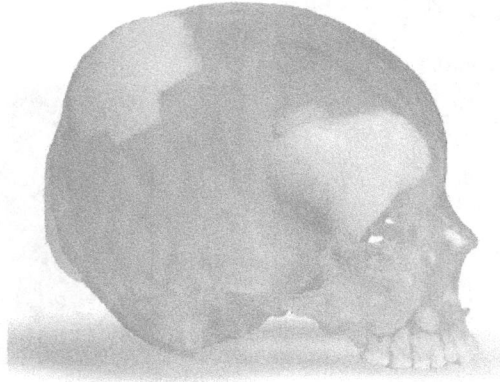

FIGURE 14.5   Transparent PolyJet skull model with the pathology defined in opaque white color.

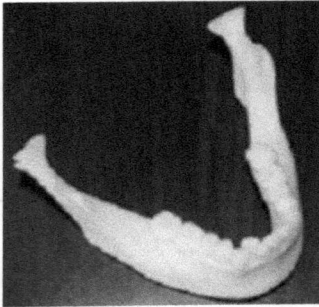

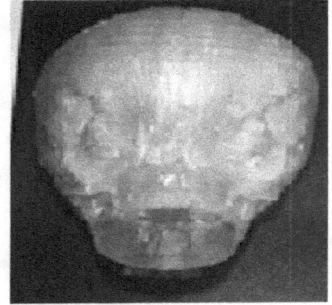

FIGURE 14.6   Mandible model made with FDM technology.   **FIGURE 14.7**   SLA model of a skull.

Stereolithography (SLA) models are as accurate as the PolyJet models. The process is capable of producing highly complex thin-walled parts. Thin vasculature that is very essential in highly complex surgeries can be replicated very precisely in SLA models. Although the process produces single-color transparent models, selective modification of the curing process in regions of interest results in color differences depicting affected areas. Models are built in a vat of liquid resin and the platform rises at the end of the build process. Parts are then removed from the platform by heating the platform to 70°C. Additional curing in an oven is required prior to post-processing. Post-processing can be achieved by bead blasting. Figure 14.7 shows an SLA model of a skull. SLA models change color over time when exposed to the atmosphere.

Starch models made from Z Corp. machines are less expensive compared to PolyJet, SLA, and FDM models. The models can be used as visualization tools, but they are hygroscopic and change dimensions very quickly. Starch models post-processed by the injection of epoxy resins are stable and do not dissolve in water. Figure 14.8 shows a starch model of a mandible made with a lattice-design structure.

14.2.2.2   Surgical Guides

Surgical guides are devices that transfer a virtual surgical plan to real life. They are of two different types: the marking guide and the cutting guide. These products can be placed in a hierarchy between less-complex models and implantable devices, which are quite complex. The guides come in direct contact with the surgical site as they are placed over the bone and markings for the cuts are made (marking guide), or the guides themselves are used as devices for making the cut with the appropriate surgical tools (cutting guide). The guides should be made from materials that are

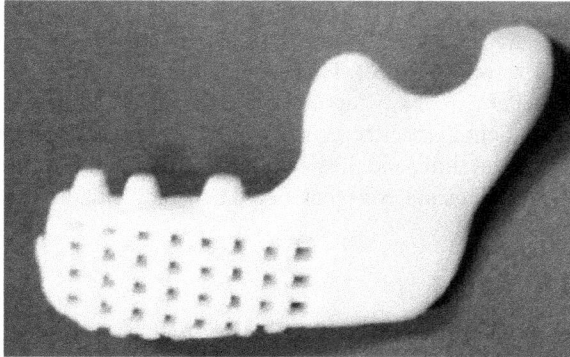

FIGURE 14.8   Starch model of a mandible with lattice design.

FIGURE 14.9   ULTEM 9085 FDM process: fibula guide placed on PolyJet model of fibula bone.

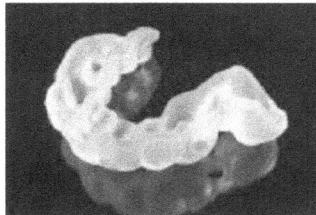

FIGURE 14.10   VisiJet dental implant surgical guide (www.3Dsystems.com).

biocompatible, can be sterilized, and are approved for short-time skin, blood, or mucosal contact. Cutting guides should not be abrasive and should be able to be cut easily without the use of surgical tools that produce powder particles that could get into the operative site and cause infection or into the bloodstream and block blood vessels.

Surgical guides can be made of ABS-M30i or ULTEM 9085 (FDM); MED 610, FullCure® 630, or FullCure® 655 (PolyJet); or VisiJet® Crystal, StonePlast, and Clear plastic materials (SLA), which are USP Class VI certified materials for the production of surgical guides. Figures 14.9 and 14.10 show surgical guides made of ULTEM 9085 and VisiJet, respectively.

14.2.2.3   Implants

Implants are highest in the product complexity hierarchy. The final product requirements for implantable products include the following:

1. Made of biocompatible materials
2. Proven for long-term implantation in the body
3. Sterilizable
4. Easily handled or altered during surgery
5. Either resorbable or non-resorbable
6. Have the same mechanical properties as those of the neighboring tissues in the region of implantation
7. Possibly integrate with the body system or remain inert at the minimum

Replacement or reconstruction of skeletal structures can be performed using autografts, allografts, and xenografts. Autografts are tissues grafted from the same individual and are considered the gold standard from an immune response aspect. However, autografts are limited by the availability of suitable donor sites, additional expensive surgeries, tissue harvesting problems, secondary sites of morbidity with additional patient discomfort, chances of infection both at the site of surgery and at the donor site, increased surgical time, and the need for an additionally skilled surgical team.[2–6] This has led to the search for extraneous material that would be suitable without the inherent problems.[7]

14.2.3 MATERIALS

Biocompatible materials available for the fabrication of implants include bioceramics, polymers, and metals and metal alloys. Their advantages and disadvantages are discussed from the perspective that some of these materials may not be currently utilized but could be included in the spectrum of AM materials in the future as a result of ongoing research.

14.2.3.1 Bioceramics

Bioceramics used for the fabrication of implants include calcium phosphate materials such as hydroxyapatite (HA) and tricalcium phosphate (TCP). These materials have been successfully used in orthopedic and dental applications for decades. TCP is readily resorbed, but HA is nearly a permanent material. These materials have good osteoconductive properties but limited osteoinductive properties and mechanical strength,[8] which restricts their usage to non-weight-bearing regions. Alumina ceramics are very hard, with a high modulus of elasticity of 380 GPa, almost twice that of metal alloys, which can cause stresses at the implant–bone interface that lead to bone resorption and aseptic loosening. Because their resistance to flexion is low these materials can be used only as osteosynthesis plates. Zirconia (ZrO_2) has excellent mechanical properties with regard to flexion and resistance to wear, but fracture of the implanted femoral heads has been reported.[9] Moreover, fabrication of the materials to the complicated anatomical geometry is also a limiting factor.

14.2.3.2 Polymers

Polymers used in biomedical applications include polymethylmethacrylate (PMMA), polylactide (PLA), polyglycolide (PGA), polycaprolactone (PCL), ultra-high-molecular-weight polyethylene (UHMWPE), polyetheretherketone (PEEK), and polyetherketoneketone (PEKK). Heat-cured PMMA has been used in dentistry for the fabrication of dentures for several years. Cold-cured PMMA (bone cements), however, is not an ideal implant material as the heat produced during polymerization could exceed 70°C, which is higher than the coagulation temperature of proteins (56°C) and bone collagen (70°C). Increased rate of infection is seen with PMMA cranioplasty plates as compared to other materials such as titanium and ceramic implant materials.[10] UHMWPE is used for making friction components for prostheses of the hip, knee, and elbow due to its mechanical properties. PLA, poly(lactic-co-glycolic acid) (PLGA), polyethylene glycol (PEG), and PCL are being used to a limited extent as tissue-engineered implant materials due to their programmable biodegradable properties. Bioresorbable scaffolds are engineered to dissolve inside the body via hydrolysis and enzymatic activity and have a range of mechanical and physical properties that can be altered to suit specific applications. The degradation characteristics depend on several parameters, including molecular structure, crystallinity, and copolymer ratio. Resorbable properties can be tailored to match the healing or new bone formation rate. These materials are currently being used in clinical scenarios as absorbable suture materials and osteosynthesis plates. They are also being used for closure of trephination holes and small cranial defects but cannot be used for large defects and load-bearing implants. Their routine use as a clinically implantable prosthesis would be the ultimate achievement of tissue engineering.

14.2.3.3 Metals and Metal Alloys

Metals and metal alloys such as stainless steel, cobalt chromium, and titanium and titanium alloys have been successfully used as fracture fixation and as joint replacement materials, and expectations for their performance are growing. Stainless steel used for biomedical applications is designated 316LV (ASTM F138), which is austenitic, low carbon, and vacuum processed. Fracture fixation plates and syringe needles made of 316LV have been used, but removal of the plates has been required due to corrosion.[11] Moreover, the modulus of elasticity of the material is above 200 GPa, which makes the material unsuitable for reconstruction prostheses as it is intended to remain permanently inside the body. Cobalt chromium alloys used for biomedical applications include cobalt–chromium–molybdenum (CoCrMo; ASTM F75) and cobalt–nickel–chromium–molybdenum (CoNiCrMo; ASTM F562). Other alloys with the presence of nickel (25 to 30%) promise increased corrosion resistance but raise concerns about toxicity or immune-related reactions. The modulus of elasticity is around 200 GPa, ten times that of bone, which can lead to stress shielding and consequent loosening of the implants.

Zirconium (Zr) and tantalum (Ta) alloys have a high corrosion resistance due to the stability of the oxide layer that forms on the implant.[12,13] The materials also have a high wear resistance. Difficulties with forming and machining the materials restrict their use in implant fabrication. Large streaking and burst star artifacts have been noticed in spinal replacements with this metal.

Titanium and titanium alloys have found wide aerospace and medical (e.g., implant) applications. Titanium has been used as an implant material due to its high strength-to-weight ratio, corrosion resistance, biocompatibility, and osseointegration properties. Commercially pure titanium (CPTi) has been used in the manufacture of dental implants due to a ductility that allows cold working. Ti6Al4V (ASTM 136) and Ti6Al4V-ELI are used as joint replacement components due to their superior mechanical properties compared to CPTi. Ti6Al4V is composed of α+β material phases—that is, hexagonal close-packed (HCP) α phase and body-centered cubic (BCC) β phase. Aluminum (5.5 to 6.5%) stabilizes the α phase, and vanadium (3.5 to 4.5%) stabilizes the β phase. Titanium alloys are particularly preferred implant materials compared to stainless steel and CoCrMo alloys because of their high corrosion resistance due to the oxide coating (TiO_2) formed on the surface of the implant. This stable protective oxide coating protects Ti alloys from pitting corrosion, intergranular corrosion, and crevice corrosion and is mainly responsible for the excellent biocompatibility of titanium alloys. The strength of titanium alloys exceeds that of stainless steel or CoCrMo alloys. The modulus of elasticity of titanium (114 GPa) is much closer than stainless steel and CoCrMo alloys to that of bone; thus, less stress shielding is seen with titanium alloys. The above-mentioned property, their excellent biocompatibility, and their corrosion resistance make titanium alloys the best available material for prosthesis fabrication. Table 14.1 shows the comparative mechanical properties of implantable biocompatible materials as compared to cortical bone. It can be observed that most of the materials used for implantation have a much higher elastic modulus compared to bone. Among the metallic materials it can be seen that Ti6Al4V has the lowest elastic modulus. A very important reason for titanium being the preferred implant material is its unique property of osseointegration, which is the direct structural and functional connection between living bone and the surface of a load-bearing implant, combining the implant and adjoining bone function in unison. The process was first reported by Brånemark et al.[14] and Albrektsson et al.[15]

Titanium implants built using machining, such as forming, casting, and swaging, are often heavy and can cause discomfort to the patients. The Young's modulus of titanium is almost five times that of cortical bone and results in stress shielding effects.[16,17] Adding further complexity to their geometry, implants are required to have varying mechanical properties in different regions of the single implant.

TABLE 14.1

Comparison of Mechanical Properties of Cortical Bone and Implantable Materials

Property	Cortical Bone	Bioceramics (Hydroxyapatite, Tricalcium Phosphate)	Resorbable Polymers (PCL, PLLA, PLGA)	Non-Resorbable Polymers (PMMA, PEEK, PEKK)	Metals Titanium and Ti Alloys	SS 316	Chrome Cobalt
Osseoconduction and osseoinduction	Inherent process	Yes	Yes	No; requires coating	Yes; forms TiO_2 layer due to surface oxidation	No; requires coating	No; requires coating
Osseointegration	Natural healing process	Yes	Replaced with natural tissue	No	Yes	No	No
Elastic modulus (GPa)	3–30	80–110	Varies depending on the design	1.8–3.6	114	200	200–210
Rockwell hardness (GPa)	—	6	—	93–99	33	80	25–35
Corrosion resistance	—	—	—		TiO_2 high	Cr_2O_3 low	Cr_2O_3 low
Radiological imaging	Radiopaque	Radiopaque	Radiolucent	Radiolucent	Radiopaque; fewer artifacts compared to other metals	Radiopaque; metal artifacts	Radiopaque; metal artifacts

Source: Adapted from Katti, K.S., *Coll. Surf. B Biointerfaces*, 39(3), 133–142, 2004; Ravaglioli, A. and Krajewski, A., *Bioceramics: Materials, Properties, Applications*, Chapman & Hall, London, 1992.

Note: PCL, polycaprolactone; PEEK, polyetheretherketone; PEKK, poly(ether-ketone-ketone); PLGA, poly(lactic-co-glycolic acid); PLLA, poly(lactide); PMMA, poly(methyl methacrylate).

14.2.4 NEED FOR A NEW GENERATION OF IMPLANTS

Skeletal structure is the first and foremost mechanical structure, the major functions of which are to transmit forces from one part of the body to another (limbs, mandible) and to provide protection to internal organs such as the heart, lungs, and brain. Bone is a calcified structure that is continuously being remodeled dynamically by bone formation (osteoblastic) and bone resorption (osteoclastic) activity, depending on the quantum of mechanical forces transmitted. Balancing this osteoblastic and osteoclastic activity allows healthy bone and muscles to perform their necessary activities. Disuse, atrophy, and hypertrophy due to increased usage are well known in physiological and pathological situations. The process of bone adaptation continues with the surgical placement of implants and is a key factor for successful retention and performance of the implant. Adaptation to the newly placed material at the bone–implant interface is both chemical and mechanical. Chemical factors affecting the cellular reaction are directly related to the chemical composition of the implant material. The mechanical factors affecting the cellular reaction are the forces at the implant–host tissue interface. These factors are directly dependent on the mechanical properties of the implant material and the bone. Further, the structure of bone is not uniform. It has an inner, less dense region (cancellous bone) and an outer, more dense region (cortical bone) with a modulus of elasticity ranging from 0.5 to 20 GPa. A reduction in or even a total lack of stress transmission at the interface due to the high modulus of elasticity of the implant material leads to an imbalance in the process of bone resorption and apposition. This causes increased bone resorption at the bone–implant interface, leading to loosening of the implant. This is the process known as stress shielding. Increased mechanical forces at the interface also lead to failure of the implant, emphasizing the need for a new generation of implants with the appropriate mechanical properties.

14.2.4.1 Requirements for the Next Generation of Implants

The next generation of implants will have to be porous to enable the growth of healthy bone and tissue for implant fixation and stabilization. Implants will have to conform to the external shape of the site that is to be replaced. More importantly, the effective elastic modulus of the implant should match that of surrounding tissue. Ideally, the weight of the implant should be equal to the weight of the tissue being replaced, leading to increased patient comfort. All of the above requirements would have to be met at an affordable cost and within a reasonable time frame.

Some examples of load-bearing implants wherein the mechanical properties could be varied to effectively reduce weight without compromising function are hip implants and mandible implants. The load-bearing parts of a hip implant must be dense for effective function, while the stem and mandible body could be more porous, as seen in Figure 14.11. Therefore, an ideal design strategy for any implant would be to determine the percent of porosity that would provide an elastic modulus of 5 to 20 GPa, equivalent to that of the adjoining bone, and a weight equal to that of bone (density of cranial bone, 2 g/cm^3). Pore sizes ranging from 500 to 2000 µm would facilitate tissue ingrowth.

It is imperative to explore manufacturing methodologies for building net-shaped porous parts with repeatable mechanical properties. The advantages lie in the reduction of weight of the implant, bringing it close to that of natural bone, and in the reduction of effective stiffness of the material,

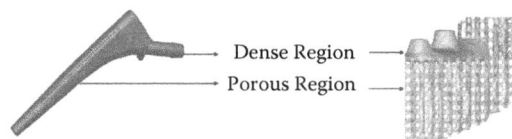

Dense Region

Porous Region

FIGURE 14.11 (Left) Hip implant with porous stem and dense head region. (Right) Part of a mandible with a dense dental abutment structure and porous body.

thereby reducing the stress shielding effects and increasing the longevity of implants.[18-21] This design strategy can only be achieved through the use of AM processes. Advantages of using AM processes to manufacture implants include the following:

- Direct fabrication of the implant without the need for a template
- Conformance of the implant to the normalized shape of the part that it replaces
- Mechanical properties close to those of the region of implantation and relevant to function
- Variation of mechanical properties depending on the implant
- Repeatability of implant properties

Table 14.2 summarizes the materials and processes for AM of implantable and non-implantable medical devices.

TABLE 14.2

Materials, Processes, and Applications

Material	Approval	Process	OEM	Applications	Sterilization Process
MED610 clear biocompatible	ISO 10993 USP Class VI[a]	PolyJet	Stratasys	Orthodontic appliances, delivery and positioning trays, full and partial denture try-ins, surgical guides	Cold
MED620	NA	PolyJet	Stratasys	Veneer models, diagnostic wax-ups	Cold
MED670	NA	PolyJet	Stratasys	Models	Cold
MED690	NA	PolyJet	Stratasys	Models	Cold
FullCure® 630 Clear	NA	PolyJet	Stratasys	Hearing aids	Cold
FullCure® 655 Rose Clear	NA	PolyJet	Stratasys	Hearing aids	Cold
PC-ISO™ (polycarbonate-ISO)	ISO 10993 USP Class VI	FDM	Stratasys	Medical models	Gamma radiation and EtO
ABS-M30i	ISO 10993 USP Class VI	FDM	Stratasys	Medical models	Gamma radiation and EtO
ULTEM 9085 flame retardant	NA	FDM	Stratasys	Medical models and surgical guides	Autoclave
VisiJet® crystal plastic material	USP Class VI	SLA	3D Systems	Dental, orthopedic surgical, crown preparation guides	Cold
VisiJet® StonePlast plastic material	USP Class VI	SLA	3D Systems	Dental, orthopedic surgical, crown preparation guides	Cold
VisiJet® clear plastic material	USP Class VI	SLA	3D Systems	Dental, orthopedic surgical, crown preparation guides	Cold
VisiJet® EX200 plastic material	USP Class VI	SLA	3D Systems	Dental, orthopedic surgical, crown preparation guides	Cold

(Continued)

TABLE 14.2 (Continued)
Materials, Processes, and Applications

Material	Approval	Process	OEM	Applications	Sterilization Process
VisiJet® MP200 plastic material	USP Class VI	SLA	3D Systems	Dental, orthopedic surgical, crown preparation guides	Cold
DSM Somos	USP Class VI ISO 10993	SLA	DSM	Medical models	Cold
E 600	USP Class VI	Bioplotter	Envisiontec	Surgical guides	Gamma radiation, e-beam radiation, and EtO
e-Dent 100	FDA	DLPM	Envisiontec	Crowns for short- or long-term placement	—
Bioresorbable polymers (PLLA, PLGA, PCL)	—	Laser sintering, bioplotting, extrusion	Envisiontec, RepRap, Makerbot	Tissue engineering scaffolds	Gamma radiation, e-beam radiation, and EtO; sterilization with nitrogen dioxide (NO_2)
PEKK	FDA	Laser sintering	EOS	Cranial, facial, and spinal implants	Gamma radiation, and EtO
Titanium and titanium alloys	FDA CE mark	EBM, DMLS	Arcam, EOS	Cranial, facial, spinal, and orthopedic implants	Autoclave

Note: DLPM, direct light projection manufacturing; DMLS, direct metal laser sintering; EBM, electron beam melting; EtO, ethylene oxide; FDA, Food and Drug Administration; FDM, fusion deposition modeling; SLA, stereolithography.

a Prolonged skin contact more than 30 days; mucosal contact up to 24 hr.

14.2.5 IMPLANTABLE PRODUCTS

14.2.5.1 Bioresorbable Polymer Scaffolds

Selective laser sintering, fused deposition modeling, and extrusion technologies use low-temperature melting polymers such as PLA and PCL to make implantable products. OsteoNexus provides 3D printed PCL meshes, burr hole covers, and custom craniofacial solutions and scaffolds that are osteoconductive. The biomimetic design allows for blood clotting and healing within the pores. The material slowly degrades in the body, allowing itself to be replaced by bone healing. The products are cleared through FDA 510(k) regulations. Similarly, a PCL 3D tracheal stent has been cleared on an emergency basis for a surgery. Figure 14.12 shows a burr hole cover from OsteoNexus. These scaffolds, when combined with tissue engineering and implanted in the body, have osteoconductive properties that encourage bone growth within the scaffold and over time are replaced by natural bone.

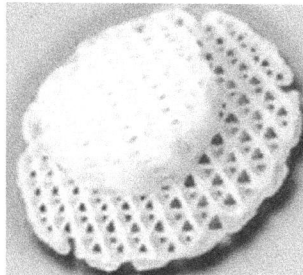

FIGURE 14.12 OsteoNexus burr hole cover (www.osteonexus.com).

14.2.5.2 Non-Resorbable Polymer Implants

High-performance polymers, or metallic polymers such as PEEK HP3 and PEKK, are non-resorbable polymers that are biocompatible and approved for long-term implantation in the body. They have higher strength compared to the resorbable polymers. The polymers are light, unlike metals, and have elastic modulus values closest to that of bone. The polymers lend themselves well to AM. High-temperature laser sintering (HTLS) is a suitable process for processing higher end polymers. EOS PEEK HP3 is a material for AM that is compliant with Federal Aviation Regulation (FAR) 25.853 and Underwriters Laboratory (UL) 94 V0, and it has excellent chemical and hydrolysis resistance. PEEK parts exhibit high potential for biocompatibility. They can be sterilized for medical applications and are light weight with high wear resistance. The lighter weight, strength, and biocompatibility of PEEK make it a desirable alternative to titanium. PEEK HP3 is processed using the EOSINT P 800 process. Oxford Performance Materials is a supplier of PEKK. Figure 14.13 shows their OsteoFab® PEKK laser sintered implant.

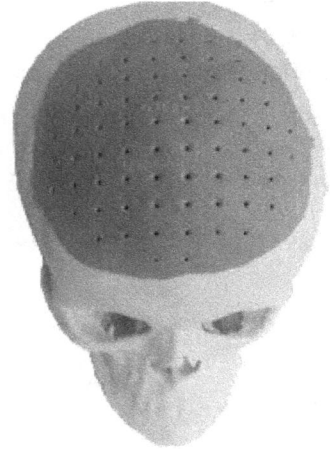

FIGURE 14.13 OsteoFab® PEKK laser sintered implant (www.oxfordpm.com).

14.2.5.3 Metallic Implants

Metallic materials—titanium, titanium alloys (e.g., Ti6Al4V), CoCr, and stainless steel—have been used for several years. Titanium and its alloys have been the materials of choice but problems with their use have been discussed earlier. The necessity of building lattice structures for implants has also been discussed. This section discusses the AM processes that can be used for producing implantable products such as implants and custom-designed fixation plates.

14.2.5.3.1 Electron Beam Melting

A 3D digital model of the cranium can be generated from CT data. The virtual model is then used to create the implant design either by mirroring from the contralateral side or by generating curves based on the anatomical region with CAD-based/haptic devices. Figure 14.14 shows the road map

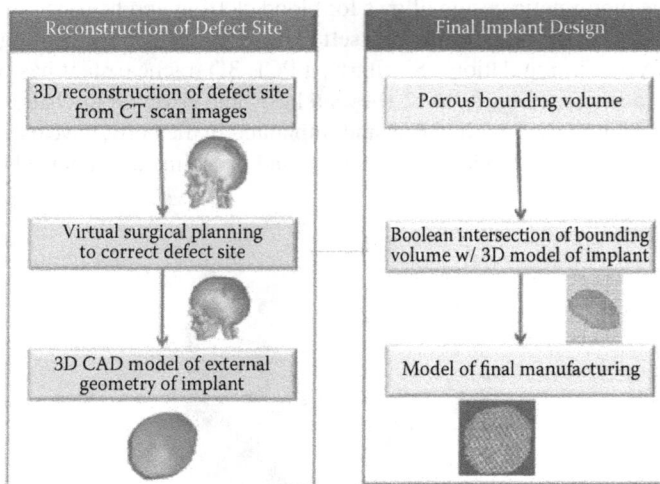

FIGURE 14.14 Road map for arriving at the final design of the implant.

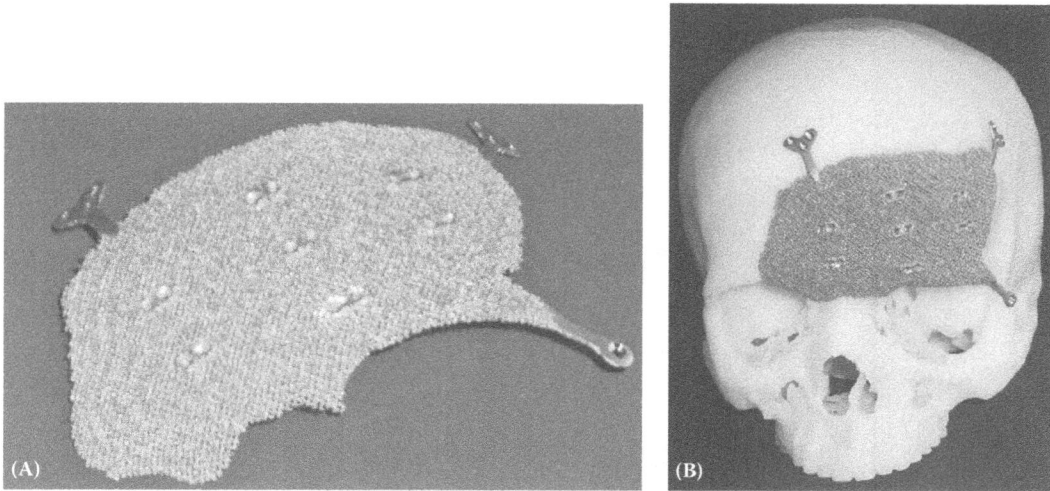

FIGURE 14.15   (A) Patient-specific porous titanium implant made using EBM; and (B) same fitting to 3D printed skull model.

for arriving at the final design of the implant. The implant model is sent to an Arcam EBM or EOS DMLS machine. The software creates layers of 2D images that are used for solidification of the part from a bed of Ti6Al4V powder, layer by layer, to create an implant ready for implantation. EBM and DMLS technologies alleviate the need for a skull model or a secondary process to create a custom implant. Figure 14.15 shows a patient-specific porous titanium implant made using EBM and the same fitting to the 3D printed skull model. The EBM process builds parts by melting Ti6Al4V powder a layer at a time at high temperatures above 1600°C. The minimum wall thickness that can be built with EBM or DMLS is 0.2 mm.

Scanning electron microscopy studies of EBM parts show no interlayer differentiation when probed on the exterior face of the part (Figure 14.16A). This indicates complete melting of the powder metal and good metallurgical bonding between layers during the fabrication process. Figure

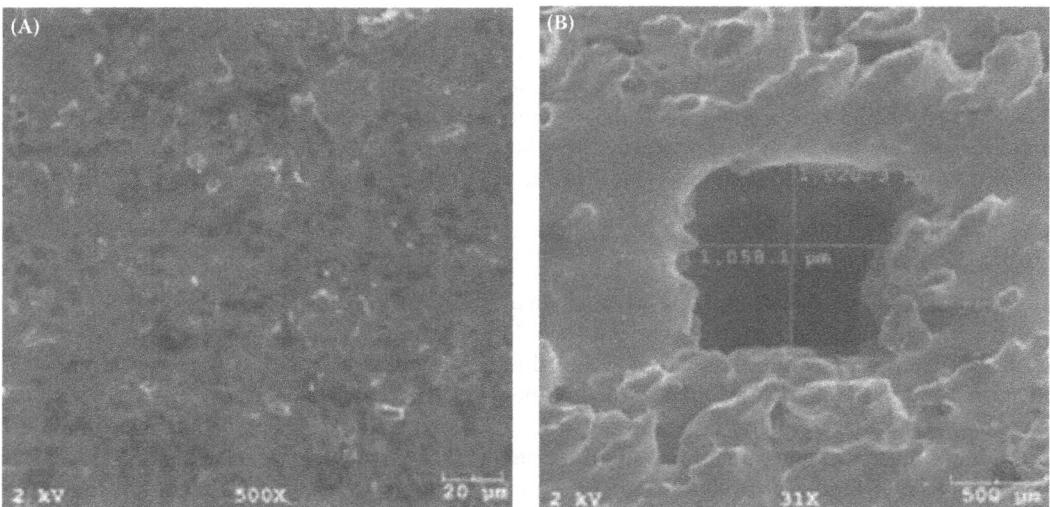

FIGURE 14.16   (A) SEM image of surface microstructure showing complete melting of powder particles, and (B) SEM image showing pore size close to that of design parameters.

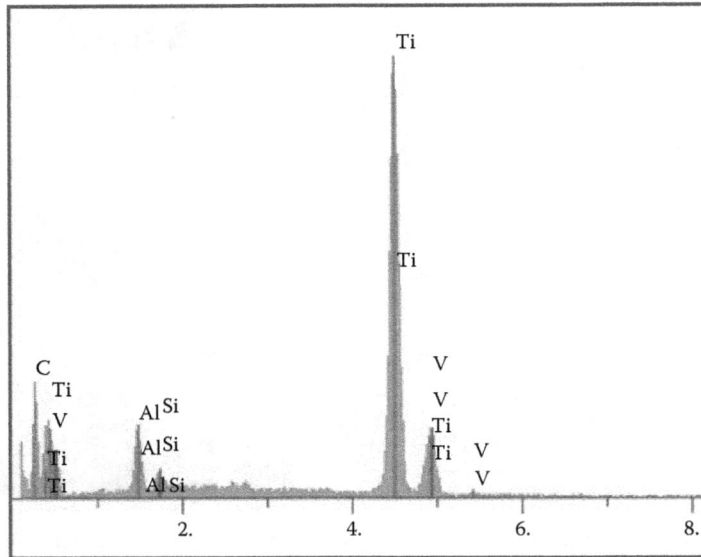

FIGURE 14.17   X-ray diffraction studies of EBM-fabricated Ti6Al4V parts: cleaning, finishing, and polishing.

14.16B shows the pore dimensions in the x and y directions to be equal. X-ray diffraction studies show the presence of mainly titanium, aluminum, and vanadium; small amounts of carbon, oxygen, and silica are also seen (Figure 14.17).

14.2.6   CLEANING, FINISHING, AND POLISHING

14.2.6.1   Anatomical Models and Surgical Guides

In the SLA process the support structures and the model material are the same, and support structures are broken off. Because models are made in a vat of resin, uncured resin can be found on the surface. The models are oven cured in a finisher for 20 to 30 minutes, depending on the size of the part. The models are then bead blasted for a fine finish. PolyJet models are washed with a water jet, soaked in lye, and then washed again with a water jet to completely remove the support material. The models can then be sanded and buffed before a polyurethane coating is applied. FDM models utilize ultrasonic baths with a proprietary solution to remove supports. Starch models from Z Corp. are made in a bed of powder and the powder is blown out. These models can also be infused with epoxy for additional strength. Cleaning is performed as recommended by the equipment manufacturers.

14.2.6.2   Implants

Both metallic and polymer implants are made in a bed of the implant material. Upon removal, the powder particles on the surface of the part should be completely blown away. Implants made of lattice structures and solid parts have the added issue of powder particles becoming entrapped in the pores and in the undulations of the struts. The parts are therefore bead blasted to ensure complete removal of the raw material. As noted, x-ray diffraction studies show the presence of mainly titanium, aluminum, and vanadium; small amounts of carbon, oxygen, and silica are also seen (Figure 14.17). Silica is a residue of post-processing wherein silica beads are blasted to remove the entrapped Ti6Al4V powder particles present inside the pores that act as the support material during the fabrication process. The presence of silica is also observed during high-resolution

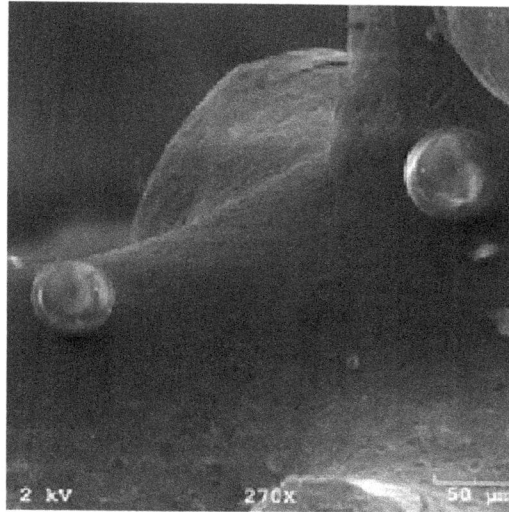

FIGURE 14.18 High-resolution scanning showing the presence of silica particles.

scanning (Figure 14.18). A validated process of blowing and bead blasting needs to be established to ensure complete removal of the raw material and silica beads. The surface finishing and smoothening are done using traditional methods. A typical surface roughness of 5 to 50 µm (Ra) can be achieved. Parts with lattice structures present a greater challenge in terms of cleaning, finishing, and polishing. Voids that are less than 500 µm pose greater challenges. For any device manufactured with voids, systems and processes have to be established for complete removal of residual support material.

14.2.7 STERILIZATION OF AM DEVICES

Patient-specific anatomical models are used mainly as presurgical planning tools and are found in the operatory only as reference models. The models themselves do not come in contact with the operatory field; however, the models do require a standard sterilization process. Models made of low-melting-point resins are sterilized using gluteraldehyde, sterrad, or ethylene oxide sterilization or gamma radiation. Metallic implants can be autoclaved along with other surgical requisites. Table 14.2 provides a summary of the sterilization processes for all AM parts. Parts with lattice structures pose a greater challenge for sterilization, requiring the establishment of validated processes for the specific device.

14.3 FUTURE PROSPECTS OF AM IN THE MEDICAL INDUSTRY

The future prospects of additive manufacturing in the medical device industry are promising. Research in biomaterials for AM will introduce new classes of resorbable and non-resorbable materials and associated equipment. Multi-material 3D printing equipment, especially equipment capable of the simultaneous printing of metals and polymers combined with cells, will produce improved implants with increased longevity. The polymers will resorb, being replaced by natural tissues, while the metals will provide strength and improved load-bearing capacity. Organ printing is the ultimate goal of biofabrication. Additive manufacturing in the near future will lead to on-demand production of patient-specific body parts with optimized properties that meet the biological and structural needs of the individual.

REFERENCES

1. FDA. (2015). *Guidance Documents (Medical Devices and Radiation-Emitting Products)*. U.S. Food and Drug Administration, Washington, DC (http://www.fda.gov/MedicalDevices/deviceregulationandguidance/guidancedocuments/default.htm).

2. Schlickewei, W. and Schlickewei, C. (2007). The use of bone substitutes in the treatment of bone defects—the clinical view and history. *Macromol. Symp.*, 253: 10–23.

3. Shimko, D.A. and Nauman, E.A. (2007). Development and characterization of a porous poly(methyl methacrylate) scaffold with controllable modulus and permeability. *J. Biomed. Mater. Res. Part B Appl. Biomater.*, 80(2): 360–369.

4. Silber, J.S., Anderson, G.D., Daffner, S.D., Brislin, T.B., Leland, J.M., Hilibrand, A.S. et al. (2003). Donor site morbidity after anterior iliac crest bone harvest for single-level anterior cervical discectomy and fusion. *Spine*, 28: 134–139.

5. St. John, T.A., Vaccaro, A.R., Sah, A.P., Schaefer, M., Berta, S.C., Albert, T., and Hilibrand, A. (2003). Physical and monetary costs associated with autogenous bone graft harvesting. *Am. J. Orthoped.*, 32: 18–23.

6. Lane, H.S. and Sandhu, J.M. (1987). Current approaches to experimental bone grafting. *Orthop. Clin. N. Am.*, 18: 213–225.

7. Martin, P.J., O'Leary, M.J., and Hayden, R.E. (1994). Free tissue transfer in oromandibular reconstruction. Necessity or extravagance? *Otolaryngol. Clin. N. Am.*, 27(6): 1141–1150.

8. Khan, S.N., Tomin, E., and Lane, J.M. (2000). Clinical applications of bone grafts substitutes. *Tissue Eng. Orthop. Surg.*, 31: 389–398.

9. Kraay, M.J. et al. (2006). Zirconia versus Co–Cr femoral heads in total hip arthroplasty: early assessment of wear. *Clin. Orthop. Relat. Res.*, 453: 86–90.

10. Matsuno, A., Tanaka, H., Iwamuro, H., Takanashi, S., Miyawaki, S. et al. (2006). Analyses of the factors influencing bone graft infection after delayed cranioplasty. *Acta Neurochir. (Wien)*, 148: 535–540.

11. Urban, R.M., Jacobs, J.J., Gilbert, J.L., Skipor, A.K., Hallab, N.J. et al. (2003). Corrosion products generated from mechanically assisted crevice corrosion of stainless steel orthopedic implants. In: *Stainless Steels for Medical and Surgical Applications* (Winters, G.L. and Nutt, M.J., Eds.), pp. 262–272. ASTM, West Conshohocken, PA.

12. Jacobs, J.J., Gilbert, J.L., and Urban, R.M. (1998). Corrosion of metal orthopedic implants. *J. Bone Joint Surg.*, 80: 268–282.

13. Levine, B.R., Sporer, S., Poggie, R.A., Valle, C.J., and Jacobs, J.J. (2006). Experimental and clinical performance of porous tantalum on orthopedic surgery. *Biomaterials*, 27(27): 4671–4681.

14. Brånemark, P.-I., Hansson, B.O., Adell, R., Breine, U., Lindström, J., Hallén, O. et al. (1977). *Osseointegrated Implants in the Treatment of the Edentulous Jaw*. Almqvist & Wiksell, Stockholm, p. 132.

15. Albrektsson, T., Brånemark, P.-I., Hansson, H.A., Kasemo, B., Larsson, K., Lundstrom, I. et al. (1983). The interface zone of inorganic implants *in vivo*: titanium implants in bone. *Ann. Biomed. Eng.*, 11: 1–27.

16. Robertson, D.M., Pierre, L., and Chahal, R. (1976). Preliminary observations of bone ingrowth into porous materials. *J. Biomed. Mater. Res.*, 10: 335–344.

17. Ryan, G., Pandit, A., and Apatsidis, D. (2006). Fabrication methods of porous metals for use in orthopedic applications. *Biomaterials*, 27: 2651–2670.

18. Parthasarathy, J., Starly, B., and Raman, S. (2009). Computer aided bio-modeling and analysis of patient specific porous titanium mandibular implants. *J. Med. Devices*, 3(3): 031007.

19. Parthasarathy, J., Starly, B., and Raman, S. (2008). Design of Patient-Specific Porous Titanium Implants for Craniofacial Applications, paper presented at RAPID 2008 Conference & Exposition, Society of Manufacturing Engineers, Lake Buena Vista, FL, May 20–22.

20. Parthasarathy, J. (2014). 3D modeling, custom implants and its future perspectives in craniofacial surgery. *Ann. Maxillofac. Surg.*, 4: 9–18.

21. Parthasarathy, J., Starly, B., and Raman, S. (2011). A design for the additive manufacture of functionally graded porous structures with tailored mechanical properties for biomedical applications. *J. Manuf. Proc.*, 13(2): 160–170.

15 Medical Applications of Additive Manufacturing

Jayanthi Parthasarathy

CONTENTS

ABSTRACT

Rapid advances in computer technology and additive manufacturing (AM) are changing the way in which surgeons diagnose, analyze, plan, and treat patients. AM, with its wide range of processes and biocompatible and implantable material choices, is continuing to find more applications. Patient-specific models and surgical guides have come to be widely used in craniofacial and orthopedic surgery. Dental laboratory procedures are using AM to produce more precisely fitted parts. Metallic implants with mechanical properties equivalent to the region of implantation and surfaces designed for better adhesion to native bone are being fabricated using electron beam melting, direct metal laser sintering, and selective laser melting. Bioprinting and the printing of resorbable implants are gaining in importance and finding wider applications that will become routine in the near future. On-demand organ printing is the ultimate aim of tissue engineering. This chapter discusses the current and future applications of AM in the healthcare industry.

15.1 INTRODUCTION

Rapid advances in computer technology and additive manufacturing (AM) are changing the way surgeons diagnose, analyze, plan, and treat patients. Improved imaging with computed tomography (CT), magnetic resonance imaging (MRI), and ultrasound (US) has made it possible to reformat data as three-dimensional (3D) images. The role of radiologic imaging is to demonstrate

TABLE 15.1

AM Processes and Their Applications in Medicine

Process	Materials	Application	Device Type
Stereolithography	Photopolymers	Template models Surgical guides	Non-implantable
Fused deposition modeling (FDM)	Thermoplastic materials (e.g., ABS, ABS-M30, ABS-M30i, ULTEM 9085)	Template models Surgical guides	Non-implantable
PolyJet™	Resins (e.g., VeroClear; Objet MED610, MED690, MED655)	Template models Surgical guides	Non-implantable
Selective laser sintering (SLS)	Polymers (e.g., polyethylene, PEKK)	Surgical guides Implants	Non-implantable and implantable
Electron beam melting (EBM)	Titanium Ti6Al4V Chrome cobalt Stainless steel	Craniofacial implants Orthopedic implants	Implantable
Direct metal laser sintering (DMLS)	Titanium Ti6Al4V Chrome cobalt Stainless steel	Craniofacial implants Orthopedic implants	Implantable
Bioprinting	Resorbable polymers Chitosan Hydrogels Cells	Craniofacial implants Orthopedic implants Plastic and reconstructive surgery	Implantable

the existence or absence of pathology in the human body. Just as important is communicating these findings as digital or physical models using the proper display, presentation, and transmission. Advances in computer graphics and image processing have revolutionized imaging such that displays of 3D objects can be viewed at different angles and in varying colors, lighting, and surface properties. 3D reconstruction from serial images of CT, MRI, and US allows identification of objects of interest in 3D and clarifies relationships among these objects. 3D post-processing is useful to both the radiologist and the clinician, as it is a useful tool for the display, interpretation, and communication of complex data. The surgeon is a hands-on individual who needs to operate on patients to achieve the ultimate goal of treatment. The complexity of the human anatomy and regions that are not visible from the exterior complicate surgery. Also, each individual is unique and varies in age, sex, and race. It would be very beneficial for surgeons to have a physical model that they could touch, cut, and use as a template in the operating room. In this regard, additive manufacturing—with its unlimited capability to manufacture complex shapes in varying colors and transparency directly from a computer-aided design (CAD) model—is probably the only option currently available.

For the past three decades, additive manufacturing has allowed the recreation of patient-specific anatomical objects as 3D physical models, thus giving surgeons a realistic view of complex structures before surgical intervention. Improvements in speed, dimensional accuracy, and choice of materials, including biocompatible and implantable resorbable and non-resorbable materials, have transformed AM from a model-making process to a solution-providing technology. Some current applications of AM include the fabrication of patient-specific models, surgical guides, implants, prostheses, and related instrumentation and tissue engineering. This chapter discusses the various medical applications of AM and the benefits they offer to surgeons and patients. Table 15.1 shows the AM processes and medical applications for which they can be used.

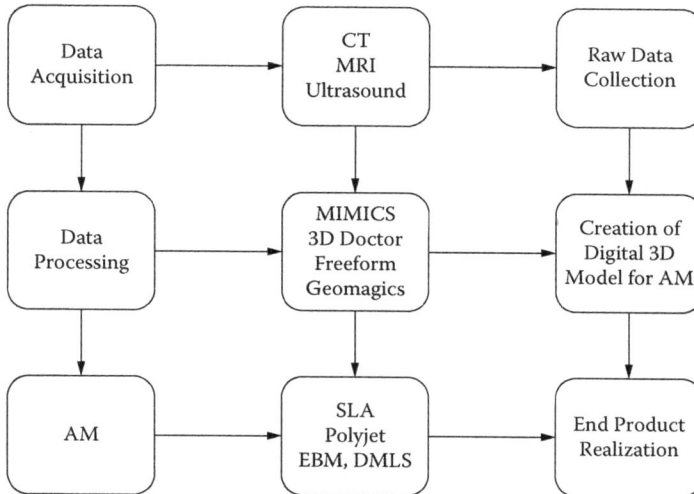

FIGURE 15.1 Process flow for product realization process for a patient-specific additive manufactured product.

15.1.1 Process Flow

The common process flow for all medical applications for patient-specific devices involves three steps: data acquisition, data processing and manufacturing. The first two steps are akin to reverse engineering and involve design processes in preparation of a printable file format. The third step is the actual manufacturing of the device. Figure 15.1 shows the process flow for the product realization process for a patient-specific additive manufactured product.

15.1.2 Adaptability of Additive Manufacturing

Computed tomography scans, magnetic resonance imaging, and ultrasound are human reverse-engineering methodologies in the truest sense. AM is a similar manufacturing technology that takes a 3D model and creates slice data to manufacture the final product from the bottom up. The commonality of layer-based methods of processing enables straightforward data transfer between the imaging and the manufacturing process. Human anatomy is very complex, involving bones, muscles, nerves, and vasculature which all have different physical and radiological properties. Also, additional differences are seen based on gender, age, and race. The complex anatomical shapes cannot be restricted to the confines of definable geometry. AM empowers the designer with geometric freedom. Any part that is designed in a CAD environment can be manufactured using AM. Mechanical property requirements vary according to the end use of the product. Implants placed in load-bearing regions, particularly orthopedic and some craniofacial applications, are subject to stresses and strains during normal use. Surgical devices and implants also experience chemical stresses due to body fluids. In emergencies, the parts must be produced within a short time frame. All of these characteristics of medical devices require using biocompatible materials for the production of unique patient-specific parts that may require long-term implantation in the body. Current additive manufacturing technologies are well suited for manufacturing complex shapes, as they can eliminate expensive tooling and offer a wide range of material choices. Parts can be made in a very short time (in some cases within a few hours), thus making the process invaluable in emergency situations. The models and implants so produced can be sterilized using common hospital sterilization protocols such as steam, ethylene oxide, gamma radiation, or the STERRAD® system.

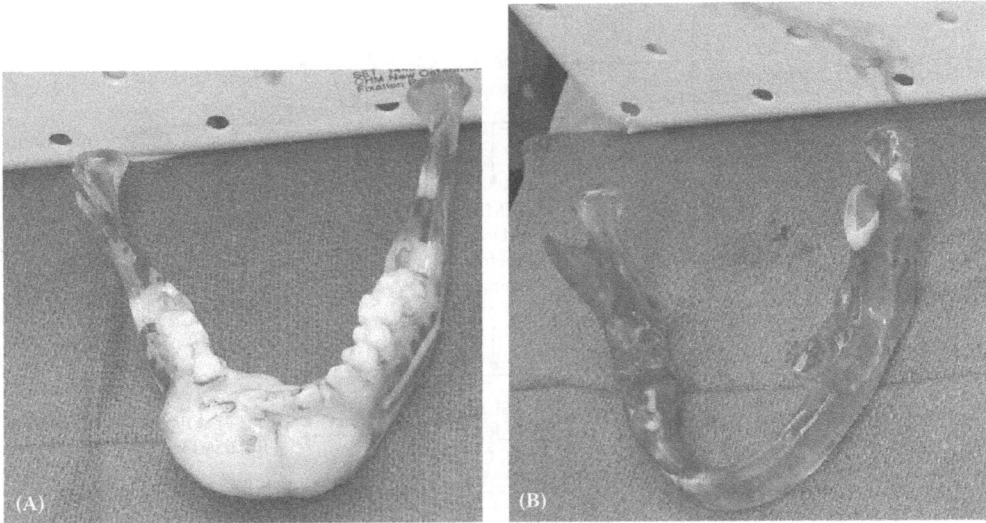

FIGURE 15.2   (A) PolyJet model of a mandible model of a tumor, and (B) reconstructed mandible reference model.

15.2   APPLICATIONS OF ADDITIVE MANUFACTURING IN MEDICINE

15.2.1   ANATOMICAL MODELS

Additive manufacturing models are being used by surgeons to plan and explain complex surgeries, especially in the craniofacial, maxillofacial, orthopedic, cardiac, and reconstructive surgical specialties. Surgery is an art and a science. As with any manual art, surgery requires great precision and dexterity, and much of the outcome depends on the surgeon's skills. The current imaging modalities of CT, MRI, or ultrasound scans do not fully reveal vital anatomical structures in the vicinity of the surgical region as well as the surgeon might prefer. Many surgeries are staged, with the decision for the second stage being based on the outcome of the previous surgery. There is no substitute for rehearsing complex surgeries to determine the course and progress of the operation. The haptic sensation the surgeon can experience when using a model to plan a surgery can make the actual procedure feel more familiar. Models are frequently present in the operating room, where they are used as templates and guides. Figure 15.2 shows images of a PolyJet™ mandible model of a tumor and a reconstructed mandible reference model in the operatory.

Transparency of the models, color printing capabilities, and printing with dual material textures allow distinct visualization of normal blood vessels, nerves, and abnormal structures such as tumors within and surrounding the region of interest. Many surgeries involve a team and are performed in stages. Patient-specific anatomical models that are made from the CT of the patient are very useful tools for discussing and rehearsing the surgery as a team and for staging and optimizing the surgical procedure to effectively reduce operating time and achieve the best outcome for the patient. Anatomical patient replica models are used as templates in the manufacture of such medical devices as cranial implants and joint reconstruction. Temporomandibular joint reconstruction is a typical example of such an application (Figure 15.3A). The models are used during the design process and as templates to check the fitting of the devices (Figure 15.3B). The models can also be used to custom bend plates for complex fracture fixation and in reconstructive surgery. This application typically reduces operating and healing times, as it repositions the fractured reconstructed bone in an optimized position and the plate is supplied to the surgeon in the desired position (Figure 15.4).[1]

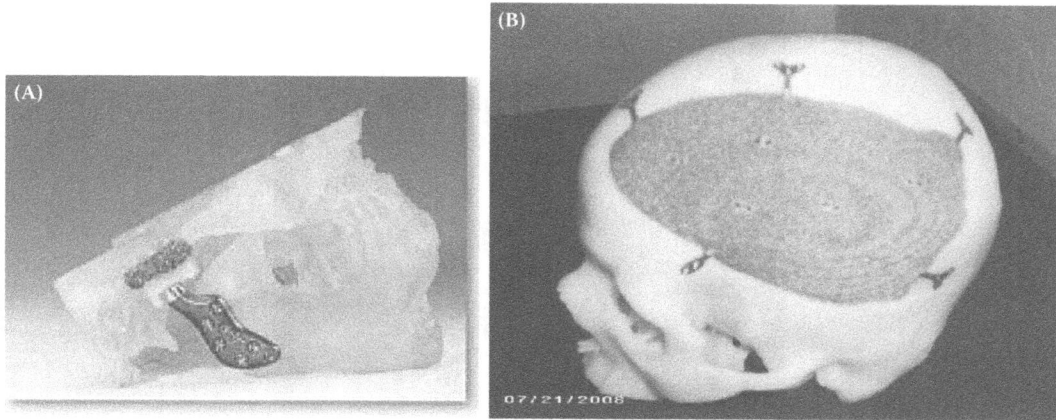

FIGURE 15.3 (A) TMJ implant fitted to SLA maxilla and mandible model (www.tmjconcepts.com). (B) Patient-specific cranial implant fitted to the skull model for verification of fit.

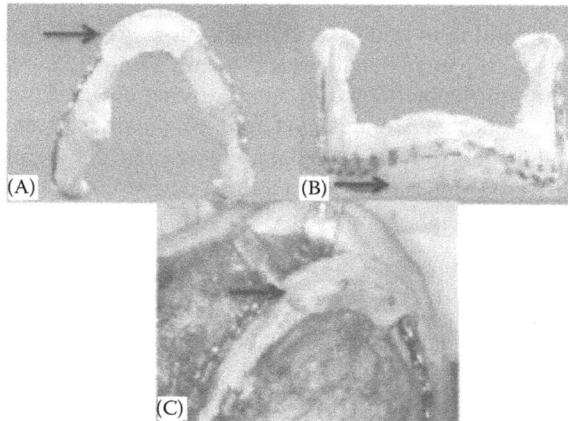

FIGURE 15.4 (A,B) 3D-printed mandible model as a custom plate-bending device. (C) Bone plate installed.[1]

15.2.2 CRANIOFACIAL SURGERY

Surgeons operating on the craniofacial region have been the earliest adapters of technology for enhanced visualization due to the complexity of the anatomy involved, deep-seated pathology in certain scenarios, proximity to vital structures such as brain and major blood vessels, need for staged and progressive multiple surgeries to treat congenital defects, and the relationship of the treatment outcome to the personality and societal integration of the patient. AM models have been used by surgeons for two decades now, since the first SLA models were first used in 1994.[2] 3D-printed models have served as a great communication tool between the operating team and the patient's family with regard to developing an understanding of the treatment and its outcome. Surgical guides for resection and reconstruction have significant drawbacks, but AM can help reduce surgical time and enhance surgical precision. Reducing anesthesia time is of utmost importance for any patient, more so for pediatric, geriatric, and medically compromised patients, and can be effectively achieved by the use of patient-specific models and guides. Figure 15.5 shows a reconstructed maxilla model printed using a Connex™ 3D printer (Objet500) to adapt a preoperative titanium mesh ridge for use in surgery.

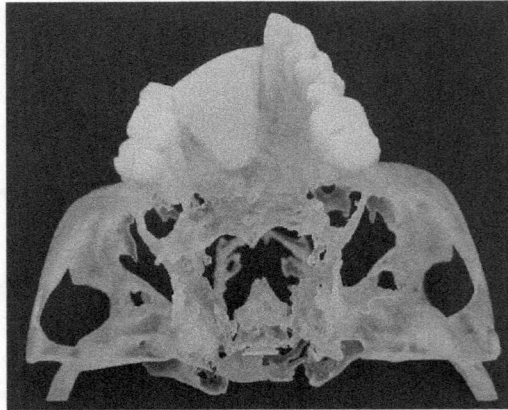

FIGURE 15.5   Reconstructed maxilla model printed on the Objet500 Connex™ printer.

15.2.3 CARDIOLOGY

Printing cardiac models is gaining in importance among surgeons wishing to evaluate and discuss existing cardiac problems with patients and surgical teams. Realistic models of cardiac anatomy and related vasculature are invaluable tools for demonstrating device fit and bench-top testing of endovascular stents and graft design. Complicated congenital cardiac defects can be better understood using an anatomical model from CT scan data. Figures 15.6 shows images of 3D reconstruction using Mimics™ software (Materialise) and physical device testing of the AbioCor® artificial heart model (Abiomed).

15.2.4 PATIENT-SPECIFIC IMPLANTS

The success and longevity of implants depend on such factors as material characteristics, design of the implant, region of implantation, patient-specific response, and the surgeon's skill. The skeletal structure is the first and foremost mechanical structure, the major functions of which are to transmit forces from one part of the body to another (limbs, mandible) and provide protection to internal organs such as the heart, lungs, and brain. Bone is a calcified structure that is continuously being remodeled dynamically by bone formation (osteoblastic) and bone resorption (osteoclastic) activity, depending on the quantum of mechanical forces transmitted. Balancing this osteoblastic and osteoclastic activity allows healthy bone and muscles to perform their necessary activities. Disuse, atrophy, and hypertrophy due to increased usage are well known in physiological and pathological situations. The process of bone adaptation continues with the surgical placement of implants and is a key factor for successful retention and performance of the implant. Adaptation to the newly placed

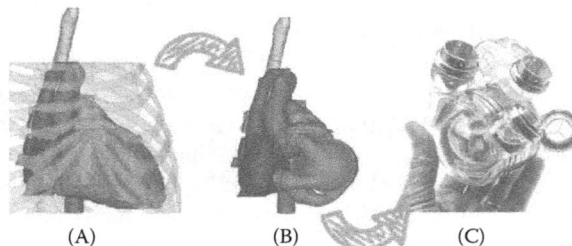

(A)   (B)   (C)

FIGURE 15.6   (A) Heart segmentation, (B) replacement of ventricles with artificial heart, and (C) AbioCor® total heart replacement. (Courtesy of Abiomed, Danvers, CO; Materialise, Leuven, Belgium.)

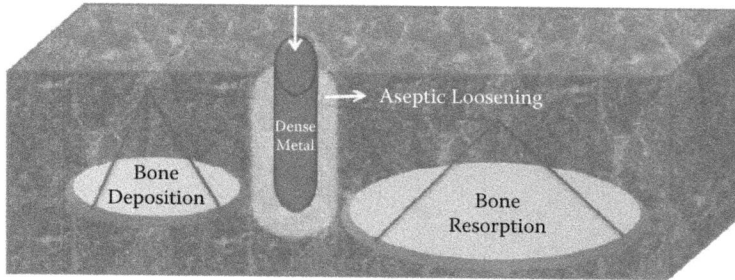

FIGURE 15.7 Aseptic loosening due to stress shielding.

material at the bone–implant interface is both chemical and mechanical. Chemical factors affecting the cellular reaction are directly related to the chemical composition of the implant material. The mechanical factors affecting the cellular reaction are the forces at the host tissue–implant interface. These factors are directly dependent on the mechanical properties of the implant material and the bone. Further, the structure of bone is not uniform and has an inner, less dense region (cancellous bone) and an outer, more dense region (cortical bone) with a modulus of elasticity ranging from 0.5 to 20 GPa. In accordance with Wolf's law, bone in a healthy individual remodels in response to the mechanical stress it is subjected to during use. A reduction in or even a total lack of stress transmission at the bone–implant interface due to the high modulus of elasticity of the implant material (titanium, cobalt chrome, stainless steel) leads to an imbalance in the process of bone resorption and apposition. This causes increased bone resorption at the bone–implant interface, leading to loosening of the implant. This is the process known as stress shielding, which reduces the longevity of the implant as shown in Figure 15.7.[3] Biomechanical forces also vary with age, sex, and body weight.

It is imperative to build lighter implants with weights closer to those of the replaced bone. In addition, the effective stiffness of the implant must be comparable to that of healthy surrounding bone tissue to reduce stress shielding effects, thus improving the longevity of implants. To reduce the weight and effective stiffness of titanium implants, the implant structure must be made porous.[4–7] The implants are designed as a combination of lattice structures and solid regions with appropriate mechanical properties for load-bearing requirements.[8] The lattice structures also facilitate increased adhesion of the implant to the adjoining bone by enhancing tissue ingrowth into the implant and increasing the surface area for cell adhesion. The solid regions are finished as smooth parts to prevent abrasion due to wear and tear. The only manufacturing process capable of producing such complicated geometries consistently is additive manufacturing. Direct manufacturing of such end-use patient devices as craniofacial, orthopedic, and spinal implants can be achieved only by AM. This is the future of patient-specific devices.

Patient-specific implants that take into consideration real-life conditions can be manufactured using AM. The ability of AM to produce parts that have the dual configuration of lattice structures providing a surface texture conducive to tissue ingrowth and solid, smooth joint surfaces to prevent wear and tear makes the technology most suitable for the production of patient-specific implants, including acetabular cups, femoral stems, cranial implants, and mandible reconstruction. Custom or patient-specific implants have the advantages of better fit, reduced operative time, better functionality, faster healing, and less chance of failure compared to generic implants, as they are precisely made to fit a particular patient's anatomy and physical requirements. The processes of electron beam melting (EBM), direct metal laser sintering (DMLS), and laser sintering of polymers have led to a series of U.S. Food and Drug Administration (FDA) 510(k)-cleared and CE-marked products being available on the market. Oxford Performance Materials uses high-melting-point polymers such as polyetherketoneketone (PEKK) to make craniofacial implants that are FDA 510(k) cleared. Figure 15.8 shows a patient-specific, EBM-printed cranial implant designed with a porous structure and osseo fixation plates designed as solid parts fitted onto the 3D-printed resin skull model.

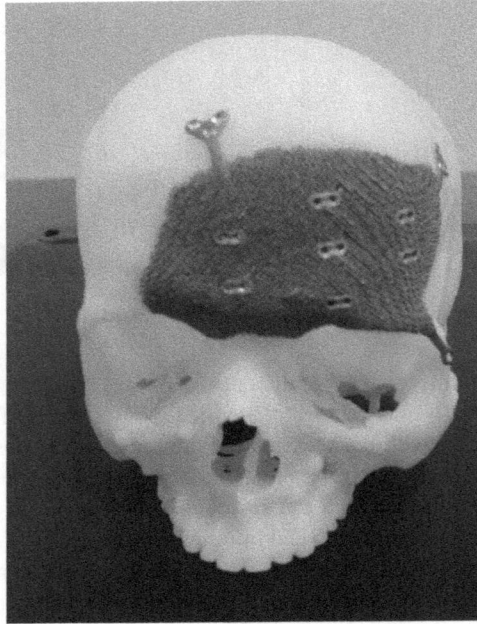

FIGURE 15.8 Patient-specific cranial implant printed with the EBM process and fitted to the 3D-printed resin skull model.

15.2.5 ORTHOPEDIC

Conventional orthopedic implants made of titanium and its alloys (Ti, Ti6Al4V, and Ti6Al4V-ELI), cobalt chrome (CoCr), and stainless steel (316L) have been in use in the orthopedic world for more than half a century. The chemical composition and mechanical properties of the implant affect the acceptance and longevity of the implant. Dense titanium implants are approximately twice as heavy as natural dense cortical bone equivalent. More importantly, the elastic modulus of Ti6Al4V is about 114 GPa, while that of cancellous and cortical bone ranges from 0.5 GPa to a maximum of 20 GPa. Electron beam melting and direct metal laser sintering are technologies that lend themselves to the manufacturing of such implants. OEM manufacturers such as Stryker use AM to manufacture titanium acetabular cups. Trabecular structures can be designed and optimized to improve fit and osseointegration by varying pore geometry, pore size, density, and surface roughness. Acetabular cups manufactured with electron beam melting by Arcam and spinal cages manufactured with direct metal laser sintering by EOS are shown in Figure 15.9.

15.2.6 DENTAL APPLICATIONS

Dentistry was an early adopter of CAD and computer-aided manufacturing (CAM), because dentistry, being an art and a science, is related to the person's facial form and personality. Being able to visualize nearby vital structures, develop a treatment plan, and determine the probable outcome can greatly empower surgeons to achieve the desired treatment goals. AM is being used in the manufacture of patient-specific crown and bridge restorations. The large metal sprue that requires removal in the traditional casting process is completely eliminated in metal additive manufacturing of cast crowns and bridges. Laser sintering is the most common technology used for this purpose.

    Implantology procedures are increasing in numbers every day. Precise placement of the implants in the right direction and at the proper angulation is key to the success of an implant. Cone beam computed tomography (CBCT) is a low-radiation CT scan currently available in

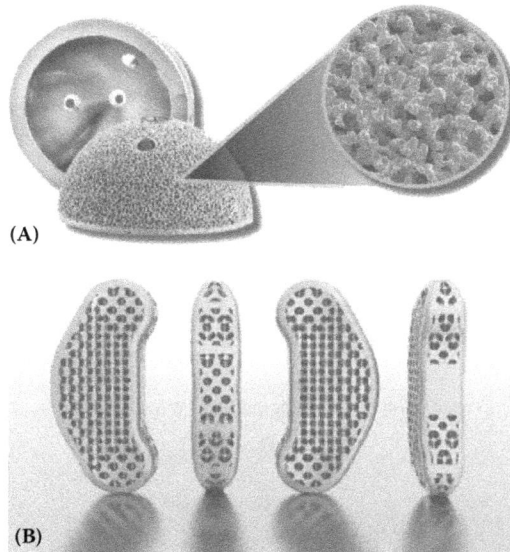

FIGURE 15.9 (A) Arcam acetabular cup manufactured with electron beam melting (EBM) (www.arcam.com). (B) EOS spinal cages manufactured using direct metal laser sintering (DMLS) (www.eos.info/en)

many dental offices. Various software (e.g., NobelGuide®, Dental Wings, Anatomage's Invivo, Materialise's SimPlant®) is used by surgeons and service centers to convert the CT data to 3D models. The software allows for precise dental implant positioning and avoiding the maxillary sinus and nerve canals during placement of the implant. Transferring the virtual surgical plan to the clinical scenario is achieved by using drill guides made with AM technologies. Guides are made by SLA or PolyJet using such materials as MED 610 and VisiJet® materials that have been approved for mucosal contact. Figure 15.10 shows a dental implant positioning guide made with MED 610 from Stratasys.

Within the orthodontic specialty, Invisalign® is a procedure for correcting malocclusion where the patient wears a series of appliances that reposition the teeth. Intraoral scans and CBCT are combined to create a composite 3D model of the tooth alignment. The teeth are then moved virtually to a new position and the malocclusion corrected. The stage models of the corrected dentition are 3D printed using SLA or PolyJet AM technologies. A series of appliances is then molded in clear plastic material from the 3D-printed models depicting the different stages of tooth movement. In recent years, a greater number of dental labs have adapted to 3D printing due to its increased precision and

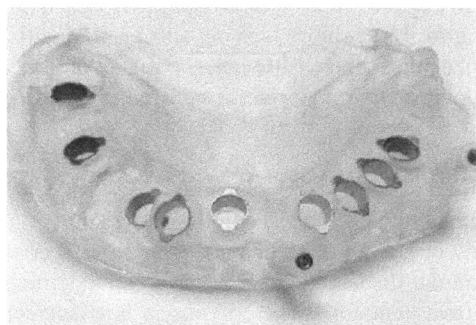

FIGURE 15.10 Dental implant placement guide (www.stratasys.com).

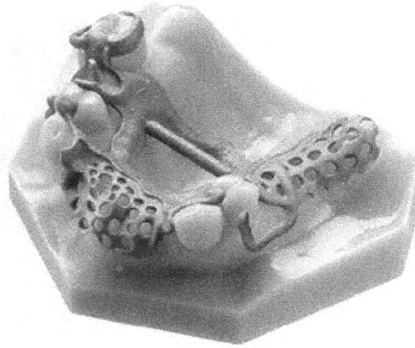

FIGURE 15.11   Partial wax up ready for investment casting (www.stratasys.com).

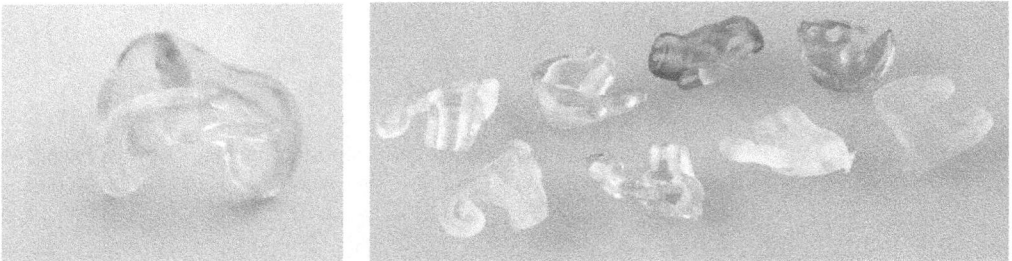

FIGURE 15.12   3D hearing aids printed with (left) Stratasys (www.stratasys.com) and (right) EnvisionTEC materials (www.envisiontec.com).

efficiency compared to the traditional stone model and wax-up procedures. Wax-up procedures such as crowns and partial frameworks can be printed using wax deposition method (WDM) technology from Stratasys. Figure 15.11 shows a partial wax-up that is ready for investment casting.

15.2.7   HEARING AIDS

No two ear canals are the same. Traditionally, producing acceptable hearing aids has been a labor-intensive process, and the results have left room for improvement. Customized, precisely fitted hearing aids and accessories are now being made with the E-Shell® line of liquid photoreactive acrylates. The materials are CE certified and classified as Class II biocompatible according to ISO 10993 (Medical Product Law) for hearing aids. Objet's FullCure® 630 Clear and FullCure® 655 Rose Clear materials are FDA approved and biocompatible. The two materials produce smooth, well-molded hearing aids. Figure 15.12 shows hearing aids printed with Stratasys and EnvisionTEC materials. The customized size and shape ensure better patient comfort and improved functionality, in addition to the advantage of reducing manufacturing time.

15.2.8   ADDITIVE MANUFACTURING IN TISSUE ENGINEERING

Structural defects of the human body can be classified as congenital or acquired. Congenital defects are caused by malformations or the absence of particular tissues or organs. In acquired defects, tissue loss occurs due to trauma, disease processes, or ablative surgical resection to contain diseases such as cancer. The treatment for these conditions is tissue replacement. Allografts have been and are being currently used but have some limitations, such as the potential for infection, lifelong intake of immunosuppressants to prevent rejection, and differences in the mechanical and physicochemical properties of the allograft. Scientists are therefore turning to tissue engineering, which entails the growth of the patient's own tissues to the required shape within a scaffold made of bioresorbable materials that degrade in the body after implantation. Ideally, scaffolds should have the following attributes:[9]

- Provide optimal surface texture and porosity for cell adhesion.
- Promote cell growth.
- Retain differentiated cell functions.
- Promote extracellular matrix (ECM) regeneration.
- Allow homogeneous tissue formation.
- Have adequate mechanical strength to function as the host tissue.
- Reproduce with the same predicted properties into 3D complex anatomic shapes.
- Offer variable properties depending on the anatomic site of implantation.

Several researchers have worked on scaffold design and internal architecture,[6,10–15] and the designs have been adopted for fabricating tissue engineering scaffolds.

Biodegradable materials have been researched, including polycaprolactone (PCL), polylactide (PLA), poly(lactic-co-glycolic acid) (PLGA), chitosan, and extracellular matrix. The scaffold itself can be fabricated using AM technologies. Initially, cells were seeded manually but currently scaffold seeding is being done using a bioplotter/bioprinter. Stem cells are also being used, with cues being given for appropriate cell differentiation depending on the region of implantation.

15.2.8.1 3D Printing of Bone Tissue

Three-dimensional printing of bone has attracted great interest. Table 15.2 shows AM techniques that have been used for scaffold fabrication for bone tissue engineering, materials used, and advantages and disadvantages.[16] Advantages of 3D printing include one-of-a-kind or low-volume production of scaffolds in patient-specific shapes with highly interconnected pores. The process is versatile and offers a wide choice of materials; however, one challenge is the selection of a suitable binder. As for the sintering process, pore and strut sizes have been a major issue. Post-processing is invariably required for densification as sintered parts could cause shrinkage, affecting the dimensional accuracy.

15.2.8.2 Organ Printing

Organ replacement has been a remedy for persons with nonfunctioning organs or end-stage disease. Kidney and liver transplants have been in vogue for many years, but the lack of suitable donors has led to the loss of patients awaiting organs, and rejection has been a major contributor to failure of transplanted organs. These factors have spawned research on bioprinting of organs. Figure 15.13 shows the schematic of an organ printing process as proposed in Murphy and Atala.[17] The authors described three basic approaches to 3D bioprinting: biomimicry, autonomous self-assembly, and mini-tissues. Biomimicry entails replicating the structural components of the cellular structure including the blood vessels and extracellular matrix (ECM). Autonomous self-assembly may be used without a bioprinted scaffold. In the mini-tissue approach, smaller components of the organ are printed and then either assembled into larger units or maintained as smaller units for use as organs on a chip for drug screening and testing.

Three types of bioprinting processes are described in the literature: inkjet, microextrusion, and laser-assisted. Some examples of inket bioprinting are direct deposition of skin[18] and cartilage[19] *in situ*. Microextrusion bioprinting involves the use of temperature-controlled printers. Print resolution and speed are a major challenge for this type of printer. Microextrusion printers have been

TABLE 15.2
AM Techniques for Scaffold Fabrication for Bone Tissue Engineering

Technique	Process Details	Processed Materials for Bone Tissue Engineering	Advantages	Disadvantages
3D plotting/ direct ink writing	Extrusion of strands of paste/viscous material (in solution form) based on the predesigned structure Layer-by-layer deposition of strands at a constant rate under specific pressure Disruption of strands according to tear-off speed	PCL HA Bioactive glasses Mesoporous bioactive glass/ alginate composite PLA/PEG PLA/PEG/G5 glass PHMGCL Bioactive 6P53B glass	Mild condition of process allows drug and biomolecule (proteins and living cells) plotting.	Heating/ post-processing that is required for some materials restricts the biomolecule incorporation.
Laser-assisted bioprinting (LAB)	Coating the desired material on transparent quartz disk (ribbon) Deposition control by laser pulse energy Resolution control by distance between ribbon and substrate, spot size, and stage movement	HA Zirconia HA/MG63 osteoblast-like cell Nanohydroxyapatite Human osteoprogenitor cell Human umbilical vein endothelial cell	Ambient conditions Applicable for organic and inorganic materials; cells Quantitatively controlled 3D stage movement	Homogeneous ribbons needed

Source: Adapted from Bose, S. et al., *Mater. Today*, 16(12), 496–504, 2013.

Note: HA, hydroxyapatite; PCL, polycaprolactone; PEG, polyethylene glycol; PHMGCL, poly(hydroxymethylglycolide-*co*-ε-caprolactone); PLA, polylactic acid.

used to print aortic valves,[20] branched vascular trees,[21] and tumor-model tissues.[22] Laser-assisted bioprinting, initially developed for printing metals, has now been adapted for bioprinting using nanohydroxyapatite and hydrogels. The process has been used to produce tracheobronchial stents in patients.[23] Among all of the resorbable printing processes, the degradation kinetics, required mechanical strength, and byproducts of degradation play major roles in material selection and design of the implantable device.

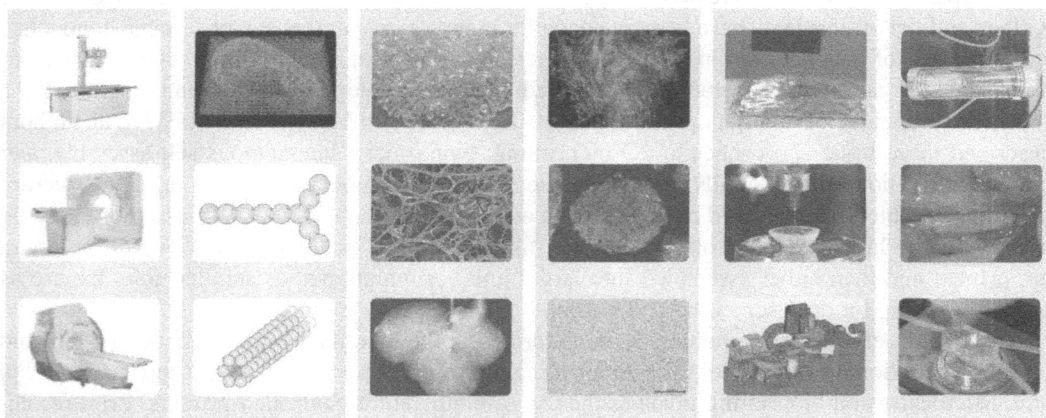

FIGURE 15.13 Schematic of organ printing.[18]

15.3 FUTURE PERSPECTIVES

Medicine has lengthened the human life span, but with this come the problems of aging and an associated need for orthopedic implants resulting from trauma in old age and organ transplants due to failing organs. On-demand replacement of body parts with a patient's own tissue is the ultimate goal of tissue engineering and regenerative medicine. Patients in the future will not have to wait on long lists for compatible donors, and they will not have to be on immune suppressants for the rest of their lives. Patient-specific, anatomically precise organs will be printed using AM. Progress in regenerative medicine, the future for organ replacement, depends on further development in biomaterials, scaffold manufacturing, cell regeneration, and vascularization. A good number of resorbable materials (e.g., polylactides, polyglycolides, polycaprolactone, bioglass) have been developed for 3D printing; however, we still have to find the perfect 3D-printable resorbable biomaterials for load-bearing regions. Printing cells along with the scaffold is a challenge currently being researched. Cues for cell differentiation from basic stem cells, cell multiplication, sustaining the multiplied cells, and vascularization are significant problems related to biofabrication and on-demand organ printing. Despite all of these challenges, skin, tracheas, and bladders have been successfully printed using AM processes and implanted in patients. Printing of vital organs such as the heart and kidney and tubular structures such as intestines will be seen the future.

REFERENCES

1. Lauria, A., Mayrink, G., Moreira, R.W.F., Asprino, L., and De Moraes, M. (2013). Evaluation of the use of biomodels in sequelae of maxillofacial trauma. *Int. J. Odontostomat.*, 7(1): 113–116.
2. Mankovich, N.J., Samson, D., Pratt, W., Lew, D., and Beumer III, J. (1994). Surgical planning using three-dimensional imaging and computer modeling. *Otolaryngol. Clin. North Am.*, 27(5): 875–889.
3. Park, J.B. and Lakes, R.S. (1992). *Biomaterials and Introduction*. Plenum, New York.
4. Parthasarathy, J., Starly, B., and Raman, S. (2008). Design of Patient-Specific Porous Titanium Implants for Craniofacial Applications, paper presented at RAPID 2008 Conference & Exposition, Society of Manufacturing Engineers, Lake Buena Vista, FL, May 20–22.
5. Parthasarathy, J., Starly, B., and Raman, S. (2009). Computer aided bio-modeling and analysis of patient specific porous titanium mandibular implants. *J. Med. Devices*, 3(3): 031007.
6. Starly, B., Gomez, C., Darling, A., Fang, Z., Lau, A. et al. (2003). Computer-aided bone scaffold design: a biomimetic approach. In: *Proceedings of the 2003 IEEE 29th Annual Bioengineering Conference*, pp. 172–173.
7. Sun, W., Starly, B., Nam, J., and Darling, A. (2005). Bio-CAD modeling and its application in computer-aided tissue engineering. *Comput.-Aided Design*, 37(11): 1097–1114.
8. Parthasarathy, J., Starly, B., and Raman, S. (2011). A design for the additive manufacture of functionally graded porous structures with tailored mechanical properties for biomedical applications. *J. Manuf. Proc.*, 13(2): 160–170.
9. Rajagopalan, S. and Robb, R.A. (2006). Schwarz meets Schwan: design and fabrication of biomorphic and durataxic tissue engineering scaffolds. *Med. Image Anal.*, 10(5): 693–712.
10. Hutmacher, D. (2000). Scaffolds in tissue engineering bone and cartilage. *Biomaterials*, 21(24): 2529–2543.
11. Wei, S., Darling, A., Starl, B., Nam, J., and Darling, A. (2005). Bio-CAD modeling and its applications in computer-aided tissue engineering. *Comput.-Aided Design*, 37: 1097–1114.
12. Wei, S., Starly, B., Darling, A., and Gomez, C. (2004). Computer aided tissue engineering and application to biomimetic modeling and design of tissue scaffolds. *Biotechnol. Appl. Biochem.*, 39: 49–58.
13. Wettergreen, M.A., Bucklena, B.S., Starly, B., Yuksel, E., Wei, S., and Liebschnera, M.A. (2005). Creation of a unit block library of architectures for use in assembled scaffold engineering. *Comput.-Aided Design*, 37: 1141–1149.
14. Starly, B., Fang, Z., Wei, S., Shokoufandeh, A., and Regli, W. (2005). Three dimensional reconstruction for medical CAD modelling. *Comput.-Aided Design Appl.*, 2(1–4): 431-438.
15. Starly, B., Lau, W., Bradbury, T., and Sun, W. (2006). Internal architecture design and freeform fabrication of tissue replacement structures. *Comput.-Aided Design*, 38(2): 115–124.
16. Bose, S., Vahabzadeh, S., and Bandyopadhyay, A. (2013). Bone tissue engineering using 3D printing. *Mater. Today*, 16(12): 496–504.

17. Murphy, S. and Atala, A. (2014). 3D bioprinting of tissues and organs. *Nat. Biotechnol.*, 32: 773–785.

18 Skardal, A., Mack, D., Kapetanovic, E., Atala, A., Jackson, J.D., Yoo, J., and Soker, S. (2012). Bioprinted amniotic fluid-derived stem cells accelerate healing of large skin wounds. *Stem Cells Transl. Med.*, 1: 792–802.

19. Cui, X., Breitenkamp, K., Finn, M.G., Lotz, M., and D'Lima, D.D. (2012) Direct human cartilage repair using three-dimensional bioprinting technology. *Tissue Eng. Part A*, 18: 1304–1312.

20. Duan, B., Hockaday, L.A., Kang, K.H., and Butcher, J.T. (2013). 3D bioprinting of heterogeneous aortic valve conduits with alginate/gelatin hydrogels. *J. Biomed. Mater. Res. A*, 101: 1255–1264.

21. Norotte, C., Marga, F.S., Niklason, L.E., and Forgacs, G. (2009). Scaffold-free vascular tissue engineering using bioprinting. *Biomaterials*, 30: 5910–5917.

22. Xu, F., Celli, J., Rizvi, I., Moon, S., Hasan, T., and Demirci, U. (2011). A three-dimensional *in vitro* ovarian cancer coculture model using a high-throughput cell patterning platform. *Biotechnol. J.*, 6: 204–212.

23. Zopf, D.A., Hollister, S.J., Nelson, M.E., Ohye, R.G., and Green, G.E. (2013) Bioresorbable airway splint created with a three-dimensional printer. *N. Engl. J. Med.*, 368, 2043–2045.

16 Additive Manufacturing of Pluronic/Alginate Composite Thermogels for Drug and Cell Delivery

Sara Maria Giannitelli, Pamela Mozetic,
Marcella Trombetta, and Alberto Rainer

CONTENTS

ABSTRACT

The use of biocompatible hydrogels has widely extended the potential of additive manufacturing (AM) in the biomedical field. Indeed, AM hydrogel scaffolds can be used for controlled administration and delivery of bioactive molecules and even living cells. To reach this ultimate goal, however, gel formulations with a proper set of physicochemical properties must be developed in order to be combined with AM setups. In this work, the pressure-assisted deposition of a biocompatible thermosensitive gel has been explored for administration of both drugs and cells.

16.1 INTRODUCTION

In the last decade, tissue engineering (TE) has benefited from the development of several additive manufacturing (AM) techniques, which have led to the production of freeform porous scaffolds with custom-tailored architectures. Advances introduced by additive manufacturing have significantly improved the ability to control porosity, pore size, and interconnectivity as well as the mechanical performance of TE scaffolds.[1]

One of the major obstacles to the diffusion of AM in TE has long been represented by the limited set of biomaterials that could be directly processed. However, recent advances in printing technologies, together with the synthesis of novel composite biomaterials, have enabled the fabrication of various scaffolds with defined shape and controlled *in vitro* behavior.[2] In this regard, the processing of hydrogels can be considered the main challenge, as these materials enable cell encapsulation during the deposition process, thus representing an intriguing candidate for soft tissue engineering applications. Furthermore, direct manufacturing of tissue precursors with a cell density similar to native tissue has the potential to overcome the extensive *in vitro* culture required by seeding prefabricated three-dimensional (3D) scaffolds.

As some AM processes operate at room temperature, hydrogel scaffolds containing cells and growth factors have been successfully fabricated without significantly affecting cell viability. Thus, several research groups have adapted different AM techniques to assemble living cell-laden constructs directly from computer-generated design models with high resolution, aiming to demonstrate their abilities in the area of complex tissue and organ manufacturing.[3] As a result, scaffolds containing multiple cell types, as well as heterogeneous distribution of drugs and growth factors, have been fabricated using novel multi-syringe AM techniques.[4,5] A further advancement has recently been made by Andersen et al.,[6] who developed an additive manufacturing technique to produce an implant with compartmentalized siRNAs in the locations corresponding to distinct tissues. However, although a set of AM techniques has been used for hydrogel processing, only a few of them are fully suitable for the generation of clinically relevant structures with self-supporting networks. A comprehensive overview on current trends and limitations of hydrogel-friendly AM techniques has recently been provided by Billiet et al.[7] Furthermore, to achieve an accurate reproduction of the designed architecture, hydrogels have to meet specific requirements in terms of viscosity and gelation rate, which limit the number of formulations that can be processed by AM.[8] Moreover, the gel must be non-cytotoxic and have adequate structural integrity and mechanical properties for *in vitro* culture and *in vivo* implantation.

The majority of hydrogels used for organ printing, mostly represented by sodium alginates,[9,10] collagen, and poloxamers (e.g., Pluronic®),[11] suffer from a lack of adhesive/biomimetic sequences, inadequate mechanical properties, and instability in culture conditions. Furthermore, when performing crosslinking of cell-printed constructs, the toxicity of the crosslinker should be carefully considered. For example, several common crosslinking agents, such as glutaraldehyde and EDC/NHS (1-ethyl-3-(3-dimethylaminopropyl) carbodiimide hydrochloride and *N*-hydroxysuccinimide), as well as some photoinitiators, have been demonstrated to negatively affect cell viability.[7] Thus, the development of suitable, tailored hydrogel matrices with cytocompatible gelation processes and tunable chemomechanical behavior represents a highly desirable goal.

In this regard, innovative solutions to improve the mechanical stability of hydrogel matrices propose to reinforce the construct by integration with secondary thermoplastic compartments, such as electrospun polymer networks and rigid thermoplastic AM architectures in a multi-compartmental structure.[12,13] To enhance the interface-binding strength of reinforced constructs, covalent grafting between the 3D printed thermoplastic network and the hydrogel has also been proposed.[14] Moreover, hybrid structures consisting of interdigitated struts of different materials in a layer-by-layer strand configuration have been fabricated with the goal to fully integrate the reinforcing skeleton and the cell-supportive hydrogel in the same microstructure.[15]

Alternatively, one of the most popular approaches to obtain functional soft structures takes advantage of the combination of different materials by using hydrogel blends.[16] In this manner, tunable mechanical and degradation properties can be achieved through the integration of multiple solidification/gelling mechanisms and multi-step crosslinking systems.[17–19] A notable application is represented by the work of Xu et al.,[20,21] who developed a gel system using gelatin, sodium alginate, and fibrinogen, characterized by a double crosslinking mechanism (alginate with bivalent ions and fibrinogen with thrombin). Another recent study reported the use of silk–gelatin-based bioink for the fabrication of 3D cell-laden constructs and the combination of enzymatic crosslinking (by mushroom tyrosinase) and physical crosslinking (by sonication).[22]

16.2 PLURONIC COMPOSITE THERMOGELS

Thermoresponsive polymers represent one of the best candidates for *in situ* generation of advanced polymeric systems.[23] Such hydrogels use a temperature change as the trigger that determines their gelling behavior without any additional factor. This unique feature can be advantageous for biomedical applications, as these materials can transform from a liquid to a gel when administered topically. Among the available temperature-responsive polymers, poly(ethylene oxide)–poly(propylene oxide)–poly(ethylene oxide) (PEO/PPO/PEO) triblocks, also known as Pluronic® or poloxamers, constitute a group of important reverse thermal gelation materials that are capable of producing low-viscosity aqueous solutions at ambient temperature and forming a gel at higher temperatures. The gelation mechanism of PEO/PPO/PEO block copolymers in aqueous solutions has been a topic of extensive investigation, and the more accredited explanation of this phenomenon could be related to changes in the micellar arrangement as a function of both concentration and temperature.[24]

Although such gels can preserve their shape very well during printing,[25] they lose their structural integrity upon dilution with other aqueous fluids due to an immediate decrease in the copolymer concentration required for the packing of Pluronic micelles. Injection of Pluronic sols into living tissues leads to the immediate obtaining of a thermosetting gel structure at the site, but the gel rapidly dissolves. This problem at the injection site severely limits the use of Pluronic scaffolds for the sustained release of various drugs. To overcome this issue, stable thermosensitive hydrogels based on Pluronic triblock copolymers were prepared by combining the temperature-induced physical sol–gel transition with the irreversible light-activated chemical crosslinking (photopolymerization).[26,27] However, low photoinitiator concentrations and low-intensity ultraviolet light should be used to minimize possible adverse effects of exposure of the embedded cells to free radicals, so that only a few studies have succeeded in demonstrating the feasibility of printing viable cell-laden scaffolds through Pluronic photopolymerization.[28]

Pluronic has been physically and/or chemically blended with other polymers, mostly chitosan[29] and alginates.[16] In particular, alginates represent one of the key components of the biocompatible blends due to their capability of forming ionotropic gel, pH sensitiveness, high biocompatibility, and relatively low cost. They belong to a family of linear block polyanionic copolymers composed of two monomeric units, namely β-D-mannuronic (M) and α-L-guluronic (G) acid, and form stable hydrogels in the presence of certain divalent cations (e.g., Ca^{++}, Cu^{++}) through the ionic interaction between the cation and the carboxyl functional group of G units located on the polymer chain.

Among the wide number of alginate formulations, Pluronic/alginate composite thermogels have been widely used for the sustained release of various macromolecular drugs in transdermal, ophthalmic,[30,31] and artery endoluminal delivery.[32] This combination gives rise to superior biocompatibility and enhanced cell viability during encapsulation compared to the use of Pluronic alone. Indeed, although Pluronic is usually regarded as nontoxic, systemic side effects have been reported after its administration in several animal models.[33] However, the majority of these studies have focused on the effect of different blending mechanisms on the *in vitro* gel stability, not taking into account the role of scaffold architecture.

In the present work, the Pluronic/alginate gel system has been used to fabricate cell-laden constructs with controlled 3D structure. This combination of polymers with distinct phase-transition mechanisms has been chosen with the aim of integrating the advantages of Pluronic gel in terms of printability with the enhancement of mechanical strength of the temperature-insensitive, strong alginate component.

16.3 METHODS

16.3.1 HYDROGEL SYNTHESIS AND CHARACTERIZATION

Pluronic/alginate solutions of different compositions were prepared to identify the most suitable blend ratio for the printing processing. The polymeric blends were prepared by using the so-called cold method proposed by Schmolka.[34] Briefly, a proper amount of sodium alginate powder (PROTANAL® XP 3499, FMC BioPolymer, Philadelphia, PA) was slowly added to a dilute solution of Dulbecco's Modified Eagle Medium (DMEM; Lonza, Basel, Germany) in distilled water (1:5 v/v) and maintained in an ice bath. Subsequently, the desired amount of Pluronic flakes (PF127; Sigma-Aldrich, St. Louis, MO) was slowly dispersed in the alginate solution under the same conditions. The system was stirred until complete polymer dissolution and kept at 4°C for 12 hr before use. The thermosensitive sol–gel phase transition of the copolymers was assessed using a tube inverting test with a 4-mL tightly screw-capped vial with an inner diameter of 10 mm.[35] The sol–gel transition was observed by tilting the vials, and the sol and gel conditions were defined as "flow" or "no flow" at 1 min, respectively. The effect of temperature was evaluated in the range of 5 to 80°C.

16.3.2 SCAFFOLD FABRICATION AND CHARACTERIZATION

Porous Pluronic/alginate scaffolds were fabricated using custom-designed AM equipment[36,37] consisting of a heated dispensing head terminating with a nozzle, an x-y motorized stage for positioning the dispensing head, and a z-axis for controlling its distance from the stage. The extrusion process was performed by pressure-assisted dispensing, feeding pressurized argon gas by means of a pressure line connected to a control electrovalve. Generation of the process tool-path was performed beginning from a computer-aided design input geometry using a dedicated software interface. A proper amount of Pluronic/alginate solution (20 wt% PF127, 2 wt% sodium alginate in diluted DMEM) was loaded into a 5-mL gel-dispensing syringe (Nordson, Westlake, OH) at 4°C. The syringe was then brought to 37°C, leading to gelation. The blend was extruded in the gel state through a 250-μm nozzle at a pressure of 1.2 bar with a relative speed between the nozzle and the x-y table of $10 \text{ mm} \cdot \text{s}^{-1}$. Squared scaffolds with a lattice homogeneous fiber spacing were obtained by depositing layers of fibers laminated in a 0°/90° pattern. The surface of the scaffold was then briefly exposed to a 25-mM $CaCl_2$ aqueous solution to induce alginate crosslinking. After 5 minutes, the crosslinking solution was removed, and the crosslinked scaffold was gently washed in deionized water to terminate the gelation process. Inducing gelation of the alginate on the scaffold surface is expected to prevent the premature erosion of Pluronic and favors modulation of the scaffold degradation according to specific requirements. Furthermore, mechanically stable structures with enhanced handling properties can be obtained.

16.3.3 MECHANICAL CHARACTERIZATION

Multilayer composite scaffolds (12 × 12 × 2 mm) were additively manufactured and crosslinked as previously described. Stiffness of the scaffolds was measured in compression using a universal tester (Instron® Model 3365; Norwood, MA) equipped with a 10-N load cell. Scaffolds were placed between compression plates, and force was applied at a constant crosshead speed of 0.5 mm/min until a deformation of 25% was reached. Young's modulus was calculated from the linear portion of each stress–strain curve.

16.3.4 *In Vitro* Degradation and Drug Release

In order to study the degradation behavior of the AM scaffolds, four-layered hydrogel scaffolds were incubated in 1 mL of prewarmed saline solution (0.9% NaCl) at 37°C. At selected time points, specimens were retrieved, dried under vacuum for 12 hr, and weighed using a precision microbalance (Sartorius M2P; Elk Grove, IL). Mass erosion was analyzed in triplicate according to the following formula:

$$\text{Weight loss} = (W_0 - W_t)/W_0 \ (\%) \tag{16.1}$$

where W_0 and W_t are the weight before and after incubation, respectively.[26]

To determine the suitability of the developed scaffolds for drug release, vitamin E has been chosen as a clinically relevant bioactive component. Indeed, epidemiologic evidence supports the notion that vitamin E, especially α-tocopherol, can reduce the risk of coronary heart disease. Animal studies supporting the cardioprotective effects of α-tocopherol have also been published.[38] Due to its good water solubility and higher stability against oxidation, α-tocopheryl phosphate was selected as an alternative natural source of tocopherol for *in vitro* tests.

The α-tocopheryl phosphate disodium salt (αTP, Sigma-Aldrich) was predissolved in deionized water and added to the cold Pluronic/alginate aqueous solution to a final concentration of 1 mM. Drug-loaded scaffolds were produced as previously described. Drug release tests were performed in triplicate by incubating scaffolds at 37°C in mQ water at a constant gel:buffer ratio. Then, 200 μL aliquots of the incubation buffer were collected and the amount of released αTP was determined by fluorescence intensity (Tecan Infinite® M200 microplate reader, ex/em 290/330 nm) using an αTP standard curve in the same buffer.

16.3.5 Cell-Laden Constructs

Mouse embryonic fibroblasts (Balb/3T3 cell line) were grown to a confluent monolayer, trypsinized, and sedimented into a compact pellet by centrifugation. The pellet was homogeneously mixed with the ice-cold Pluronic/alginate solution at a final concentration of 1×10^6 cells per mL. To minimize the risk of scaffold contamination during the deposition, the AM equipment was placed in a biological safety cabinet, and all the materials in direct contact with the cells were autoclaved before processing. The hydrogel–cell mixture was put into the printing syringe and kept at 37°C until gelation. The mixture was extruded into a sterile Petri dish according to the previously described printing conditions. The obtained constructs were assessed in terms of cell viability and distribution. Hydrogel disks obtained by casting were also prepared and used as controls.

Cell viability was determined by Live/Dead® cell viability assays (Invitrogen, San Diego, CA) based on cellular membrane intactness. Soon after manufacturing, specimens containing cells were dissolved in sodium citrate buffer (50-mM sodium citrate, 100-mM sodium chloride, pH 7.4). Cells were collected by centrifugation (200× g, 5 min), processed for the Live/Dead assay according to the manufacturer's recommendations, dispensed on a microscope slide, and viewed under an inverted fluorescence microscope (Nikon Ti-E equipped with NIS Elements AR software; Nikon, Tokyo, Japan) with excitation/emission filters at 488/530 nm to detect living (green) cells and at 530/580 nm to detect dead (red) cells. Cell viability was calculated as the average ratio of vital to total cells in a sample; three randomly chosen fields per sample were considered.

To assess cell distribution within the scaffolds, cells were previously labeled with green-fluorescent 5-chloromethylfluorescein diacetate (CellTracker™ Green CMFDA; Molecular Probes, Eugene, OR), at a final concentration of 10 μM. Labeled cells were mixed with the ice-cold Pluronic/alginate solution and loaded in the dispensing syringe of the AM equipment. The final constructs were observed under an inverted fluorescence microscope directly after deposition.

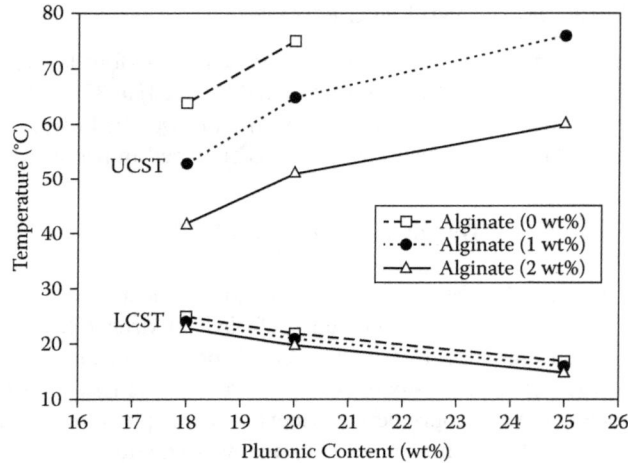

FIGURE 16.1 Sol–gel–sol transition curves for Pluronic® F127/alginate system at different compositions. LCST, lower critical solution temperature; UCST, upper critical solution temperature.

16.4 RESULTS AND DISCUSSION

16.4.1 HYDROGEL SYNTHESIS AND CHARACTERIZATION

An ideal gel for printing cell-laden constructs should be a free-flowing liquid with low viscosity under non-physiological conditions and undergo a rapid phase transition to form a stable gel capable of withstanding shear forces under physiological conditions.[7,8] As shown in Figure 16.1, at Pluronic concentrations higher than 18 wt%, the described blend systems undergo a thermodynamically reversible transition upon heating. Furthermore, the addition of alginate to Pluronic slightly lowered the sol–gel transition temperature compared to Pluronic F127 alone. As already experienced by several authors,[39,40] the presence of salts in Pluronic aqueous solution determines a decrease of structural transition temperature proportional to the concentration of salts. In this regard, the addition of sodium alginate exerted a strong influence on the structural process by hindering the initial caging and subsequent structural rearrangement of micelles.[30] However, Pluronic continued to dictate the transition behavior of the thermogel, as Pluronic micelles still underwent temperature-driven rearrangement. Taking into account these considerations, a ratio of Pluronic/alginate of 10:1 w/w was chosen as the most adequate for the aim of the study due to a beneficial equilibrium among conflicting requirements of gel structural integrity and biocompatibility.

16.4.2 SCAFFOLD FABRICATION AND MECHANICAL CHARACTERIZATION

Representative macroscopic images of a four-layered AM scaffold obtained are shown in Figure 16.2. The uniformity of the layered pattern, with interconnected regularly spaced pores, demonstrates the suitability of the selected gel system to be processed by the described AM technique. As usually observed in hydrogel processing,[28,41] transversal pores partially fused during stacking of layers because of the softness of the material. Nevertheless, in our settings, only a limited discrepancy between the computer-generated geometry and the obtained scaffold has been achieved, with a mean fiber diameter of 303.68 ± 5.17 μm, as calculated from optical micrographs (Figures 16.2C,D) by image analysis tools (ImageJ; National Institutes of Health, Bethesda, MD). Mechanical properties of the AM scaffolds were characterized by unconfined compression tests. Although Pluronic represents the most abundant component of the gel (20 wt%), a Young's modulus value close to the

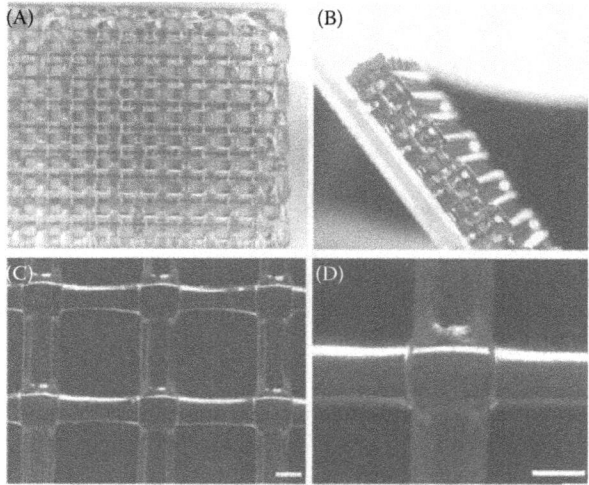

FIGURE 16.2   Macroscopic images of a four-layered hydrogel scaffold: (A) top and (B) side view. (C) Optical micrographs of scaffold microarchitecture (needle Ø 250 μm; fiber spacing, 1 mm; scale bar, 250 μm). (D) Higher magnification of the intersection between fibers (scale bar, 250 μm).

stiffness of pristine alginate AM scaffolds[42] was reported (6.4 ± 0.9 kPa, $n = 3$). As the low mechanical strength of the Pluronic hydrogels is widely recognized,[24,33] the reported enhanced mechanical behavior can be ascribed to the crosslinking of the alginate phase.

16.4.3   DEGRADATION AND DRUG RELEASE

Scaffold degradation rate was studied in terms of weight loss following incubation at 37°C. Figure 16.3A shows the remaining weight of the hydrogel in a long-term incubation period (up to 30 days). Addition of alginate to the gel composition resulted in a lower mass erosion with respect to pristine Pluronic hydrogels. In particular, while Pluronic scaffolds are highly unstable, beginning to dissolve within minutes after incubation, a burst degradation pattern followed by a much slower sustained

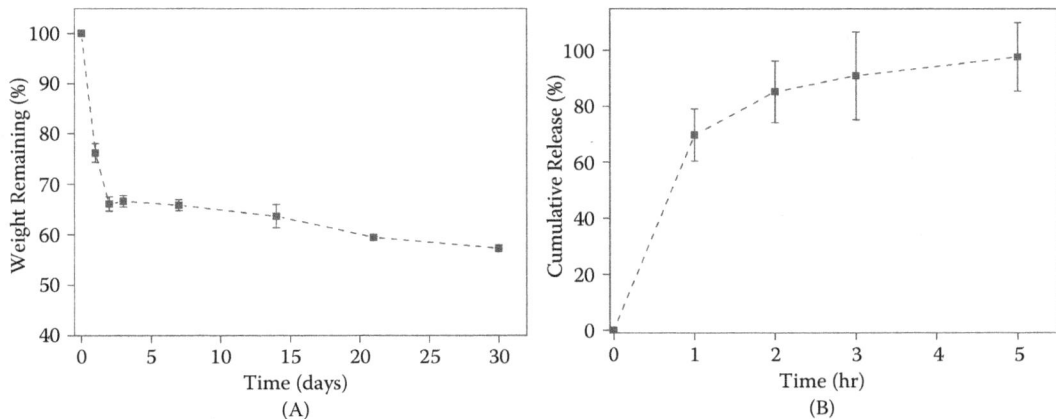

FIGURE 16.3   (A) Mass erosion of Pluronic/alginate hydrogel (20 wt% Pluronic, 2 wt% alginate) following incubation in saline solution at 37°C. (B) Release profile of αTP from hydrogel scaffolds as function of time. Results are presented as mean ± standard deviation ($n = 3$).

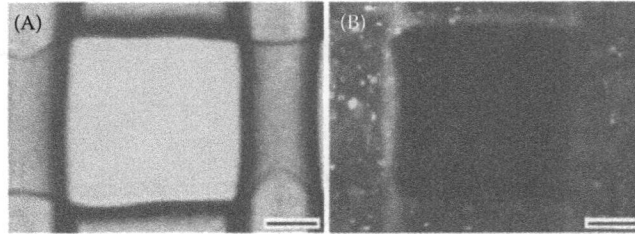

FIGURE 16.4 Encapsulation of living cells during the deposition process: (A) optical and (B) fluorescence micrographs of the Pluronic/alginate cell-laden scaffolds. Cells were fluorescently labeled with CFMDA live cell imaging probe (scale bar, 200 μm)

profile was registered in the case of Pluronic/alginate composite hydrogels. Furthermore, the continuous 3D architecture detectable at the end of the experiment suggested that homogeneous mass erosion had occurred. Scaffolds were loaded with αTP and characterized in terms of release profile to evaluate their potential as drug delivery systems. Figure 16.3B reports the cumulative release of αTP from Pluronic/alginate scaffolds as a function of time. Although the Pluronic/alginate solution performs better than the individual components in retaining pharmaceutical active molecules,[30,31] almost all of the drugs were released in the first few hours. According to the literature, both drug diffusion and polymer erosion can be considered as predominant factors governing drug release in such gel systems.[16]

16.4.4 CELL-LADEN HYDROGELS

In the design of viable cell-laden structures, survival of the printed cells and uniform cell distribution throughout the construct are paramount. Thus, the AM hydrogel scaffolds were studied in combination with Balb/3T3 cells to evaluate their potential for cell delivery. Results of Live/Dead assays performed after printing suggested that the printing process itself did not induce cell death (cell viability higher than 85%). As shown in Figure 16.4, a homogeneous distribution of cells microencapsulated within the hydrogel scaffold was obtained, thereby demonstrating the applicability of the AM equipment to obtain viable Pluronic/alginate cell-laden hydrogel.

16.5 CONCLUSIONS

Three-dimensional scaffolds with controlled architectures were manufactured beginning from a composite Pluronic/alginate gel system. This polymeric blend represents a likely candidate for additive manufacturing, as Pluronic F127 leads to thermosensitive structured systems, while alginate endows the formation of a strong gel upon exposure to bivalent cations in aqueous solution. This positive combination led to scaffolds with precisely defined microarchitectures and enhanced mechanical behavior. A more sustained degradation profile with respect to pristine Pluronic scaffolds was also observed. Furthermore, the use of this gel system for the printing of cell-laden, spatially organized constructs has been also demonstrated. Embedded cells survived the deposition process and spread homogeneously inside the gel phase. In light of the obtained results, the proposed system is expected to find applicability in the fabrication of customized 3D constructs for soft tissue engineering applications.

ACKNOWLEDGMENT

This work was partially supported by MIUR–FIRB program (Grant #RBFR10L0GK).

REFERENCES

1. Giannitelli, S.M., Accoto, D., Trombetta, M., and Rainer, A. (2014). Current trends in the design of scaffolds for computer-aided tissue engineering. *Acta Biomater.*, 10: 580–594.
2. Giannitelli, S.M., Rainer, A., Accoto, D., De Porcellinis, S., De-Juan-Pardo, E.M. et al. (2014). Optimization approaches for the design of additively manufactured scaffolds. In: *Tissue Engineering* (Fernandes, P.R. and Bartolo, P.J., Eds.), pp. 113–128. Springer, Netherlands.
3. Wang, X. (2012). Intelligent freeform manufacturing of complex organs. *Artif. Organs*, 36: 951–961.
4. Chimate, C. and Koc, B. (2014). Pressure assisted multi-syringe single nozzle deposition system for manufacturing of heterogeneous tissue scaffolds. *Int. J. Adv. Manuf. Technol.*, 75: 317–330.
5. Li, S., Yan, Y., Xiong, Z., Weng, C., Zhang, R., and Wang, X. (2009). Gradient hydrogel construct based on an improved cell assembling system. *J. Bioact. Compat. Polym.*, 24: 84–99.
6. Andersen, M.Ø., Le, D.Q.S., Chen, M., Nygaard, J.V., Kassem, M., Bünger, C., and Kjems, J. (2013). Spatially controlled delivery of siRNAs to stem cells in implants generated by multi-component additive manufacturing. *Adv. Funct. Mater.*, 23: 5599–5607.
7. Billiet, T., Vandenhaute, M., Schelfhout, J., Van Vlierberghe, S., and Dubruel, P. (2012). A review of trends and limitations in hydrogel-rapid prototyping for tissue engineering. *Biomaterials*, 33: 6020–6041.
8. Melchels, F.P., Domingos, M.A., Klein, T.J., Malda, J., Bartolo, P.J., and Hutmacher, D.W. (2012). Additive manufacturing of tissues and organs. *Prog. Polym. Sci.*, 37: 1079–1104.
9. Ahn, S., Lee, H., Puetzer, J., Bonassar, L.J., and Kim, G. (2012). Fabrication of cell-laden three-dimensional alginate-scaffolds with an aerosol cross-linking process. *J. Mater. Chem.*, 22: 18735–18740.
10. Cohen, D.L., Malone, E., Lipson, H., and Bonassar, L.J. (2006). Direct freeform fabrication of seeded hydrogels in arbitrary geometries. *Tissue Eng.*, 12: 1325–1335.
11. Smith, C.M., Stone, A.L., Parkhill, R.L., Stewart, R.L., Simpkins, M.W. et al. (2004). Three-dimensional bioassembly tool for generating viable tissue-engineered constructs. *Tissue Eng.*, 10: 1566–1576.
12. Lee, H., Ahn, S., Bonassar, L.J., and Kim, G. (2013). Cell (MC3T3-E1)-printed poly (ε-caprolactone)/alginate hybrid scaffolds for tissue regeneration. *Macromol. Rapid Comm.*, 34: 142–149.
13. Xu, T., Binder, K.W., Albanna, M.Z., Dice, D., Zhao, W., Yoo, J.J., and Atala, A. (2013). Hybrid printing of mechanically and biologically improved constructs for cartilage tissue engineering applications. *Biofabrication*, 5: 015001.
14. Boere, K.W., Visser, J., Seyednejad, H., Rahimian, S., Gawlitta, D. et al. (2014). Covalent attachment of a three-dimensionally printed thermoplast to a gelatin hydrogel for mechanically enhanced cartilage constructs. *Acta Biomater.*, 10: 2602–2611.
15. Ahn, S., Kim, Y., Lee, H., and Kim, G. (2012). A new hybrid scaffold constructed of solid freeform-fabricated PCL struts and collagen struts for bone tissue regeneration: fabrication, mechanical properties, and cellular activity. *J. Mater. Chem.*, 22: 15901–15909.
16. Dalmoro, A., Barba, A.A., Lamberti, G., Grassi, M., and d'Amore, M. (2012). Pharmaceutical applications of biocompatible polymer blends containing sodium alginate. *Adv. Polym. Technol.*, 31: 219–230.
17. Das, S., Pati, F., Chameettachal, S., Pahwa, S., Ray, A.R., Dhara, S., and Ghosh, S. (2013). Enhanced redifferentiation of chondrocytes on microperiodic silk/gelatin scaffolds: toward tailor-made tissue engineering. *Biomacromolecules*, 14: 311–321.
18. Yan, Y., Wang, X., Xiong, Z., Liu, H., Liu, F. et al. (2005). Direct construction of a three-dimensional structure with cells and hydrogel. *J. Bioact. Compat. Polym.*, 20: 259–269.
19. Zhang, T., Yan, Y., Wang, X., Xiong, Z., Lin, F. et al. (2007). Three-dimensional gelatin and gelatin/hyaluronan hydrogel structures for traumatic brain injury. *J. Bioact. Compat. Polym.*, 22: 19–29.
20. Xu, M., Wang, X., Yan, Y., Yao, R., and Ge, Y. (2010). A cell-assembly-derived physiological 3D model of the metabolic syndrome, based on adipose-derived stromal cells and a gelatin/alginate/fibrinogen matrix. *Biomaterials*, 31: 3868–3877.
21. Xu, W., Wang, X., Yan, Y., Zheng, W., Xiong, Z. et al. (2007). Rapid prototyping three-dimensional cell/gelatin/fibrinogen constructs for medical regeneration. *J. Bioact. Compat. Polym.*, 22: 363–377.
22. Das, S., Pati, F., Choi, Y.-J., Rijal, G., Shim, J.-H. et al. (2015). Bioprintable, cell-laden silk fibroin-gelatin hydrogel supporting multilineage differentiation of stem cells for fabrication of 3D tissue constructs. *Acta Biomater.*, 11: 233–246.
23. Cohn, D., Sosnik, A., and Garty, S. (2005). Smart hydrogels for *in situ* generated implants. *Biomacromolecules*, 6: 1168–1175.
24. Klouda, L. and Mikos, A.G. (2008). Thermoresponsive hydrogels in biomedical applications. *Eur. J. Pharm. Biopharm.*, 68: 34–45.

25. Park, S.A., Lee, S.H., and Kim, W. (2011). Fabrication of hydrogel scaffolds using rapid prototyping for soft tissue engineering. *Macromol. Res.*, 19: 694–698.

26. Lee, J.B., Yoon, J.J., Lee, D.S., and Park, T.G. (2004). Photo-crosslinkable, thermo-sensitive and biodegradable Pluronic hydrogels for sustained release of protein. *J. Biomater. Sci. Polymer Ed.*, 15: 1571–1583.

27. Shachaf, Y., Gonen-Wadmany, M., and Seliktar, D. (2010). The biocompatibility of Pluronic® F127 fibrinogen-based hydrogels. *Biomaterials*, 31: 2836–2847.

28. Fedorovich, N.E., Swennen, I., Girones, J., Moroni, L., van Blitterswijk, C.A. et al. (2009). Evaluation of photocrosslinked lutrol hydrogel for tissue printing applications. *Biomacromolecules*, 10: 1689–1696.

29. Choi, J.S. and Yoo, H.S. (2013). Chitosan/pluronic hydrogel containing bFGF/heparin for encapsulation of human dermal fibroblasts. *J. Biomater. Sci. Polymer Ed.*, 24: 210–223.

30. Chen, C.-C., Fang, C.-L., Al-Suwayeh, S.A., Leu, Y.-L., and Fang, J.-Y. (2011). Transdermal delivery of selegiline from alginate–Pluronic composite thermogels. *Int. J. Pharm.*, 415: 119–128.

31. Lin, H.-R., Sung, K., and Vong, W.-J. (2004). *In situ* gelling of alginate/pluronic solutions for ophthalmic delivery of pilocarpine. *Biomacromolecules*, 5: 2358–2365.

32. Abrami, M., D'Agostino, I., Milcovich, G., Fiorentino, S., Farra, R. et al. (2014). Physical characterization of alginate–Pluronic F127 gel for endoluminal NABDs delivery. *Soft Matter*, 10: 729–737.

33. Ruel-Gariépy, E. and Leroux, J.-C. (2004). *In situ*-forming hydrogels—review of temperature-sensitive systems. *Eur. J. Pharm. Biopharm.*, 58: 409–426.

34. Schmolka, I.R. (1977). A review of block polymer surfactants. *J. Am. Oil Chem. Soc.*, 54: 110–116.

35. Gong, C.Y., Shi, S., Dong, P.W., Yang, B., Qi, X.R. et al. (2009). Biodegradable *in situ* gel-forming controlled drug delivery system based on thermosensitive PCL–PEG–PCL hydrogel. Part 1. Synthesis, characterization, and acute toxicity evaluation. *J. Pharm. Sci.*, 98: 4684–4694.

36. Chiono, V., Mozetic, P., Boffito, M., Sartori, S., Gioffredi, E. et al. (2014). Polyurethane-based scaffolds for myocardial tissue engineering. *Interface Focus*, 4: 20130045.

37. Rainer, A., Giannitelli, S.M., Accoto, D., De Porcellinis, S., Guglielmelli, E., and Trombetta, M. (2012). Load-adaptive scaffold architecturing: a bioinspired approach to the design of porous additively manufactured scaffolds with optimized mechanical properties. *Ann. Biomed. Eng.*, 40: 966–975.

38. Mukherjee, S., Lekli, I., Das, M., Azzi, A., and Das, D.K. (2008). Cardioprotection with α-tocopheryl phosphate: amelioration of myocardial ischemia reperfusion injury is linked with its ability to generate a survival signal through Akt activation. *Biochim. Biophys. Acta*, 1782: 498–503.

39. Grassi, G., Crevatin, A., Farra, R., Guarnieri, G., Pascotto, A. et al. (2006). Rheological properties of aqueous Pluronic–alginate systems containing liposomes. *J. Colloid Interface Sci.*, 301: 282–290.

40. Pandit, N.K. and Kisaka, J. (1996). Loss of gelation ability of Pluronic® F127 in the presence of some salts. *Int. J. Pharm.*, 145: 129–136.

41. Fedorovich, N.E., De Wijn, J.R., Verbout, A.J., Alblas, J., and Dhert, W.J. (2008). Three-dimensional fiber deposition of cell-laden, viable, patterned constructs for bone tissue printing. *Tissue Eng. Part A*, 14: 127–133.

42. Fedorovich, N.E., Schuurman, W., Wijnberg, H.M., Prins, H.-J., Van Weeren, P.R. et al. (2011). Biofabrication of osteochondral tissue equivalents by printing topologically defined, cell-laden hydrogel scaffolds. *Tissue Eng. Part C Meth.*, 18: 33–44.

17 Additive Manufacturing of Rare Earth Permanent Magnets

Vemuru V. Krishnamurthy

CONTENTS

ABSTRACT

A review of currently pursued additive manufacturing methods of permanent magnets for application at different length scales ranging from thin films for microelectromechanical systems to bulk magnets for electrical motors is presented. The scope for additive manufacturing of permanent magnets using electron beam melting and selective laser melting systems is also outlined. The case for designing a new additive manufacturing system by combining magnetron sputtering and computer-aided design for high-performance permanent magnet thin films is presented. Recent progress in deposition methods for improving the magnetic properties such as a high coercivity field (H_c), large perpendicular magnetic anisotropy, and a high magnetic energy density product of $Nd_2Fe_{14}B$-based rare earth magnets is reviewed so as to provide a standard framework for permanent magnet development using additive manufacturing technologies. Overall, additive manufacturing is expected to change the landscape by offering new ways for more efficient use of rare earths in permanent magnets and by combining the two processes of microstructure development and design in desired geometries to minimize manual intervention.

17.1 INTRODUCTION

Magnetic materials are essential components in energy applications such as electric power generation, renewable energy, transportation, magnetic recording, and magnetic refrigeration. Recently, there has been an increasing demand for permanent magnets due to information technology advances and new consumer electronic products. New technologies often require that electronic devices containing permanent magnets be smaller, lighter, and highly reliable, yet consume little power and be less expensive to manufacture. In this context, permanent magnets made out of coercive $Nd_2Fe_{14}B$-based powders are of high importance for present and future applications.[1–3] The annual growth rate of all permanent magnets is estimated to be about 12%.[1] However, rare earth elements used in the manufacturing of permanent magnets are expensive. Table 17.1 shows the cost per kilogram for rare earth metals that are used in the synthesis of permanent magnetic compounds. As the natural resources for energy, critical rare earth metals, are limited, there is an increasing need for efficient manufacturing of rare-earth-based magnets as well as for discovering new hard magnetic materials with minimal or preferably without rare earth metals.[1–8]

Several permanent magnets have been discovered in the past 100 years, and the methods of manufacturing and characterizing these magnets have been established.[1–10] The magnetic energy density product (BH_{max}), which is a key figure of merit for permanent magnets, has steadily increased from 8 kJ/m^3 for steel to ~450 kJ/m^3 for $Nd_2Fe_{14}B$ magnets.[1] Figure 17.1 shows the development of magnetic energy density products with different permanent magnetic materials during the last century. Currently, the $Nd_2Fe_{14}B$ magnet is the best permanent magnet available for industrial applications. Recent investigations have been dedicated to better understanding the origin of magnetism in this compound and to develop bulk magnet manufacturing methods with superior magnetic energy density products. Site-specific magnetic investigations using x-ray magnetic circular dichroism experiments near neodymium (Nd) atomic resonances revealed that the intrinsic magnetic stability of $Nd_2Fe_{14}B$ has its atomic origins predominately at the Nd g sites, which strongly prefer c-axis alignment at ambient temperature and dictate the macroscopic easy-axis direction.[11] At present, there is not much progress in finding novel hard magnetic materials with higher remanent magnetization, which is now defined as the relevant parameter, and further

TABLE 17.1
Comparison of Cost of Rare Earth Metals Used in Permanent Magnet Manufacturing

Rare Earth Metal	Cost (US$/kg)
Lanthanum	10
Cerium	11
Praseodymium	150
Neodymium	85
Samarium	26.5
Dysprosium	475
Gadolinium	132.5
Europium	1000

Source: http://www.metal-pages.com.

Note: All metals have a purity of 99% and are supplied from China. The prices are as of October 2014.

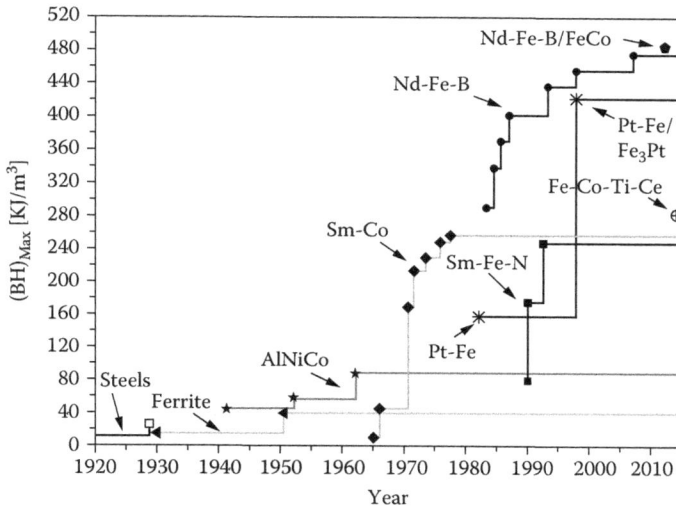

FIGURE 17.1 Development of magnetic energy density (BH_{max}) with different types of materials during the last 100 years. Each magnet is designed such that a field of 100 mT is produced at a distance of 5 mm from the magnet pole as indicated by the size/volume plots. (Plot was made by modifying and combining the data from Gutfleisch et al.,[1] Goll et al.,[13] Cui et al.,[33] and Liu et al.[14])

research is highly desirable. Only a few ternary and quaternary compounds have been investigated and only few new hard magnetic compounds have been discovered. Research is being pursued on nanocomposites to texture the hard magnets for higher anisotropies and for exchange-coupling with soft magnetic materials. The exchange interaction between the rare earth 4f-electrons and transition metal 3d-electrons is expected to improve the ferromagnetic properties, such as the Curie temperature.[12] Advanced manufacturing methods, especially additive manufacturing using multi-electron beam melting or selective laser melting of bulk magnetic materials or other layer-by-layer additive thin-film deposition methods based on magnetron sputtering, have the potential and can be explored further to develop permanent magnets with superior magnetic properties for a wide range of applications. Presenting an overview of various layer-by-layer growth methods for either bulk magnetic manufacturing or thin-film development and reviewing the current status of experimental results on the structural and magnetic properties are the main goals of this chapter.

General interest in the research is associated with the need for rare earth magnets with a high magnetic energy density product at various operating temperatures. $Pr_2Fe_{14}B$ magnets are used for applications with high critical temperature superconductors at 77 K. $Nd_2Fe_{14}B$ magnets with less dysprosium (Dy) content have the necessary temperature stability around 450 K for motor applications. Sm_2Co_{17} magnets are suitable for applications at temperatures around 670 K.[1] For applications in high-power microelectromechanical systems (MEMS) as high-speed magnetic generators, highly textured and thick films of rare earth magnets are desirable. There is growing interest in research on materials chemistry, the structure of grain boundary phases, structure of the interfaces for a better understanding of the mechanisms of coercivity field enhancement, and critical magnetic element magnetization reversal processes. Hard magnet–soft magnet nanocomposites with textured hard magnets are projected to play a crucial role in future technologies. This requires the fabrication of highly mixed multi-phases and well-oriented nanocomposite magnets. Physical vapor deposition techniques, such as magnetron sputtering or triode sputter deposition, have been used to make textured films of $Nd_2Fe_{14}B$-type magnets on metal and silicon (Si) substrates.

Most of the currently employed manufacturing methods are based on the production of aniso-tropic permanent magnets by aligning the particles so that magnetic domains are pointed in a specific direction.[2] One of these methods is called *uniaxial pressing.* It makes use of an aligning field that is either (1) parallel to the applied pressure (axial pressing) or (2) perpendicular to the applied pressure (transverse pressing) for powder compaction. The transverse pressing yields a better alignment and a higher magnetic energy density product compared to axial pressing. However, the compaction of powders in mechanical or hydraulic presses limits the shapes of the particles to simple cross-sections so they can be easily pushed out from the die cavity. This limits the texture and thus the anisotropy. Another method is called *isostatic pressing.* In this method, the powder is sealed in a flexible container, an aligning field is applied, and then a hydraulic fluid such as water is used to apply equal pressure from all sides on the outside of the container to compact the powder. This method has the advantage over uniaxial pressing of being able to produce larger magnets with higher magnetic energy density product. Typically, 60 to 70% of the theoretical density is achieved in this type of compacting process; hence, a sintering process is also necessary to achieve the maximum density.[2] For rare earth materials, the sintering process is carried out in the presence of a liquid phase. This type of liquid phase sintering results in inevitable grain growth and hence changes the properties associated with the nanostructures. The conventional techniques do not meet the fabrication needs of nanocomposite magnets as each process can further modify the nanocomposite properties; thus, new techniques are needed to make the necessary progress.

Hard magnets for applications such as motors should have high saturation magnetization ($\mu_0 M_s$), which is required to generate higher magnetic flux, and high coercivity field (H_c) or coercivity expressed as $\mu_0 H_c$, which determines the resistance to field demagnetization and thermal demag-netization. A higher Curie temperature (T_C) is also required to resist thermal demagnetization. The magnetic hardness of a material is determined by the intrinsic coercivity and the normal coerciv-ity. Higher coercivities give rise to higher magnetic hardness. Another requirement for a stable operation of hard magnets is that the demagnetization induction (B) should be linear in the second quadrant of the hysteresis curve, and it requires that the intrinsic coercivity be higher than the normal coercivity. Currently, there are mainly three types of materials with such properties: SmCo magnets, $Nd_2Fe_{14}B$ magnets, and the hard ferrites. There is a need to improve the manufacturing processes of these materials and to search for potential new materials with a wide range of operat-ing temperatures.

17.2 INDUSTRIAL MANUFACTURING OF POWDER-BASED RARE EARTH MAGNETS

The current magnet manufacturing process is comprised primarily of three steps: powder prepa-ration, forming and consolidation, and grinding and coating.[2] Figure 17.2 illustrates the typical processing steps of an energy-critical rare-earth-based permanent magnet. The magnetic alloys are produced using (1) mechanical techniques including jaw crushing, jet milling, and mechanical alloying, or (2) chemical fabrication techniques such as calciothermic reduction, which is often used for rare earth magnets. The key to producing a good magnet is to align the magnetic particles such that their magnetic moments are also aligned in a specific direction. The manufacturing process involving either cold isostatic pressing or axial die pressing of the particles meets this alignment requirement for particles with sizes in the range of 1 to 7 µm. In the case of epoxy bonded magnets, the particle size is larger (i.e., 5 to 150 µm); therefore, an additional thermal treatment is done to complete the alignment. After the compacting to maximum packing density and the thermal treat-ment, the magnets are machined to finish dimensions and then finish coatings are applied as neces-sary. The current methods therefore involve a lot of steps, and there can be some irrecoverable loss of crucial metals during the manufacturing process.

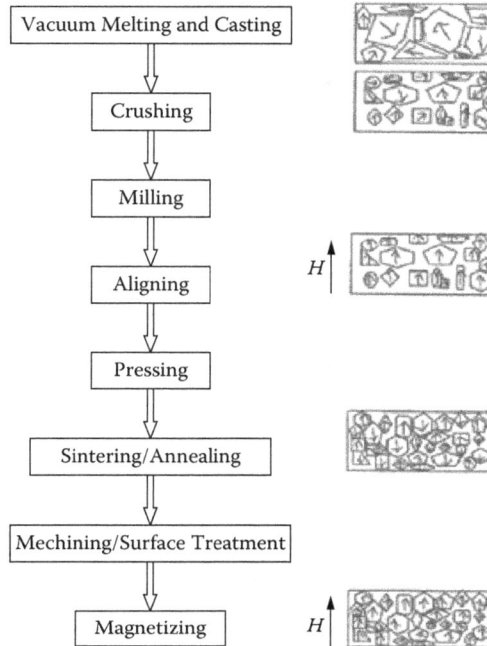

FIGURE 17.2 A typical manufacturing process of rare-earth-based permanent magnets used in industrial applications. (Adapted from Yin, W. and Constantinides, S., *Energy Critical Magnetic Material Manufacturing Processes*, technical paper, Arnold Magnetic Technologies, Marengo, IL, 2014 (http://www.arnoldmagnetics.com/Content1.aspx?id=4828.)

17.3 HIGH-RATE SPUTTERING GROWTH OF THICK $Nd_2Fe_{14}B$ FILMS

Recent work on a two-step thick-film growth method (deposition at temperatures $\leq 500°C$ and annealing at $750°C$ for 10 min) of $Nd_2Fe_{14}B$ magnetic films using high-rate sputtering (~30 µm/hr) on Si substrates shows that the out-of-plane remanent magnetization increases with the deposition temperature, reaching a maximum of 1.4 T, while the coercivity remains constant at 1.6 T.[15] The magnetic energy density product of these films was found to be 400 kJ/m³, which is comparable to that of high-quality sintered $Nd_2Fe_{14}B$ magnets. Figure 17.3A and Figure 17.3B compare the structural properties from scanning electron microscopy (SEM), x-ray diffraction (XRD), and magnetic hysteresis curves of these films as a function of the growth temperature.[15] Such films can be suitable for applications at a microscopic scale such as MEMS. Applications such as motors, power generators, windmills, and magnetic refrigeration would benefit from energy-critical permanent magnets with higher magnetic energy density products in the bulk form. Therefore, it is strongly desirable to pursue further research in the growth and synthesis of self-supported bulk magnetic films and bulk $Nd_2Fe_{14}B$ and related hard magnets.

17.4 PROSPECTS FOR PERMANENT MAGNET SYNTHESIS USING ELECTRON BEAM MELTING-BASED ADDITIVE MANUFACTURING WITH 3D SCREEN PRINTERS

Technological developments in 3D screen printing give rise to a new method for the growth of $Nd_2Fe_{14}B$ magnets using additive manufacturing combined with electron beam melting.[16,17] For example, the additive manufacturing approach can be used to fabricate self-supported $Nd_2Fe_{14}B$-based

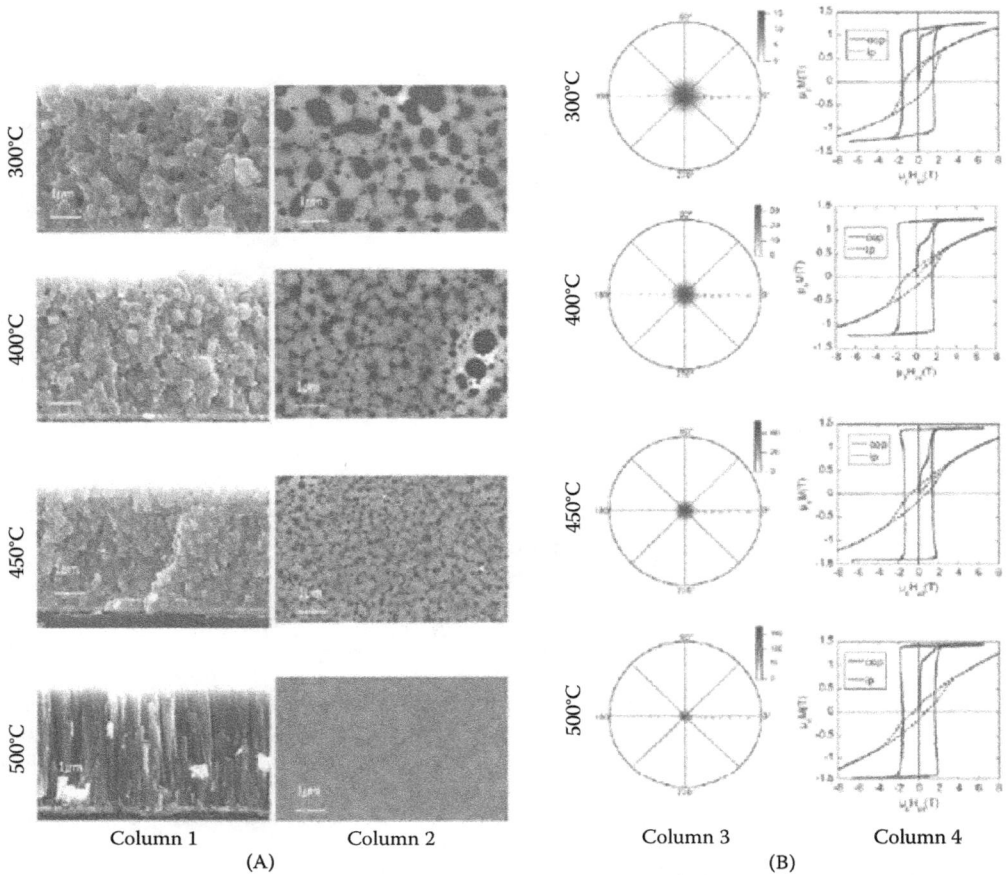

FIGURE 17.3 (A) Structural and magnetic properties of thick NdFeB films grown at high sputtering rates. Column 1 shows cross-sectional scanning electron microscope (SEM) images, and Column 2 shows plane-view images from SEM. (B) Structural and magnetic properties of thick NdFeB films grown at high sputtering rates. Column 3 shows x-ray pole figures, and Column 4 shows magnetic hysteresis curves. (Adapted from Dempsey, N.M. et al., *Appl. Phys. Lett.*, 90, 092509, 2007. With permission.)

nanocomposite magnets under various growth condition options available for the Arcam A2XX system. In this process, the magnetic layers can be grown layer by layer by melting using an electron multi-beam. The grains are expected to be aligned within layers grown by this method. By altering the beam and scan parameters, new, unusual, and even non-equilibrium local structures can be produced in the material. Thin films of $Nd_2Fe_{14}B$ and $Nd_2Fe_{14}B/\alpha$-Fe or $Nd_2Fe_{14}B/FeCo$ type of magnetic nanocomposites can be fabricated through natural grading of the material in additive manufacturing. The objective for such an additive manufacturing process is, therefore, to identify the synthesis conditions, the nanocomposite materials, growth parameters, grain boundary structures, and a combination of these so as to maximize the magnetic energy density product of the synthesized magnets. If this can be achieved, it will allow rapid and direct production of high-performance permanent magnets in a cost-effective way. Such a method will make use of much smaller amounts of critical rare earth elements due to the incorporation of soft magnetic alloys such as α-Fe or FeCo alloy into the magnet.

The main challenges that need to be addressed are the following: (1) achieving a nanocomposite structure so that an exchange coupling exists between the hard magnet and the soft magnet to form an exchange spring magnet; (2) alignment of easy axes; and (3) dense packing of nanoparticle assemblies. The magnetic coupling can be helpful for combining the magnetic hardness of the rare earth magnets and the high magnetization of the soft magnetic material and can result in a high magnetic energy density product in nanoparticle composites.[18-21] Such an exchange coupling can be realized in nanocomposites of $Nd_2Fe_{14}B$ with either α-Fe or FeCo.[22-33]

17.4.1 Synthesis of Permanent Magnet FeNi₃ by Selective Laser Melting

Selective laser melting enables the melting of metal powders such as aluminum and its alloys, titanium, cobalt, steel, nickel, and rare earth magnetic compounds such as $Nd_2Fe_{14}B$ to build parts in the desired shapes through a layer-by-layer approach. The scanning laser beam uses its energy to locally melt the supplied metal powder and fuses it onto a layer that was previously melted and solidified. Zhang et al.[34] recently used selective laser melting for *in situ* synthesis of the Permalloy FeNi from Fe–80wt%Ni metal powders. Permalloy is a soft magnetic material used in sensors, magnetic recording, transformers, and motors. Iron powder with an average particle size of 35 μm was blended with nickel powder with a particle size of 30 μm in a tumbling mixer for 45 minutes. Then, the MCP Realizer II SLM (MCP-HEK Tooling GmBH, Lübeck, Germany) was used to make 5 × 5 × 5-mm specimens from the powders by selective laser melting with a beam diameter of 50 μm. The effect of laser melting conditions on the synthesis was investigated by varying the laser power in the range of 50 to 100 W and scan velocity in the range of 0.1 to 1.6 m/s with a layer thickness of 0.05 mm. An alternative scanning pattern from layer to layer with equal spacing in x and y directions was used in the melting. The chamber was filled with argon and the powder bed was kept at 80°C.

Microstructure determined from SEM showed that dense structures without any porosity were formed. Crystal structure was determined from XRD, and elemental composition analysis was carried out using energy dispersive spectroscopy (EDS). These analyses showed the formation of an $FeNi_3$ intermetallic phase with fcc structure for the laser scan velocity range of 0.1 to 0.3 m/s. The lattice parameter was found to decrease from 0.2876 nm at 0.1 m/s to 0.2872 nm at 0.4 m/s. The average grain size was found to increase from 58 nm at 0.1 m/s to 115 nm at 0.4 m/s, showing a strong dependence on the laser scanning velocity. Both saturation magnetization and coercivity field showed a weak dependence on the scanning velocity, with values in the ranges of 1.19 to 1.24 m³/kg and 2.78 to 3.18 kA/m, respectively. The saturation magnetization of $FeNi_3$ synthesized by selective laser melting was lower, but the coercivity field was higher when compared to the properties of the $FeNi_3$ phase made by cold compacting or sintering. These results show that additive manufacturing using selective laser melting clearly offers a new route to the synthesis of permanent magnets in 3D shapes with some control over the microstructure and magnetic properties using the laser parameters. Synthesis of both intermetallic phases and rare earth compound-based permanent magnets by selective laser melting is a newly developing field of research. At present, research is ongoing on the synthesis of rare earth permanent magnets such as $Nd_2Fe_{14}B$ by selective laser melting. It is anticipated that new work will be aimed at $Nd_2Fe_{14}B$ as well as other permanent materials such as SmCo and nanocomposite magnets.

17.5 MAGNETRON SPUTTERING OF Nd₂Fe₁₄B THIN FILMS

Magnetron sputtering is a widely used plasma-based physical vapor deposition method to deposit thin films.[35-39] The concept is based on the vaporization of a material from a solid target by sputtering and condensation of the vaporized form of a material onto any substrate. Target materials can be an element, an alloy, or a compound. Multiple targets can be loaded and can be used either simultaneously or sequentially to grow a desired material component of the film. Thin films can be grown by layer-by-layer deposition from multiple target materials, enabling additive manufacturing of alloys, compounds, and multilayers. The technique is both robust and upscalable. Magnetron sputtering uses a diode sputtering configuration with a series of magnets placed behind the cathode (i.e., the target) to better confine the plasma in the sputtering region. Several recent developments in plasma magnetron sputtering have resulted in new deposition techniques such as unbalanced magnetron sputtering, pulsed magnetron sputtering, and ion-assisted magnetron sputtering.[38] High-power pulsed magnetron sputtering combines sputtering from pulsed plasma discharges to generate highly ionized plasma with large quantities of ions of the target material. This method enables the growth of smooth and dense films with control over the phase in reactively deposited compounds.[39] Ion-assisted magnetron sputtering involves periodical ion bombardment of the thin film during its growth, usually by adding nitrogen or carbon to the deposition atmosphere to improve the structural and physical properties.[40,41] These developments make magnetron sputtering an easily applicable technique for additive manufacturing of a wide variety of thin films.

Because thin-film growth in these methods is achieved layer by layer, often using multiple targets in a vacuum chamber with a typical base pressure of 0.1 to 10^{-5} Pa, magnetron sputtering is also an effective additive manufacturing method. A commercial disk magnet of $Nd_2Fe_{14}B$ with the desired chemical composition is often used for deposition of the magnetic layer. At the start of the deposition, each target is pre-sputtered for a time period ranging from 10 min to an hour. In this method, the substrate is heated to an elevated temperature, often in the range of 100 to ~700°C, depending on the combination of materials and the substrate. Lower temperature growth usually results in polycrystalline films, while higher temperature growth is often necessary for epitaxial growth. Other conditions, such as lattice match or lattice mismatch, also play a role in the grain size, texture, and buffer layers.

17.5.1 Growth of Textured Films

The growth and magnetic properties of anisotropic $Nd_2Fe_{14}B$ thin films with high magnetic energy density products have been investigated for applications such as MEMS and other micromagnetic devices. The key to producing large perpendicular magnetic anisotropy is to grow $Nd_2Fe_{14}B$ thin films with large c-axis texture. Tang et al.[42] used the DC magnetron sputtering to deposit or additively manufacture $Nd_2Fe_{14}B$ thin films on niobium (Nb) buffer layers on thermally oxidized Si substrates.[42] The x-ray diffraction patterns and the magnetic hysteresis loops from these $Nd_2Fe_{14}B$ films are shown in Figure 17.4. In the x-ray diffraction from $Nd_2Fe_{14}B$, the (004), (006), and (008) peaks indicate the c-axis texture with the c-axis perpendicular to the substrate surface. The (410) peak is the most intense peak from the randomly oriented grains of the tetragonal $Nd_2Fe_{14}B$ phase. By comparing the intensities of the (004), (006), and (008) peaks with the intensity of the (410) peak from films grown at different deposition temperatures in the range of 300 to 470°C with annealing at 590°C, the authors were able to show that high substrate temperatures result in a strong c-axis texture and increased perpendicular anisotropy. The texture of the $Nd_2Fe_{14}B$ phase or the amorphous phase created in the first deposition process of films was suggested as the origin for the strong texture in the epitaxial films. The magnetic hysteresis loops for these films measured at room temperature clearly indicate the perpendicular magnetic anisotropy. The ratio of the perpendicular to in-plane remanence increases with the substrate temperature, indicating that the c-axis texture is strengthened. Further analysis of the domains in these films was carried out using atomic force microscopy (AFM) and magnetic force microscopy (MFM). Figure 17.5 shows the AFM and MFM images from the $Nd_2Fe_{14}B$ films at different growth temperatures in the range of 300 to 470°C. AFM and MFM indicate that the amount of the stripe domain phase often observed in the amorphous phase decreases and the spike domain phase increases with the substrate temperature, resulting in the increased c-axis texture of the thin films.

17.5.2 Coercivity Enhancement Using the Doping of Heavy Rare Earths

Highly coercive $Nd_2Fe_{14}B$-based permanent magnets are in increasing demand for automotive applications. In this context, many recent studies have been dedicated to enhancing the mechanism of coercivity in $Nd_2Fe_{14}B$ thin films. A standard way to enhance the coercivity of these films is to add a heavy rare earth element such as Dy or terbium (Tb) to the rare earth site, resulting

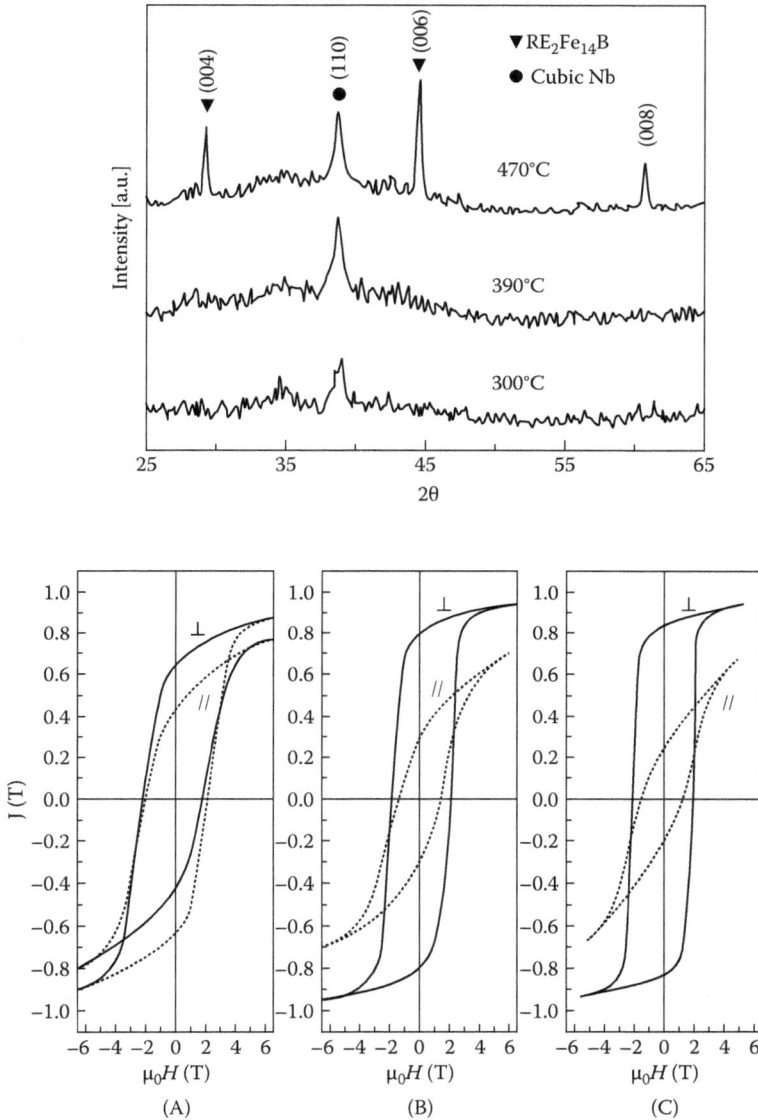

FIGURE 17.4 (Top) XRD pattern. (Bottom) Hysteresis loops from DC magnetron sputtering-deposited $Nd_2Fe_{14}B$ films. (From Tang, S.L. et al., *J. Appl. Phys.*, 103, 07E113, 2008. With permission.)

in (Nd,Dy)–Fe–B or (Nd,Tb)–Fe–B compounds. However, the remanence (B_r) and the BH_{max} are reported to decrease with an increase in Dy and Tb doping in $Nd_2Fe_{14}B$ phases. The cost of the magnet is also expected to go up with the addition of heavy rare earth elements to the magnet (see Table 17.1 for a comparison of the prices of rare earth metals). An alternative way to achieve a higher coercivity without sacrificing the B_r and BH_{max} is by modification of grain boundaries. Recent work shows that surface treatment-based addition (i.e., vapor sorption of Dy and Tb elements to the sintered $Nd_2Fe_{14}B$ magnets) results in the formation of a thin and continuous wetting layer between the $Nd_2Fe_{14}B$ grains, resulting in an increased coercivity field without affecting either B_r or BH_{max}.[43–45] Recently, You et al.[46] examined the Dy layer deposition at different temperatures on $Nd_2Fe_{14}B$ films and the mechanism of coercivity field enhancement through surface diffusion using magnetization

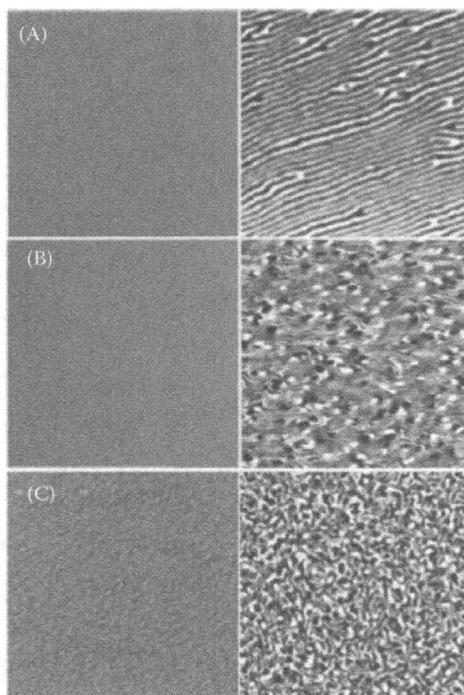

FIGURE 17.5   AFM (left) and MFM (right) 20×20 μm^2 scan size images of $Nd_2Fe_{14}B$ films grown at various deposition temperatures: (A) 300°C, (B) 390°C, and (C) 470°C. (From Tang, S.L. et al., *J. Appl. Phys.*, 103, 07E113, 2008. With permission.)

and transmission electron microscopy (TEM) analysis. A high deposition temperature of 460°C was found to result in an increase of the coercivity field—from 1297 kA/m without the Dy capping layer to 2005 kA/m with the Dy capping layer. TEM analysis indicated that the higher coercivity field resulted from enrichment of the grain boundaries with the Nd and structural modifications of the Dy-doped $Nd_2Fe_{14}B$ phase. There is an initial drop of 17°C when the Dy layer is deposited at room temperature compared to the $Nd_2Fe_{14}B$ without a Dy deposition layer. Higher deposition temperatures of the Dy capping layer in the range of 250 to 575°C were found to have no effect on the spin reorientation temperature of $Nd_2Fe_{14}B$ phase.

17.5.3   COERCIVITY ENHANCEMENT IN THICK FILMS USING SUPERFERRIMAGNETISM

Although the coercivity can be enhanced using Dy, it is an expensive and critical rare earth material. Moreover, the introduction of Dy into NdFeB films results in the reduction of remanent magnetization. Recently, Akadogan et al.[47] demonstrated a new approach for coercivity enhancement without using expensive rare earth materials.[47] In this approach, a soft magnetic layer (a Gd/Fe bilayer) is deposited on the surface of the grains of the $Nd_2Fe_{14}B$ hard magnet using triode sputtering. In this method, a stack of tantalum (Ta) (0.1 μm)/gadolinium (Gd) (0.3 μm)/Fe (0.23 μm)/$Nd_2Fe_{14}B$ (1 μm)/ Fe (1 μm)/Gd (1 μm)/Ta (1 μm) layers are deposited on Si/SiO$_2$(1 0 0) substrates that are 100-mm diameter and 500-μm-thick wafers. In these films, only the $Nd_2Fe_{14}B$ layer was deposited at 450°C as an amorphous layer. A subsequent annealing at 450°C for 10 min changed the microstructure of the film, resulting in the $Nd_2Fe_{14}B$ crystalline phase and formation of the $GdFe_2$ phase. The new $GdFe_2$ layer is then exchange-coupled with its magnetization antiparallel to the magnetization of

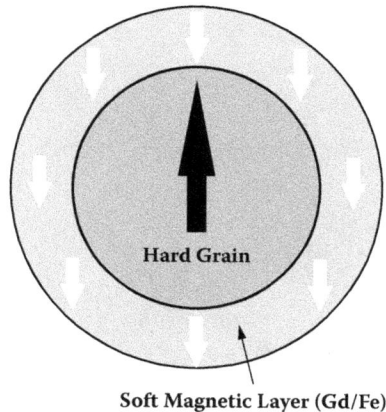

Soft Magnetic Layer (Gd/Fe)

FIGURE 17.6 The concept of superferrimagnetism for enhancing the coercivity of the hard magnetic layers of NdFeB using a soft magnetic layer (Gd/Fe), with the magnetization of Gd antiparallel to the magnetization of the grains of the hard magnet $Nd_2Fe_{14}B$. (Adapted from Akadogan, O. et al., *J. Appl. Phys.*, 115, 17A764, 2014. With permission.)

the $Nd_2Fe_{14}B$ hard magnetic layer, thus inducing superferrimagnetism in the films. Figure 17.6 schematically illustrates the exchange coupling that occurs between the hard magnetic grain core and the soft magnetic layer that forms the shell.

The top panel of Figure 17.7 shows a comparison of the microstructures obtained from SEM images for films with (A) Ta/Nd-Fe-B/Ta layers and films in which the (B) NdFeB layer sandwiched between Gd/Fe layers and film was not annealed, and (C) NdFeB layer sandwiched between Gd/Fe layers after annealing at 450°C for 10 min. The annealing also results in formation of the $GdFe_2$ phase. The differences in the microstructure can be clearly noted from the figure. The features in SEM images indicate that the as-deposited NdFeB layer is amorphous and it crystallizes following the annealing. The bottom panel compares the magnetic hysteresis curves. It is apparent that the coercivity of the NdFeB layer sandwiched between the Gd/Fe layers is significantly enhanced compared to the other two cases, indicating that the antiparallel exchange coupling between NdFeB and Gd in $GdFe_2$ plays a role.

17.5.4 HARD AND SOFT MAGNETIC NANOCOMPOSITES

Their increasing cost combined with the scarce resources of rare earth elements have resulted in a strong interest to develop alternatives to rare earth permanent magnets or permanent magnets that use smaller amounts of rare earth elements with equivalent or superior magnetic energy density products. Hard and soft magnetic nanocomposite magnets are based on the idea of combining two types of grains: a $Nd_2Fe_{14}B$ type of rare earth hard magnetic material component and a soft magnetic material such as Fe_3B or α-Fe or FeCo alloys that are exchange coupled.[33,48–50] Such an exchange-coupled magnet can combine the high coercivity of the hard magnetization of the soft component to produce high remanence and high magnetic energy density product. Using melt-spinning, Coehoorn et al.[48] developed the first nanocomposite magnet $Fe_3B/Nd_2Fe_{14}B$ with isotropic microstructure and relatively low amount of rare earths of 4.5 at% compared to sintered magnets that have 14 at% of rare earths.[48]

Many studies have been dedicated to isotropic nanocomposite bulk magnets. Bonded magnets made of $Fe_3B/Nd_2Fe_{14}B$ have been used as medium-performance magnets in commercial applications. A superior magnetic energy density product is expected in anisotropic nanocomposite magnets. Liu et al.[50–52] have investigated the structural and magnetic properties of rare earth nanocomposite

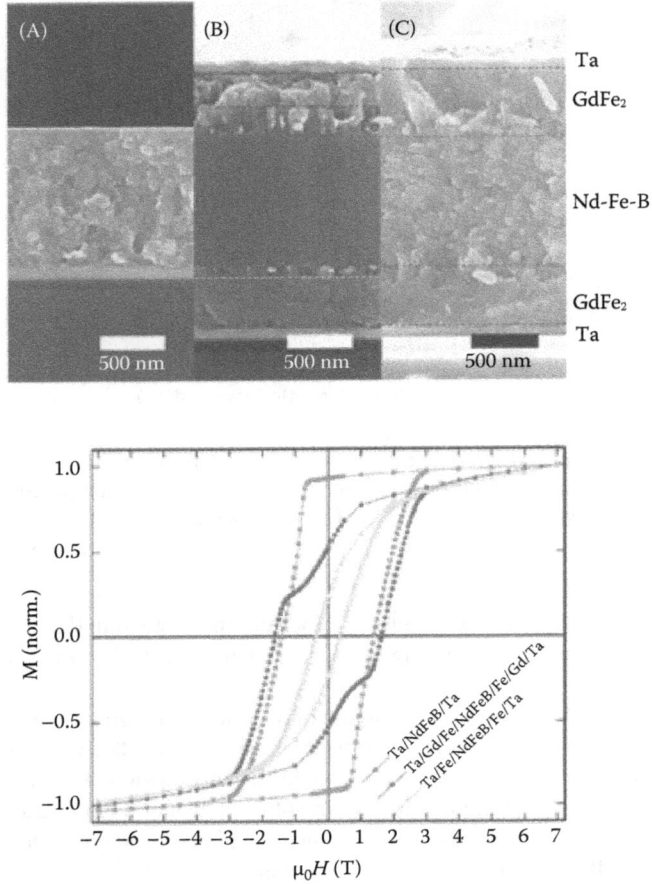

FIGURE 17.7 (Top) SEM images of (A) 1-μm-thick $Nd_2Fe_{14}B$ film after annealing, (B) 1-μm-thick $Nd_2Fe_{14}B$ layer sandwiched between Gd/Fe layers, and (C) 1-μm-thick $Nd_2Fe_{14}B$ layer sandwiched between Gd/Fe layers after annealing. (Bottom) Magnetization hysteresis curves for the three films: films with Ta/NdFeB/Ta, films with Ta/Gd/Fe/NdFeB/Fe/Gd/Ta layers, and films with Ta/Fe/NdFeB/Fe/Ta layers. (From Akadogan, O. et al., *J. Appl. Phys.*, 115, 17A764, 2014. With permission.)

magnets of a $(Nd,Dy)(Fe,Co,Nb,B)_{5.5}$ single layer and a $(Nd,Dy)(Fe,Co,Nb,B)_{5.5}/\alpha$-Fe multilayer prepared by sputtering and heat treatment. They achieved magnetic properties with $B_r = 1.11$ T, $\mu_0H_c = 0.88$ T, and $BH_{max} = 192$ kJ/m³ in the multilayer magnet annealed at 550°C for 30 min. These authors suggested that multilayer films with thinner soft magnetic layers may favor the formation of more ideal nanostructures for exchange coupling between soft and hard magnetic phases with an improved magnetic energy density product.

Typically, anisotropic nanocomposite magnets are made in thin-film form by controlling the crystallographic texture. However, in $Nd_2Fe_{14}B$/Fe thin films, due to the lack of pinning centers in $Nd_2Fe_{14}B$, the coercivity becomes much lower than one half of μ_0M_r, where M_r is the remanent magnetization. Cui et al.[33] achieved improved growth by modifying the interfaces between the hard and soft magnetic composites. They were able to grow high-coercivity anisotropic $[NdFeB/FeCo/Ta]_N$ composite multilayer thin films (N is the number of multilayers) with coercivity of 1.38 T and BH_{max} of 486 kJ/m³, which is the largest value found in thin films so far. The multilayers were deposited by magnetron sputtering with a chamber base pressure less than 10^{-6} Pa while maintaining the Ar gas pressure for sputtering at 13 Pa. The $Nd_2Fe_{14}B$ layers have the (0 0 1) texture. To suppress oxygen, a 50-nm Ta underlayer and a 20-nm Ta cover layer were deposited at room temperature. Both $Nd_2Fe_{14}B$ and Nd layers were deposited at 600°C and were annealed at 650°C for 30 min. After cooling the substrates, 1 nm Ta/10 nm FeCo/1 nm Ta were deposited at 200°C. The films were subsequently annealed at 650°C to allow the diffusion of Nd into the $Nd_2Fe_{14}B$ layers.

The coercivity, saturation magnetization, and magnetic energy density product have been extracted from in-plane and out-of-plane magnetic hysteresis curves and as a function of the number of multilayers for these films. Notable features include the gradual decrease of coercivity from about ~1.8 T with 4 multilayers to ~0.8 T with 19 multilayers. The magnetization $\mu_0 M$ has a peak with a maximum of 1.9 T in the film with 10 multilayers in which a peak in the BH_{max} is also observed with a value of ~460 kJ/m³. Microstructure analysis of these thin-film multilayers using Lorentz transmission electron microscopy in the Fresnel mode shows the formation of domain walls normal to the film plane. In addition, domain wall motion in an external field shows both pinning and movable features that are consistent with the initial magnetization curves. It would be interesting if this type of microstructure can be realized in thicker films or in bulk in order to manufacture bulk nanocomposite $Nd_2Fe_{14}B$ magnets with superior magnetic energy density product compared to sintered $Nd_2Fe_{14}B$ magnets.

17.6 PLASMA-ASSISTED PULSED LASER DEPOSITION OF NdFeB MAGNET THIN FILMS

Plasma-assisted pulsed laser deposition is another technique that is effective for the additive manufacturing of $Nd_2Fe_{14}B$ thin films with the desired c-axis texture. The structural and magnetic properties of the film can be affected by the conditions during the deposition, such as target material, deposition temperature, laser wavelength, laser fluence, deposition rate, and presence of the radio-frequency plasma in a reactive or inert gas in the deposition chamber. Constantinescu et al.[53,54] investigated the role of the deposition rate, laser wavelength, and effect of nitrogen environment in the growth chamber on the structural and magnetic properties of $Nd_2Fe_{14}B$ thin films.

17.6.1 Corrosion Resistance Using a Nitrogen Environment

The $Nd_2Fe_{14}B$ magnets have a poor corrosion resistance in humid environments due to the high tendency of oxidation and higher capacity of hydrogen absorption of rare earth elements. The multiphase microstructure of these magnets gives rise to a galvanic coupling effect between the ferromagnetic matrix phase and the Nd-rich or B-rich intragranular phase regions.[53,54] The addition (doping) of other rare earth elements such as Dy and praseodymium (Pr), transition metal elements such as cobalt and nickel, and sp-metal elements such as galium and aluminum is known to improve the corrosion resistance of $Nd_2Fe_{14}B$ magnets. Constantinescu et al.[53,54] compared $Nd_2Fe_{14}B$ films deposited on silicon- and platinum-covered Si substrates in vacuum to those depositions in nitrogen plasma. The sample temperature was maintained at a temperature in the range of room temperature to 850°C during the deposition. Nitrogen gas pressure was maintained at 6 Pa for the nitrogen-doped samples as compared to a pressure of about 10^{-3} Pa for the vacuum-deposited samples. The structural and magnetic properties of the films were characterized using AFM, x-ray diffraction, and magnetometry.

Atomic force microscopy of thin films was used to compare the RMS roughness of thin films grown at different wavelengths of the laser: 355 nm (4 ω) and 266 nm (3 ω) with a RF laser power of 75 or 100 W (figure not shown here). The use of lower wavelength lasers results in the compact growth of films with fewer droplets. Further, nitrogen RF plasma-assisted pulsed laser deposition decreases the RMS surface roughness for films as compared to films grown in argon RF plasma.[53] An RMS roughness of 65 nm was obtained for films deposited at 600°C or higher deposition temperatures. Magnetic hysteresis has revealed that films deposited in vacuum have a coercivity field of 20 to 50 kA/m, whereas films deposited in the presence of nitrogen have an enhanced coercivity field of 100 kA/m. Crystal structure analysis using x-ray diffraction indicated the likely formation of the $Nd_2Fe_{17}N_3$ phase for films containing nitrogen[53,55] in samples made at 600°C or higher growth temperatures using nitrogen RF-plasma-assisted pulsed laser deposition. The RF laser power of 75 or 100 W did not seem to affect the composition of the thin films. The degradation of films, which can affect the chemical structure at the grain boundaries, was investigated by exposing the films to normal atmospheric conditions and enhanced by irradiation with the x-rays. Samples without nitrogen are observed to have a tendency for oxidation and to lose superficial reflectivity, whereas samples containing nitrogen are stable with a tendency to maintain superficial reflectivity. Further, thermogravimetric analysis has indicated that the magnetic films are stable up to 341°C, where nitriding of the materials begins to take place and continues until 1000°C. Nitriding of the magnets results in a 3.2% increase of the mass. Nitrogen inclusion in the structure thus provides an alternative to protective coatings while improving the magnetic and surface morphological properties of $Nd_2Fe_{14}B$ thin films.

17.6.2 HIGHLY TEXTURED NDFEB FILMS ON TA(1 0 0) BUFFER LAYERS

Future applications of NdFeB films in MEMS such as motors and actuators require a high remanence along the direction of application of the magnetic field. This requires a texture along the easy axis of magnetization. Pulsed laser deposition has also proved to be a successful growth technique to deposit textured NdFeB thin films on Cr/Ta buffer layers on MgO(1 0 0) substrate. NdFeB-phase thin films grow epitaxially with the c-axis out of plane. Textured NdFeB film growth was reported by Kwon et al.,[56] who used an excimer laser (KrF, 246 nm, 25 ns, 5 J/cm², 10 Hz) in an ultrahigh vacuum of ~10⁻⁷ Pa. The first buffer layer (i.e., 50-nm thick Cr layer) was deposited at 245°C, followed by deposition of the 50-nm-thick Ta second buffer layer and the $Nd_2Fe_{14}B$ layer on MgO(0 0 1) substrate (both Ta and $Nd_2Fe_{14}B$ layers were deposited at the same temperature) by alternatively ablating Nd, Fe, and FeB targets by adjusting the number of pulses to obtain the desired stoichiometry. AFM of the films shows irregular morphologies, and MFM shows stripe domain formation, a typical feature expected for films with large perpendicular magnetic anisotropy. Magnetic hysteresis measurements show a coercivity of about 1.3 T at growth temperatures in the range of 430 to 600°C. This method offers a growth method for NdFeB films at a relatively low temperature of 420°C with an out-of-plane coercivity of 1 T. These results show that NdFeB thin films with suitable texture along the easy axis of magnetization can be grown by using the proper combination of buffer layer and substrate.

17.7 SUMMARY AND CONCLUSIONS

Additive manufacturing is a well-established and widely used physical vapor deposition method for the layer-by-layer growth of magnetic thin films and offers scalable manufacturing options for thin films. Particularly, magnetron sputtering and pulsed laser deposition methods offer unique routes to the synthesis of rare earth permanent magnets based on $Nd_2Fe_{14}B$, SmCo, and exchange-coupled magnetic nanocomposite magnets. With some additional research, the recently developed multi-electron beam melting and selective laser melting types of 3D screen printing or additive manufacturing methods can be combined with the well-developed additive synthesis methods for thin films to further develop the growth technologies to scale up and fabricate bulk permanent magnets in industrially desired shapes and structures. Currently, research efforts are focused on additive manufacturing of bonded permanent magnets using electron beam melting with anisotropic magnetic $Nd_2Fe_{14}B$ particles and various polymer binders. This method is expected to reduce the waste and offer flexibility in the choice of polymers for use as binders, with potential to produce a higher packing density of the magnetic particles. Additive manufacturing using selective laser melting and selective laser annealing is also expected to emerge as a new alternative for permanent magnet development in the coming years. It has the potential to allow growing desired nanocomposites from metallic starting materials in the powder form with control over the degree of crystalline order being enabled by rapid thermal processing in successive layers. The laser intensity, beam profile, processing time, and region of selection offer possibilities to fine-tune the structural and magnetic properties of amorphous NdFeB thin films.[57] Further research is desirable to understand the role played by grain boundary magnetic phases in magnets made by additive manufacturing. There is also a need to develop methods that can reduce the internal stresses associated with selective laser melting. It is expected that additive manufacturing equipment will be developed by combining computer-aided design (CAD) with other methods used for thin-film growth such as magnetron sputtering to manufacture magnetically anisotropic particles or grains with desired texture in successive layers. The field of permanent magnetic manufacturing is therefore expected to open up new avenues for developing high-quality end-product magnetic components for a wide range of applications, including permanent magnets for use in windmills for renewable energy, magnetic hard disk drives, and industrial motors.

ACKNOWLEDGMENTS

The author acknowledges helpful discussions with Dr. S.G. Sankar, Dr. F. Rothwarf, Dr. M.P. Paranthaman, Dr. Gopi Somayajula, and Mr. Fred Forstner.

REFERENCES

1. Gutfleisch, O., Willard, M.A., Brück, E., Chen, C.H., Sankar, S.G., and Liu, J.P. (2011). Magnetic materials for 21st century: stronger, lighter, and more energy efficient. *Adv. Mater.*, 23(7): 821–842.
2. Yin, W. and Constantinides S. (2014). *Energy Critical Magnetic Material Manufacturing Processes*, technical paper. Arnold Magnetic Technologies, Marengo, IL (http://www.arnoldmagnetics.com/Content1. aspx?id=4828).
3. Coey, J.M.D. (2012). Permanent magnets: plugging the gap. *Scripta Mater.*, 67(6): 524–529.
4. Kramer, M.J., MeCallum, R., Anderson, I.E., and Constantinidis, S. (2012). Prospects for non-rare earth permanent magnets for traction motors and generators. *JOM*, 64(7): 752–763.
5. Zhu, L., Nie, S. Meng, K. Pan, D. Zhao, J., and Zheng, H. (2013). Multifunctional $L1_0$-$Mn_{1.5}Ga$ films with ultrahigh coercivity, giant perpendicular magnetocrystalline anisotropy and large magnetic energy product. *Adv. Mater.*, 24(23): 4547–4551.
6. Carroll, K.J., Huba, Z.J., Spurgeon, S.R., Qian, M., Khanna, S.N., Hudgins, D.M., Taheri, M.L., and Carpenter, E.E. (2012). Magnetic properties of Co_2C and Co_3C nanoparticles and their assemblies. *Appl. Phys. Lett.*, 101(1): 012409.
7. Herbst, J.F., Meyer, M.S., and Pinkerton, F.E. (2012). Magnetic hardening of $Ce_2Fe_{14}B$. *J. Appl. Phys.*, 111(1): 07A718.
8. Brown, D.N., Wu, Z., He, F., Miller, D.J., and Herchenroder, J.W. (2014). Dysprosium-free melt-spun permanent magnets. *J. Phys. Condens. Matter*, 26(6): 064202.
9. Sagawa, M., Fujimura, S., Togawa, N., Yamamoto, H., Matsuura, Y., and Hiraga, K. (1984). Permanent magnet materials based on the rare earth-iron-boron tetragonal compounds. *IEEE Trans. Magn.*, MAG-20(5): 1584–1589.
10. Croat, J.J., Herbst, J.F., Lee, R.W., and Pinkerton, F.E. (1984). Pr–Fe and Nd–Fe based permanent magnets: a new class of high performance permanent magnets. *J. Appl. Phys.*, 55(6): 2078–2082.
11. Haskel, D., Lang, J.C., Islam, Z., Cady, A., Srajer, G., van Veenendaal, M., and Canfield, P.C. (2005). Atomic origin of magnetocrystalline anisotropy in $Nd_2Fe_{14}B$. *Phys. Rev. Lett.*, 95(21): 217207.
12. Krishnamurthy, V.V., Lang, J.C., Haskel, D., Keavney, D.J., Srajer, G. et al. (2007). Ferrimagnetism in $EuFe_4Sb_{12}$ due to the interplay of f-electron moments and a nearly ferromagnetic host. *Phys. Rev. Lett.*, 98(12): 126403.
13. Goll, D., Loeffler, R., Stein, R., Pflanz, U., Goeb, S., Karimi, R., and Schneider, G. (2014). Temperature dependent magnetic properties and application potential of intermetallic $Fe_{11-x}Co_xTiCe$. *Phys. Status Solidi*, 8(10): 862–865.
14. Liu, J.P., Luo, C.P., Liu, Y., and Sellmyer, D.J. (1998). High energy products in rapidly annealed nanoscale Fe/Pt multilayers. *Appl. Phys. Lett.*, 72(4): 483–485.
15. Dempsey, N.M., Walther, A., May, F., and Givord, D. (2007). High performance hard magnetic thick films of NdFeB for integration into micro-electro-mechanical-systems. *Appl. Phys. Lett.*, 90(9): 092509.

16. Arcam. (2014). *Setting the Standard for Additive Manufacturing*, technical brochure. Arcam AB, Mölndal, Sweden (http://www.arcam.com).

17. Murr, L.E., Gaytan, S.E., Ramirez, D., Martinez, E., Hernandez, J. et al. (2012). Metal fabrication by additive manufacturing using laser and electron beam melting technologies. *J. Mater. Sci. Technol.*, 28(1): 1–14.

18. Betancourt, I. and Davies, H.A. (2010). Exchange coupled hard magnetic nanocomposite alloys. *Mater. Sci. Technol. Lond.*, 26(1): 5–19.

19. Balamurugan, B., Sellmyer, D.J., Hadjipanayis, G.C., and Skomski, R. (2012). Prospects for nanoparticle based permanent magnets. *Scripta Mater.*, 67(6): 542–547.

20. Yu, Y., Sun, K., Tian, Y., Li, X.-J., Kramer, M.J. et al. (2013). One-pot synthesis of Urchin like FePt–Fe_3O_4 and their conversion into exchange coupled L10 FePd–Fe nanocomposite magnets. *Nano Lett.*, 13(10): 4975–4979.

21. Choi, Y., Jiang, J.S., Ding, Y., Rosenberg, R.A., Pearson, J.E. et al. (2007). Role of diffused Co atoms in improving effective exchange coupling in Sm-Co/Fe spring magnets. *Phys. Rev. B*, 75(10): 104432.

22. Bader, S.D. (2006). *Colloquium*: Opportunities in nanomagnetism. *Rev. Mod. Phys.*, 78(1): 1–15.

23. Yue, M., Niu, N.P. Li, Y.L. Zhang, D.T. Liu, W.Q. et al. (2008). Higgins, structure and magnetic properties of bulk isotropic and anisotropic $Nd_2Fe_{12}B/\alpha$-Fe nanocomposite permanent magnets with different α-Fe contents. *J. Appl. Phys.*, 103(7): 01E101.

24. Su, K.P., Liu, Z.W., Zeng, D.C., Huo, D.X., Li, W.L., and Zhang, G.Q. (2013). Structure and size-dependent properties of Nd-Fe-B nanoparticles and textured nano-flakes prepared from nanocrystalline ribbons. *J. Phys. D: Appl. Phys.*, 46(24): 245003.

25. Hussain, A., Jhadav, A.P., Baek, Y.K., Choi, H.J., Lee, J., and Kang, Y.S. (2013). One pot synthesis of exchange coupled $Nd_2Fe_{14}B$/alpha-Fe by Pechini type sol–gel method. *J. Nanosci. Nanotechnol.*, 13(11): 7717–7722.

26. Xia, J., Zhang, X.C., and Zhao, G.P. (2013). Micromagnetic analysis of the effect of the easy axis orientation on demagnetization process in $Nd_2Fe_{14}B$/alpha-Fe bilayers. *Acta Phys. Sinica*, 62(22): 227502.

27. Rong, C.B., Wang, D. Nguyen, V.V. Daniil, M. Willard, M.A. et al. (2012). Effect of selective Co addition on magnetic properties of $Nd_2(FeCo)_{14}B/\alpha$-Fe nanocomposite magnets. *J. Phys. D: Appl. Phys.*, 46(4): 045001.

28. Poudyal, N. and Liu, A.P. (2013). Advances in nanostructured permanent magnet research, *J. Phys. D: Appl. Phys.*, 46(4): 043001.

29. Lou, L., Hou, F.C., Wang, Y.N., Cheng, Y., Li, H.L. et al. (2013). Texturing α-Fe/$Nd_2Fe_{14}B$ nanocomposites with enhanced magnetic properties. *J. Magn. Magn. Mater.*, 352: 45–48.

30. Zhang, P.Y., Pan, M.X., Ge, H.L., Yue, M., and Liu, W.Q. (2013). Study of magnetization reversal behavior for annealed $Nd_2Fe_{14}B/\alpha$-Fe nanocomposite alloys. *J. Rare Earths*, 31(8): 759–764.

31. Hadjipanayis, G.C., Liu, J., Gabay, A., and Marinescu, M. (2006). Current status of rare-earth permanent magnet research in USA. *J. Iron Steel Res. Int.*, 13(Suppl. 1): 12–22.

32. Magnfält, D., Abadias, G., and Sarakinos, K. (2013). Atom insertion into grain boundaries and stress generation in physically vapor deposited films. *Appl. Phys. Lett.*, 103(5): 051910.

33. Cui, W.B., Takahaski, Y.K., and Hono, K. (2012). $Nd_2Fe_{14}B$/FeCo anisotropic nanocomposite films with high magnetic energy density product. *Adv. Mater.*, 24(48): 6530–6535.

34. Zhang, B., Fenineche, N.-E., Liao, H., and Coddet, C. (2013). Magnetic properties of *in-situ* synthesized $FeNi_3$ by selective laser melting Fe-80wt.%Ni powders. *J. Magn. Magn. Mater.*, 336: 49–54.

35. Ohring, M. (2002). *Materials Science of Thin Films*. Academic Press, San Diego, CA.

36. Penning, F.M. (1939). Coating by Cathode Disintegration, U.S. Patent 2,146,025.

37. Glocker, D.A., Romach, M.M., and Lindberg, V.W. (2001). Recent developments in inverted cylindrical magnetron sputtering. *Surf. Coat. Technol.*, 146–147: 457–462.

38. Xing, Y., Chengbiao, W., Yang, L., Deyang, Y., and Tingyan, X. (2006). Recent developments in magnetron sputtering. *Plasma Sci. Technol.*, 8(3): 337–343.

39. Lundin, D. and Sarakinos, K. (2012). An introduction to thin film processing using high-power impulse magnetron sputtering. *J. Mater. Res.*, 27(5): 780–792.

40. Kouznetsov, V., Macák, K., Schneider, J.M., Helmersson, U., and Petrov, I. (1999). A novel pulsed magnetron sputter technique utilizing very high target power densities. *Surf. Coat. Technol.*, 122(2–3): 290–293.

41. Helmersson, U., Lattemann, M., Bohlmark, J., Ehiasarian, A.P., and Gudmundsson, J.T. (2006). Ionized physical vapor deposition (IPVD): a review of technology and applications. *Thin Solid Films*, 513(1–2): 1–24.

42. Tang, S.L., Gibbs, M.R.J., Davis, H.A., Mateen, N.E., Nie, B., and Du, Y.W. (2008). The possible origin of RE–Fe–B thin films with *c*-axis texture. *J. Appl. Phys.*, 103(7): 07E113.
43 Komuro, M. and Satsu, Y. (2008). Structure and magnetic properties of NdFeB powder surrounded with layer of rare-earth fluorides. *J. Appl. Phys.*, 103(7): 07E142.
44. Kianvash, A., Mottram, R.S., and Harris, I.R. (1999). Densification of $Nd_{13}Fe_{78}NbCoB_7$-type sintered magnet by (Nd, Dy)-hydride additions using a powder blending technique. *J. Alloys Compd.*, 287(1–2): 206–214.
45. Liu, Q.Z., Zhang, L.T., Dong, X.P., Xu, F., and Kumoro, M. (2009). Increased coercivity in sintered Nd–Fe–B magnets with NdF_3 additions and the related grain boundary phase. *Scripta Mater.*, 61(11): 1048–1051.
46. You, C.Y., Zhu, J., Tian, N., and Lu, X.J. (2011). Coercivity enhancement of thin film magnets through Dy surface diffusion process. *J. Mater. Sci. Technol.*, 27(9): 826–830.
47. Akadogan, O., Dobyrynin, A., Le Roy, D., Dempsey, N.M., and Givord, D. (2014). Superferrimagnetism in hard Nd–Fe–B thick films, an original concept for coercivity enhancement. *J. Appl. Phys.*, 115(17): 17A764.
48. Coehoorn, R., de Mooji, D.B., and Waard, C.D.E. (1989). Meltspun permanent magnet materials containing Fe_3B as the main phase. *J. Magn. Magn. Mater.*, 80: 101–104.
49. Skomski, R. and Coey, J.M.D. (1993). Giant energy product in nanostructured two-phase magnets. *Phys. Rev. B*, 48(21): 15812–15816.
50. Schrefl, T., Kromueller, H., and Fidler, J. (1993). Exchange hardening in nano-structured 2-phase permanent-magnets. *J. Magn. Magn. Mater.*, 127: L273–277.
51. Liu, W., Zhang, Z.D., Liu, J.P., Li, X.Z., Sun, X.K., and Sellmyer, D.J. (2002). Structure and magnetic properties of sputtered (Nd,Dy) $(Fe,Co,Nb,B)_{5.5}/M$ ($M = FeCo,Co$) multilayer magnets. *J. Appl. Phys.*, 91(10): 7890–7892.
52. Liu, W., Zhang, Z.D., Liu, J.P., Dai, Z.R., Wang, Z.L., Sun, X.K., and Sellmyer, D.J. (2003). Nanocomposite $(Nd,Dy)(Fe,Co,Nb,B)_{5.5}$/alpha-Fe multilayer magnets with high performance. *J. Phys. D: Appl. Phys.*, 36(17): L63–66.
53. Constantinescu, C., Patroi, E., Codescu, M., and Dinescu, M. (2013). Effect of nitrogen environment on NdFeB thin films grown by radio frequency plasma beam assisted pulsed laser deposition. *Mater. Sci. Eng. B*, 178(4): 267–271.
54. Constantinescu, C., Ion, V., Codescu, M., Rotaru, P., and Dinescu, M. (2013). Optical, morphological and thermal behavior of NdFeB magnetic thin films grown by radiofrequency plasma-assisted pulsed laser deposition. *Curr. Appl. Phys.*, 13(9): 2019–2025.
55. Kajitani, T., Morii, Y., Funabashi, S., Iriyama, T., Kobayashi, K. et al. (1993). High resolution neutron powder diffraction study on nitrogenated Nd_2Fe_{17}. *J. Appl. Phys.*, 73(10): 6032–6034.
56. Kwon, A.R., Hanneann, U., Neu, V., Faehler, S., and Schutz, L. (2005). Microstructure and magnetic properties of highly textured Nd–Fe–B films grown on Ta(1 0 0). *J. Magn. Magn. Mater.*, 290–291: 1247–1250.
57. Hopkinson, D., Cockburn, A., and O'Neill, W. (2011). Fiber laser processing of amorphous rare earth NdFeB magnetic materials. In: *Proceedings of the 30th International Congress on Applications of Lasers and Electro-Optics*, pp. 1233–1240.

Index

For Product Safety Concerns and Information please contact our EU
representative GPSR@taylorandfrancis.com
Taylor & Francis Verlag GmbH, Kaufingerstraße 24, 80331 München, Germany